Methods in Cell Biology

VOLUME 88

Introduction to Electron Microscopy for Biologists

Series Editors

Leslie Wilson
Department of Molecular, Cellular and Developmental Biology
University of California
Santa Barbara, California

Paul Matsudaira
Whitehead Institute for Biomedical Research
Department of Biology
Division of Biological Engineering
Massachusetts Institute of Technology
Cambridge, Massachusetts

Methods in Cell Biology

VOLUME 88

Introduction to Electron Microscopy for Biologists

Edited by

Prof. Terence D. Allen

Structural Cell Biology
Paterson Institute for Cancer Research
Manchester

AMSTERDAM • BOSTON • HEIDELBERG • LONDON
NEW YORK • OXFORD • PARIS • SAN DIEGO
SAN FRANCISCO • SINGAPORE • SYDNEY • TOKYO
Academic Press is an imprint of Elsevier

Cover Illustration: Field Emission Scanning EM of a folded region of an isolated nuclear envelope from the oocyte of Xenopus, displaying both outer (left) and inner (right) surfaces with nuclear pore complexes. Magnification 80,000X.
Courtesy of S Bagley and T D Allen, University of Manchester.

Academic Press is an imprint of Elsevier
30 Corporate Drive, Suite 400, Burlington, MA 01803, USA
525 B Street, Suite 1900, San Diego, CA 92101-4495, USA

First edition 2008

Copyright © 2008 Elsevier Inc. All rights reserved

No part of this publication may be reproduced, stored in a retrieval system or transmitted in any form or by any means electronic, mechanical, photocopying, recording or otherwise without the prior written permission of the publisher

Permissions may be sought directly from Elsevier's Science & Technology Rights Department in Oxford, UK: phone (+44) (0) 1865 843830; fax (+44) (0) 1865 853333; email: permissions@elsevier.com. Alternatively you can submit your request online by visiting the Elsevier web site at http://elsevier.com/locate/permissions, and selecting *Obtaining permission to use Elsevier material*

Notice
No responsibility is assumed by the publisher for any injury and/or damage to persons or property as a matter of products liability, negligence or otherwise, or from any use or operation of any methods, products, instructions or ideas contained in the material herein. Because of rapid advances in the medical sciences, in particular, independent verification of diagnoses and drug dosages should be made

ISBN: 978-0-12-374320-6
ISSN: 0091-679X

For information on all Academic Press publications
visit our website at books.elsevier.com

Printed and bound in USA

Transferred to Digital Printing, 2011

Working together to grow libraries in developing countries

www.elsevier.com | www.bookaid.org | www.sabre.org

ELSEVIER   BOOK AID International   Sabre Foundation

I should like to dedicate this volume to my wife, Gabrielle, for her unstinting support and encouragement not only through the production of this book, but throughout my entire scientific career. Secondly, to our children Neil and Debbie, for their help in later years.

Terence D. Allen

CONTENTS

Contributors   xv

Preface   xix

PART I   Exploring the Organisation of the Cell by Electron Microscopy

SECTION 1   Basic Transmission and Scanning Electron Microscopy

1. High Pressure Freezing and Freeze Substitution of *Schizosaccharomyces pombe* and *Saccharomyces cerevisiae* for TEM
   Stephen Murray

I. Introduction	4
II. Materials and Instrumentation	4
III. Procedures	7
IV. Comments and Problems	10
References	17

2. Electron Probe X-ray Microanalysis for the Study of Cell Physiology
   E. Fernandez-Segura and Alice Warley

I. Introduction	20
II. Rationale	21
III. Methods	22
IV. Equipment	37
V. Discussion	38
References	40

3. Preparation of Cells and Tissues for Immuno EM
   Paul Webster, Heinz Schwarz, and Gareth Griffiths

I. Preparing Biological Specimens for Examination by Electron Microscopy	46
II. Vitrification and Chemical Fixation for Immunolocalization	48

III. Vitrification Followed by Freeze Substitution — 48
IV. Chemical Cross-linking (or Fixation) — 49
V. Embedding in Resin for Sectioning — 50
VI. Cryosectioning for Immunocytochemistry: The Tokuyasu Method — 50
VII. The Starting Material — 51
VIII. Protocols — 51
IX. Alternative Approaches to Freeze Substitution — 53
X. Correlative Microscopy — 54
References — 55

4. Quantification of Structures and Gold Labeling in Transmission Electron Microscopy

John Lucocq

I. Introduction — 59
II. Sampling and Stereology — 60
III. Quantities Displayed on Sections—Gold Labeling and Profile Data — 62
IV. Quantities in Three Dimension — 69
V. Spatial Analysis — 79
References — 80

5. Combined Video Fluorescence and 3D Electron Microscopy

Alexander A. Mironov, Roman S. Polishchuk, and Galina V. Beznoussenko

I. Introduction — 84
II. Rationale — 84
III. Method Steps — 85
IV. Immunolabeling for EM with NANOGOLD — 87
V. Immunolabeling for EM with HRP — 88
References — 95

6. From Live-Cell Imaging to Scanning Electron Microscopy (SEM): The Use of Green Fluorescent Protein (GFP) as a Common Label

Sheona P. Drummond and Terence D. Allen

I. Introduction — 98
II. Rationale — 99
III. Methods — 99
IV. Summary — 103
V. Concluding Remarks — 104
References — 107

7. Immunolabeling for Scanning Electron Microscopy (SEM) and Field Emission SEM
Martin W. Goldberg

    I. Introduction — 110
    II. Methods — 120
    III. Procedure — 122
    Conclusions — 129
    References — 129

8. Immunogold Labeling of Thawed Cryosections
Peter J. Peters and Jason Pierson

    I. Introduction — 132
    II. Preparation of Carbon- and Formvar-Coated Copper Girds — 132
    III. Aldehyde Fixation of Cells for Cryoimmuno Labeling — 134
    IV. Embedding Samples for Cryoimmuno Labeling — 136
    V. Cryosectioning for Cryoimmuno Labeling — 138
    VI. Immunogold Labeling — 140
    VII. Reagents and Solutions — 143
    VIII. Future Outlooks — 146
    References — 148

9. Close-to-Native Ultrastructural Preservation by High Pressure Freezing
Dimitri Vanhecke, Werner Graber, and Daniel Studer

    I. Introduction — 152
    II. Rational — 155
    III. Material — 156
    IV. Methods — 158
    V. High Pressure Frozen Samples — 159
    VI. Discussion — 160
    References — 162

10. High-Pressure Freezing and Low-Temperature Fixation of Cell Monolayers Grown on Sapphire Coverslips
Siegfried Reipert and Gerhard Wiche

    I. Introduction — 166
    II. Materials and Instrumentation — 167
    III. Procedures — 167
    IV. Comments and Pitfalls — 174
    References — 180

11. Freeze-Fracture Cytochemistry in Cell Biology
   Nicholas J. Severs and Horst Robenek

I. Introduction	182
II. Basic Rationale: Solving the Problems of Combining Freeze Fracture with Cytochemistry	183
III. Methods	185
IV. Discussion: Impact on Topical Questions in Cell Biology	191
V. Concluding Comment	201
References	202

PART II Electron Microscopy of Specific Cellular Structure

SECTION 1 The Cell Membrane

12. Three-Dimensional Molecular Architecture of the Plasma-Membrane-Associated Cytoskeleton as Reconstructed by Freeze-Etch Electron Tomography
    Nobuhiro Morone, Chieko Nakada, Yasuhiro Umemura, Jiro Usukura, and Akihiro Kusumi

I. Introduction	208
II. Protocol for Visualization of the Three-Dimensional Structure of the MSK of the Cytoplasmic Surface of the Plasma Membrane	212
III. 3D Structure of the Cytoskeleton-Plasma Membrane Interface	219
IV. Electron Tomography Clarified that Some of the Actin Filaments are Laterally Bound to the Cytoplasmic Surface of the Plasma Membrane	231
References	232

13. Visualization of Dynamins
    Jason A. Mears and Jenny E. Hinshaw

I. Introduction	237
II. Methods and Materials	240
III. Discussion	251
IV. Summary	253
References	254

SECTION 2 The Cytoskeleton

14. Correlated Light and Electron Microscopy of the Cytoskeleton
Sonja Auinger and J. Victor Small

I. Introduction	258
II. Materials and Methods	259
III. Methods	260
IV. Results and Discussion	266
V. Summary	271
References	271

15. Electron Microscopy of Intermediate Filaments: Teaming up with Atomic Force and Confocal Laser Scanning Microscopy
Laurent Kreplak, Karsten Richter, Ueli Aebi, and Harald Herrmann

I. Introduction	274
II. Rationale	277
III. Visualization of Intermediate Filaments *in vitro* and in Cultured Cells	278
IV. Materials	293
V. Conclusions and Outlook	294
References	295

16. Studying Microtubules by Electron Microscopy
Carolyn Moores

I. Introduction	300
II. Molecular Electron Microscopy	302
III. Cellular Electron Microscopy	311
IV. Outlook	314
References	315

SECTION 3 Extracellular Matrix and Cell Junctions

17. Electron Microscopy of Collagen Fibril Structure *In Vitro* and *In Vivo* Including Three-Dimensional Reconstruction
Tobias Starborg, Yinhui Lu, Karl E. Kadler, and David F. Holmes

I. Introduction	320
II. Electron Microscopy of Isolated Collagen Fibrils	327
III. Fibroblast/Fibril Interface in Developing Tendon	334
IV. Discussion	340
V. Conclusions	341
References	341

18. Visualization of Desmosomes in the Electron Microscope
   Anthea Scothern and David Garrod

I. Introduction	348
II. Rationale	351
III. Methods	351
IV. Materials	360
V. Discussion	364
VI. Summary	364
References	365

SECTION 4 The Nucleus

19. A Protocol for Isolation and Visualization of Yeast Nuclei by Scanning Electron Microscopy
    Stephen Murray and Elena Kiseleva

I. Introduction	368
II. Materials and Instrumentation	370
III. Procedures	372
IV. Comments and Problems	385
References	386

20. Scanning Electron Microscopy of Nuclear Structure
    Terence D. Allen, Sandra A. Rutherford, Stephen Murray, Sheona P. Drummond, Martin W. Goldberg, and Elena Kiseleva

I. Introduction	390
II. Rationale	393
III. Methods	393
IV. Colloidal Gold in the SEM	400
V. Immunolabeling Protocol	403
VI. CPD for High Resolution SEM	404
VII. Discussion	407
References	408

21. Electron Microscopy of Lamin and the Nuclear Lamina in *Caenorhabditis elegans*
    Merav Cohen, Rachel Santarella, Naama Wiesel, Iain Mattaj, and Yosef Gruenbaum

I. General Introduction	412
II. *In Vitro* Assembly of Ce-Lamin Filaments	414
III. Preparation of Embryos and Adults for Transmission Electron Microscopy Using Microwave Fixation	416

IV. Preparation of *C. elegans* Embryos and Adults for Conventional Transmission Electron Microscopy by High Pressure Freezing Combined with Freeze Substitution   419
V. Preembedding Immunogold EM Staining of Lamina Proteins in *C. elegans* Embryos   423
VI. Postembedding Immunogold EM Staining of Lamina Proteins in *C. elegans* Embryos   425
VII. Summary   427
References   427

22. Visualization of Nuclear Organization by Ultrastructural Cytochemistry

Marco Biggiogera and Stanislav Fakan

I. Introduction   432
II. Cytochemical Contrasting Approaches   432
III. High Resolution Autoradiography   441
IV. Immunocytochemistry   441
V. Molecular *in situ* Hybridization   443
VI. Identification of Nucleic Acids by Means of Enzymatic Reactions   444
VII. Targeting of Intranuclear Substrates using Enzyme-Colloidal Gold Complexes   446
VIII. Concluding Remarks   446
References   447

23. Scanning Electron Microscopy of Chromosomes

Gerhard Wanner and Elizabeth Schroeder-Reiter

I. Introduction   452
II. Materials and Methods   454
III. Chromosome Preparation   456
IV. Chromosome Structure in SEM   458
V. Chromosome Analysis in SEM   460
VI. Outlook   470
References   473

PART III Cells and Infectious Agents

24. Infection at the Cellular Level

Christian Goosmann, Ulrike Abu Abed, and Volker Brinkmann

I. Introduction   477
II. Methods and Materials   478
III. Overview and Conclusion   494
References   495

25. Electron Microscopy of Viruses and Virus–Cell Interactions

Peter Wild

I. Introduction	498
II. Methods	500
III. Material	515
IV. Discussion	516
V. Summary	520
References	521

Index	525
Volumes in Series	537

CONTRIBUTORS

Numbers in parentheses indicate the pages on which the authors' contributions begin.

Ulrike Abu Abed (477), Max-Planck-Institut für Infektionsbiologie, Charitéplatz 1, 10117 Berlin, Germany

Ueli Aebi (273), M. E. Müller Institute for Structural Biology, Biozentrum, University of Basel, Klingelbergstrasse 70, 4056 Basel, Switzerland

Terence D. Allen (97, 389), Department of Structural Cell Biology, Paterson Institute for Cancer Research, University of Manchester, Manchester M20 4BX, United Kingdom

Sonja Auinger (257), Institute of Molecular Biotechnology, Dr Bohr-Gasse 3 1030, Vienna, Austria

Galina V. Beznoussenko (83), Department of Cell Biology and Oncology, Consorzio Mario Negri Sud, 66030 Santa Maria Imbaro, Chieti, Italy

Marco Biggiogera (431), Laboratorio di Biologia Cellulare e Neurobiologia, Dipartimento di Biologia Animale, Università di Pavia, Piazza Botta 10, I-27100 Pavia, Italy, and Centre of Electron Microscopy, University of Lausanne, Bugnon 27, CH-1005 Lausanne, Switzerland

Volker Brinkmann (477), Max-Planck-Institut für Infektionsbiologie, Charitéplatz 1, 10117 Berlin, Germany

Merav Cohen (411), Division of Cell Biology, MRC-Laboratory of Molecular Biology, Cambridge CB2 0QH, United Kingdom, and Department of Genetics, The Institute of Life Sciences, The Hebrew University of Jerusalem, Jerusalem 91904, Israel

Sheona P. Drummond (97, 389), Post-transcriptional Research Group, Manchester Interdisciplinary Biocentre, University of Manchester, Manchester M1 7ND, United Kingdom

Stanislav Fakan (431), Centre of Electron Microscopy, University of Lausanne, Bugnon 27, CH-1005 Lausanne, Switzerland

E. Fernandez-Segura (19), Department of Histology, Faculty of Medicine, University of Granada, E-10871, Granada, Spain

David Garrod (347), Faculty of Life Sciences, University of Manchester, Manchester, UK

Martin W. Goldberg (109, 389), Department of Biological and Biomedical Sciences, University of Durham, Durham DH1 3LE, United Kingdom

Christian Goosmann (477), Max-Planck-Institut für Infektionsbiologie, Charitéplatz 1, 10117 Berlin, Germany

Werner Graber (151), Institute of Anatomy, University of Bern, 3000 Bern 9, Switzerland

Gareth Griffiths (45), Cell Biology and Biophysics Unit, EMBL, Heidelberg, Meyerhofstrasse 69117

Yosef Gruenbaum (411), Department of Genetics, The Institute of Life Sciences, The Hebrew University of Jerusalem, Jerusalem 91904, Israel

Harald Herrmann (273), Functional Architecture of the Cell, German Cancer Research Center (DKFZ), 69120 Heidelberg, Germany

Jenny E. Hinshaw (237), Laboratory of Cell Biochemistry and Biology, NIDDK, NIH, Bethesda, Maryland 20892

David F. Holmes (319), Wellcome Trust Centre for Cell-Matrix Research, Faculty of Life Sciences, University of Manchester, Manchester M13 9PT, United Kingdom

Karl E. Kadler (319), Wellcome Trust Centre for Cell-Matrix Research, Faculty of Life Sciences, University of Manchester, Manchester M13 9PT, United Kingdom

Elena Kiseleva (367, 389), Institute of Cytology and Genetics, Russian Academy of Science, Novosibirsk 630090, Russia

Laurent Kreplak (273), M. E. Müller Institute for Structural Biology, Biozentrum, University of Basel, Klingelbergstrasse 70, 4056 Basel, Switzerland

Akihiro Kusumi (207), Membrane Mechanisms Project, International Cooperative Research Project (ICORP), Japan Science and Technology Agency, Institute for Integrated Cell-Material Sciences and Institute for Frontier Medical Sciences, Kyoto University, Shougoin, Kyoto 606-8507, Japan

Yinhui Lu (319), Wellcome Trust Centre for Cell-Matrix Research, Faculty of Life Sciences, University of Manchester, Manchester M13 9PT, United Kingdom

John Lucocq (59), School of Life Sciences, University of Dundee, Dundee DDI 4HN, United Kingdom

Iain Mattaj (411), European Molecular Biology Laboratory, Heidelberg, Germany

Jason A. Mears (237), Laboratory of Cell Biochemistry and Biology, NIDDK, NIH, Bethesda, Maryland 20892

Alexander A. Mironov (83), Department of Cell Biology and Oncology, Consorzio Mario Negri Sud, 66030 Santa Maria Imbaro, Chieti, Italy

Carolyn Moores (299), School of Crystallography, Birkbeck College, London WC1E 7HX, United Kingdom

Nobuhiro Morone (207), Department of Ultrastructural Research, National Institute of Neuroscience, National Center of Neurology and Psychiatry, Kodaira 187-8502, Japan

Stephen Murray (3, 367, 389), TEM Service Facility and Department of Structural Cell Biology, Paterson Institute for Cancer Research, University of Manchester, Manchester M20 4BX, United Kingdom

Chieko Nakada (207), Membrane Mechanisms Project, International Cooperative Research Project (ICORP), Japan Science and Technology Agency, Institute for Integrated Cell-Material Sciences and Institute for Frontier Medical Sciences, Kyoto University, Shougoin, Kyoto 606-8507, Japan

Peter J. Peters (131), Department Tumor Biology: H4, The Netherlands Cancer Institute, Amsterdam, The Netherlands 1066 CX

Jason Pierson (131), Department Tumor Biology: H4, The Netherlands Cancer Institute, Amsterdam, The Netherlands 1066 CX

Roman S. Polishchuk (83), Department of Cell Biology and Oncology, Consorzio Mario Negri Sud, 66030 Santa Maria Imbaro, Chieti, Italy

Contributors

Siegfried Reipert (165), Department of Molecular Cell Biology, Max F. Perutz Laboratories, University of Vienna, Dr. Bohr-Gasse 9, A-1030 Vienna, Austria

Karsten Richter (273), Department of Molecular Genetics, German Cancer Research Center (DKFZ), 69120 Heidelberg, Germany

Horst Robenek (181), Institute for Arteriosclerosis Research, University of Münster, Münster, Germany

Sandra A. Rutherford (389), Department of Structural Cell Biology, Paterson Institute for Cancer Research, University of Manchester, Manchester M20 4BX, United Kingdom

Rachel Santarella (411), European Molecular Biology Laboratory, Heidelberg, Germany

Elizabeth Schroeder-Reiter (451), Department of Biology I, Ludwig-Maximilians-Universität München, Menzinger Straße 67, 80638 Munich, Germany

Heinz Schwarz (45), Max Planck Institute for Developmental Biology, Tubingen Germany

Anthea Scothern (347), Faculty of Life Sciences, University of Manchester, Manchester, UK

Nicholas J. Severs (181), Imperial College London, National Heart and Lung Division, London, United Kingdom

J. Victor Small (257), Institute of Molecular Biotechnology, Dr Bohr-Gasse 3 1030, Vienna, Austria

Tobias Starborg (319), Wellcome Trust Centre for Cell-Matrix Research, Faculty of Life Sciences, University of Manchester, Manchester M13 9PT, United Kingdom

Daniel Studer (151), Institute of Anatomy, University of Bern, 3000 Bern 9, Switzerland

Yasuhiro Umemura (207), Membrane Mechanisms Project, International Cooperative Research Project (ICORP), Japan Science and Technology Agency, Institute for Integrated Cell-Material Sciences and Institute for Frontier Medical Sciences, Kyoto University, Shougoin, Kyoto 606-8507, Japan

Jiro Usukura (207), Division of Integrated Projects, EcoTopia Science Institute, Nagoya University, Furo-cho, Chikusa, Nagoya 464-8603, Japan

Dimitri Vanhecke (151), Institute of Anatomy, University of Bern, 3000 Bern 9, Switzerland

Gerhard Wanner (451), Department of Biology I, Ludwig-Maximilians-Universität München, Menzinger Straße 67, 80638 Munich, Germany

Alice Warley (19), Centre for Ultrastructural Imaging, King's College London, Guy's Campus London SE1 1UL, United Kingdom

Paul Webster (45), Cell Biology and Biophysics Unit, EMBL, Heidelberg, Meyerhofstrasse 69117

Gerhard Wiche (165), Department of Molecular Cell Biology, Max F. Perutz Laboratories, University of Vienna, Dr. Bohr-Gasse 9, A-1030 Vienna, Austria

Naama Wiesel (411), Department of Genetics, The Institute of Life Sciences, The Hebrew University of Jerusalem, Jerusalem 91904, Israel

Peter Wild (497), Electron Microscopy, Institutes of Veterinary Anatomy and of Virology, University of Zürich, CH-8057 Zürich, Switzerland

PREFACE

Visualisation of the mechanics of cell activity is crucial to our understanding of normal cellular processes, their variations throughout the cell cycle and responses to altered circumstances such as stress and disease. The majority of biologists (at least of my generation) were most likely initially attracted to the subject by their earliest views down a microscope, however rudimentary in nature. Even for the current generation who may prefer the intellectual satisfaction (but not beauty) of the presence and absence of specific proteins in gels, or tables showing variation in the genome, proteome, metabolome etc; we must still allow for the fact that a cells response to change is usually in the form of production of new protein, and that protein has to be in the right place at the right time to affect the desired response. Consequently there will always be a major role for microscopy in cell and molecular biology.

The role of electron microscopy in cell biology has shown the typical phases of any novel technology, at the start, (in the late '40s and subsequently) Porter, Sjostrand, Palade *et al.*, began to map out cellular ultrastructure in terms of morphological features, to be followed by characterisation of the structures involved by various labelling techniques. As with all new techniques, the phase of intensive application for the fresh information generated is often followed by a relative reduction in use of the technique for its own sake, and a redirection of effort to address specific problems. Over the last few years, the tools that have been generated by molecular biology to visualise specific molecules within the living cell have led to massive investment in light microscopy, with matching results in terms of our understanding of dynamic events within the cell. The nature of electron microscopy does not allow live cell imaging, and EM observations will always be a snapshot of the events occurring at the point of initial intervention for EM preservation. This intervention can occur during a live series of observations, or a series of time course EM preparations can be run in parallel with live cell imaging. Optimal EM preservation involves rapid freezing, when the cell and its contents become immobilised *in situ* in amorphous ice within milliseconds. More traditional fixation methods are slower, and therefore have a greater potential for induced change, but there are circumstances where freezing is not an accessible option for a particular experimental system, and chemical fixation still generates unique (and useful) information. As freezing alone (via frozen sections) may fail to produce sufficient contrast for visualisation of some subcellular elements the choice of approach for EM observation needs to consider a spectrum of approaches with a view to the required endpoint. A combination of 'instant' stabilisation by rapid freezing is often followed by resin embedding via freeze substitution to allow the production of thin sections, which form a more easily managed starting point for such activities as tomographic 3D reconstruction.

Given the exquisite sensitivity of light microscopic detection of fluorescence from a single molecule within the cell, the uncommitted experimenter may well ask why EM should be used at all. There are two overwhelming reasons for EM usage, firstly resolution, and secondly the macromolecular context of the molecule of interest. Resolution in electron microscopy is at the Angstrom level, and in light microscopy the figure is generally around 200 nanometres (disputed by some!) resulting in a 40 fold improvement in the EM (McIntosh, 2006). Secondly, the nature of fluorescence in LM is to present a light dot against a dark background, and consequently there is no indication of the macromolecular context of the molecule in question. Co-localisation of other fluorophores may partially address this limitation, but this approach cannot compare with the ability to produce 3D reconstruction at molecular resolution available from EM tomography. At this point, it is worth stressing that LM and EM should not be thought of as mutually exclusive, as they are truly complementary, and there are chapters in this volume where one and the same cell is imaged firstly live in LM, then subsequently imaged in EM. This approach incorporates a fluorescent LM tag (such as green fluorescent protein) allowing dynamic visualisation in live cell fluorescence microscopy, then fixation for EM and relocation with specific antibodies in the EM (marked with colloidal gold particles) (see Mironov chapter for TEM relocation, and Drummond chapter for SEM relocation).

EM preservation protocols have always been the source of healthy debate among those who practise them. The spectre of artefact has always been in the background, despite a 50 year contribution of information that has subsequently been confirmed by a variety of complementary approaches. More sophistication in preservation protocols approach has extended previous information, but this is the very nature of scientific advance, and does not render previous efforts either inaccuracte or insignificant. One current area of controversy is the view which tends to overemphasize shortcomings in 'conventional' preparation protocols (i.e. chemical fixation) and suggests that material that has reached the electron beam without an initial rapid freezing stabilisation is hardly worth observation. There are however some circumstances in which an initial freezing step is precluded, or simply not required as in the case of purified macromolecular complexes or isolated organelles, where negative staining EM can be performed in minutes, producing high resolution information. Negative staining may be used as a precursor to the more sophisticated approach of low angle rotary shadowing, but is an incredibly useful short cut to allow concentrations etc to be quickly and easily optimised. As macromolecular complexes are the universal building block at the sub-organelle level, visualisation at this level of biological structure is clearly of great importance. (see Mears chapter) Also, as these complexes may have been purified as the result of an extended biochemical purification procedure, or even assembled *in vitro* from recombinant proteins, clearly the requirement for rapid freezing *'in situ'* is irrelevant. Freezing can also exclude some antibody tagging approaches, so that the full range of EM methodologies should be considered and a choice made that is relevant to the desired structural information.

A final consideration in the application of EM to sub-cellular structure is also the type of EM utilised. In a recent and extremely comprehensive MCB volume (vol 79, 2006) (which should be consulted in tandem with this volume) where nearly 900 pages are devoted to 'Cellular Electron Microscopy' there is no mention of scanning EM. (SEM) Since the advent of field emission SEM, resolution is in the same order as TEM for biological material. Although SEM for subcellular ultrastructure therefore might be considered a 'minority' interest, it has great advantages in the visualisation of surfaces within the cell, which are the interfaces at which the majority of significant events occur. Scanning EM has the advantage of easy and wide coverage of material, (very similar to light microscopy) providing the ability to scan large areas for relatively rare events which would require unfeasibly labour intensive efforts to locate in thin sections for TEM investigation. Surface imaging in SEM also allows direct 3D visualisation of membrane events in a way that TEM sections cannot replicate. For 30 years, the nuclear pore complex was thought to be symmetrical around the plane of the nuclear envelope, until Hans Ris (1991) and others (Goldberg and Allen, 1992) showed otherwise by the use of Field Emission SEM. A further example of the advantage of surface imaging is in the formation of nuclear pore complexes, which are initiated by a point fusion between inner and outer nuclear envelope membranes, occurring initially as a 5 nm diameter 'dimple' in the outer membrane at the point of fusion. Given that the thinnest section for TEM is around 50 nms, to see a variation in the membrane contours that occurs within only one tenth of the full thickness of the section is unlikely. Surface imaging within the highly fibrous environment of the nucleus in SEM allows direct visualisation in 3D of the interaction of fibres with substructures such a Cajal bodies or nucleoli in a way that is difficult in the extreme to replicate from thin section TEM (Kiseleva *et al.*, 2004). Exposure of internal cellular surfaces, the nuclear interior or extraction of membranes for SEM imaging can be much simpler than anticipated (see Allen chapter).

This volume contains what might be considered a somewhat eclectic mix of chapters. What I tried to do in the commissioning was to cover a wide variety of both subcellular, intracellular and extracellular sites with the idea that this diversity would have a potential example or relationship to an area of cell biology that might be deemed useful by the reader considering the use of Electron microscopy as a new element in his/her research profile. Consequently there are chapters that address the cytoskeleton, with EM imaging of actin (Small); intermediate filaments (Aebi) and microtubules (Moore). The plasma membrane skeleton and associated structures (Kusumi; Mears) moves to the cell boundary; then extracellular collagen (Kadler) and junctional complexes (Scothern) between cells. Moving back inside the cell, there are methods for the characterisation of nuclei in yeast (Murray) nuclear organisation by EM cytochemistry (Fakan), nuclear lamins (Grunebaum), and SEM in the study of the nucleus (Allen) and chromosomes (Wanner). These are complemented by methodologies for visualisation of macromolecular complexes (Mears), high pressure freezing (Studer; Peters; Webster; Reipert, Murray) in variety of applications, including tissue culture cells and yeast. This approach to freezing is complemented by freeze-fracture and immunocytochemistry (Severs). High resolution SEM immunolabelling (Goldberg) and analytical EM (Warley) complete the mainly methodology approaches, but I thought it was also important to cover viral interactions (Wild) and infection at the cellular level (Brinkmann) in view of the novel structural alterations in these instances. I would also urge the reader to look beyond the titles of the chapters, lest they miss such gems as the best protocols yet published for *C.elegans*, (Cohen) an organism which will continue to swim in standard fixative for up to 30 mins! Best of all is to browse the illusrations and see which type of presentation would best fit the intended use.

I should like to thank all the contributing authors, who have made the effort to lay out their methodologies in a way that is useful to those joining or already working in the field. Thanks to Paul Matsudiara, who thought that a more basic introduction would stand as a useful addition to the recent 'Cellular Electron Microscopy' (vol 79 in this series). Thanks also to Luna Han, Patricia Gonzalez, Kirsten Funk and Charles Neelakandan for their encouragement and help in the production of this book.

T. D. Allen

References

McIntosh, J. R. (2006). 'Cellular Electron Microscopy'. *Methods in Cell Biol.* **79**, 1–5.

Ris, H. (1991). The three dimensional structure of the nuclear pore complex as seen by high voltage electron microscopy and high resolution low voltage scanning electron microscopy. *EMSA Bull.* **21**, 54–56.

Goldberg, M. W., and Allen, T. D. (1992). High resolution scanning electron microscopy of the nuclear envelope: demonstration of a new, regular, fibrous lattice attached to the baskets of the nucleoplasmic face of the nuclear pores. *J. Cell Biol.* **119**, 429–440.

Kiseleva, E., Drummond, S. P., Goldberg, M. W., Rutherford, S. A., Allen, T. D., and Wilson, K. L. (2004). Actin- and protein-4.1-containing filaments link nuclear pore complexes to subnuclear organelles in Xenopus oocyte nuclei. *J. Cell Sci.* **117**, 2481–2490.

PART I

Exploring the Organisation of the Cell by Electron Microscopy

SECTION 1

Basic Transmission and Scanning Electron Microscopy

CHAPTER 1

High Pressure Freezing and Freeze Substitution of *Schizosaccharomyces pombe* and *Saccharomyces cerevisiae* for TEM

Stephen Murray

TEM Service Facility
Paterson Institute for Cancer Research
University of Manchester
Manchester, United Kingdom

    Abstract
  I. Introduction
 II. Materials and Instrumentation
III. Procedures
      A. Preparation of Sample Holders
      B. Filtration of Yeast Sample
      C. High-Pressure Freezing of the Sample
      D. Freeze Substitution of Yeast
IV. Comments and Problems
    References

Abstract

The use of standard room temperature chemical fixation protocols for the ultrastructural preservation of yeast and subsequent observation under the electron microscope is fraught with difficulties. Many protocols require the use of enzymatic digestion of the cell wall in order to facilitate the entry of fixatives into the cell interior. Others rely on the use of permanganate-based fixative solutions, which whilst enabling overall preservation of the cell, does require multiple centrifugation, washing, and resuspension steps. This often results in the significant loss of sample volume whilst the use of permanganate can cause extraction of cytoplasmic components. The use of low temperature techniques and in particular high pressure freezing (HPF) and freeze substitution (FS) overcomes many of these problems.

With the recent advances in cryotechnologies and in particular the development of commercially available equipment such as the high pressure freezer, the level of ultrastructural preservation attainable in electron microscopy has increased markedly. It is now possible to capture dynamic time sensitive events and to place them in their ultrastructural context with a level of resolution which at the present time can only be achieved with electron microscopy.

I. Introduction

The use of high pressure freezing (HPF) and freeze substitution (FS) as a method of preserving yeast can now be considered a mainstream technique. As with all procedures in electron microscopy, there are numerous HPF and FS protocols published in the scientific literature, with each individual worker having a preference for a particular processing regime. Some are designed to with one specific aim e.g. preservation of epitopes (Monaghan and Robertson, 1990; Monaghan *et al.*, 1998; Neuhaus *et al.*, 1998) while others would seem to be counterintuitive e.g. addition of water to the FS cocktail (van Donselaar, *et al.*, 2007; Walther and Ziegler, 2002). All of the protocols have one single aim, the preservation of the tissue or cell being investigated in as close to the native state as possible. The ultimate is of course to observe the sample in a frozen fully hydrated state with out addition of chemical fixatives (Al-Amoudi *et al.*, 2004; McDowall *et al.*, 1983, 1984; Michel, 1991). However, for the purposes of this chapter, the author deals purely with a HPF and FS protocol he has found to work consistently well with both wild type and mutant strains of *Schizosaccharomyces pombe* and *Schizosaccharomyces cerevisae*.

II. Materials and Instrumentation

HPF was performed using a Bal-Tec HPM010 high pressure freezer (Bal-Tec AG Principality of Liechtenstein) and using interlocking brass hats as the specimen carrier (Swiss Precision, Inc., Palo Alto, CA, USA). Subsequent FS was performed using an automatic FS unit (Leica AFS; Leica Microsystems, Vienna) with the FS solution being contained in 1.5-ml conical Eppendorf centrifuge tubes (catalogue 0030 120.086; Eppendorf UK, Ltd., Cambridge, UK).

Yeast cultures were vacuum concentrated onto 0.45 µm membrane filters (catalogue # VLP02500; Millipore UK Ltd., Watford, UK) using a KNF Laboport N86KT.18 dry vacuum pump (KNF Neuberger UK Ltd.) and Sartorius suction flask and 25 mm glass filter holder assembly (catalogue 16672 & 16306; Sartorius Ltd., Epsom, UK). To facilitate quick vacuum release, an in line Edwards AV10K manual air admit valve and T-piece assembly were placed between the vacuum pump and the suction flask. An image of the filtration system can be seen in Fig. 2. To aid sample loading into the specimen carriers, any standard stereo zoom dissecting microscope and cold light source may be used. Removal of the yeast sample from the filter and its subsequent loading into the brass hat was achieved using a pointed cocktail stick. Fig. 1 shows additional useful tools for the handling, manipulation, removal, and separation of the sample carriers following HPF include two pairs of Dumont Dumoxel #3 medical grade tweezers catalogue T5272 (Agar Scientific, Essex, UK) used to manipulate the brass hats and load them into the specimen rod. Two flat bladed screwdrivers with fine tips, CK Xonic 4880X, one with 1.5 mm tip, the other with 2.5 mm tip (catalogue 2508619214 & 2508618794; RS Components, Corby, UK). The screwdrivers are used to pry the brass hat assembly apart once frozen.

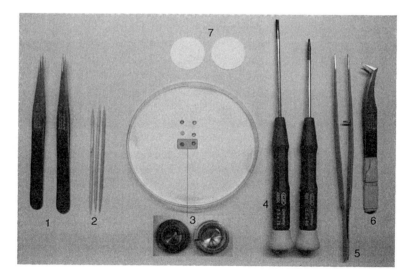

Fig. 1 Tools used during loading and manipulation: (1) Dumont No. 3 forceps, (2) Pointed applicator sticks for paste scraping/loading, (3) Interlocking brass sample carriers, (4) Fine blade screwdrivers for separating hats following freezing, (5) Insulated cryoforceps for handling frozen sample, (6) Dumont No. 6 forceps for pushing sample carrier from specimen rod, (7) Millipore membrane filter 0.45 µm, 22 mm.

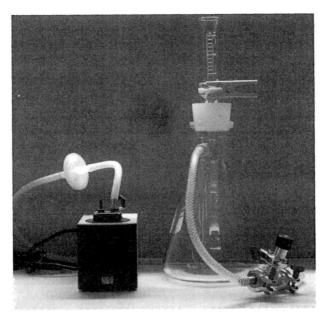

Fig. 2 Filtration system for sample harvesting (manual air inlet valve on the right).

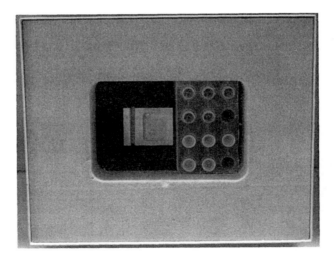

Fig. 3 Cryowork box with cryotubes and holder.

A pair of Dumont Dumostar #6 medical grade tweezers (catalogue T5277; Agar Scientific, Essex, UK) is useful to push the frozen brass hat assembly from the HPF specimen rod should it be required. The frozen samples are placed into prelabeled 1 ml CryoPlus tubes held in the work chamber of the HPF (catalogue 72.377;

Sarstedt, Leicester, UK). In order to hold the vials and allow easy attachment of the lids, the author has found it invaluable to use a Sarsted Cryorack 40 cut to size and fixed firmly to the bottom of the HPF cryo work chamber (see Fig. 3) using Araldite or epoxy resin (catalogue 93.856.040; Sarstedt, Leicester, UK).

Other chemicals and consumables mentioned in this article are commonly available from electron microscopy supply companies.

III. Procedures

A. Preparation of Sample Holders

1. Clean the Brass Hats as Follows

    - Ultrasonicate for 5 min in acetone
    - Rinse in double distilled water
    - Ultrasonicate in 70% ethanol
    - Dry using a hair dryer
    - Place in a plastic Petri dish lined with hardened filter paper ready for use

B. Filtration of Yeast Sample

The cells are grown to mid log phase equivalent to $0.4-1.0 \times 10^7$ cells/ml. At this density approximately 50–100 ml of culture will be required to produce a paste which should be slightly glossy and have an apple sauce consistency. It should be noted that the degree of filtration required to produce the paste varies from strain to strain. If upon removal of the membrane filter the paste is found to be to wet, excess moisture can be removed by carefully dabbing the underside of the membrane filter onto hardened filter paper. If the paste is too dry, nothing can be done. Too much fluid in the paste will result in poor freezing, too little will result in poor FS and embedding.

1. Place a fresh membrane filter into the filter assembly. Turn on the vacuum pump and ensure that the vacuum release valve is fully closed.
2. Pour the yeast culture into the filter assembly and top up as required.
3. At the moment the last visible trace of solution disappears, turn off the pump and open the vacuum release valve.
4. Unclamp the filter funnel and remove the membrane filter.
5. Using a cocktail stick carefully scrape some of the paste from the filter and load it into the bottom of the brass hat pair. The hat should be overfilled and no air bubbles should be seen. Air bubbles will result in poor freezing.
6. Place the top part of the brass hat onto the bottom and press down using the forceps.
7. Quickly place the complete assembly into the specimen rod of the HPF and clamp into position.

C. High-Pressure Freezing of the Sample

Following filtration, the filling of the specimen carrier and the loading and freezing of the sample should take between 20 and 30 s. While this procedure can be carried out by a single operator, it is much better to have an additional person to assist in the filtration and sample loading of the brass hats. Additionally, the HPF sample rod will need to be defrosted between each freezing cycle. The author has found the purchase of an additional specimen rod invaluable in speeding up the process. While one rod defrosts and dries out, the other can be put into use. All tools used subsequent to the freezing process must be precooled to LN2 temperature before handling the sample.

1. Insert the specimen rod into the freezing chamber of the HPF and secure with the locking bolt.
2. Initiate the freezing cycle by pressing the "Jet" button.
3. As soon as the Jet button has been pressed, allow a couple of seconds to pass before removing the locking bolt and rapidly transferring the specimen rod into the cryo work chamber of the HPF.
4. Allow the end of the rod to rest for a few seconds in the liquid nitrogen before unclamping the specimen holder.
5. Under nitrogen and on the metal working platform of the cryowork chamber, push the brass hat from the specimen holder using the precooled No. 6 forceps if necessary.
6. Using the two precooled screwdrivers, separate the two halves of the brass hat by prising them apart. Take care at this stage, it is very easy to apply too much pressure and catapult the brass hat out of the cryowork station.
7. Place the separated halves of the specimen carrier into the relevant prelabeled vial using precooled No. 6 forceps.
8. Once all samples have been frozen place the lids on the cryovials and transfer them either to a storage Dewar or the FS unit.

D. Freeze Substitution of Yeast

After much experimentation, the author has found that the addition of water to the FS cocktail has resulted in the optimal preservation of his particular samples. It was of particular importance that the nuclear membranes, the spindle pole bodies and microtubules be well delineated and preserved. The protocol which follows fulfills these requirements and has the added bonus that subsequent staining of ultrathin sections requires the use of Reynolds lead citrate only (Reynolds, 1963).

1. Programming of the AFS

    1. Set the first temperature step to −90 °C and the time to 72 h
    2. Set the first slope to +5 °C/h

3. Set the second temperature step to $-20\ ^\circ C$ and the time to 2 h
4. Set the second slope to $+5\ ^\circ C/h$
5. Set the final temperature step to $+4\ ^\circ C$ and the time to 4 h
6. Place the AFS into pause mode and allow the temperature to reach $-90\ ^\circ C$

2. Prepare the Freeze Substitution Cocktail

   1. Prelabel the 1.5 ml Eppendorf tubes as required. Up to ten 1.5 ml Ependorfs can be accommodated at one time.
   2. Dissolve 0.2 g of crystalline osmium tetroxide in 9.5 ml of acetone.
   3. Add 500 µl of 2% aqueous uranyl acetate to the acetone/osmium mix.
   4. Dispense 1 ml into each of the Eppendorfs.
   5. Place the Eppendorfs into the AFS and allow the temperature to equilibrate.
   6. Due to the toxicity of osmium tetroxide and uranyl acetate Steps 2–4 of the above must be carried out in a fume hood.

3. Transfer of Specimens to AFS

   1. Place the cryovials containing the samples into a small transfer container filled with liquid N_2.
   2. Using precooled No. 3 forceps, quickly transfer the specimen carrier from the cryotube to the relevant pre labeled Eppendorf tube in the AFS.
   3. Once all samples have been placed into the AFS, cancel the pause function to begin FS cycle.

4. Resin Infiltration and Embedding of Samples

   During FS, the samples will readily separate from the specimen carrier. After 2 h at $+4\ ^\circ C$, the FS cocktail is rinsed out with fresh acetone. This can be done with the Eppendorf carrier in the AFS or, it can be transferred to a fume hood and placed over ice.

   1. Wash the specimens three times with fresh acetone over a 1 h period at $+4\ ^\circ C$.
   2. During the third wash, remove the empty specimen carriers from the Eppendorfs.
   3. Carefully transfer each sample from the Eppendorf into a prelabeled 7-ml glass vial containing 2 ml of acetone. The transfer can be effected either by using fine point forceps (No. 3) or a wide bore disposable plastic Pasteur pipette.
   4. Infiltrate the sample with Spurr's (Spurr, 1969) resin/acetone mixtures. If possible, degas the resin under vacuum prior to dilution with acetone
      - 1:7 resin/acetone 3 h
      - 1:3 resin/acetone overnight

- 1:1 resin/acetone 4 h
- 3:1 resin/acetone 4 h or overnight
- 100% resin 4 h (with no lid on sample vials)
- 100% resin 4 h (under vacuum if available)
- 100% resin overnight (under vacuum if available)
- Place a small volume of fresh resin into prelabeled Beem capsules
- Carefully transfer each sample from the glass vial to the Beem capsule using a pointed wooden stick or wide bore plastic Pasteur pipette and centralize in the capsule
- Fill the capsules with resin
- Polymerize the resin for 48 h at 60 °C or 24 h at 70 °C under vacuum if possible.

5. Once polymerized trim a standard trapezoid block face and proceed to cut 50–70-nm ultrathin sections and collect on formvar/carbon coated grids.
6. Post stain the sections for 5 min using Reynold's lead citrate in the presence of sodium hydroxide pellets.
7. Wash sections to remove stain.
8. Carefully remove excess water and allow to dry.
9. Image under the TEM at 80 kV.

IV. Comments and Problems

It has been reported that the presence of water in the FS medium in excess of 1% severely reduces the ability of acetone to replace the sample ice (Humbel and Mueller, 1986) Certainly the addition of water to the FS medium would seem counter intuitive but it has been demonstrated in numerous publications that addition of water to the FS medium dramatically enhances the visibility of membranes and in particular the nuclear envelope in yeast (Fig. 4). It is possible to achieve improved membrane visibility using other protocols (Giddings, 2003), but this often requires the use of two different FS mediums and has the considerable challenge of washing and replacement steps being carried out at −90 °C.

It has been reported that the use of epoxy resin as a fixative during FS of *Caenorhabditis elegans* resulted in a more complete preservation of cellular proteins and membranes (Matsko and Mueller, 2005). Results with yeast show some improvement in membrane visibility (Fig. 5) when compared to a standard osmium/acetone protocol (Fig. 6), but are not as remarkable as those in which water is added to the FS medium (Fig. 4). It should also be noted that the concentration of water in the FS medium is important. There appears to be virtually no difference in terms of membrane visibility in yeast when only 1% water is added, compared to FS medium containing no water (Fig. 6). Another intriguing observation is that in the presence of high water concentration (10%) *S. cerevisiae* appears to be well

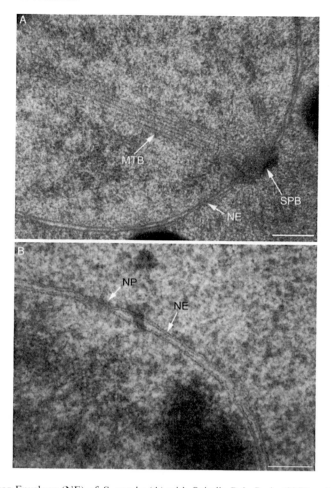

Fig. 4 Nuclear Envelope (NE) of *S. pombe* (A) with Spindle Pole Body (SPB) and Microtubules (MTB) and (B) Nuclear Pore (NP). Sample freeze substituted in medium containing 2% osmium tetroxide/0.1% uranyl acetate/5% water in acetone. Each element of the NE is clearly visible, demonstrating typical membrane structure. Bars (A) 200 nm and (B) 100 nm.

preserved while *S. pombe* is very poorly preserved and shows distinct ice damage (Fig. 7), which would seem to imply that effects of water addition are not only concentration dependant but also sample dependant. It is recommended that a water concentration of 5% be used initially and if necessary adjustments made during subsequent FS runs.

One common artifact induced by HPF particularly when freezing yeast is the rupture of the cell wall and/or the nuclear envelope (Fig. 8). In the author's experience, this occurs in 5–10% of yeast cells both wild type and mutant (an observation also made by M. Morphew, personal communication).

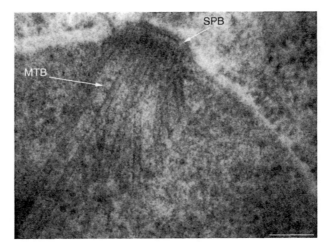

Fig. 5 SPB and MTB of *S. cerevisiae* freeze substituted in epoxy/araldite embedding mix. Bar 100 nm.

1. Safety

It is extremely important when using cryogenic equipment that the manufacturer's recommendations and safety instructions are followed. It is essential that the safe handling and use of liquid nitrogen and a thorough understanding of the very specific hazards associated with it are clearly understood before any work commences. Many of the chemicals used for FS and embedding are at the very least irritants and many are extremely toxic and even carcinogenic. They should therefore always be handled in a fume hood with the operator wearing suitable protective clothing and gloves. When using resins, standard latex or nitrile gloves do not afford sufficient protection. Instead, the use of vinyl gloves is recommended. Once contaminated, gloves should be changed immediately. The exhaust from the AFS should be placed into the fume hood. Care should be exercised when handling the polymerized resin blocks. If it is necessary to file or hacksaw the blocks, the operator should where gloves and a mask and all debris and dust removed immediately.

2. Vacuum Concentration of the Yeast Culture and Loading of Specimen Carriers

It is very important that the yeast paste produced by vacuum filtration is neither too wet nor too dry. The ideal paste should have an apple sauce consistency. When scraping the paste from the Millipore filter, care should be taken not to rip it with the tip of the cocktail stick. When loading the sample carriers, it is essential for good freezing that no air pockets are present in the sample or the carrier. The paste will dry very quickly so speed is important when loading the sample carrier. The paste is also quite sticky and during handling with forceps, the specimen carriers will quite happily attach themselves to the tips. This can provide the operator with minutes

1. High Pressure Freezing and Freeze Substitution

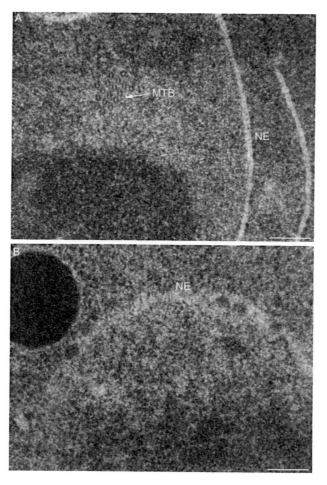

Fig. 6 Comparison of the nuclear envelope (NE of *S. pombe* freeze substituted in medium ± water: (A) Sample freeze substituted in medium containing 2% osmium tetroxide/0.1% uranyl acetate/1% water in acetone showing MTB in cross section. (B) 2% osmium tetroxide/0.1% uranyl acetate in acetone. Bars 200 nm.

of endless amusement as they try to place the sample into the specimen freezing rod. It is best to have two pairs of forceps in order to overcome such difficulties.

3. High Pressure Freezing with the Bal-Tec HPM010

- Ensure that the alcohol reservoir contains sufficient isopropanol for the run.
- Ensure that the instrument has been thoroughly dried by running the air function for at least an hour prior to cool down.

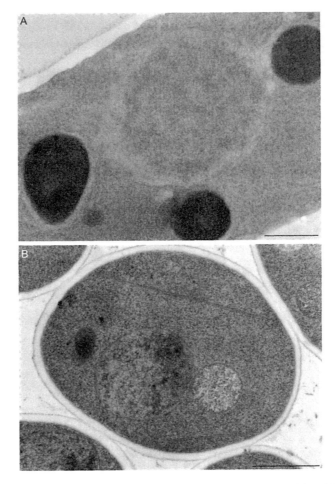

Fig. 7 Tolerance of (A) *S. pombe* and (B) *S. cerevisiae* to FS medium containing 10% added water. Samples freeze substituted in medium containing 2% osmium tetroxide/0.1% uranyl acetate/10% water in acetone. Bars 500 nm.

- Ensure that the compressor outlet pressure and alcohol pressure settings are correct.
- Ensure that there is sufficient LN2 in the supply dewar for the instrument cool down and freezing run.
- Ensure that the output pressure from the supply dewar is correct.
- Fill the cryowork box with liquid nitrogen to the correct level and allow to equilibrate.
- Replace and grease the O-rings on the sample freezing rods and the recording rod.

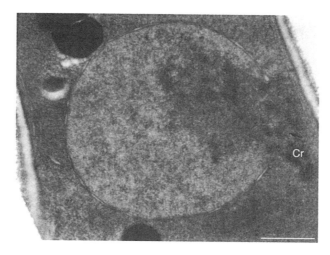

Fig. 8 Ruptured nuclear envelope (Arrow) of high pressure frozen *S. pombe*. Note the presence of chromatin (Cr) in the cytoplasm of the cell. Bar 500 nm.

- Carry out three test shots using the recording rod and ensure that the recorded values are within range.
- Grease the O-ring on the sample freezing rods every five shots and replace every 50 or at the start of each run.
- Precool all instruments to be used in handling the frozen specimen in the cryowork box.
- Try and insulate the end of the forceps with tape to prevent cold burns when handling.
- Allow 90 s to pass between each shot.
- Always defrost and dry the rod between each shot.
- Take care when manipulating the frozen sample and handling the precooled forceps/tools; wear two pairs of nitrile gloves, this affords only minimal protection from the extremely cold tools but it is better than nothing. Cryogloves cannot be worn because of the severe restriction to dexterity.
- Take care when screwing the specimen carrier clamp into position on the freezing rod, it is very easy to trap the nitrile gloves in the assembly.
- As soon as the "Jet" button has been pressed, count quickly to four and then remove the locking pin and specimen rod.
- When removing the sample from the rod always keep it within the work box and at LN2 temperature.
- When separating the sample carrier halves with the screwdrivers, take care not to apply too much force—the whole sample can be catapulted out of the box and lost for good.

4. Freeze Substitution, AFS and Embedding Issues

On the whole FS and the use of the AFS is fairly trouble free. Major issues include poor preservation and/or ice damage. This can result from a number of causes all of which are operator related

- Failure to pre cool instruments used for handling of the frozen specimen.
- Poor transfer or handling of specimen after freezing or during FS.
- Yeast paste too wet or too dry prior to freezing.
- TF setting on the AFS too high or AFS lid left open for long periods causing freezing of the FS medium.

In its grossest form ice damage can manifest itself as lattice type pattern or large holes within the cell. Subtle damage is more difficult to interpret. In general, segregation or granularity of the chromatin or within the cytoplasm is an indicator that ice damage has occurred (Fig. 9).

Resin embedding of yeast can be problematic and poor infiltration is not uncommon. If the initial paste was too dry prior to freezing, subsequent embedding is adversely affected. A slow incremental infiltration protocol together with vacuum infiltration of the samples once in pure resin is a major help in reducing these problems. It is, however, important to stress that continuous pumping of the sample during vacuum infiltration can be detrimental to good polymerization. This is because the individual resin components have different vapor pressures and continuous pumping reduces the effective concentration of the more volatile components. As a result the original formulation of the resin is changed and with it, the polymerization and hardening characteristics. It is better to attain a specific level (100 mb) of vacuum in the chamber and then isolate it from the pump.

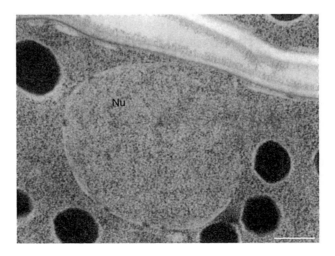

Fig. 9 Ice damage in the nucleus (Nu) of *S. pombe*. Bar 500 nm.

The use of a slow speed angled rotator is also beneficial during the infiltration steps.

The original formulation of Spurr's resin is no longer available due to the toxicity of one of its major components, ERL4206. However, a modified formulation which replaces the original ERL4206 with ERL4221D is available from TAAB laboratory supplies, UK. This particular formulation is only slightly more viscous than the original and is to be recommended.

Acknowledgment

I thank Victor Alvarez and Iain Hagan of the Cell Division group, for the provision *S. pombe* and *S. cerevisiae* cultures.

I also thank Sandra Rutherford for expert technical assistance. This work was supported by CRUK and carried out at The Paterson Institute for Cancer Research, University of Manchester, UK.

References

Al-Amoudi, A., Chang, J. J., Leforestier, A., McDowall, A., Salamin, L.-A., Norlén, L. P. O., Richter, K., Sartori Blanc, N., Studer, D., and Jacques Dubochet, J. (2004). Cryo-electron microscopy of vitreous sections. *EMBO J.* **23**(18), 3583–3588.

Giddings, T. H. (2003). Freeze-substitution protocols for improved visualization of membranes in high-pressure frozen samples. *J. Microsc.* **212**(1), 53–61.

Humbel, B. M., and Müller, M. (1986). Freeze substitution and low temperature embedding. In "The Science of Biological Specimen Preparation" (M. Müller, R. P. Becker, A. Boyde, and J. J. Wolosewick, eds.), pp. 175–183. SEM, AMF O'Hare, IL.

Matsko, N., and Mueller, M. (2005). Epoxy resin as fixative during freeze-substitution. *J. Struct. Biol.* **152**(2), 92–103.

McDowall, A. W., Chang, J. J., *et al.* (1983). Electron microscopy of frozen hydrated sections of vitreous ice and vitrified biological samples. *J. Microsc.* **131**(Pt 1), 1–9.

McDowall, A. W., Hofmann, W., *et al.* (1984). Cryo-electron microscopy of vitrified insect flight muscle. *J. Mol. Biol.* **178**(1), 105–111.

Michel, M. M. (1991). Cryosectioning of plant material frozen at high pressure. *J. Microsc.* **163**(1), 3–18.

Monaghan, P., *et al.* (1998). High-pressure freezing for immunocytochemistry. *J. Microsc.* **192**(3), 248–258.

Monaghan, P., and Robertson, D. (1990). Freeze-substitution without aldehyde or osmium fixatives: ultrastructure and implications for immunocytochemistry. *J. Microsc.* **158**, 355–363.

Neuhaus, E. M., Horstmann, H., *et al.* (1998). Ethane-freezing/methanol-fixation of cell monolayers: A procedure for improved preservation of structure and antigenicity for light and electron microscopies. *J. Struct. Biol.* **121**(3), 326–342.

Reynolds, E. S. (1963). The use of lead citrate at high pH as an electron-opaque stain in electron microscopy. *J. Cell Biol.* **17**, 208–212.

Spurr, A. R. (1969). A low-viscosity epoxy resin embedding medium for electron microscopy. *J. Ultrastruct. Res.* **26**(1–2), 31–43.

van Donselaar, E., Posthuma, G., *et al.* (2007). Immunogold labeling of cryosections from high-pressure frozen cells. *Traffic* **8**(5), 471–485.

Walther, P., and Ziegler, A. (2002). Freeze substitution of high-pressure frozen samples: The visibility of biological membranes is improved when the substitution medium contains water. *J. Microsc.* **208**(1), 3–10.

CHAPTER 2

Electron Probe X-ray Microanalysis for the Study of Cell Physiology

E. Fernandez-Segura* and Alice Warley[†]

*Department of Histology
Faculty of Medicine
University of Granada
E-10871, Granada, Spain

[†]Centre for Ultrastructural Imaging
King's College London
Guy's Campus London SE1 1UL
United Kingdom

    Abstract
I. Introduction
II. Rationale
III. Methods
    A. Specimen Preparation
    B. Treatment Post Cryofixation
    C. Analysis
    D. Deriving Qualitative and Quantitative Information from the Spectrum
IV. Equipment
    A. Electron Microscopes
    B. X-ray Detection Systems
V. Discussion
    References

Abstract

Of the analytical electron microscopy techniques available, electron probe X-ray microanalysis has been most widely used for the study of biological specimens. This technique is able to identify, localize, and quantify elements both at the whole cell and at the intracellular level. The use SEM or TEM to analyze individual whole cells gives a simple and rapid method to study changes in ion transport after stimulation, whereas the analysis of thin sections of cryoprepared cell sections, although technically more difficult, allows details about ionic content in intracellullar compartments, such as mitochondria, ER, and lysosomes, to be obtained. In this chapter the principles underlying X-ray emission are briefly outlined, step-by-step methods for specimen preparation of whole cells and cell sections for microanalysis are given, as are the methods used for deriving quantitative information from spectra. Areas where problems might occur have been highlighted. The different areas in which X-ray microanalysis is being used in the study of cell physiology are briefly reviewed.

I. Introduction

In biological sciences electron microscope (EM) imaging is being used to its fullest but the capability for element detection is one of the less well-known techniques. There are three methods for detection of elements using EM, wavelength dispersive spectroscopy (WDS), electron energy-loss spectroscopy (EELS), and energy dispersive spectroscopy (EDS), more commonly known as electron probe X-ray microanalysis (EPXMA). Of these, WDS is now only rarely used for biological specimens and EELS is generally considered to be more demanding but with technological developments holds tremendous potential for element detection and imaging at the molecular level (Leapman, 2003). EPXMA has been widely applied in the biological sciences and will be the focus of this chapter.

When used at low resolution, EPXMA bridges the gap between bulk chemical techniques (such as atomic adsorption spectroscopy) and fluorescent imaging at the light microscope level by providing information about large numbers of individual cells, thus allowing changes in subsets in a population of cells to be determined. At high resolution, EPXMA is one of the very few techniques capable of providing information about element concentrations in intracellular organelles, or gradients of ions within cells. The major advantage of EPXMA is that it allows direct correlation of element composition with structural information. In addition, unlike other techniques, all elements present in the area of analysis are detected simultaneously. The technique is not, however, suitable for the detection of elements present in trace concentrations, minimal detectable levels are in the order of a few millimoles per kilogram ($\sim$100 ppm), and it should be realized that EPXMA measures total element content and does not discriminate between free and bound forms. For elements such as Na, K, and Cl that are generally considered to be free within the cell this does not matter, but EPXMA measurements of Ca yield results that give information mainly about the intracellular stores.

EPXMA was first applied to biological specimens over 40 years ago and since then has been used for a wide variety of applications with a varying methodology for specimen preparation (Ingram *et al.*, 1999a,b). As an introductory text, this chapter is restricted to methods in general use for the study of cells in culture, freshly isolated cells, or single cellular organisms.

II. Rationale

When a specimen is analyzed in the EM, interactions of the incident electron beam with atomic nuclei of the specimen result in the generation of X-rays, the energy of which depends on the atomic number of the element in which the interaction is occurring and on the transitions that occur. The technique of X-ray microanalysis exploits these interactions. A detector situated close to the specimen collects the X-rays and associated electronic equipment sorts them according to energy resulting in the production of a spectrum (Fig. 1), which is a histogram plot of number of counts against X-ray energy. The spectrum contains both qualitative and quantitative information. The position of a peak in the spectrum, its energy, identifies the element; the area under-the-peak is proportional to the number of atoms of the element in the irradiated area. X-rays are also produced when the electron beam is slowed by the electrostatic fields of the atomic nuclei of elements present in the specimen. These X-rays form a continuous radiation that appears below the peaks in the spectrum (see Fig. 1). The production of continuum X-rays depends on the total number of atoms in the irradiated area and is used as a measure of the total mass of the specimen. The ratio of peak-to-continuum counts is used as a basis for quantification.

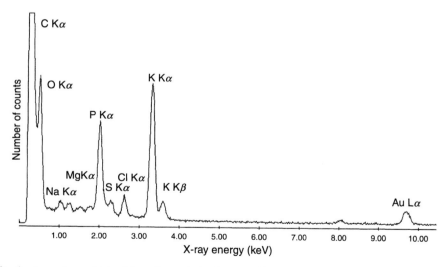

Fig. 1 A spectrum obtained from the analysis of an osteoblast cell, using a thin window detector. The spectrum is displayed from 0–10 keV and consists of characteristic peaks of the elements present in the analysed area including the low atomic number elements C and O. Characteristic peaks from the supporting grid (Au) are also seen. The characteristic peaks are superimposed over a background contributed by the continuum radiation.

III. Methods

A. Specimen Preparation

When using X-ray microanalysis, the analytical results are obtained from the specimen as it exists in the EM. Since biological specimens undergo some specimen preparation to remove water, the major goal in preparation for biological microanalysis is to maintain elements at their physiologically active site so that analysis gives true and meaningful results. Diffusible elements and small molecules are quickly lost during the dehydration and embedding steps used for the routine preparation of biological specimens (see Chapter 1 by Stephen Murray, this volume), and so these methods are only relevant if it is already known that the elements of interest are tightly bound (see e.g. Fig. 2). Instead cryotechniques are used for the study of diffusible elements. These rely on the rapid lowering of temperature to below $-160\,°C$ to achieve fixation at a known time and arrest diffusible elements at their intracellular sites of action. To prevent movement of elements, specimens are then maintained at low temperature throughout any further steps until they are either stabilized by freeze-drying or analyzed in the frozen-hydrated state. Cryofixation techniques developed for the study of antigens using immunocytochemistry, especially those used in the Tokuyasu technique

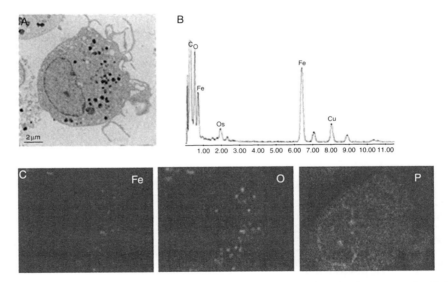

Fig. 2 (A) Electron micrograph of a thin section from a dendritic cell that has been exposed to FeO_2. The dendritic cell has engulfed the FeO_2 particles that are enclosed in the lysosomes. (B) a spectrum from an inclusion confirming the localisation of Fe and O in the lysosomes. (C) 2-D maps showing that the distribution of Fe and O coincides with the distribution of the particulate material whereas P is associated with chromatin in the margins of the nucleus. The specimen was prepared by dehydration in alcohol followed by infiltration and embedding in resin. These procedures could be used because the particles are sequestered within the lysosomes and are not lost. (See Plate no. 1 in the Color Plate Section.)

(see Chapter 10 by Siegfried Reipert and Gerhard Wiche, Chapter 12 by Nobuhiro Morone *et al.*, and Chapter 13 by Jason A. Mears and Jenny E. Hinshaw, this Volume), cannot be used for X-ray microanalytical studies, they use aldehyde fixation and cryoprotection with sucrose both of which lead to redistribution of mobile elements. The preparation steps that are required before high pressure freezing (Chapter 11, by Nicholas J. Servers and Horst Robenek, this volume) is carried out also make this technique less than optimal, particularly for physiological studies. Instead, either plunge-freezing into cooled liquid cryogens such as propane or ethane, liquid nitrogen is not suitable for high-resolution studies (Zierold, 1991), or impact against a precooled polished metal surface (Somlyo *et al.*, 1989) are used. With these techniques, minimal ice crystal damage is only achieved within the first few microns of a specimen surface, and this limits the size and type of specimen than can be used for microanalysis. However, vitrification of water, as is needed for morphological studies, is not absolutely necessary; ice crystal damage is tolerable provided that it remains below the size of the focused probe.

1. Treatment before Cryofixation

The way in which cells or tissues are handled before cryofixation affects elemental distribution or content, so care must be taken at every step in the process. Good laboratory practice should be followed to ensure cleanliness and minimize risk of contamination of specimens from environmental sources. When preparing cells, the following general considerations need to be taken into account:

(a) Adherent cells should not be removed from their substrate before cryofixation, both scraping and trypsinization damage the cells causing leakage of elements.
(b) Take care not to allow cells to cool, lowering of temperature decreases Na/K pump activity and may cause element redistribution.
(c) Do not expose concentrated cell pellets to a dry atmosphere as this may cause desiccation and alter element concentrations.
(d) If nonpenetrating cryoprotectants, such as dextran, are added to the external medium their effect on element concentrations needs to be determined.

2. Safety Considerations for Cryopreparation

Cryopreparation procedures involve known hazards (Sitte *et al.*, 1987). The evaporation of large quantities of liquid nitrogen can cause asphyxiation, liquid cryogens such as cooled propane and ethane are potentially explosive and contact with cooled metal surfaces causes low temperature burns. It is essential that anyone considering using these techniques consult with their local safety officer and make themselves aware of local safety regulations before beginning work.

3. Cryofixation of Cells for SEM Analysis

Analysis in the SEM provides a rapid and easy method for determining element composition in whole cells. The method was first described by Abraham *et al.* (1985) and has been further developed by a number of different groups. The preparation method is illustrated in Fig. 3A. Cells are grown on a suitable substrate such as graphite discs, Thermanox coverslips, Millicell inserts, or plastic-coated EM grids. For growth on EM grids, gold or titanium grids are used as these are not toxic for the cells (this preparation is also suitable for TEM analysis, Warley *et al.*, 1994). The method has also been adapted for use with suspended cells after entrapment by centrifugation onto a Milicell insert (Fernandez-Segura *et al.*, 1999, see Fig. 3B)

The method is as follows:

(a) Remove support + cells from culture medium.
(b) Dip support rapidly several times into an ice-cold washing medium.
(c) Blot the edges of the support gently against a filter paper to remove excess washing solution.

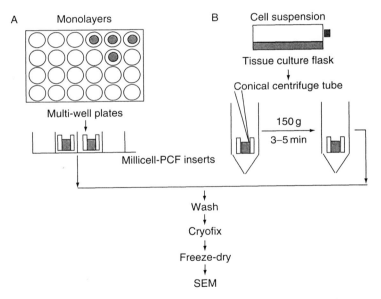

Fig. 3 (A) Preparation of whole cells for analysis. Adherent cells are grown on a suitable substrate here millicell inserts are used. The supports are removed from culture, overlying medium is removed by washing in an ice-cold solution, the cells are then cryofixed by dipping into liquid nitrogen and freeze-dried before analysis. (B) adaptaion of the method for suspended cells. An aliquot of the cell suspension is placed on a Millicell insert which is then placed into a conical centrifuge tube and the tube centrifuged to concentrate the cells onto the surface of the filter. In both procedures the filter is cut from the holder and overlying medium removed by rapidly dipping the excised filter into a washing solution. (See Plate no. 2 in the Color Plate Section.)

(d) Cryofix by plunging the support + cells into liquid nitrogen.
(e) Freeze-dry and carbon coat before analysis (Section III.B.3).

The appearance of cells prepared by this method is shown in Figs. 4 and 5.

The preparation procedure requires a washing step to remove overlying medium that could possibly cause a disturbance of the elemental balance in the cells. Of the solutions that are used, ammonium acetate (150 mM) is isotonic to most

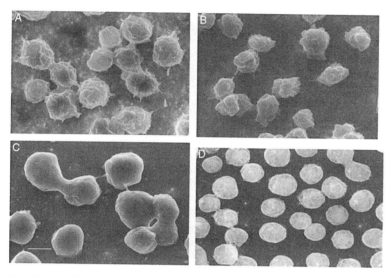

Fig. 4 The effects of different washing solutions on the morphology of U937 cells (A) unwashed cells, (B) 150 mM ammonium acetate, (C) 300 mM sucrose, (D) distilled water. Marker = 10 μm and is same fro all images.

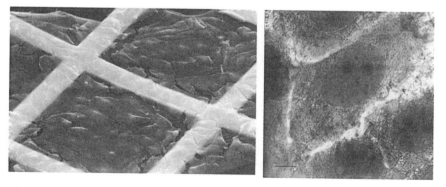

Fig. 5 Cells grown on. Pioloform-coated EM grids left LLC-PK cells seen in scanning EM, right Soas osteoblasts in TEM, marker = 1μm

cells and volatilizes during freeze-drying, but the ammonium ion (NH_4^+) could exchange for K^+, Na^+, or H^+. Mannitol (300 mM) or sucrose (300 mM) are isotonic but they form a dense deposit over the cells during freeze-drying obscuring detail and provide the possibility of absorption of emitted X-rays. Despite its lack of isotonicity, distilled water has been found to be effective for a number of cell types, but was shown to cause an increase in volume in cultured renal cells (Borgmann et al., 1994). The effects of different washing solutions on the morphology of U937 cells are shown in Fig. 4. The choice of washing solution needs to be determined experimentally for the system under consideration. This has been achieved by comparison of washed cells with unwashed controls (Fernandez-Segura et al., 1999; Warley et al., 1994), and by comparisons between whole cells that have been washed with cryosections prepared from cells in which the culture medium has not been removed (Dragomir et al., 2001; Warley et al., 1994).

In the method outlined above, liquid nitrogen is used as a cryogen, this is perfectly adequate since individual cells are analyzed using a large area probe and any ice crystal damage does not affect the analysis.

4. Cryofixation of Cells for TEM Analysis

Cells prepared by the methods outlined above provide information about elemental content at the whole cell level, at the best discrimination between cytoplasm and nucleus can be made (Fernandez-Segura et al., 1997). For the study of element distributions within intracellular organelles cells need to be cryofixed and then cryosectioned before analysis. The approach most frequently used for such studies is immersion fixation of a cell pellet using a commercially available apparatus that uses either liquefied propane or liquefied ethane as the cryogen. Immediately before cryofixation suspended cells need to be concentrated by centrifugation to reduce the volume of extracellular medium in which ice crystals form making subsequent cryosectioning difficult. This is achieved by use of either in a microfuge at ~10,000 rpm for periods of up to 1 min, or in a bench centrifuge for up to 5 min, both methods have proved reliable. The supernatant is removed and droplets of the thick cell suspension are cryofixed using the protocol given below:

(a) Cool the cryofixation apparatus down to operating temperature and allow to equilibrate, fill with cryogenic liquid and allow to cool to just above freezing point.
(b) Place a clearly labeled cryostorage tube into the chamber of the cryopreparation apparatus and allow to cool.
(c) Rapidly transfer a droplet of the thick cell suspension onto a carrier suitable for cryofixation.
(d) Mount into the plunging device and plunge immediately into the cold cryogen.

(e) Remove the specimen from the arm of the crofixation apparatus and place within the cooled cryotube, keeping the cryofixed specimen within the cooled atmosphere at all times.

(f) Cap the cryotube and transfer to a storage Dewar under liquid nitrogen.

Although a commercial cryofixation device is recommended this apparatus is not absolutely necessary, home-made devices can be used (see e.g. Somlyo *et al.*, 1989; Warley and Gupta, 1991). However, if a home-made apparatus is used and specimens are plunge frozen by hand, it must be remembered that there is a cool layer of gas above the cryogen that precools the specimen promoting ice crystal formation. The specimen must be moved through this layer as quickly as possible. The specimen support also influences the cooling rate and thus the quality of the resulting cryofixation. Metal pins supplied by manufacturers are not appropriate. Instead pieces of wooden orange stick can be used or the cells can be supported on wedges of filter paper. Cells cryofixed on filter paper are subsequently attached to stubs for sectioning in the chamber of the cryoultramicrotome using a cryoglue such as *n*-heptane (Steinbrecht and Zierold, 1984). Cell pellets can also be cryofixed by impact against a cooled metal block using a proprietary metal mirror fixation device.

The method for preparing cells in suspension as outlined above is relatively routine, but the thinness of cells grown as monolayers makes handling them much more difficult, so cryofixation and cryosectioning of monolayers is not in general use and will not be discussed in great detail here. Zierold (1997) constructed a custom-made apparatus for immersion cryofixation of monolayers of rat hepatocytes, others have exploited growth of cells on plastic film (James-Kracke *et al.*, 1980; Warley *et al.*, 1994), or on a layer of gelatine (Buja *et al.*, 1985) to allow removal of cells from culture without disruption. Interested readers are referred to these publications. Similarly, more complex methods have been developed to allow the cryofixation of cells at defined physiological time points (see e.g., Hardt and Plattman, 1999; Wendt-Gallitelli and Isenberg, 1991; Zierold, 1991).

B. Treatment Post Cryofixation

Frozen specimens need to be maintained at temperatures of $<-150\,°C$ to prevent ice crystal growth with consequent movement of elements. Liquid nitrogen cooled storage vessels designed for the storage of tissue culture cells are suitable for storing cryofixed specimens. Careful note should be made of the place of storage so that specimens can be retrieved without disturbing other specimens. Specimens should always be transferred between the cryofixation apparatus and storage Dewars, and between storage Dewars and the cryoultramicrotome under liquid nitrogen.

1. Cryofixation Followed by Embedding Techniques

If cryosectioning or cryotransfer are not available, then low temperature embedding or freeze-drying and embedding can be considered as alternative methods for specimen preparation but these are not in general use because of concerns about movement of ions during either substitution or embedding, and also because of ambiguities in quantified results (Hall, 1989; Warley, 1993). If these methods are used, there is a requirement to show that element redistribution does not occur; a detailed approach is given in Hardt and Plattman (1999).

2. Cryosectioning

Thin cryosections are required for the study of element concentrations in intracellular compartments; these are cut at low temperatures using a cryoultramicrotome. All tools and equipment needed for sectioning should be assembled beforehand. In addition to the usual tweezers and eyelash probes, the following are also needed (see Fig. 6A):

(a) A supply of glass knives and a cryodiamond knife if available, if glass knives are used they should be prepared using the balanced break method.

(b) A supply of grids. Grids made from copper or nickel are generally used. The grids should have high transmission properties; a mesh size of 100 or 150 is advantageous to minimize the contribution of the grid to the X-ray spectrum. Single slotted or multislotted grids are also suitable but these are fragile in use. The grids need to be coated with a support film; carbon, Pioloform, or Formvar have all been used. The grids should be checked for cleanliness before use.

(c) A copper rod or proprietary device for pressing sections onto the grids.

(d) A cryotransfer device for carrying grids with sections under liquid nitrogen.

The following is a general guideline for cutting cryosections for microanalysis, when using a cryoultramicrotome, the manufacturer's instructions for operation must be followed.

(a) Cool the microtome to the required temperature, this is usually between $-125\,°C$ and $-140\,°C$.

(b) Place knives into the cryochamber during cooling but do not clamp in place.

(c) Once the desired temperature is reached transfer the specimen into the cryochamber and clamp firmly in the specimen holder.

(d) Clamp knives into place.

(e) Trim the specimen only if necessary. Trimming removes the best cryofixed part of the specimen.

(f) Cut sections, a thickness of 100–150 nm is generally used, and transfer to coated grids using an eyelash probe (see Fig. 6B).

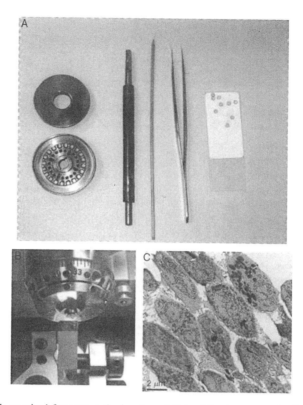

Fig. 6 (A) Tools required for cryosectioning from left to right A cryotransfer holder for carrying section covered grids between the cryoultramicrotome and freeze-drier, a copper rod used for pressing the sections onto the grids, an eyelash for picking up sections from the knife edge and transferring them to the grids, antimagnetic tweezers for handling Ni grids and Pioloform-covered Ni grids. (B) Interior of cryosectioning chamber ($-130\ °C$) showing, the blockface a knife with a grid containing sections, the grid is supported on the knife by a tape strip. Sections are from the knife to the grid using an eyelash. (C) Freeze-dried cryosections of human neutrophils. The sections are compressed in the direction of cutting (arrow). The sections show high contrast despite the lack of staining, the contrast is due to differences in dry mass between the different organelles. (See Plate no. 3 in the Color Plate Section.)

(g) Once sufficient sections have been cut they need to be pressed onto the grids. Several methods are used for this purpose. Sections are either pressed between two overlying single grids or oyster grids are used.

(h) Place the grids bearing the sections into a precooled cryotransfer device and transfer to the freeze-dryer under liquid nitrogen.

The major problem encountered with frozen sectioning is the development of static electricity, which causes sections to fly away from the edge of the knife. In a modern microtome this is neutralized using a static line deioniser, alternatively an antistatic gun can be used.

3. Freeze-Drying

Thin frozen sections are not analyzed in the TEM in their fully hydrated form since, at the probe currents that are required to elicit X-ray formation, interaction of incident electrons with ice in the specimens leads to the generation of free radicals that destroy the specimen (Zierold, 1988). Instead they are either transferred to a proprietary cryoworkstation to allow placement in a cryoholder and transfer to the EM column where they are freeze-dried by raising the temperature of the holder to $-110\,°C$ for a period of 20–30 min (Scott *et al.*, 1997), or freeze-dried externally using either a carbon coating unit (Di Francesco *et al.*, 1998; LeFurgey *et al.*, 2001) or a specially designed freeze-drier with temperature controlled stage (Warley and Skepper, 2000). The grids are transferred to the freeze-drying apparatus in a covered cryotransfer device under liquid nitrogen. The chamber of the freeze-drying apparatus is then evacuated and the sections allowed to freeze-dry overnight. If a carbon coater is used for freeze-drying, insulating materials such a piece of Perspex should be placed between the cryotransfer device and the base plate of the coater to allow the transfer device to warm up slowly. There has been some discussion about the length of time needed for freeze-drying of biological specimens for microanalysis (Edelman, 1994) but it has been shown experimentally (Warley and Skepper, 2000) that, using the conditions described above, element redistribution does not occur.

4. Treatment after Freeze-Drying

Freeze-dried cryosections are hygroscopic and care needs to be taken to ensure that the sections are not exposed to water vapor. Sections are usually coated with a thin layer of carbon to protect the sections from the atmosphere this also has the benefit that their conductivity in the electron beam is increased making the sections more stable during analysis. Ideally coating should be done without breaking the vacuum of the freeze-drying apparatus. If the vacuum of the freeze-drier or coating apparatus needs to be broken, then the chamber should be flushed with dry nitrogen gas rather than room air.

Even when coated with carbon, freeze-dried sections are easily contaminated from the environment. The best policy is to ensure that sections are analyzed as soon as possible, preferably within 24–48 h after preparation. Although storage within desiccators is sometimes recommended sections easily pick up contamination from desiccants such as silica gel, or if stored under vacuum from O-rings. If long-term storage is envisaged, the conditions should be checked thoroughly.

C. Analysis

In its simplest form analysis is achieved by focusing the incident beam of the EM onto the area of interest in the specimen and producing a spectrum. The sensitivity of analysis is determined by the ratio of counts in the characteristic peak divided by the number of counts in the background under-the-peak, so whatever system is used conditions must be selected that maximize this ratio. The factors that affect P/B include the total number of counts in the characteristic peaks, the number of counts in the background and the calibration of the detector. The detector should be calibrated on a regular basis to ensure correct assignment of peaks and accurate deconvolution of overlapping peaks during spectrum processing.

1. Analysis in the SEM

When cells are analyzed in the SEM, the main problem that affects sensitivity is over-penetration of the specimen by the electron beam causing excitation of X-rays from the underlying substrate on which the cells are grown (Fig. 7). This results in the possibility of characteristic X-rays from the substrate, for example Si, if the cells are grown on glass appearing as spurious peaks within the spectrum, and also the generation of unknown amounts of continuum radiation that will affect quantification. To avoid this problem, lower accelerating voltages need to be used, but this reduces characteristic X-ray production in the specimen with consequent lowering of sensitivity. The optimum accelerating voltage can be determined experimentally by coating the substrate with a substance, such as gold, that emits a characteristic X-ray, and then growing the cells on the coated substrate. When analyzing the cells, the highest accelerating voltage that does not cause the appearance of characteristic X-rays is the one that should be used for analysis (Fig. 7; Fernandez-Segura *et al.*, 1999).

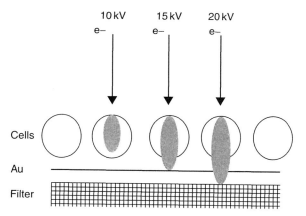

Fig. 7 A schematic representation of over-penetration of the electron beam caused by higher voltages in the SEM. Cells are grown on a gold coated filter. At a lower voltage here 10 kV, the electron beam remains within the cell so that X-ray excitation occurs only from this source. As the voltage is increased the primary electron beam penetrates deeper so that X-rays are produced from the underlying gold layer and from the filter as well as from the cell.

2. Analysis in the TEM

Analysis in the TEM is generally used to obtain high-resolution information from small areas of a thin section. The success of analysis is affected by the limited number of counts that are available from the thin specimen and by the generation of extraneous X-rays from the materials surrounding the specimen in the restricted space of the specimen area. The success of analysis can be improved by ensuring that the microscope is set up correctly (see Section V.A) and by carrying out analysis in a careful manner.

When carrying out analysis in the TEM, higher accelerating voltages are used as this leads to less scattering of electrons, reducing the unwanted background radiation. Analysis in the TEM is also affected by the geometric relationship between the incident electron beam, the specimen and the detector. In the TEM, the detector is usually mounted at a low or zero angle to the specimen stage and the specimen needs to be tilted towards the detector to allow spectrum collection. A take off angle of 35° is generally used to maximize collection of X-rays and minimize absorption effects. Since the specimen needs to be tilted it is important that the specimen stage is set at its eucentric height before tilting so that the specimen and emitted X-rays are in the direct line of sight of the detector.

The way in which the analysis is carried out is also very important. Analysis should always take place in the centre of the grid and in the central areas between grid bars. Analysis close to the edge of the holder will lead to unacceptable levels of X-ray emission from this source and in some cases to absorption of X-rays. Analysis close to a grid bar, especially when the specimen is tilted, can also lead to excessive production of X-rays from the grid and possibly to absorption of X-rays (see Roomans, 1988a).

Loss of specimen mass during the course of analysis also affects quantification. Loss from specific peaks (S and Cl are particularly susceptible; Hall and Gupta, 1986) causes underestimation of these elements, whilst loss from the organic matrix causes an underestimate of total specimen mass. Mass loss can be minimized by cooling the specimen during analysis using a cold stage; if this is done, use of an anticontaminator is required to prevent deposition of contamination onto the specimen. Mass loss can also be compensated for during quantification procedures (Section III.D.4).

D. Deriving Qualitative and Quantitative Information from the Spectrum

Details about the requirements of analytical systems are given in Section V. Qualitative analysis, i.e. identification of elements in the spectrum, is usually achieved using manufacturer's software. These systems are not foolproof, they are designed for materials applications rather than biology and misidentification of a peak by the software is a distinct possibility, e.g., the Na K peak is often labeled as Zn L. It is essential to look for expected accompanying X-ray lines, the Zn L line should be accompanied by a Zn K at 8.63 keV. Similarly, if the program is set to label only K_α lines, then the accompanying K_β lines are often mislabeled. A case in point is that for the potassium K_β peak (see Fig. 1), this overlaps the Ca K_α peak, and often appears in spectra wrongly labeled as Ca K_α. All analysts need to be aware of the principles involved in X-ray generation and the lines that are generated (Section III) full details are outside the scope of this chapter but can be found in Ingram *et al.* (1999a,b) and Warley (1997).

Direct comparison of net counts in a peak between spectra is not valid because differences in specimen thickness and fluctuations in probe current between analyses affect the number of counts collected; in addition, in thicker specimens, absorption of generated X-rays and secondary X-ray fluorescence become important. Some means of quantification is therefore necessary. In materials science, thick specimens with polished surfaces are analyzed in the SEM and using well-documented ZAF (atomic number, absorption, and fluorescence) procedures to obtain quantitative results. These methods are not applicable in biology where the surface of the specimen is rarely smooth and the composition of the matrix is of low atomic number and not necessarily known. The two methods in general use for obtaining quantitative results from biological specimens are given below.

1. Continuum Normalization

The method most often used for quantification of elements in thin sections is the continuum normalization method developed by Hall (see Hall, 1989; Hall and Gupta, 1986). This method assumes that in thin sections composed of low Z elements (biological specimens) absorption and fluorescence corrections are negligible and are not applied. The concentration of a given element x is expressed in terms of mass x/total mass in the area analyzed. The mass of x in the area of analysis is proportional to the net number of counts in its characteristic peak (P_x) and the total mass in the area of analysis is proportional to the continuum radiation (W see Fig. 8) generated by the specimen itself. The basic equation is:

$$C_{x,\mathrm{sp}} = \frac{(P_x/W)_{\mathrm{sp}}}{(P_x/W)_{\mathrm{std}}} \cdot \frac{\overline{Z^2/A_{\mathrm{sp}}}}{\overline{Z^2/A_{\mathrm{std}}}} \cdot C_{x,\mathrm{st}} \qquad (1)$$

where C is the concentration (mmole/kg), the subscripts sp and std refer to specimen and standard respectively, and the value of Z^2/A is the weighted mean value of atomic number squared divided by the atomic weight for all constituent elements of the sample.

The method automatically corrects for variations in thickness of the specimens and fluctuations in probe current since they both affect P_x and W to the same extent.

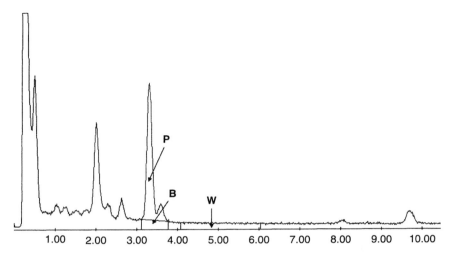

Fig. 8 Spectrum showing the areas used in different quantification regimes the net peak integral (P), the background under-the-peak (B) and a continuum region W.

a. Estimation of Continuum Counts

The use of Eq. (1) requires an estimation of continuum counts. This is obtained by setting a region of interest in a peak-free region of the spectrum (see Fig. 6). Two regions are in general use, either 1.34–1.64 or 4.2–6.2 keV. If the manufacturer's software does not give an option for setting a continuum region, this can be done simply by including an element that is not present in the specimen in the list of elements to be analyzed, and using the counts under-the-peak for this element as the continuum region. The counts in the continuum need to be those contributed by the specimen itself. Freeze-dried sections are supported on a plastic film on an EM grid, both of which contribute to the continuum region of the spectrum from the specimen so that the total counts in the continuum region (W_t) are the sum of those from the specimen itself (W_{sp}) plus those from the film (W_f) and those from the grid (W_g). The counts contributed by the specimen are estimated as follows:

(a) The grid correction factor is calculated by obtaining a spectrum from an analysis close to the grid bar from a bare grid (i.e., no film, no specimen), the position needs to be adjusted to give a count rate of ~2000 cps. From this the ratio of counts in the continuum to counts in the grid peak is determined. This procedure is repeated several times. The ratio W_g/P_g is constant, and forms the grid correction factor that is used to estimate the counts contributed by the grid in subsequent spectra.

(b) A spectrum is obtained from the analysis of the film close to the area of interest in the cell. The contribution of the grid to the continuum region of this spectrum is obtained by multiplying the number of counts in the grid peak by the grid correction factor, obtained as outlined above, and this value is subtracted from the total counts in the continuum region to give the counts contributed by the film. ($W_t - W_g = W_f$).

(c) In the spectrum obtained from the specimen the contribution of the grid is determined by multiplying the counts in the grid peak by the grid correction factor, these are then subtracted from the total continuum counts along with the film correction as obtained in (b) to give the number of continuum counts produced by the specimen.

Continuum normalization tends to be inaccurate when the corrections to the continuum exceed the counts generated by the specimen itself. This occurs if the sections are very thin, or if analysis takes in an inappropriate position (Roomans, 1988a).

2. Quantification for Semithick Specimens

When semithick specimens such as whole cells are analyzed in the SEM, the assumption that absorption and secondary fluorescence do not occur cannot be made and a different quantitative approach is required. Statham and Pawley (1978) proposed the use of the ratio of peak to background under-the-peak, rather than a distant continuum region, as a basis for quantitative analysis of particulate material. The reasoning is that, since both the peak and the continuum are produced from the same area of the specimen and are of the same energy, absorption and fluorescence effects should be equivalent. The equation used is

$$C_x = k_x P_x / B_x \qquad (2)$$

This method has found wide application in SEM analysis in biology. As with continuum normalization, standards are required for quantification and these should closely resemble the specimen in composition so that absorption and fluorescence are similar and a Z^2/A correction is not required. Standards are made as described in Section III.D.4.

3. Thin Widow Detectors and Quantification

Much of the literature about quantitative analysis of biological specimens was written in the late 1980s early 1990s. Instrumentation has changed since then. In particular, Be-window detectors have been replaced by thin window detectors so that the statement that biological specimens generally consist of an organic matrix of elements, C, N, O, and H, that are not detectable with conventional EDS detectors no longer holds true (Zierold et al., 2005). The ability to analyze light elements makes standardless quantification a possibility (Laquerriere et al., 2001; Marshall, 1994); however, very few groups are using this approach. If this method is considered specimens need to be supported on a holey or lacy carbon film since standardless analysis does not allow for the contribution of the film to the characteristic peaks for C and O. Laquerriere et al. (2001) have suggested that, with thin window detectors the presence of an oxygen sum peak at 1.04 keV, the energy of the sodium K peak might cause problems for quantification; however, Marshall and Patak (1993) and experience in the author's laboratory suggest that this would not be a problem in the analysis of thin sections.

4. Standards

All biological specimens suffer from loss of low atomic number elements under the electron beam; this can be reduced by analysis at low temperatures using a cold stage and is also compensated for by the choice of an appropriate standard. The standard should be of similar overall composition to the specimen with the assumption that mass loss is similar in both specimen and standard. This strategy has further advantages. In Eq. (1) the term Z^2/A (sometimes called the G factor) is used to correct for differences in continuum generation between the specimen and standard. If a standard is chosen that closely resembles the specimen in composition, then the value of $\frac{Z^2/A_{sp}}{Z^2/A_{std}}$ cancels out and the equation becomes

$$C_{x,\text{sp}} = \frac{(P_x/W)_{\text{sp}} \cdot C_{x,\text{st}}}{(P_x/W)_{\text{std}}} \qquad (3)$$

where the term $C_{x,\text{std}}/(P_x/W)_{\text{std}}$ is the constant for the element x determined from the analysis of standards.

Standards are made by dissolving known concentrations of salts in a solution of either 15–20% protein such as gelatin or albumin, or 15–20% dextran, freezing droplets of the solution and taking them through exactly the same procedure used for specimen preparation. For each element a number of standards of different concentration is analyzed; the mean value of $C_{x,\text{std}}/(P_x/W)_{\text{std}}$ gives the constant for that element. A plot of $(P_x/W)_{\text{std}}$ against C_x gives a straight line. The accuracy of standardization is checked either by the analysis of a specimen of known composition such as red blood cells or by analyzing a standard of known composition that was not used in the standardization procedure. Further information about the preparation and analysis of standards can be found in Roomans (1988b) and Warley (1990, 1997).

5. Estimation of Water Content

When analyzing biological specimens an attempt at estimation of local water content should be considered for two reasons. First, analysis of freeze-dried sections produces results as mmoles/kg dry weight, units that are unfamiliar to the majority of scientists, and second, because differences in dry mass between intracellular compartments can give the impression of concentration differences where nonexists in the hydrated state (see e.g. Zierold, 1997). It is recommended that some attempt is made to convert the dry weight values to millimoles concentrations to allow comparisons with the literature. If local values for dry mass are available in the literature these can be used, otherwise local dry mass can be estimated. Methods for achieving this are outlined in Roomans (1988b), Warley (1997), and Zierold (1988).

6. Mapping Techniques

In a STEM instrument, the ability to scan the electron beam across the specimen in a 2D raster allows the collection of elemental information at each dwell point. The data collected is presented as an element map that provides a 2D map of element distribution that can be directly related to the image acquired at the same time. Mapping software is generally included in analysis software, and maps are usually acquired by first collecting a spectrum from the area of interest using a wide probe then setting windows over the energy regions of interest. During the collection of the map the number of counts in each region of interest is stored for subsequent processing (see Fig. 2). Since no corrections are applied such maps are only qualitative. In the spectral imaging techniques, a full spectrum is collected at each pixel that can then be processed by a variety of methods to produce fully quantitative maps. These methods have been developed and exploited by the group working at Duke University (Ingram *et al.*, 1999a,b; LeFurgey *et al.*, 1992, 2005; Simm *et al.*, 2007).

IV. Equipment

A. Electron Microscopes

SEMs are used for the study of whole cells prepared as in Section III.A.3. High resolution is not required for such studies so any SEM that is fitted with a detection system should be suitable. The SEM can be made more versatile by fitting a STEM detector allowing more detail to be imaged and also analysis at higher kilovolts with the benefit of increased count rates. If the SEM is fitted with a back scatter detector this can be used to image major cellular compartments such as nucleus and cytoplasm, allowing differentiation between these two compartments to be made (Fernandez-Segura *et al.*, 1997).

The requirements of a TEM for analysis are more stringent. A TEM can be used for low resolution analysis such as whole cells, but is most useful for the analysis of intracellular compartments. This requires the ability to form a highly focused electron probe so the TEM should be equipped with the ability to operate in the nanoprobe mode or to have STEM capability. The number of X-ray emitted from the specimen during irradiation depends directly on the current in the focused probe which itself depends on the type of emitter used. Conventional tungsten filaments are adequate, but the use of LaB_6 filaments gives higher X-ray count rates, these are useful when small spot sizes are used. Field emission TEMs have not yet gained the widespread use in analysis of biological specimens that they have in materials science. Although in earlier publications, cleanliness of the EM vacuum system was emphasized as a factor in EPXMA this is unlikely to be a problem with modern TEMs, especially if an anticontaminator is used as a routine. The main problem encountered with TEM analysis is the production of extraneous X-rays from various sources including apertures in the column, the specimen holder and the grid carrying the specimen. These produce spurious peaks in the spectrum and also contribute unwanted continuum radiation that affects the accuracy of quantification (Section III.D). The production of extraneous X-rays from the column is minimized by having a well-aligned electron beam and by using thick apertures to prevent X-rays generated higher in the column from reaching the specimen area. The objective aperture must be removed during analysis. To reduce the generation of X-rays from the specimen surroundings manufacturers produce specially designed holders for analytical applications these are shielded with low atomic number elements such as carbon or beryllium to minimize X-ray production from this source although it should be noted that carbon will be detected by the atmospheric thin window (ATW) detectors that are now in general use.

B. X-ray Detection Systems

Energy dispersive detection systems (EDS) are used for the analysis of biological specimens because of their multielement capability and their greater effectiveness at low probe currents. The detectors associated with these systems are now usually thin windowed and detect elements from carbon ($Z = 6$) upwards. Full details of detection systems are usually available from the manufacturers. Quantitative analysis has several requirements from the associated computer software. Reliable peak deconvolution routines are necessary to separate peaks from the underlying continuum and to deconvolve overlapping peaks, this is especially important for the K K_β Ca K_α overlap (Somlyo *et al.*, 1989). The ability to obtain basic data (net peak counts, counts in the background under-the-peak, and continuum counts) is essential for quantification using P/B or continuum normalization. To the author's knowledge only one manufacturer includes a program for biological analysis in their software but this should not be a deterrent; independent systems are available. NIST provides a desktop spectrum analyzer for Mac computers (www.cstl.nist.gov/div837/division/outputs/DTSA/DTSA.wm) and software is also available for spectrum processing, quantification, as well as qualitative and quantitative analysis (see Simm *et al.*, 2007). In the absence of suitable software from the manufacturer, the calculations outlined in Sections III.D.1 and III.D.2 can easily be carried out using a spreadsheet on a personal computer provided that the basic data is available.

V. Discussion

The procedures that we have described here for preparing whole cells for analysis by either SEM or TEM provide simple and rapid methods for measuring element concentrations at the whole cell level. The use of cryosections is more demanding but gives unambiguous information about concentrations within intracellular organelles. Neither method should be considered in any way "static," as EM experiments are often described. Samples can be taken for analysis at time points throughout an experiment, a strategy that is used in many biochemical investigations, allowing temporal information to be acquired (Arrebola et al., Salido et al., 2001; 2006; Zierold, 1997). Tracer elements, e.g., Rb in exchange for K, Sr for Ca, Br for Cl, can be used to study the uptake of specific elements into the cell and when coupled with specific inhibitors allow the study of ion transport pathways (Arrebola et al., 2005a,b). In such studies, the multielement capability of X-ray microanalysis has the added advantage of being able to detect the effects of inhibiting one ion transporter on the intracellular concentrations of other ions.

X-ray microanalysis is now being used in a variety of applications (Table I) with emphasis on epithelial transport, the toxic effects of environmental pollutants and in the mechanisms of apoptotic and oncotic cell death. The work on epithelial transport has already been well summarized (Roomans, 2002). In the studies on toxicity and cell death a common finding has been that K/Na ratios are changed often very early after treatment and this has lead various authors to suggest that a lowering of the K/Na ratio can be used as a measure of cell viability. Whilst there is no doubt that high values for K/Na are an indication of healthy cells a lowering of the ratio needs to be interpreted with caution and with due consideration both of experimental techniques and of the concentrations of other elements. Lowering of the K/Na can occur due to harsh handling of cells before cryofixation (Section III.A.1) this has been well documented when isolated tissue preparations are used

Table I
Some Applications of X-ray Microanalysis in Cell Biology

Biological problem	Preparation technique	System studied	References
Apoptotic cell death	Whole cells SEM	U937, staurosporine	Arrebola et al. (2005a,b)
	Whole cells SEM	Prostatic cell line etoposide	Salido et al. (2001)
	Whole cells TEM	Monocyte/macrophages LDL	Skepper et al. (1999)
	Cell sections TEM	U937, UV light	Arrebola et al. (2006)
Epithelial transport	Whole cells SEM	Nasal epithelium	Dragomir et al. (2001), Serretnyk et al. (2006)
Metal toxicity	Cell sections TEM	Hepatocytes Cd, Zn	Zierold (1997)
	Cell sections TEM	Hepatocytes	Zierold (2000)
	Cell sections TEM	Macrophages, Al, Zr	Nkamgeue et al. (2000)
Trypanosome composition	Cell sections TEM	Trypanosomes	Scott et al. (1997)
Leishmania	Cell sections TEM	Leishmania volume regulation	LeFurgey et al. (2001, 2005)
Cardiac myocytes	Cell sections TEM	Isolated myocytes ATP	Zhang et al. (2001)
Cell growth/differentiation	Cell sections TEM	Smooth muscle	Warley (2001)
	Cell sections TEM	HL60 cells, retinoic acid	Di Francesco et al. (1998)
Tissue engineering	Whole cells SEM	Oral mucosa	Sanchez-Quevedo et al. (2007)
	Whole cells SEM	Corneal endothelium	Alaminos et al. (2007)

(see e.g. Dragomir and Roomans, 2000) and needs to be considered for freshly isolated cells such as leucocytes when a period of incubation may be needed to restore the ratio to normal. Adherent cells are known to lose K as they come out of active growth (Warley, 2001) with a consequent reduction in the K/Na ratio, but without loss of viability. Similarly differentiation in HL60 cells has been shown to result in a lowering of K/Na in the nucleus (Di Francesco et al., 1998) again without affecting viability. So the growth state of cells in culture always needs to be considered.

One area where a lowering of K/Na has been shown to be important is in studies of cell death pathways. Both apoptotic and oncotic cell death are characterized by a decrease in K/Na ratio. The mechanisms underlying the change are, however, quite different. Apoptosis is characterized by an early loss of K that occurs well before the concentration of Na begins to increase and is coupled with a loss of Cl, whereas oncotic cell death is characterized by a loss of K with a concomitant increase in Na and an increase in Cl. It is therefore important not just to look at the ratio itself, but how the individual elements are changing as this gives insight into the pathways involved. When interpreting the results from such studies it should be remembered that X-ray microanalysis couples element content with the information provided in an image that can provide further information about the status of the cell such as nuclear condensation in the later stages of apoptosis (Arrebola et al., 2005b). However, it is increasingly clear both from studies in apoptosis and metal toxicity that changes in element concentration occur well before changes in structure either of the surface, or of intracellular organelles become apparent, and well before the uptake of trypan blue or the release of LDH the markers usually used to demonstrate cell death. Also, in studies of metal toxicity, the changes in diffusible elements occur when cells are exposed to levels of toxicant that are well below levels that can be detected by this EPXMA. Such early changes in K/Na reflect changes in the cell itself, making EPXMA an ideal method for detecting the changes in ionic pathways that accompany signaling events in cells.

Acknowledgements

The authors would like to thank Dr. F. Arrebola of the University of Granada for help with the diagrams Ms. Jane Storey of the CUI King's College for the photographs. AW would like to thank Prof. A. Campos, Department of Histology, University of Granada, for the stay in his Department that allowed a large part of this work to be written.

References

Abraham, E. H., Breslow, J. L., Epstein, J., Chang-Sing, P., and Lechene, C. (1985). Preparation of individual human diploid fibroblasts and study of ion transport. *Am. J. Physiol. Cell Physiol.* **248,** C145–C164.

Alaminos, M., Sanchez-Quevedo, M. C., Munoz-Avila, J. I., Garcia, J. M., Crespo, P. V., Gonzalez-Andrades, M., and Campos, A. (2007). Evaluation of the viability of cultured corneal endothelial cells by quantitative electron-probe X-ray microanalysis. *J. Cell. Physiol.* **211,** 692–698.

Arrebola, F., Canizares, J., Cubero, M. A., Crespo, P. V., Warley, A., and Fernandez-Segura, E. (2005a). Biphasic behaviour of changes in elemental composition during staurosporine- induced apoptosis. *Apoptosis* **10**, 1317–1331.

Arrebola, F., Zabiti, S., Canizares, F. J., Cubero, M. A., Crespo, P. V., and Fernandez-Segura, E. (2005b). Changes in intracellular sodium,chlorine, and potassium concentrations in staurosporine-induced apoptosis. *J. Cell. Physiol.* **204**, 500–507.

Arrebola, F., Fernandez-Segura, E., Campos, A., Crespo, P. V., Skepper, J. N., and Warley, A. (2006). Changes in intracellular electrolyte concentrations during apoptosis induced by UV irradiation of human myeloblastic cells. *Am. J. Physiol. Cell Physiol.* **290**, C638–C649.

Borgmann, S., Granitzer, M., Crabbé, J., Beck, F. X., Nagel, W., and Dörge, A. (1994). Electron microprobe analysis of electrolytes in whole cultured eptithelial cells. *Scanning Microsc. Suppl.* **8**, 139–148.

Buja, L. M., Hagler, H. K., Parsons, D. B., Chien, K., Reynolds, R. C., and Willerson, J. T. (1985). Alterations of ultrastructure and elemental composition in cultured neonatal rat cardiac myocytes after metabolic inhibition with iodoacetic acid. *Lab. Invest.* **53**, 397–412.

Di Francesco, A., Desnoyer, R. W., Covacci, V., Wolf, F. I., Romani, A., Cittadini, A., and Bond, M. (1998). Changes in magnesium content and subcellular distribution during retinoic acid-induced differentiation of HL60 cells. *Arch. Biochem. Biophys.* **360**(2), 149–157.

Dragomir, A., and Roomans, G. M. (2000). X-ray microanalysis of intestinal chloride transport *in vitro* and *in vivo*. *Microsc. Res. Tech.* **50**, 176–181.

Dragomir, A., Andersson, C., Aslund, M., Hjelte, L., and Roomans, G. M. (2001). Assessment of chloride secretion in human nasal epithelial cells by X-ray microanalysis. *J. Microsc.* **203**, 277–284.

Edelman, L. (1994). Optimal freeze-drying of cryosections and bulk specimens for X-ray microanalysis. *Scanning Microsc. Suppl.* **8**, 67–81.

Fernandez-Segura, E., Canizares, F. J., Cubero, M. A., Revelles, F., and Campos, A. (1997). Backscattered electron imaging of cultured cells: Application to electron probe X-ray microanalysis using a scanning electron microscope. *J. Microsc.* **188**, 72–78.

Fernandez-Segura, E., Canizares, F. J., Cubero, M. A., Campos, A., and Warley, A. (1999). A procedure to prepare cultured cells in suspension for electron probe X-ray microanalysis: Application to scanning and transmission electron microscopy. *J. Microsc.* **196**, 19–25.

Hall, T. A. (1989). Quantitative electron probe X-ray microanalysis in biology. *Scanning Microsc.* **3**, 461–466.

Hall, T. A., and Gupta, B. L. (1986). EDS quantitation and application to biology. In "Principles of Analytical Electron Microscopy" (D. C. Joy, A. D. Romig, Jr., and J. R. Goldstein, eds.). Plenum Press, New York.

Hardt, M., and Plattman, H. (1999). Quantitative energy-dispersive X-ray microanalayasis of calcium dynamics in cell suspensions during stimulation on a subsecond time scale: Preparation and analytical aspects as exemplified with *Paramecium* cells. *J. Struct. Biol.* **128**, 187–199.

Ingram, P., Shelburne, J. D., Roggli, V. L., and LeFurgey, A. (1999a). "Biomedical Applications of Microprobe Analysis." Academic Press, San Diego, CA.

Ingram, P., Shelburne, J. D., and LeFurgey, A. (1999b). Principles and instrumentation. In "Biomedical Applications of Microprobe Analysis" (P. Ingram, J. Shelburne, V. Roggli, and A. LeFurgey, eds.), pp. 1–57. Academic Press, San Diego, CA.

James-Kracke, M. R., Sloane, B. F., Shuman, H., Karp, R., and Somlyo, A. P. (1980). Electron probe analysis of cultured vascular smooth muscle. *J. Cell. Physiol.* **103**, 313–322.

Laquerriere, P., Banchet, V., Michel, J., Zierold, K., Balsossier, G., and Bonhomme, P. (2001). X-ray microanalysis of organic thin sectionsin TEM using an UTW Si(Li) detector: Comparison of quantification methods. *Microsc. Res. Tech.* **52**, 231–238.

Leapman, R. D. (2003). Detecting single atoms of calcium and iron in biological structures by electron energy-loss spectrum imaging. *J. Microsc.* **210**, 5–15.

LeFurgey, A., Davilla, S. D., Kopf, D. A., Sommer, J. R., and Ingram, P. (1992). Real time quantitative elemental analysis and mapping: Microchemical imaging in cell physiology. *J. Microsc.* **165**, 191–223.

LeFurgey, A., Ingram, P., and Blum, J. J. (2001). Compartmental responses to acute osmotic stress in leishmania major result in rapid loss of Na^+ and Cl^-. *Comp. Biochem. Physiol. A* **128,** 385–394.

LeFurgey, A., Gannon, M., Blum, J., and Ingram, P. (2005). Leishmania donovani amastigotes mobilize organic and inorganic osmolytes during regulatory volume decrease. *J. Eukaryotic Microbiol.* **52,** 277–289.

Marshall, A. T. (1994). Light element X-ray microanalysis in Biology. *Scanning Microsc. Suppl.* **8,** 187–201.

Marshall, A. T., and Patak, A. (1993). Use of ultra-thin window detectors for biological microanalysis. *Scanning Microsc.* **7,** 677–691.

Nkamgeue, E. M., Adnet, J.-J., Bernard, J., Zierold, K., Kilian, L., Jallot, E., Benhayoune, H., and Bonhomme, P. (2000). *In vitro* effects of zirconia and alumina particles on human blood monocyte-derived macrophages: X-ray microanalysis and flow cytometric studies. *J. Biomed. Mater. Res.* **52,** 587–594.

Roomans, G. M. (1988a). The correction for extraneous background in quantitative X-ray microanalysis of biological thin sections: Some practical aspects. *Scanning Microsc.* **2,** 311–318.

Roomans, G. M. (1988b). Quantitative X-ray microanalysis of biological specimens. *J. Electron Microsc Tech.* **9,** 19–43.

Roomans, G. M. (2002). X-ray microanalysis of epithelial cells in culture. *In* "Epithelial Cell Culture Protocols" (C. Wise, ed.). Humana Press, Totowa, NJ.

Salido, M., Vilches, J., Lopez, A., and Roomans, G. M. (2001). X-ray microanalysis of etoposide-induced apoptosis in the PC-3 prostatic cancer cell line. *Cell Biol. Int.* **25,** 499–508.

Sanchez-Quevedo, M. C., Alaminos, M., Capitan, L. M., Moreau, G., Gorzon, I., Crespo, P. V., and campos, A. (2007). Histological and histochemical evaluation of human oral mucosa constructs developed by tissue engineering. *Histol. Histopathol.* **22,** 631–640.

Scott, D. A., Docampo, R., Dvorak, J. A., Shi, S., and Leapman, R. D. (1997). *In situ* compositional analysis of acidocalciosomes in *Trypanosomi cruzi. J. Biol. Chem.* **272**(4), 28020–28029.

Serretnyk, Z., Krjukova, J., Gaston, B., Zaman, K., Hjelte, L., Roomans, G. M., and Dragomir, A. (2006). Activation of chloride transport in CF airway epithelial cell lines and primary CF nasal epithelial cells by *S*-nitrosoglutathione. *Resp. Res.* DOI: 10.11861/1645-9921-7-124.

Simm, C., Lahner, B., Salt, D., LeFurgey, A., Ingram, P., Yandell, B., and Eide, D. J. (2007). The yeast vacuole in zinc storage and intracellular zinc distribution. *Eukaryotic Cell.* DOI: 10.1128/EC.00077-07.

Sitte, H., Neumann, K., and Edelman, L. (1987). Safety rules for cryopreparation. *In* "Cryotechniques in Biological Electron Microscopy" (R. A. Steinbrecht, and K. Zierold, eds.), pp. 285–289. Springer-Verlag, Berlin, Heidelberg.

Skepper, J. N., Karydis, I., Garnett, M. R., Hegyi, L., Hardwick, S. J., Warley, A., Mitchinson, M. J., and Cary, N. R. B. (1999). Changes in element concentrations are associated with early stages of apoptosis in human monocyte-macrophages exposed to oxidised low-density lipoprotein: An X-ray microanalysis study. *J. Pathol.* **188,** 100–106.

Somlyo, A. V., Shuman, H., and Somlyo, A. P. (1989). Electron probe X-ray microanalysis of Ca^{2+}, Mg^{2+}, and other ions in rapidly frozen cells. *Methods Enzymol.* **172,** 203–229.

Statham, P. J., and Pawley, J. B. (1978). A new method for particle X-ray microanalysis based on peak to background measurements. *Scanning Electron Microsc.* **I,** 469–477.

Steinbrecht, R. A., and Zierold, K. (1984). A cryoembedding method for cutting ultrathin cryosections from small frozen specimens. *J. Microsc.* **136,** 69–75.

Warley, A. (1990). Standards for the application of X-ray microanalysis to biological specimens. *J. Microsc.* **157,** 135–147.

Warley, A. (1993). Quantitative X-ray microanalysis of thin sections in biology: Appraisal and interpretation of results. *In* "X-ray Microanalysis in Biology. Experimental Techniques and Applications" (D. C. Sigee, A. J. Morgan, A. T. Sumner, and A. Warley, eds.), pp. 47–57. Cambridge University Press, Cambridge.

Warley, A. (1997). X-ray microanalysis for biologists. *In* "Practical Methods in Electron Microscopy" (A. M. Glauert, ed.), Vol. 16. Portland Press, London.

Warley, A. (2001). Potassium concentration is reduced in cultured rabbit tracheal smooth muscle cells after withdrawal of serum. *Cell Biol. Int.* **25,** 691–695.

Warley, A., and Gupta, B. L. (1991). Quantitative biological X-ray microanalysis. *In* "Electron Microscopy in Biology: A practical Approach" (J. R. Harris, ed.), pp. 243–281. IRL Press, Oxford.

Warley, A., and Skepper, J. N. (2000). Long freeze-drying times are not necessary during the preparation of thin sections for X-ray microanalysis. *J. Microsc.* **198,** 116–123.

Warley, A., Cracknell, K. P. B., Cammish, H. B., Twort, C. H. C., Ward, J. P. T., and Hirst, S. J. (1994). Preparation of cultured airway smooth muscle for study of intracellular element concentrations by X-ray microanalysis:Comparisons of whole cells with cryosections. *J. Microsc.* **175,** 143–145.

Wendt-Gallitelli, M. F., and Isenberg, G. (1991). Total and free myoplasmic calcium during a contraction cycle: X-raymicroanalysis in guinea-pig ventricular myocytes. *J. Physiol.* **435,** 349–352.

Zhang, B.-X., Desnoyer, R. W., and Bond, M. (2001). Extracellular adenosine triphosphate triggers arrythmias and element redistribution in electrically stimulated rat cardiac myocytes. *Microsc. Microanal.* **7,** 48–55.

Zierold, K. (1988). X-ray microanalysis of freeze-dried and frozen-hydrated cryosections. *J. Electron Microsc. Tech.* **9,** 65–82.

Zierold, K. (1991). Cryofixation methods for ion localization in cells by electron probe X-ray microanalysis: A review. *J. Microsc.* **161,** 357–366.

Zierold, K. (1997). Effects of cadmium on electrolyte ions in cultured rat hepatocytes studied by X-ray microanalysis of cryosections. *Toxicol. Appl. Pharmacol.* **144,** 70–76.

Zierold, K. (2000). Heavy metal cytotoxicity studied by electron probe X-ray microanalysis of cultured rat hepatocytes. *Toxicology in vitro* **14,** 557–563.

Zierold, K., Michel, J., Terryn, C., and Balossier, G. (2005). The distribution of light elements in biological cells measured by electron probe X-ray microanalysis of cryosections. *Microsc. Microanal.* **11,** 138–145.

CHAPTER 3

Preparation of Cells and Tissues for Immuno EM

Paul Webster,* Heinz Schwarz,[†] and Gareth Griffiths*

*Cell Biology and Biophysics Unit, EMBL
Heidelberg, Meyerhofstrasse 69117

[†]Max Planck Institute for Developmental Biology
Tubingen, Germany

Abstract
I. Preparing Biological Specimens for Examination by Electron Microscopy
II. Vitrification and Chemical Fixation for Immunolocalization
III. Vitrification Followed by Freeze Substitution
IV. Chemical Cross-linking (or Fixation)
V. Embedding in Resin for Sectioning
VI. Cryosectioning for Immunocytochemistry: The Tokuyasu Method
VII. The Starting Material
VIII. Protocols
  A. High Pressure Freezing, Freeze Substitution, and Resin Embedding
  B. Chemical Fixation, Cryoprotection, Freezing, Freeze Substitution, and Resin Embedding
IX. Alternative Approaches to Freeze Substitution
X. Correlative Microscopy
References

Abstract

Transmission electron microscopy (TEM) provides a powerful set of methods to investigate cellular and subcellular structures using thin sections. In this article we summarize some of the different approaches available for researchers interested in using these methods. The essential details involved in specimen preparation for immunolabelling are covered.

The best sectioning approach for preserving specimens for structural analysis is **Cryo EM of Vitrified Sections** (CEMOVIS), a method where still frozen sections are examined in the transmission electron microscope. Because the specimens are kept at low temperature during sectioning and examination, this method is not amenable for immunolabelling, where antibodies are applied to sections at ambient temperature.

To combine structural analysis with immunocytochemical analysis of antigens, the approach of freeze-substitution without chemical fixative is the method of choice, at least from a theoretical point of view. In practice, however, the vast majority of electron microscopic (EM) immunocytochemical analyses are carried out using chemically-fixed specimens that have been embedded in specialized resins (such as the Lowicryls) using freeze-substitution or ambient temperature methods. Antibody labelling of thawed cryosections through chemically-fixed specimens (the Tokuyasu method) is also a popular method for preparing cells and tissues for TEM analysis.

Here, we provide an overview of all these sectioning methods for EM, focusing mostly on the practical details. Given the space limitation, the fine details necessary to apply these methods have been successfully omitted and will have to be obtained from the technical references we provide.

I. Preparing Biological Specimens for Examination by Electron Microscopy

The eventual aim of biological microscopy is to observe cellular processes as they occur in living cells. In light microscopy this is now routine for cultured cells and is increasingly possible with live tissues (Huisken *et al.*, 2004). Improvements in biomarkers, molecules expressed by living cells that can be detected by this approach, now make it possible to follow subcellular processes and organelles *in vivo*. Improvements in instrumentation enable researchers to observe subcellular events as they occur in real time in a variety of different ways.

While much advances as these have been welcomed, the inherent limitations of light microscopy must also be acknowledged. The main limitation for light microscopy is the resolving power available to examine cells. The resolving power of a microscope is a measure of how well a microscope is able to separate two points and therefore to show specimen detail. In the light microscope the smallest detectable details are, at best, in the size of about 200 nm. So, although the light microscope is an excellent tool for examining living cells in real time, it cannot give the detailed view of structures that is readily achievable by EM.

Electron microscopy (EM) is a technology ideally suited to examining small details in cells and has been in the forefront of dissecting cellular structures and their functions. However, the need for high vacuums and dead specimens meant that examination by electron microscopes have mostly been restricted to specimens that have been altered in some way. The most usual alteration performed on biological specimens for EM examination is to dehydrate them. Drying is a process that can induce many structural alterations inside the cells.

Another alteration, usually performed to enable biological specimens to withstand the drying process, is chemical cross-linking of cellular components.

Classical images of cells, tissues, and organelles usually consist of images taken from material that has been chemically cross-linked (or fixed) using glutaraldehyde, stained with heavy metals, and dehydrated. Chemical cross-linking followed by dehydration form the basis of modern subcellular morphology studies. The possibility of structural alterations occurring during these specimen preparation methods has led to the idea that a lot of EM data in cell biology is questionable.

The discovery that water can be frozen so rapidly that ice crystals are unable to form (the vitreous state) has stimulated a revolution in electron microscopy (Dubochet et al., 1988; Saibil, 2000a,b; Steven and Aebi, 2003). Cryo-EM specimen preparation methods have been used to examine biological material and, in most instances, validate much of what had been observed using chemical fixation and dehydration specimen preparation methods.

At the most sophisticated end of the cryo-EM specimen preparation spectrum is the approach where rapidly vitrified specimens are examined, while still solid and in a fully hydrated state in the electron microscope. Although vitrification of hydrated biological specimens can only be judged to be successful by examining diffraction patterns in the TEM, we will use the term to apply to all specimens that have been rapidly frozen using methods previously shown to produce vitrification.

Vitrified specimens that can be easily examined while still fully hydrated and a low temperature can consist of small particles such as viruses and sub-cellular organelles, frozen in thin films of water. Multiple images, which have been obtained while the particles are in a frozen, fully hydrated state, are now used routinely for 3D reconstruction (Baumeister et al., 1999; Cyrklaff et al., 2005).

The most marked use of this approach has been described in the recent works investigating whole, intact cells (Beck et al., 2004; Braet et al., 2003; Kurner et al., 2004; Medalia et al., 2002). Cells on coated specimen grids are rapidly vitrified by immersion in liquid ethane and examined, while still solid at low temperature, in the TEM. In the first publication (Medalia et al., 2002), *Dictyostelium discoideum* were shown to move actively on an EM grid prior to rapid freezing. This approach makes it possible to look inside thinner part of the cells (less than 500 nm). In a later study, the organization of actin filaments in the filopodia of *Dictyostelium* were elucidated using similar preparation and imaging methods (Medalia et al., 2007).

To examine the thicker regions of cells and tissues, the technique of CEMOVIS (**C**ryo-**E**lectron **M**icroscopy **o**f **V**itreous **S**ections) must be used (Al-Amoudi et al., 2002, 2004a,b, 2005). In this method, vitrified (rapidly frozen) biological material is sectioned at low temperature and the still-solid sections are examined at low temperature in the TEM. Such thin sections, through rapidly vitrified cell pellets and biological tissues, have also been examined to reveal structures familiar to classical morphologists (McIntosh et al., 2005).

Although it is possible to perform on-grid labeling of particle surfaces or cells prior to freezing (Roos et al., 1996), fully hydrated, frozen specimens cannot be immunolabeled using sequential applications of reagents once they are frozen. One possibility for examining structures in frozen material is to incorporate markers into living cells, and there is much interest in developing such biomarkers for use with cryo-EM methods. The ideal biomarkers are tags attached to normally expressing proteins that can be easily visualized by electron microscopy in frozen specimens. So far there are only a few such biomarkers available for use and their application is not widespread (Gaietta *et al.*, 2002, 2006; Sosinsky *et al.*, 2007).

II. Vitrification and Chemical Fixation for Immunolocalization

If the location of molecules within cells is to be studied using affinity markers to cells or sections, then some form of immobilization of the biological material must take place to preserve the location of the molecules of interest. This immobilization process must also preserve the structures in which the molecules are operating, and as many related structures and molecules as possible. The immobilization must withstand subsequent processing steps such as dehydration, sectioning, and immunolabeling.

Molecules that are not immobilized may be displaced from their normal location, resulting in either a false (positive) localization or a spurious (negative) result, such as when molecules are washed away (van Genderen *et al.*, 1991). Two approaches can be considered for immobilizing specimens for EM: these are (1) the vitrification of biological material, and (2) chemical fixation.

III. Vitrification Followed by Freeze Substitution

Rapid heat removal from biological material so as to place specimens into a vitreous or microcrystalline state is currently the optimal way of preserving ultrastructural features. Vitrification is the process where fully hydrated specimens are cooled rapidly enough to prevent ice crystal formation. Frozen specimens can be dehydrated at low temperature ($-90\ °C$) by dissolving the ice with an organic solvent using a process called freeze-substitution (Kellenberger, 1991; Walther and Ziegler, 2002). Chemical cross-linkers, fixatives, or contrasting chemicals, such as uranyl acetate, osmium tetroxide, or glutaraldehyde, can be added to the solvent. In this approach, the specimens are chemically-fixed (or stabilized) while still in a cryoimmobilized state, at low temperature, a condition that is much less prone to extraction and reorganization of sub-cellular architecture (Bohrmann and Kellenberger, 2001; Kellenberger, 1991). However, alterations in specimen morphology can occur if vitrification is not successfully carried out (Han and Bischof, 2004). In most cases the freeze substituted specimen is finally infiltrated with resin and polymerized (Giddings, 2003).

Lowicryl resins belong to a group of resins that can be polymerized at low temperature by UV light, making them useful for embedding freeze substituted material while they are still in a solid state. Alternatively, the specimens can be slowly warmed to ambient temperature and embedded in a resin also designed for immunolabelling such as LR White, an acrylic resin. The resin can be heat polymerized either in a conventional oven or by using a microwave processor (Webster, 2007). Other resins are available that can be polymerized by heat or catalytic action.

IV. Chemical Cross-linking (or Fixation)

It is generally accepted that the best approach for immunolabeling, resulting in minimal alterations in morphology, is cryo-immobilization followed by freeze-substitution and embedding at low temperatures (−50 to −90 °C) (Hess, 2007; Humbel and Schwarz, 1989a; Schwarz et al., 1993). However, although high-pressure freezing (McDonald, 1999; Moor, 1987) is generally acknowledged to be the method of choice for immobilizing relatively large biological specimens (up to 200 μm thickness and 2 mm in diameter), the technology and skills required to prepare specimens are demanding, especially when attempting to sample material in its native state without causing changes due to initial specimen preparation. Other, technically less-challenging rapid freezing methods are available that are able to vitrify biological specimens (Gilkey and Staehelin, 1986; Roos and Morgan, 1990) but the limited freezing efficiency at ambient pressure restricts their use to small specimens up to 10 μm.

Although freeze-substitution is understandably the preferred method for immobilizing specimens, most users of electron microscopy still prefer the technically easier immobilization method of chemical fixing material, using aldehydes to cross-link cellular components. Chemical fixatives are convenient to use, help to preserve specimen morphology, and immobilize antigens. Despite the reasonable concerns about fixation artifacts, EM immunocytochemistry is widely and successfully used in many laboratories around the world.

In cases where structural reorganization due to the processing method are suspected (artifacts), the only alternative is to examine the morphology using the CEMOVIS technique (which is not applicable for immunolabelling), or freeze-substitution. The theoretical background to chemical fixation has been extensively covered elsewhere (Griffiths, 1993). It is important to note that chemical cross-linking as a process for immobilizing subcellular structures involves the relatively slow (many seconds) conversion of a live cell state to a dead one. Significant physiological changes are inevitable in the period before most cross-links induced by the fixative occurs.

For immunocytochemistry, tissues or cells can be fixed by immersion in a buffered solution of aldehyde using the same methods that are used for many years to prepare specimens for epoxy resin embedding. The specimen should be small for rapid penetration of aldehyde and care should be taken not to damage the sample by over-manipulation (i.e., squeezing, pulling, cutting, etc.). Large organs are best fixed by perfusing with warm fixative.

The buffers used with fixatives will affect the morphology of biological material and although the exact role of pH in fixation is not known, it seems that all commonly used buffers are charged molecules, and are thus likely to display their effects outside the cells. This is potentially worrisome given the ability of aqueous solutions of glutaraldehyde and formaldehyde to lower the cytoplasmic pH from 7.0 to perhaps 6.0 (Griffiths, 1993). However, it has been empirically determined that the need for a robust buffer with the chemical fixative generally improves structural preservation.

V. Embedding in Resin for Sectioning

Chemically fixed biological specimens can be prepared for immunocytochemistry in a variety of ways. Perhaps the easiest way is to embed the material in a resin that can be sectioned. However, there are multiple choices for which resin to use and how to embed the specimen in resin.

A popular method used for embedding in methacrylates such as Lowicryl resins (Armbruster et al., 1982; Carlemalm et al., 1985) introduced by the group of Kellenberger in the early 1980s, called the progressive lowering of temperature (PLT) method (Hobot, 1989; Robertson et al., 1992), subjected aldehyde-fixed material to increasing concentrations of dehydrating solvent (usually ethanol or acetone) while progressively lowering the temperature of the solvent. Dehydrated specimens were subsequently infiltrated with resin at low temperature (around −50 °C) and polymerized under UV light. Several researchers have successfully used this approach for immunolabelling (Altman et al., 1984; Bayer et al., 1985; Lucocq and Roth, 1984; Oprins et al., 1994; Warhol et al., 1985). However, the Lowicryl resins, and some related resins, must be handled with care because of their tendency to cause skin allergies.

Chemically fixed biological material can also be dehydrated using freeze-substitution methods if the material is cryoprotected before immersion freezing in liquid nitrogen (Humbel and Schwarz, 1989b; Oprins et al., 1994; van Genderen et al., 1991). Chemically fixed specimens are trimmed to a suitable size, infiltrated with cryoprotectant and frozen by immersion in liquid nitrogen when sucrose is used. This has been demonstrated to vitrify the specimens. The specimens are then soaked in freeze-substitution medium at low temperature in a way similar to that of rapidly frozen specimens.

Chemically fixed biological specimens have also been processed for immunocytochemistry by rapid dehydration at ambient temperature and resin embedded using heat polymerization (Gocht, 1992; Osamura et al., 2000; Waller et al., 1998).

VI. Cryosectioning for Immunocytochemistry: The Tokuyasu Method

A popular, but specialized method for preparing biological material for immunolabelling is the Tokuyasu cryosectioning technique (Griffiths et al., 1983; Tokuyasu, 1980). It is often the preferred immunolabelling method because it is the only post-embedding immunolabelling approach that does not require dehydration by polar solvents before application of affinity markers. It is also the fastest approach for obtaining results on ultrastructure and immunolabelling. In general, in the absence of plastic resin, thawed cryosections tend to give the highest accessibility of antigens to the antibodies.

Biological material is chemically-fixed using buffered aldehyde solution, cryoprotected in sucrose and vitrified by immersion in liquid nitrogen (Griffiths *et al.*, 1984; Webster and Webster, 2007). The vitrified, cryoprotected specimen blocks are sectioned at low temperature (between -60 and $-120\,°C$) in a cryo-ultramicrotome using a knife, retrieved from the knife surface and thawed onto metal specimen grids. The thawed sections are labeled with specific affinity markers and colloidal gold probes and finally embedded in a thin film of plastic (Webster and Webster, 2007). The hydrated state of the sections means that the final drying step can be made in a variety of plastic films. The final embedment and drying step can be used to manipulate the final appearance of the sections in the transmission electron microscope (Himmelhoch, 1994; Keller *et al.*, 1984; Takizawa *et al.*, 2003; Tokuyasu, 1986, 1989).

VII. The Starting Material

For any EM analyses one needs access to a transmission electron microscope, specimen preparation equipment such as an ultramicrotome, a glass knife maker, or a diamond knife for sectioning and various laboratory essentials such as a chemical hood for extracting toxic fumes away from the workspace.

Metal specimen grids will be required, and if cryosections are to be used, the grids should be coated with a thin layer of plastic to support the fragile sections (Webster and Webster, 2007). Suitable antibodies and colloidal gold probes are important, as are the many buffers used.

Finally, there are multiple protocols that can be applied to the specimens under study. Choosing the best method for use may be difficult for novice microscopists as the final choice will depend on the specimens to be examined, the specimen preparation equipment available, and the expected end result. Much can be learned from talking with experts or attending specialist courses. By seeing the "tricks of the trade" from an experienced person, a beginner can rapidly learn the basics.

VIII. Protocols

A. High Pressure Freezing, Freeze Substitution, and Resin Embedding

High pressure freezing is the current method of choice for immobilizing biological specimens for subsequent examination by electron microscopy. The method requires access to a high-pressure freezer and to a freeze-substitution device and has been adequately described elsewhere (McDonald, 1999, 2007).

B. Chemical Fixation, Cryoprotection, Freezing, Freeze Substitution, and Resin Embedding

Although high-pressure freezing is considered to be the optimal method for immobilization, the technology is not routinely available. A useful compromise when preparing specimens for resin embedding is to freeze-substitute frozen, cryoprotected and chemically-fixed material. Although this method does not avoid chemical cross-linking it does omit ambient temperature dehydration, a process that is often considered to be a contributor to morphological damage (Bohrmann and Kellenberger, 2001).

Sucrose has been used successfully for cryoprotection prior to freeze-substitution (van Genderen et al., 1991), but it can only be used if methanol is the substitution medium. A precipitate may form in and around specimens if cold acetone or ethanol is used to dehydrate material cryoprotected with sucrose.

1. Method A: Preparing Specimens for Morphological Examination

Immerse tissue pieces in buffered aldehyde (e.g., 2% formaldehyde (w/v) and 2% glutaraldehyde (v/v) in 100 mM HEPES buffer, pH 7.2) and leave for 2 h. If cells are used, the fixative can be added directly to the culture medium (use double strength fixative to allow for the dilution in the medium) and the cells can be scraped off the dish after approx 5 min fixation. Pellet the cells and leave them in the fixative for the remaining time.

Infiltrate the specimens by immersion in either 30% glycerol, 2% dimethyl formamide (a solvent used previously for dehydrating prior to resin embedding (Altman et al., 1984), or 2.3 M sucrose. Place the specimens on metal pins or wire stubs (for convenient handling) and freeze by immersion in liquid propane (glycerol and dimethyl formamide) or in liquid nitrogen (sucrose).

Place tubes containing dry acetone, ethanol, or methanol in liquid nitrogen to freeze the solvents and then place the frozen specimens into tubes. Use dry acetone or ethanol for the materials cryoprotected in glycerol or dimethyl formamide and the dry methanol for sucrose cryoprotected material. When applied to rapidly frozen specimens, small amounts of glutaraldehyde, osmium tetroxide, tannic acid, or uranyl acetate have been added to the solvents to improve specimen contrast (Giddings, 2003; Walther and Ziegler, 2002). Similar approaches, although not thoroughly explored, should improve specimen contrast of the chemically-fixed and cryoprotected frozen specimens.

Place the tubes containing substitution media and specimens into a freeze-substitution device set at −90 °C and leave for 36 h. Replace the substitution medium with fresh, cold, and dry medium and gradually bring the temperature to 20 °C (5 °C per hour). Replace the substitution medium with a 1:1 mixture of uncatalyzed epoxy resin (e.g., any of the Epon substitutes, Araldite or Spurr's resin are all available commercially in kit form) and substitution medium, gradually increasing the concentration of epoxy resin over several hours. Leave in uncatalyzed resin overnight, transfer to catalyzed resin in specimen molds and polymerize by placing at 60 °C overnight. The hardened blocks of resin-embedded specimens are now ready to be sectioned using an ultramicrotome and either a glass or diamond knife (Hagler, 2007; Webster, 2007) and sections can be examined by light or transmission electron microscopy.

2. Method B: Preparing Specimens for Immunocytochemistry

Biological material prepared for immunocytochemical experiments are usually cross-linked in lower concentrations of aldehyde to preserve antigenicity. Many suitable aldehyde mixtures have been used and the final choice will depend on many factors that can only be determined empirically. The main issues to take into account are the ability of the specific antibodies to gain access to and bind the target antigen after fixation yet still obtain adequate specimen preservation. These issues are too complicated to discuss in depth here but it should be noted that fixing specimens to obtain good morphology is often at odds with the ability of the antibody to gain access to the target antigen. The more effective the cross-linking the less likely is the antigen in the section to be accessible to antibodies.

Finding the best conditions for retaining antigenicity is a matter of trial and error. However, there are many simple aldehyde solutions that have been successfully used for immunolabelling (e.g., 0.1% glutaraldehyde in 100 mM phosphate buffer (pH 7.2), or 2% formaldehyde in 100 mM PIPES buffer, pH 7.2 (van Genderen *et al.*, 1991)). As a general rule, the more glutaraldehyde used for cross-linking, the better the morphology will be but at the expense of antigenicity. Using glutaraldehyde to crosslink biological tissues is not recommended if the tissues are to be subsequently used for light microscopy using epi-fluorescent illumination. Glutaraldehyde often induces an autofluorescent signal in the tissues that can interfere with detection of fluorescent antibodies. Formaldehyde-based fixatives, which do not create an autofuorescent signal and also cross-link less effectively than glutaraldehyde, are generally used to cross-link specimens that are to be examined by light and electron microscopy.

Once the tissues or cells have been cross-linked, they can be cryoprotected as described above, frozen by immersion in cryogen, and freeze substituted in a dry, cold freeze-substitution medium. Specimens that have been freeze substituted can be infiltrated in the cold with an acrylic resin (such as one of the Lowicryl or London resins (Newman and Hobot, 1987)), and polymerized at low temperature with ultraviolet light. Commercially available freeze substitution machines have the capability to process specimens this way, and come with ample instructions for their operation.

IX. Alternative Approaches to Freeze Substitution

Laboratories that are fully equipped to prepare rapidly frozen material use commercially available machines to control the temperatures at which freeze-substitution occurs. Beginners who want to try out this method of specimen preparation before investing in instrumentation can perform freeze-substitution on dry ice held in an insulated cooler or Styrofoam box. It is highly likely that users who want to try cryo-methods will not have easy access to a high-pressure freezer so freezing chemically cross-linked, cryoprotected material is a useful alternative method.

Freeze substitution of frozen material can be performed on dry ice in a Styrofoam box (Giddings, 2003). In its simplest form, the specimen bottles are placed at the bottom of the Styrofoam box and covered with dry ice, which maintains a temperature of $-80\,°C$. The substitution medium can be changed as the dry ice evaporates, and the box can be re-filled with dry ice if necessary. Specimens can be polymerized while still at dry ice temperature but placing a UV light over the box, as described in the brochure supplied with Lowicryl resins. Alternative, specimens can be gradually warmed by placing the Styrofoam box in a large freezer and left to let the dry ice evaporate, gradually warming the specimens to $-20\,°C$ for infiltration with resin and subsequent polymerization. Specimens can also be warmed to ambient temperature and polymerized using a microwave processor (Webster, 2007).

Modifications to this simple system aimed at improving temperature stability can be incorporated into the Styrofoam box design. The specimen tubes can be inserted into metal blocks to provide uniform temperatures around the specimens (Giddings, 2003), and thermocouple probes (McDonald, 1999) can be used to monitor specimen temperatures during the freeze-substitution process. Styrofoam containers can be replaced with more solid ice boxes, and specimen shaking can be incorporated into the protocol (McDonald, 1999).

Whatever approach is used, it is important to remember that specimen warming should occur slowly to avoid rapid release of gases from the substitution medium. It should also be noted that the substances used for freeze-substitution, and the resins used for specimen embedding, are toxic and should be handled carefully. Adequate ventilation is required to remove the toxic fumes released by these chemicals to prevent users from being exposed to them.

X. Correlative Microscopy

At the beginning of this chapter we said that the eventual aim of biological microscopy is to observe cellular processes as they occur in living cells. Although it is still not possible for users of the transmission electron microscope to look inside living cells, the instrumentation and specimen preparation protocols currently available make it possible to see snapshots of cellular processes as they occur.

Cells can now be examined in the light microscope and then immediately vitrified, freeze substituted, embedded in resin and sectioned (McDonald *et al.*, 2007). These methods make it possible to identify single cells in the light microscope and then find them again in thin sections for examination in the TEM (Biel *et al.*, 2003).

A much simpler approach to correlative microscopy, where light microscopy is correlated with TEM data, involves the collection of sequential sections through the biological material. Alternate sections are then examined either by light or electron microscopy (Muller-Reichert et al., 2007; Robinson et al., 2001; Schwarz and Humbel, 2007; Takizawa and Robinson, 2003a,b, 2006). Such an approach makes it possible to correlate immunolabelling observed at the light microscope level (and thus determine optimal antibody dilutions, signal intensity and distribution) with subsequent TEM analysis.

Applying fluorescent antibodies to thin sections of biological material offers a high-resolution alternative to the confocal microscope (Mori et al., 2006; Schwarz and Humbel, 2007; Takizawa and Robinson, 2004), where z-axis resolution is determined by aperture diameter and can be in the micron range. The z-axis resolution in thin sections is determined by the section thickness and in some instances, only the surface of the section it labeled with antibody (Schwarz and Humbel, 2007; Stierhof et al., 1986). Ultra-small subcellular organelles such as caveoli can be resolved using thin sections for immunofluorescence microscopy (Mori et al., 2006).

References

Al-Amoudi, A., Chang, J. J., Leforestier, A., McDowall, A., Salamin, L. M., Norlen, L. P., Richter, K., Blanc, N. S., Studer, D., and Dubochet, J. (2004a). Cryo-electron microscopy of vitreous sections. *Embo J.* **23**, 3583–3588.

Al-Amoudi, A., Dubochet, J., and Studer, D. (2002). Amorphous solid water produced by cryosectioning of crystalline ice at 113 K. *J. Microsc.* **207**, 146–153.

Al-Amoudi, A., Norlen, L. P., and Dubochet, J. (2004b). Cryo-electron microscopy of vitreous sections of native biological cells and tissues. *J. Struct. Biol.* **148**, 131–135.

Al-Amoudi, A., Studer, D., and Dubochet, J. (2005). Cutting artefacts and cutting process in vitreous sections for cryo-electron microscopy. *J. Struct. Biol.* **150**, 109–121.

Altman, L. G., Schneider, B. G., and Papermaster, D. S. (1984). Rapid embedding of tissues in Lowicryl K4M for immunoelectron microscopy. *J. Histochem. Cytochem.* **32**, 1217–1223.

Armbruster, B. L., Carlemalm, E., Chiovetti, R., Garavito, R. M., Hobot, J. A., Kellenberger, E., and Villiger, W. (1982). Specimen preparation for electron microscopy using low temperature embedding resins. *J. Microsc.* **126**, 77–85.

Baumeister, W., Grimm, R., and Walz, J. (1999). Electron tomography of molecules and cells. *Trends Cell. Biol.* **9**, 81–85.

Bayer, M. E., Carlemalm, E., and Kellenberger, E. (1985). Capsule of *Escherichia coli* K29: Ultrastructural preservation and immunoelectron microscopy. *J. Bacteriol.* **162**, 985–991.

Beck, M., Forster, F., Ecke, M., Plitzko, J. M., Melchior, F., Gerisch, G., Baumeister, W., and Medalia, O. (2004). Nuclear pore complex structure and dynamics revealed by cryoelectron tomography. *Science* **306**, 1387–1390.

Biel, S. S., Kawaschinski, K., Wittern, K. P., Hintze, U., and Wepf, R. (2003). From tissue to cellular ultrastructure: Closing the gap between micro- and nanostructural imaging. *J. Microsc.* **212**, 91–99.

Bohrmann, B., and Kellenberger, E. (2001). Cryosubstitution of frozen biological specimens in electron microscopy: Use and application as an alternative to chemical fixation. *Micron* **32**, 11–19.

Braet, F., Bomans, P. H., Wisse, E., and Frederik, P. M. (2003). The observation of intact hepatic endothelial cells by cryo-electron microscopy. *J. Microsc.* **212**, 175–185.

Carlemalm, E., Villiger, W., Hobot, J. A., Acetarin, J. D., and Kellenberger, E. (1985). Low temperature embedding with Lowicryl resins: Two new formulations and some applications. *J. Microsc.* **140**, 55–63.

Cyrklaff, M., Risco, C., Fernandez, J. J., Jimenez, M. V., Esteban, M., Baumeister, W., and Carrascosa, J. L. (2005). Cryo-electron tomography of vaccinia virus. *Proc. Natl. Acad. Sci. USA* **102**, 2772–2777.

Dubochet, J., Adrian, M., Chang, J. J., Homo, J. C., Lepault, J., McDowall, A. W., and Schultz, P. (1988). Cryo-electron microscopy of vitrified specimens. *Q. Rev. Biophys.* **21**, 129–228.

Gaietta, G., Deerinck, T. J., Adams, S. R., Bouwer, J., Tour, O., Laird, D. W., Sosinsky, G. E., Tsien, R. Y., and Ellisman, M. H. (2002). Multicolor and electron microscopic imaging of connexin trafficking. *Science* **296**, 503–507.

Gaietta, G. M., Giepmans, B. N., Deerinck, T. J., Smith, W. B., Ngan, L., Llopis, J., Adams, S. R., Tsien, R. Y., and Ellisman, M. H. (2006). Golgi twins in late mitosis revealed by genetically encoded tags for live cell imaging and correlated electron microscopy. *Proc. Natl. Acad. Sci. USA* **103**, 17777–17782.

Giddings, T. H. (2003). Freeze-substitution protocols for improved visualization of membranes in high-pressure frozen samples. *J. Microsc.* **212**, 53–61.

Gilkey, J. C., and Staehelin, L. A. (1986). Advances in ultrarapid freezing for the preservation of cellular ultrastructure. *J. Electron Microsc. Tech.* **3**, 177–210.

Gocht, A. (1992). Use of LR white resin for post-embedding immunolabelling of brain tissue. *Acta Anat. (Basel)* **145**, 327–339.

Griffiths, G. (1993). "Fine Structure Immunocytochemistry." Springer, Berlin.

Griffiths, G., McDowall, A., Back, R., and Dubochet, J. (1984). On the preparation of cryosections for immunocytochemistry. *J. Ultrastruct. Res.* **89**, 65–78.

Griffiths, G., Simons, K., Warren, G., and Tokuyasu, K. T. (1983). Immunoelectron microscopy using thin, frozen sections: Application to studies of the intracellular transport of Semliki Forest virus spike glycoproteins. *Methods Enzymol.* **96**, 466–485.

Hagler, H. K. (2007). Ultramicrotome for biological electron microscopy. *In* "Methods in Molecular Biology. Electron microscopy: Methods and Protocols" (J. Kuo, ed.), Vol. 369, pp. 67–106. Humana Press, Totowa, NJ.

Han, B., and Bischof, J. C. (2004). Direct cell injury associated with eutectic crystallization during freezing. *Cryobiology* **48**, 8–21.

Hess, M. W. (2007). Cryopreparation methodology for plant cell biology. *In* "Cellular Electron Microscopy" (J. R. McIntosh, ed.) Elsevier, Oxford, UK.

Himmelhoch, S. (1994). A negative contrast stain for ultra-thin frozen sections. *Microsc. Res. Tech.* **29**, 23–28.

Hobot, J. A. (1989). Lowicryls and low temperature embedding for colloidal gold methods. *In* "Colloidal gold: Principles, methods, and applications" (M. A. Hayat, ed.), Vol. 2, p. 75. Academic Press, San Diego, CA.

Huisken, J., Swoger, J., Del Bene, F., Wittbrodt, J., and Stelzer, E. H. (2004). Optical sectioning deep inside live embryos by selective plane illumination microscopy. *Science* **305**, 1007–1009.

Humbel, B. M., and Schwarz, H. (1989a). Freeze substitution for immunocytochemistry. *In* "Immuno-gold labeling in cell biology" (A. J. Verkleij, J. L. M. Leunissen, and V. Insler, eds.), pp. 114–134. CRC Press, Boca Raton, FL.

Humbel, B. M., and Schwarz, H. (1989b). Freeze-substitution for immunochemistry. *In* "Immuno-gold labelling in cell biology" (A. J. Verkleij, and J. L. M. Leunissen, eds.), p. 115. CRC Press, Boca Raton, FL.

Kellenberger, E. (1991). The potential of cryofixation and freeze-substitution: Observations and theoretical considerations. *J. Microsc.* **161**, 183–203.

Keller, G. A., Tokuyasu, K. T., Dutton, A. H., and Singer, S. J. (1984). An improved procedure for immunoelectron microscopy: Ultrathin plastic embedding of immunolabeled ultrathin frozen sections. *Proc. Natl. Acad. Sci. USA* **81**, 5744–5747.

Kurner, J., Medalia, O., Linaroudis, A. A., and Baumeister, W. (2004). New insights into the structural organization of eukaryotic and prokaryotic cytoskeletons using cryo-electron tomography. *Exp. Cell Res.* **301**, 38–42.

Lucocq, J. M., and Roth, J. (1984). Applications of immunocolloids in light microscopy. III. Demonstration of antigenic and lectin-binding sites in semithin resin sections. *J. Histochem. Cytochem.* **32**, 1075–1083.

McDonald, K. (1999). High-pressure freezing for preservation of high resolution fine structure and antigenicity for immunolabeling. *Methods Mol. Biol.* **117**, 77–97.

McDonald, K. L., Morphew, M., Verkade, P., and Müller-Reichert, T. (2007). Recent advances in high-pressure freezing. *In* "Methods in Molecular Biology. Electron microscopy: Methods and protocols" (J. Kuo, ed.), Vol. 369, pp. 143–173. Humana Press, Totowa, NJ.

McIntosh, R., Nicastro, D., and Mastronarde, D. (2005). New views of cells in 3D: An introduction to electron tomography. *Trends Cell Biol.* **15**, 43–51.

Medalia, O., Beck, M., Ecke, M., Weber, I., Neujahr, R., Baumeister, W., and Gerisch, G. (2007). Organization of actin networks in intact filopodia. *Curr. Biol.* **17**, 79–84.

Medalia, O., Weber, I., Frangakis, A. S., Nicastro, D., Gerisch, G., and Baumeister, W. (2002). Macromolecular architecture in eukaryotic cells visualized by cryoelectron tomography. *Science* **298**, 1209–1213.

Moor, H. (1987). Theory and practice of high pressure freezing. *In* "Cryotechniques in biological electron microscopy" (R. A. Steinbrecht, and K. Zierold, eds.), pp. 175–191. Springer, Berlin/Heidelberg.

Mori, M., Ishikawa, G., Takeshita, T., Goto, T., Robinson, J. M., and Takizawa, T. (2006). Ultrahigh-resolution immunofluorescence microscopy using ultrathin cryosections: Subcellular distribution of caveolin-1alpha and CD31 in human placental endothelial cells. *J. Electron Microsc. (Tokyo)* **55**, 107–112.

Muller-Reichert, T., Srayko, M., Hyman, A., O'Toole, E. T., and McDonald, K. (2007). Correlative light and electron microscopy of early *Caenorhabditis elegans* embryos in mitosis. *Methods Cell. Biol.* **79**, 101–119.

Newman, G. R., and Hobot, J. A. (1987). Modern acrylics for post-embedding immunostaining techniques. *J. Histochem. Cytochem.* **35**, 971–981.

Oprins, A., Geuze, H. J., and Slot, J. W. (1994). Cryosubstitution dehydration of aldehyde-fixed tissue: A favorable approach to quantitative immunocytochemistry. *J. Histochem. Cytochem.* **42**, 497–503.

Osamura, R. Y., Itoh, Y., and Matsuno, A. (2000). Applications of plastic embedding to electron microscopic immunocytochemistry and *in situ* hybridization in observations of production and secretion of peptide hormones. *J. Histochem. Cytochem.* **48**, 885–891.

Robertson, D., Monaghan, P., Clarke, C., and Atherton, A. J. (1992). An appraisal of low-temperature embedding by progressive lowering of temperature into Lowicryl HM20 for immunocytochemical studies. *J. Microsc.* **168**, 85–100.

Robinson, J. M., Takizawa, T., Pombo, A., and Cook, P. R. (2001). Correlative fluorescence and electron microscopy on ultrathin cryosections: Bridging the resolution gap. *J. Histochem. Cytochem.* **49**, 803–808.

Roos, N., Cyrklaff, M., Cudmore, S., Blasco, R., Krijnse-Locker, J., and Griffiths, G. (1996). A novel immunogold cryoelectron microscopic approach to investigate the structure of the intracellular and extracellular forms of vaccinia virus. *EMBO J.* **15**, 2343–2355.

Roos, N., and Morgan, J. A. (1990). "Cryopreparation of thin biological specimens for electron microscopy: Methods and applications." Oxford Scientific Publications, .

Saibil, H. R. (2000a). Conformational changes studied by cryo-electron microscopy. *Nat. Struct. Biol.* **7**, 711–714.

Saibil, H. R. (2000b). Macromolecular structure determination by cryo-electron microscopy. *Acta Crystallogr. D* **56**, 1215–1222.

Schwarz, H., Hohenberg, H., and Humbel, B. M. (1993). Freeze-substitution in virus research: A preview. *In* "Immuno-gold electron microscopy" (A. D. Hyatt, and B. T. Eaton, eds.), pp. 349–376. CRC Press, Boca-Raton, FL.

Schwarz, H., and Humbel, B. M. (2007). Correlative light and electron microscopy using immunolabeled resin sections. *In* "Methods in Molecular Biology. Electron microscopy: Methods and Protocols" (J. Kuo, ed.), Vol. 369, pp. 229–256. Humana Press, Totowa, NJ.

Sosinsky, G. E., Giepmans, B. N., Deerinck, T. J., Gaietta, G. M., and Ellisman, M. H. (2007). Markers for correlated light and electron microscopy. *In* "Cellular Electron Microscopy" (J. R. McIntosh, ed.), Vol. 79, pp. 575–591. Elsevier.

Steven, A. C., and Aebi, U. (2003). The next ice age: Cryo-electron tomography of intact cells. *Trends Cell Biol.* **13**, 107–110.

Stierhof, Y. D., Schwarz, H., and Frank, H. (1986). Transverse sectioning of plastic-embedded immunolabeled cryosections: morphology and permeability to protein A-colloidal gold complexes. *J. Ultrastruct. Mol. Struct. Res.* **97**, 187–196.

Takizawa, T., Anderson, C. L., and Robinson, J. M. (2003). A new method to enhance contrast of ultrathin cryosections for immunoelectron microscopy. *J. Histochem. Cytochem.* **51**, 31–39.

Takizawa, T., and Robinson, J. M. (2003a). Correlative microscopy of ultrathin cryosections is a powerful tool for placental research. *Placenta* **24**, 557–565.

Takizawa, T., and Robinson, J. M. (2003b). Ultrathin cryosections: An important tool for immunofluorescence and correlative microscopy. *J. Histochem. Cytochem.* **51**, 707–714.

Takizawa, T., and Robinson, J. M. (2004). Thin is better! Ultrathin cryosection immunocytochemistry. *J. Nippon. Med. Sch.* **71**, 306–307.

Takizawa, T., and Robinson, J. M. (2006). Correlative microscopy of ultrathin cryosections in placental research. *Methods Mol. Med.* **121**, 351–369.

Tokuyasu, K. T. (1980). Immunochemistry on ultrathin frozen sections. *Histochem J.* **12**, 381–403.

Tokuyasu, K. T. (1986). Application of cryoultramicrotomy to immunocytochemistry. *J. Microsc.* **143**, 139–149.

Tokuyasu, K. T. (1989). Use of poly(vinylpyrrolidone) and poly(vinyl alcohol) for cryoultramicrotomy. *Histochem J.* **21**, 163–171.

van Genderen, I. L., van Meer, G., Slot, J. W., Geuze, H. J., and Voorhout, W. F. (1991). Subcellular localization of Forssman glycolipid in epithelial MDCK cells by immuno-electronmicroscopy after freeze-substitution. *J. Cell. Biol.* **115**, 1009–1019.

Waller, R. F., Keeling, P. J., Donald, R. G., Striepen, B., Handman, E., Lang-Unnasch, N., Cowman, A. F., Besra, G. S., Roos, D. S., and McFadden, G. I. (1998). Nuclear-encoded proteins target to the plastid in Toxoplasma gondii and Plasmodium falciparum. *Proc. Natl. Acad. Sci. USA* **95**, 12352–12357.

Walther, P., and Ziegler, A. (2002). Freeze substitution of high-pressure frozen samples: The visibility of biological membranes is improved when the substitution medium contains water. *J. Microsc.* **208**, 3–10.

Warhol, M. J., Lucocq, J. M., Carlemalm, E., and Roth, J. (1985). Ultrastructural localization of keratin proteins in human skin using low-temperature embedding and the protein A-gold technique. *J. Invest. Dermatol.* **84**, 69–72.

Webster, P. (2007). Microwave-assisted processing and embedding for transmission electron microscopy. *In* "Methods in Molecular Biology. Electron microscopy: Methods and Protocols" (J. Kuo, ed.), Vol. 369, pp. 47–65. Humana Press, Totowa, NJ.

Webster, P., and Webster, A. (2007). Cryosectioning fixed and cryoprotected biological material for immunocytochemistry. *In* "Methods in Molecular Biology: Electron microscopy: Methods and Protocols" (J. Kuo, ed.), Vol. 369, pp. 257–289. Humana Press, Totowa, NJ.

CHAPTER 4

Quantification of Structures and Gold Labeling in Transmission Electron Microscopy

John Lucocq

School of Life Sciences
University of Dundee
Dundee DDI 4HN
United Kingdom

I. Introduction
II. Sampling and Stereology
III. Quantities Displayed on Sections—Gold Labeling and Profile Data
    A. Gold Labeling Distribution
    B. Gold Labeling Density
    C. Statistical Assessment of Labeling
    D. Specificity
IV. Quantities in Three Dimensions
    A. Densities
    B. Volume
    C. Number
V. Spatial Analysis
References

I. Introduction

Ninety years ago in his book *On Growth and Form* the biomathematician D'Arcy (1917) wrote: "Dreams apart, numerical precision is the very soul of science, and its attainment affords the best, perhaps the only criterion of the truth of theories and the correctness of experiments". Today the sentiment of this statement is reflected in an increasing demand for reliable numerical precision as cell biologists strive to analyze increasingly complex and subtle effects *in vitro* and *in vivo*. At the level of ultrastructure, numerical precision is now provided by sophisticated methods for efficient quantitation of immunogold labeling and cell structure. This review describes how a range of quantitative tools can be applied to ultrathin sections of biological material.

II. Sampling and Stereology

In transmission electron microscopy quantitative "estimations" are generally made on images taken of ultrathin sections (see below) but the very first concern of any well designed study is the origin of these images. The reason is that any measurements must faithfully represent the bigger biological picture found in the whole animal, plant, or cell culture, and the application of appropriate sampling protocols is therefore unavoidable.

Electron microscope images are extremely small representations of intact specimens and a 20,000× micrograph has, for example, a volume that represents $1/1 \times 10^{11}$ of the volume of a rat kidney. These small samples are obtained by successive selection of animals or cultures, organs, blocks, and sections, and together this chain forms a "sampling cascade" (Fig. 1). The cascade will only faithfully represent quantities if selection is nonpreferential at each level, and the way to ensure this is to use a lottery-style selection method called simple uniform random sampling (Lucocq, 1994).[1] In EM quantitation, simple random sampling is used less often and a powerful modification called systematic uniform random (SUR) sampling is preferred. SUR sampling spreads samples at regular intervals and is quicker, easier, more precise, and therefore more efficient than random sampling especially on the mostly heterogeneous biological specimens examined in cell biological EM.[2] SUR sampling can be achieved, for example, by cutting tissue/cell pellets into parallel strips or by positioning micrographs on a thin section as a regular array or by casting probes onto a micrograph in a regular lattice; each time unbiased results are obtained by random positioning of the array with respect to the specimen.

Once an appropriate sample has been obtained, EM quantitation uses a stereological approach that applies simple geometrical probes (points, lines, or quadrats or brick shaped volumes) to estimate the size or number of structures (Howard and Reed, 1998; Weibel, 1979). The investigator counts chance encounters between the geometrical probes and structures of interest, and converts them into quantities

---

[1] To randomly select countable items such as animals/pellets/dishes/experiments, read a random number from a random number table. If the random number is 0901 then 0.0901 multiplied by the number of animals, indicates the animal to be selected. Thus, for nine animals the result is 0.8109, which is nearer to 1 than 0, so take animal number 1.

[2] SUR sampling is more efficient as long as the structures underlying the array do not happen to be in register with the array! In this case uniform random sampling will tend to be more efficient.

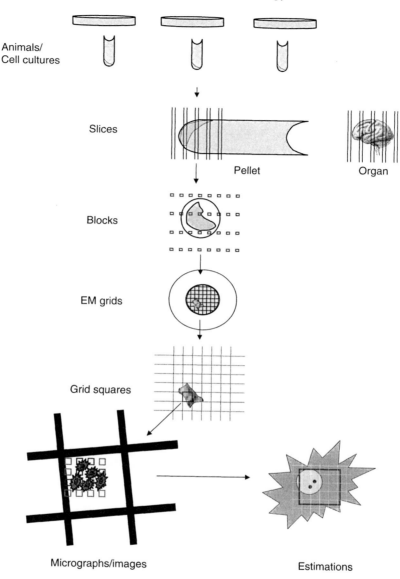

Fig. 1 Sampling "cascade" for quantitative EM analysis. The images analyzed quantitatively are derived from animals or cell cultures, organs, blocks, and sections. Unbiased results are obtained if the sampling at each level uses either simple random or systematic uniform random (SUR). SUR sampling is applied here to cell pellets, blocks, micrographs, and estimations. SUR sampling is more efficient than random sampling when samples are heterogeneous *and* structures are not periodic in register with the sampling arrays (e.g. as may be the case in striated muscle cells). Random selection using a random number table is described in footnote 1. (See Plate no. 4 in the Color Plate Section.)

using simple formulae (see examples in Figs. 3 and 4). Importantly, the method follows a strict sampling design that limits bias and optimizes precision (hence the term "design based estimators"). The use of geometrical probes has the following advantages. First, counts are made rather than measurements, making it simply a choice of "does this encounter happen or not". Second, relatively small numbers of counts are needed to make accurate estimations. And third, SUR sampling can be done easily by constructing a regular array or lattice of geometrical probes (see Figs. 1 and 2).

So how many cell cultures, pellets, blocks, and estimations should one take? It is generally more efficient to concentrate effort at the higher levels of the sampling cascade, i.e. to do more experiments or take more animals and "do more less well" (Gundersen and Osterby, 1981). This generally means at least 3 experiments/animals. At the level of sections or blocks multiple samples are preferred, especially if the specimens are known to be heterogeneous, while at the lower level of the sampling cascade a good starting point is 10–20 micrographs/images combined with a modest count of 100–200 stereological "events" (see below).

III. Quantities Displayed on Sections—Gold Labeling and Profile Data

Currently, a major use of profile data on ultrathin sections is in assessing gold particle labeling of cell components. Typically, in this approach, ultrathin sections are exposed to antibodies localized using particles of colloidal gold, and because the section "presents" the components to the gold labeling system it is crucial to follow the sampling scheme already outlined. This ensures an unbiased sample of cell components is contained in the sections and can gain access to the gold labeling system. Colloidal gold is particulate and can be quantified as a signal that represents the underlying component[3] and the two principal readouts of interest to cell biologists are the distribution and concentration of gold labeling.

A. Gold Labeling Distribution

The distribution of gold labeling over an array of cellular structures displayed in ultrathin sections can be estimated in two main ways. In the *micrograph method* (Fig. 2A; Lucocq, 1994, 2003), micrographs are evenly spaced over the whole extent of a section, an EM grid, or a randomly selected section or EM grid square.

---

[3] Not all molecules in the section will be labeled by gold particles and the number of gold particles related to the number of molecules can be expressed as ratio termed labeling efficiency (Griffiths and Hoppeler, 1986). Most workers proceed on the basis that labeling efficiency is not known but this effect should always be borne in mind because it can vary from compartment to compartment. There are ways to equalize labeling efficiency over intracellular compartments, if necessary (see Griffiths, 1993; Posthuma *et al.*, 1986, 1987, 1988; Slot *et al.*, 1989, for background to this problem).

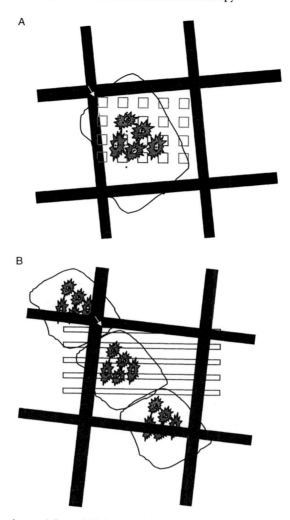

Fig. 2 Image samples used for gold labeling quantitation. (A) *Micrograph method*: An optimally contrasted and randomly selected section is photographed in a systematic array positioned according to an EM grid hole corner (white arrow). Notice it is only necessary to take photos when they contain labeling or compartments of interest (red outlines). Furthermore, it is crucial not to move the field and include (or exclude) items of interest. (B) *Scanning method*: An optimally contrasted section is selected at random and the section scanned systematically. The array is aligned with the grid hole corner to ensure random positioning (white arrow). (See Plate no. 5 in the Color Plate Section.)

Counting and assignment of gold particles to compartments is made easier by placing a quadrat over the image. In the *scanning method* (Fig. 2B; Lucocq, 2003; Watt *et al.*, 2002, 2004), parallel scans are spaced evenly across the full extent of the labeled section or the full extent of the sections found in a randomly selected EM grid square. The advantage is that scanning can be done rapidly at the electron

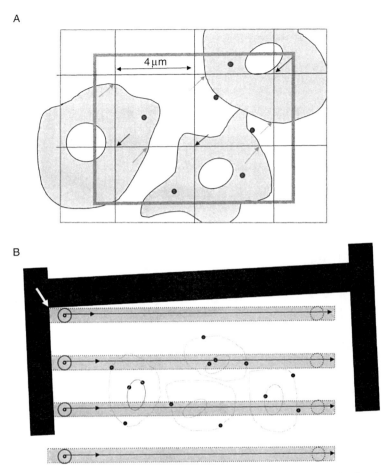

Fig. 3 Gold labeling distribution and density. (A) In the micrograph method the fraction of gold particles is counted in a quadrat placed over the micrograph. This can be placed arbitrarily with respect to the micrograph border because the micrograph has already been positioned SUR. There are six sampled gold particles of which four are over the cytoplasm and two over the plasma membrane. The estimate of the fraction of gold over the cytoplasm is 4/6. (B) Using the scanning method, the distribution of gold particles is assessed by counting gold particles over plasma membrane and nuclear envelope. Of a total of six sampled gold particles, four are over the plasma membrane and two over the nuclear envelope. The unbiased estimate of the fraction of gold over nuclear envelope is 2/6. In a real study there would be more scans (10–20), more compartments (10–15), and more gold counted (100–200 per EM grid). Both these methods can be used for gold density estimation. In the micrograph method (A) the area of the cytoplasm enclosed in the micrograph is estimated by counting the number of grid lattice corners (points, ΣP) that fall over the cytoplasm (black arrows, total 3) and multiplying this by the area, a, associated with one point, which here is 4×4 μm². Thus, 4 golds/ (SP × a) = 4/(3 ×16) = 4/48 is an estimate of the numerical density of gold over area in golds per square micron. This can also be expressed as a density per point in a labeling index, 4/3. Density over membrane length can be obtained by counting intersections of the borders of the grid lines with the plasma membrane (grey arrows). The density can be expressed as gold particles per intersection (2/5) or in absolute units using the

microscope without recording images. In these methods, the arrays can be positioned randomly by placing them at the corner of an EM grid hole. Initially, the magnification is set to a level at which both the gold particles and the compartments of interest are visible, although magnification may be increased to reduce the total gold count. When the gold particles are of finite size compared to the quadrat or scanning strip (say >1/100th of the width/length), unbiased counting rules should be applied (Gundersen, 1977; Lucocq, 1994). The simplest rule for a quadrat is to exclude all particles that touch two adjacent sides and accept all particles that are totally inside the quadrat or touch the other two sides (see Fig. 5A). In the scanning method, particles touching one of the scan borders (upper or lower) are excluded and particles within the scan or lying on the other border are accepted (Lucocq, 2003). The results can be expressed as raw frequency data or as a percentage of total particles counted. Remarkably, the shape of the frequency distribution can be assessed reproducibly by counting 100–200 particles over 10–15 identifiable compartments (Lucocq et al., 2004), although the workload will need to be increased if increased precision is required on individual compartments.

B. Gold Labeling Density

The density of gold label over structure or compartment profiles can be estimated by combining gold counts with stereological estimators of area or length (Griffiths, 1993; Lucocq, 1994).

1. Gold Density over Profile Areas

The ratio of gold particle counts to profile area is an estimate of gold density. Profile area can be estimated using a systematic array of points placed randomly over a structure profile. The number of points that fall over the profile multiplied by the area associated with each point estimates the profile area (Lucocq, 1994; see Fig. 3A). The point array can be applied to a micrograph using a transparent sheet or can be generated digitally. Again the chosen magnification should allow gold particles and structures to be identified and counting of gold and point counts are done within a quadrat placed over the image. Gold particles and point counts should be summed over all quadrats prior to calculating the ratio estimate[4] (see William, 1977). Adjustments in these totals can be made by taking more or less micrographs, adjusting the point spacing, or by sampling smaller areas

[4] This principle should be adopted for other ratio estimates discussed in this article.

---

formula $\pi/4 \times I \times d$ for membrane trace length (using intersections with vertical and horizontal lines in this type of grid lattice). In the scanning method (B), the gold particles are related to the intersections of a scanning line formed by the edge of a feature on the microscope screen. Here there are 4 plasma membrane golds and 10 intersections. Density is 4/10 as a labeling index. In comparison the ratio for nuclear envelope is 2/8 and so the labeling is more concentrated over the plasma membrane. (See Plate no. 6 in the Color Plate Section.)

(a convenient way of subsampling is to use smaller quadrats which represent a known fraction of the larger quadrat). The density can then be expressed either as an absolute value in gold per square micron or simply as density per point (labeling index). The micrograph method is most suited to this type of estimation.

2. Gold Density over Linear Profiles

Over two centuries ago Georges Louis Leclerc Comte de Buffon carried out a probability experiment by throwing sticks over his shoulder (randomly) onto a tiled floor and counting the number of times the sticks fell across the lines between the tiles (Comte de Buffon, 1777). This work led to an estimation of linear trace length which can be applied to membrane profiles displayed on ultrathin sections. In this method, a regular array of test lines is placed with random position and orientation[5] over a profile and the number of intersections between the lines and the trace are counted (Fig. 4B). If the array comprises a square grid lattice, and both vertical and horizontal lines are used, then the formula for the trace length is $\pi/4 \times I \times d$, where I is the total number of intersections and d the distance between the lines (the constant $\pi/2$ is used for a lattice in which only horizontal or vertical line intersections are counted). Intersections are defined by interaction of the test line edge with one border of the membrane trace (for example, the outer aspect of a plasma membrane profile). Gold particles are counted as associated with the trace if they lie within a preset acceptance zone, which corresponds roughly to the expected resolution of the technique. A useful rule of thumb, when using 8–10 nm gold particles, is to use an acceptance zone of two particle widths. Again, gold particles can be related to absolute units (gold per micron) or as gold particles per intersection (labeling index).

This method works with micrographs but can be adapted easily to the scanning method for more rapid results (Lucocq, 2003, 2006; Watt *et al.*, 2002, 2004). Here, the edge of a feature on the electron microscope viewing screen forms a scanning line as the specimen is translocated. Intersection counts and gold counts are made in systematic random scans and used to compare relative signal intensities along membrane traces of different compartments.

---

[5] When the position and orientation of the section/image has been assured in the sampling scheme (or can be safely assumed) the grid can be placed arbitrarily with respect to the micrograph borders. Anyway, it is advisable to ensure that the specimen already has randomized orientations as well as randomized position because both labelling density and the quantity of membranes can vary according to section orientation. Methods for obtaining randomized orientations (isotropy) are discussed in the section on surface density estimates.

C. Statistical Assessment of Labeling

New methods have been developed to evaluate labeling distributions and labeling intensities by statistical means. They enable the investigator to determine whether a compartment is preferentially labeled or whether there is a difference in gold labeling between experimental groups. These methods generate an expected distribution for the gold particles to which the observed experimental distribution is compared, and the difference between the expected distribution and the observed distribution is evaluated using Chi square statistics (Mayhew et al., 2002, 2003, 2004).

1. Preferential Labeling

This method allows preferentially labeled (or poorly labeled) compartments to be identified (Mayhew et al., 2002). First, gold particles and point counts or line intersections are counted over selected compartments. Next, an expected distribution of particles over the same set of compartments is calculated under the assumption that the density of gold is uniform over all compartments. This expected distribution serves as a reference for comparison with the "real" data. Here, the number of gold counted will vary *only* in proportion to the size of the compartments (estimated in terms of either point counts or line intersections) (see Table I). For each compartment the *observed* (real) signal is then compared

Table I
Statistical Analysis of Gold Labelling Distribution Over Endomembrane Compartments.

	Observed		Expected		
	Gold	Intersections	Gold	Relative labeling index	Chi square
ER	14	90	44.4	0.32	20.8
Golgi	6	11	5.4	1.11	0.067
Endosomes	65	13	6.4	10.2	536.6
Plasma membrane	23	105	51.8	0.44	16.01
Total	108	219			573.5
					df 3 $p < 0.005$

In this experiment gold labelling is compared over a series of membrane bound organelles sampled systematic uniform random using the scanning method (see above). Using the total gold particles (108) and total intersections (219) the expected number of golds per intersection, if gold labelling is distributed randomly, would be 108/219 = 0.493. This value multiplied by the number of intersections for each compartment gives the expected gold labelling as shown in the table. For example, over endoplasmic reticulum (ER), there would be 0.493 x 90 = 44.4 expected golds. The relative labelling index (RLI) for the ER is therefore 14/44.4 = 0.32. The total Chi square value (χ^2) = 573.5 and for 3 degrees of freedom (calculated from (r−1) x (k−1), where r = number of rows and k = number of columns), $p < 0.005$, so the distribution pattern of gold-labelling is significantly different from random. The endosomes are preferentially labelled because the RLI = 10.2 and the partial χ^2 accounts for 94% of total χ^2.

to the *expected* signal as a ratio (relative labeling index, RLI) and the Chi square statistic calculated. Those compartments with highest RLI ratios and highest partial Chi square values are labeled preferentially. The overall Chi square value can be used to assess the significance of the difference between the observed and expected distributions.

Note that recent work suggests it may be possible to mix membranes and space-filling organelles in this type of analysis (Mayhew and Lucocq, 2008). Also certain constraints apply to the use of Chi square, such that values cannot be less than 5 in more than 20% of the groups and not less than 1 in any group. Solutions include increasing the number of gold particles or fusing groups together. The number of compartments is at the discretion of the investigator and depends on the demands of the study in question, but we currently recommend any number of compartments between 3 and 15. Increasing the number of compartments increases the precision of the evaluation of the localization, but decreases the precision of the statistical analysis. One use of this technique would be, in the initial stages of an investigation, to identify compartments with preferential labeling. This method may also be useful for analyzing labeling patterns at different antibody dilutions.

Gold particles over selected membranes and intersections of systematically spaced grid lines with these membranes are counted using the micrograph or scanning method. The expected labeling density, assuming no preferential labeling, is 108/219 = 0.49 gold particles per intersection. The expected gold over each compartment membrane is given by expected labeling density × intersections with that membrane, e.g. 0.49 × 90 = 44.4, for the endoplasmic reticulum (ER) membrane. The relative labeling index is the ratio of observed to expected gold and is a measure of preferential labeling. The partial Chi square value is calculated using $(O-E)^2/E$, where O and E are observed and expected gold, respectively. The significance of the total Chi square value can be read from a table, of Chi squared statistics with degrees of freedom (df) which is calculated from the number of compartments-1; df = 3 in this case. The conclusion is that endosomes are preferentially labeled.

2. Differences in Labeling Distributions Between Groups/Experiments

This approach allows comparison of immunogold labeling patterns in the same sets of compartments in different experimental groups (Mayhew *et al.*, 2004). The procedure starts with counts of gold particles associated with the compartments that have been selected. Next, the observed numbers of gold particles in compartments are compared using a contingency table analysis. The groups are arranged in columns and compartments in rows. The data from the experimental groups are combined to generate expected gold particles for each compartment and these expected numbers are compared to the observed numbers using Chi square analysis. The individual Chi square values will show which compartments make substantial contribution to between-group differences and the total Chi square value will identify whether there is an overall difference between the groups (Mayhew *et al.*, 2004).

D. Specificity

Thus changes in either the component or reference volume will produce a change in the ratio. For example, a shrinking reference volume can increase the apparent density of a membrane without changes in membrane amounts.

The most powerful control for specificity is to change the expression or location of the antigen *in situ*. Examples would be a knockdown of protein expression using small interfering RNA, gene deletion, or mutation; the introduction of the component by gene expression, microinjection, or endocytosis; the physiological *in vivo* generation of the component; or chemical modification of the component on the section. The easiest result to interpret is when controls show very little or no signal indicating very low levels of nonspecific labeling. Distributions and concentrations of gold labeling then become quantitatively meaningful readouts. When gold signal is reduced but not ablated in controls, residual label could represent either nonspecific interactions and/or residual amounts of component. In this case, it is useful to correct quantitative data by removing the residual labeling to reveal specific signals (Watt *et al.*, 2004). Other controls for specificity analyze treatments/conditions with predictable effects on the labeling system, but are less powerful. Examples would be inhibition of labeling using purified exogenous antigen, or substituting an unrelated antibody for the specific one (see Griffiths, 1993 for more detailed discussion).[6]

IV. Quantities in Three Dimensions

Sections are excellent tools to reveal the internal structure of intact organisms and cells at all scales from gross morphology, viewed with MRI and CAT scans, down to the organelle and molecular levels examined using transmission electron microscopy. The transition from two-dimensional (2D) image to three-dimensional (3D) quantity is not an easy one to make, but careful sampling design and appropriate choice of geometrical probe has provided a wide array

---

[6] Note that prior to doing controls, an almost universal starting point in investigating a new affinity reagent is to assess the labeling pattern. A preferential distribution may be immediately obvious (which might be confirmed statistically) and this may be consistent with independent data obtained from cell fractionation, immunofluorescence or sequence information (targeting motifs/characteristic domains). On the other hand significant labeling may lie over compartments that one is fairly certain should not contain the antigen (e.g. mitochondria in the case of a secretory protein) the next step may be to analyse a range of affinity reagent concentrations to assess which pools of labeling are resistant to dilution and which ones are not. Those resistant to dilution are candidates for specific labeling.

of stereological tools with which to quantify structures[7] (Howard and Reed, 1998; Lucocq, 1993). Typically, a stereological estimate is based on two components: a density and a volume. The density is estimated easily on the micrographs and might be, for example, a packing density of number, surface, or volume inside a cell or tissue space. The volume can be estimated by a range of methods (see below) and might be the cell, organ, or tissue space. The volume allows densities to be converted into an absolute amount (e.g. V_{cell} × Density = Amount). Importantly, knowledge of the volume, or reference space as it is often called, avoids uncertainty in interpreting changes in a ratio such as a density (the so called reference trap).[8]

A. Densities

1. Volume Fraction

The concentration of organelle or structure volume inside a reference space such as a cell can be estimated easily by counting the fraction of randomly placed points that fall over the structure compared to a reference space (Fig. 4A; Howard and Reed, 1998; Weibel, 1979). This is known as volume fraction or volume density.[9] Points are most often presented as a regular (SUR) array to a set of images from a randomly positioned section, which is used to carry the probes array "into" the sample. Points are counted when they fall on the reference space and the fraction of these points falling over the component gives an estimate of the volume fraction of the component. The volume fraction has been used extensively in cell biology to monitor the relative amounts of organelles in cells such as the fraction of cytoplasm occupied by organelles such as autophagosomes, Golgi, mitochondria, or nuclei. However, because of the reference trap it is always safer to estimate the absolute volume of the component (see below).

---

[7] In simple terms, a 3D quantity "presents" itself on the thinnest of sections reduced in one dimension. For example, the volume inside an organelle or compartment (3D) displays itself as an area (2D), and, surface of an organelle (2D) presents as a length (1D), and length (1D) is displayed as a point (0D). Stereological geometric probes applied to sections supplement profile dimensions to at least three and so points (0D) can be used to estimate volume (3D), lines (1D) can be used to estimate surface (2D), and planes (2D) can be used to estimate length (1D). By this rule 2D planes (sections) cannot be used to estimate number (0D) and the solution is to use a volume probe (3D) called the disector (see text).

[8] Density is a ratio sensitive to changes in numerator and denominator. Thus changes in either the component or reference volume will produce a change in the ratio. For example, a shrinking reference volume can increase the apparent density of a membrane without changes in membrane amounts.

[9] The principle can be explained in two ways. First, the chance a randomly positioned point lands inside a component depends on the fraction of a reference space volume occupied by the component. Second, according to Delesse (1847) the profile areas of a component related to the profile area of a reference space as a ratio, reports on the fractional volume of component in the reference (the profile areas can be estimated using point counting as detailed above for gold density per area).

2. Surface and Length Density

In cell biology, membranes are functional barriers involved in transport, signal transduction, motility, and vesicular transport. Consequently, much effort has gone into quantifying their extent. The conceptual basis of surface estimation in 3D is similar to that of the Buffon stick experiment. For any line inserted into a specimen the number of intersections (I) of that line with a surface will depend on the packing density of that surface in 3D and also the total length of line applied (L). Packing density is also known as surface density or surface (S) to volume (V) ratio (S/V) and can be estimated by the formula $2I/L$ (Fig. 4B). For unbiased estimates the lines used must be positioned and orientated at random (isotropic uniform random, IUR). Random positioning can be ensured by systematic random selection/cutting of pellets/tissues and randomization orientation (isotropy) can be achieved by generating isotropic sections applying lines to these. An intuitive way to ensure isotropy is to chop a pellet or tissue into small pieces (blocks) and let them settle (see for example, Henrique et al., 2001; Stringer et al., 1982). However, as randomization is not actively assured, preferential orientations could occur that can lead to biased results. A more rigorous method suitable for cell biology is to embed the tissue fragment in a ball of gelatin, roll the ball and embed (Fig. 4B; the Isector; Nyengaard and Gundersen, 1992). Once an isotropic section at a uniform random position (IUR) has been generated, the line probes are applied as a regular (systematic) lattice to micrographs.

The need for isotropic sections for surface estimation presents a problem in the use of oriented sections through polarized cells such as epithelia or neurons. However, in 1986, Adrian Baddeley and coworkers (Baddeley et al., 1986) described the vertical section method, allowing the investigator to choose section direction but still generate isotropic lines in space (Fig. 4C). The chosen direction is orthogonal (vertical) to any preffered horizontal plane and the section is placed randomly on this plane and oriented randomly around the vertical direction. Isotropic lines in space are then presented to the image as cycloid traces that contain line elements representing all possible orientations in 3D space. In the case of electron microscopy, vertical sections might be oriented vertical to the base of polarized epithelial cells or perpendicular to the horizontal plane of a culture dish (Griffiths et al., 1989). The cycloids can be applied as a systematic array aligned along the vertical axis and positioned at random with respect to the cells. Surface density estimates can provide information on the concentration of surface/membrane functions and can also be combined with reference volume (see Section 3B) to obtain the total surface (Michel and Cruz-Orive, 1988).

Another parameter that is less often of interest is length. The length density of a linear structure in 3D can be estimated using randomly positioned and oriented planes (this is a reversal of the surface estimation just described—now using the surface as a probe for lines). The length density (L_V) is given by counting the density of intersections of the linear structure over profile area of the reference space on the section (Q_A, see Howard and Reed, 1998).

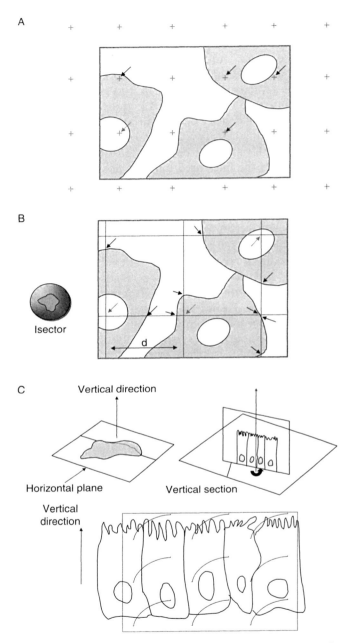

Fig. 4 Estimation of 3D densities. (A) *Volume density*: A single micrograph taken from a systematic random set obtained from a cell pellet sampled according to the scheme in Fig. 1. Random or systematic placement is important but orientation need not be controlled. A regular point lattice is applied to the micrograph and points over the structure of interest (nucleus, P_n, one hit) and reference volume (in this

The estimation of volume and surface density may become problematic when structures are small relative to the section thickness. Two effects tend to introduce bias into the estimations. First, the edges of small structures such as vesicles are lost, producing the so-called "lost-caps" effect. This will tend to reduce volume and surface density estimates. Second, volume and surface density estimations are based on the premise that images are generated from structures cut by planes with zero thickness. In reality EM sections are 50–100 nm thick and small structures enclosed in the section will tend to produce additional images when projected onto the electron microscope screen. This effect will therefore tend to increase volume and surface estimates. Correction factors for volume and surface density estimates of such small structures have been published (Weibel and Paumgartner, 1978) but should be used with caution as they are based on model shapes. A further effect of sectioning is the loss of membrane images that occurs when the membrane is tilted with respect to the section. Again there are ways to correct for this effect (Mayhew and Reith, 1988).

3. Number Density

The number of structures becomes important in studies of biogenesis, degradation and turnover of individual units (see below). Before 1984 number estimation was problematic because a 2D section alone does not report on number (in fact, the chance that a section encounters an object is related to its height in the direction of sectioning not the number!). The elegant solution was to use a 3D probe made from a set of two parallel sections (a disector; Fig. 5A; see Gundersen, 1986; Mayhew and Gundersen, 1996; Sterio, 1984). Particles sectioned by one, but not the parallel section, will have an edge situated between the sections, and if the particles have only one edge in the direction of sectioning then the particles are

---

case cell space, P_{cell}, five hits) are counted. The point counts are summed over all micrographs and the volume density (volume fraction) calculated from $\Sigma P_n / \Sigma P_{cell}$. Points are zero dimensional and defined by the corners where two lines of the cross meet (e.g. bottom left). (B) *Surface density—isotropic sections*: The micrograph is from a SUR set obtained from a cell pellet/tissue piece embedded in a ball of gelatine (illustrated; isector; Nyengaard Gundersen, 1992), and rolled prior to embedding to ensure random orientation. Otherwise, sampling is carried out as in Fig. 1. A regular array of lines (in this case square lattice) is applied and intersections (I) with membrane traces are counted. Here, plasma membrane traces intersect with horizontal and vertical lines (black arrows, nine intersections). In this grid lattice, the line length (L) applied to the cell space is given by the point hits on the cell space (P_{cell}, grey arrows, three hits) multiplied by the line length "belonging" to each point ($2d$). The surface density (S_v) is given by $2I/L$. (C) *Surface density—vertical sections*: The horizontal plane defines a vertical direction. Vertical sections are positioned at a uniform random position on the horizontal plane and randomly oriented (isotropic) around the vertical direction. In this illustration, the horizontal plane is defined by the surface of a culture dish covered by polarized columnar epithelial cells. Micrographs are taken in a SUR manner and systematic arrays of cycloid arcs are used to count intersections (I) with membrane traces. The line length (L) applied to the reference space of interest (e.g. the cell interior) is estimated from counts of points multiplied by the line length associated with each point on the grid. The surface density is calculated from the formula $S_v = 2I/L$. (See Plate no. 7 in the Color Plate Section.)

counted as individuals. Structures with single edges can include coated vesicles, peroxisomes, nuclei, nucleoli, secretion vesicles, etc. Particle profiles are first selected using the so-called reference section by applying unbiased counting rules to a quadrat. The sampled particles are then sought in the parallel, so-called look-up section, and particles that have disappeared are designated Q^-. Note that the sections should not be spaced wider than the smallest particle to avoid missing particles, but they need to be close enough to facilitate comparison of the two sections. Preparation of disectors is labor intensive and so to increase efficiency the disector can be used in both directions by reversing, one reverses the reference and the "look up" sections. As with all the stereological methods described here, precise and efficient counting depends on positioning the disectors SUR within the specimen.

The density of structures (numerical density, N_v) inside a reference volume (cell) (Fig. 5B) can be estimated by counting the number of structures in the disectors (ΣQ^-) and by estimating the volume of the reference volume by point counting ($V_{cell,dis}$): $N_v = \Sigma Q^- / V_{cell,dis}$. The product of the sum of point count hits (ΣP), the area associated with one point on a regular point lattice (a) and the interval between the sections (k) is an estimate of the volume of cell inside the disectors ($V_{cell,dis} = \Sigma P \times a \times k$). Often the reference volume is a cell space such as cytoplasm, cell, or nucleoplasm. If in parallel the reference volume (V_{cell}) is obtained by applying the methods described below, the product of numerical density (N_v) and reference volume (V_{cell}) provides the total number of structures. Note that section thickness is required to obtain these estimates of numerical density (e.g. Gunning and Hardham, 1977; Small, 1968).

B. Volume

So far the volume of tissues, cells, or organelles has been described as an important yardstick for converting densities into amounts, but volume is a fundamental biological parameter in its own right. Here are presented methods for estimating the volume as an average value for a population of items such as cells, or for estimating the volume of individual items or "particles". In each case cell volume is used as an example.

1. Average Cell Volume

If one knows the average cell volume one can use it to convert densities into absolute values per cell. A simple way of estimating cell volume is to use disectors to estimate the reciprocal of cell density in cell volume ($Q^-/V_{cell,dis}$, as described above). The cells are first sampled using SUR disectors of known size (Fig. 5B), most conveniently using semithin sections of resin or cryosectioned material. The cells (or unique components such as nuclei) are counted (ΣQ^-) and the volume of cell space ($V_{cell,dis}$) inside these disectors estimated ($V_{cell,dis} = \Sigma P \times a \times k$, where ΣP is the sum of point counts of a regular lattice grid, a is the area associated with a

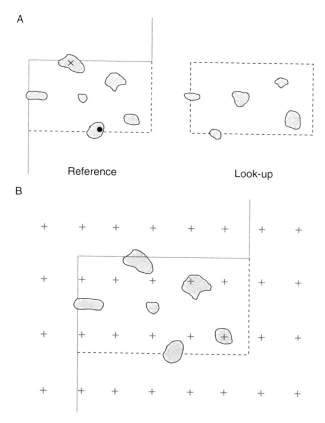

Fig. 5 Number estimation. (A) *Disector counting probe*: Number is estimated and particles are selected unbiasedly with a disector composed of two parallel sections that are situated closer than half the height of the smallest particle to be counted. One of the sections (reference section) is used to select particles and the other assesses which particles have disappeared (the look-up section). Particles can be selected on the reference section unbiasedly according to the forbidden line rule in which a particle touching the continuous line is not selected (e.g. particle marked "x" in the figure). All particles with profiles within the quadrat or over the dotted acceptance lines are selected for further consideration. Of these selected particles those that have disappeared in the look-up are counted (they have edges between the section planes; particle marked with black dots). Of these selected particles those that have disappeared in the look-up are counted (they have edges between the section planes). The disector can be used in both "directions" to increase efficiency. The disector can be used to sample particles unbiasedly for further characterization or to estimate particle size. (B) *Numerical density and particle volume using disectors*: The distance between the sections (k) is measured (e.g. using section thickness and number of sections) and combined with point counts (ΣP) and the area associated with each point (a) to estimate the volume (in this case the cell volume) in the disector ($V_{cell,dis} = \Sigma P \times a \times k$). The numerical density, $N_v = Q^-/V_{cell,dis}$ where Q^- is the number of disappearing particles. The reciprocal of this ratio estimates the mean cell volume, $V_{cell} = V_{cell,dis}/Q^-$ and can be obtained using either semithin sections or ultrathin sections. (See Plate no. 8 in the Color Plate Section.)

point on the lattice, and k is the spacing between the sections of the dissector). The average volume of the cells V_{cell} is given by $V_{cell,dis}/Q^-$. This method is suited to cells in suspension but another simple method can be applied to cells growing on culture dishes. This depends on counting cell density on the dish and estimating the average height of the cell layer orthogonal to the dish using EM (see Griffiths *et al.*, 1984 for details). Note that average cell volume may also be obtained using the methods described next for individual structures/reference spaces but these can be more labor intensive.

2. Volume of Individual Structure or Reference Space

There are a number of methods for obtaining the volume of individual identified items such as organs, tissues, cells, or organelles possessing particular characteristics. If the items are selected from a population by sectioning, this should be done using disectors before applying the volume estimator.

A powerful method for estimating volume of anything from an organelle, to cells or organs was devised by the 17th century Italian priest Bonaventura Cavalieri (Fig. 6A; Cavalieri, 1635; Gundersen and Jensen, 1987). Cavalieri's method uses a randomly placed series of regularly spaced slices through an object. The summed areas of profiles on the slices multiplied by the mean distance between the slices gives the volume, irrespective of object shape or dispersion. The area of the profiles on the slices can be estimated from the product of point hits (ΣP) of a grid lattice and the area associated with one point (a). At the EM level, distance between the sections (k) can be found from the section thickness (Gunning and Hardham, 1977; Small, 1968). A complication arises if the thickness of the slices becomes substantial (Gual-Arnau and Cruz-Orive, 1998), but for most purposes EM sections are very thin compared to the size of the object under investigation. Cavalieri's method has been used in EM to estimate the volume of mitotic cells and mitotic ER (Lucocq *et al.*, 1989; McCullough and Lucocq, 2005). In general, only 5–8 sections and point hits 100–200 per object are needed to obtain estimates considered to be precise (Gundersen and Jensen, 1987). Practically, a useful strategy is to do the estimation using semithin sections and to estimate densities from thin sections taken in between. Alternatively, thin sections are selected at intervals in a series and mounted on slot grids. The cells can be counted at low magnification and densities obtained at higher magnification. Cavalieri estimation is an important method for estimating total quantities of organelle structures in large organs such as the lung (Weibel *et al.*, 2007).

In the *rotator* method (Fig. 6B), the volume of individual cells/organelles can be estimated using sections that are isotropic around an identifiable axis passing though a central point. For example, the axis could be perpendicular to a cell monolayer and directed through the centriole or nucleolus or pass through each pole of the mitotic spindle. Note that if items at the central point are not of constant size (and therefore have a variable probability of being sampled), it will be necessary to sample them using the disector (see above). The Pappus theorum

4. Quantification of Structures and Gold Labeling in Transmission Electron Microscopy

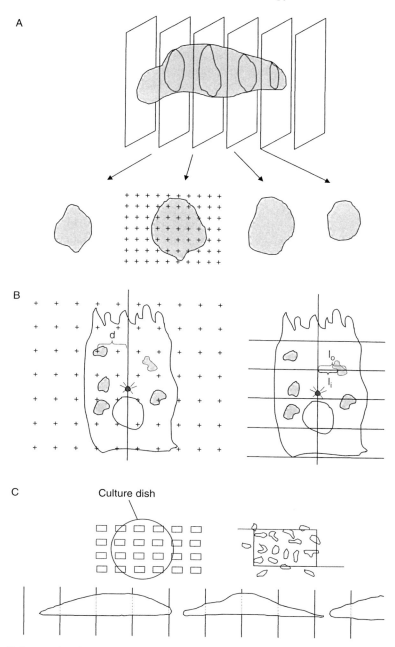

Fig. 6 Volume estimation. (A) *Cavalieri*: A stack of regularly spaced sections is placed at a uniform random location relative to the object. The spacing (k) is set for 5–10 sections to hit the object. Profile image areas are estimated by summing point hits (ΣP) of lattice corners where a is the area associated with one point (corner). The estimate of volume Est $V = \Sigma P \times a \times k$. The Cavalieri principle can be

states that the volume of any item/compartment is given by the product of the area displayed on the sections multiplied by the distance traveled by the centroid of the region in one rotation. This rotator method (Jensen and Gundersen, 1993) has been applied to both monolayer culture cells (Mironov and Mironov, 1998) and cells and organelles in mitosis (McCullough and Lucocq, 2005). If average volume is required, then the average of summed rotator estimates for each cell can be taken. An important feature of the rotator is that it allows counting or size estimations to be mapped onto locations in the cell, providing gradients of organelle size and number in different regions of the cell. This approach will be a powerful way to analyze changes in organelle distribution and cell polarity.

A related method called the *nucleator* (Gundersen *et al.*, 1988), estimates cell or organelle volume using isotropic lines emanating from unique and identifiable central locations such as nucleoli, centrioles, and centrosomes. The central locations are first identified using disectors (see below) and the volume is estimated by measuring the distance across the cell through the organelle, along a randomly oriented (isotropic) line. The isotropic lines can be generated using isotropic sections (e.g. using the isector) or on vertical sections (see Howard and Reed, 1998, for details).

C. Number

As described above the disectors can be used to obtain total numbers of an object/structure by combining the numerical density of structures (N_v) and reference volume in those disectors. However, there is a second approach that estimates total number and avoids estimation of section thickness. The principle is to count the ratio of small structures in relation to larger structures in the same stack of sections. Typically, two sections belonging to a stack of serial sections are used to form a disector to count the larger structures (e.g. nuclei) and estimate their density within any reference volume (e.g. cell space). Then, at random, a pair of sections within the stack is used to count the smaller structures and estimate their density in

---

adapted to any component contained within the stack of sections. (B) *Rotator*: (Left) An orthogonally sectioned structure through a central point sampled with uniform probability is overlaid with a set of test points (area associated with each point, *a*). A vertical axis through the point is drawn and for each point hit over a structure of interest the distance from the point to the axis measured (*d*). An estimate of the total volume of the structure is $\pi/2 \times a \times \Sigma d$. (Right) Another way of estimating volume is to use the difference in squared intercept lengths ($l_o^2 - l_i^2$) in the formula $V = \pi/2 \times h \times \Sigma(l_o^2 - l_i^2)$ where h is the distance between lines arranged SUR on the axis and l_o and l_i are the distances to the outer and inner aspects of the structure/compartment. The central point may be sampled using disectors or by using hits with a structure of constant size (e.g. centrosome). (C) *Monolayer cell volume*: Using light microscopy SUR quadrats are used to count the number of cells (ΣQ_q) and estimate the area included in these quadrats (A_q) by point counting. In randomly positioned vertical sections, the mean height of the cells over the dish area is estimated using SUR vertical intercepts. ΣQ_q is the number of cells in each quadrat. (See Plate no. 9 in the Color Plate Section.)

$$\text{Cell volume} = \frac{\text{Mean cell height} \times A_q}{\Sigma Q_q}$$

the same reference volume. The ratio of the densities gives an estimate of small to larger structures. Conveniently, both density estimates contain a section thickness term, which cancels because the estimations were done on the same section stack. This approach called the double disector (Gundersen, 1986) has been used to estimate the number of coated vesicles per cell (Smythe et al., 1989) and also the number of gold particles per cell (Lucocq, 1992).

Importantly, the disector can be used as a sampling tool as well as a counting tool. For sampling, the distance between the sampling and look-up section need not be known exactly and the particles can be characterized in any way that is relevant to the investigation, e.g. volume, length, degree of invagination, presence of a cytoplasmic coat, budding vesicles, etc.; using these measures a "number weighted" mean can be obtained and proportions of particles with a particular characteristic or size computed. As an example, coated pits sampled using disectors can be characterized according to the size of the necks, volumes or degree of invagination, etc. (Smythe et al., 1989). As mentioned above it may be necessary to sample cells or their nuclei using disectors prior to Cavalieri, rotator, or nucleator volume estimations.

When using the disector for counting small structures such as vesicles, a special situation may arise. It may be that all vesicles sectioned by one ultrathin section are not visible in the next section and so the probability of finding a disappearing profile is 100%. Therefore, under these conditions there may be no need for a look-up section to be used. This convenient adaptation of the disector must, however, be used with caution and preliminary work is needed to check the proportion of structures that are detected by one section and disappear in the next. This approach has been used to count Golgi vesicles in dividing HeLa cells (Lucocq et al., 1989).

One approach to number estimation that is less suited to electron microscopy is the fractionator (Gundersen et al., 1988). The specimen containing the complete array of particles is exhaustively divided into slices, blocks, and sections. Then at each level known fractions are sampled and disectors are used to count particles in a known fraction of the sections. If the sampling fraction at each level is f_1, f_2, etc., then the estimate of the total number of structures is $1/f_1 \times 1/f_2 \times 1/f_3$, etc. Unfortunately, the fractionator is difficult to apply in EM because exhaustive sectioning is required, although a recent modification avoids this constraint making it easier for EM analysis (Witgen et al., 2006).

V. Spatial Analysis

Most of the quantitative methods described so far estimate total or aggregate quantities and are often referred to as "first-order" stereology. However, there are also methods available for studying spatial distributions of quantities in both 2D and 3D, and these have acquired the label of "second-order" stereology. One example of a second-order problem is the statistical assessment of point processes, such as gold labeling, on linear and planar profiles (Diggle, 1983; Ripley, 1981). The first-order property of such a point pattern is the number of items per unit area, while a second-order property is characterized by the distribution of inter-particle distances. Particle distributions can be assessed using Ripley's K-function (Prior *et al.*, 2003; Ripley, 1981), which allows statistical comparison to an expected distribution obtained using randomly positioned points. This type of analysis is suited to particulate labeling located on 2D arrays of organelle membranes labeled with gold particles and can provide clues to the clustering of receptors, lipids, or signaling molecules recruited to membranes (Prior *et al.*, 2003). Methods for 3D spatial analysis are illustrated by the use of linear dipole probes which have shown some promise in assessing the spatial distribution of volume features at the EM level (Mayhew, 1999; see also Reed and Howard, 1999).

Finally, it is worth mentioning that this review covers some of the basic principles only and the reader is referred to more extensive texts (such as Howard and Reed, 1998) and to the International Society for Stereology's (ISS) web site (http://www.stereologysociety.org/) for further details. Advice from experienced workers in this field can be invaluable to anyone embarking on a quantitative EM study.

Acknowledgment

The author appreciates the critical comments of Gareth Griffiths and Christian Gawden-Bone.

References

Baddeley, A. J., Gundersen, H. J., and Cruz-Orive, L.-M. (1986). Estimation of surface area from vertical sections. *J. Microsc.* **142,** 259–276.

Cavalieri, B. (1635). "Geometria Indivisibilibus Continuorum. Typis Clemetis Feronij Bononi." Reprinted 1966 as Geometria degli Indivisibili Unione, Tipografico-Editrice Torinese, Torino.

Comte de Buffon, G. L. L. (1777). "Essai d'Arithmetique Morale." *Supplement a l'Histoire Naturelle*, Vol. 4. Imprimerie Royale Paris.

Delesse, M. A. (1847). Procédé mécanique pour déterminer la composition des roches. *C.R. Acad. Sci. Paris* **25,** 544–545.

Diggle, P. J. (1983). *Statistical Analysis of Spatial Point Patterns.* London: Academic Press.

Griffiths, G. (1993). *In* "Fine structure immunocytochemistry" (G. Griffiths, ed.) Springer-Verlag, Berlin.

Griffiths, G., Fuller, S. D., Back, R., Hollinshead, M., Pfeiffer, S., and Simons, K. (1989). The dynamic nature of the Golgi complex. *J. Cell Biol.* **108,** 277–297.

Griffiths, G., and Hoppeler, H. (1986). Quantitation in immunocytochemistry: Correlation of immunogold labeling to absolute number of membrane antigens. *J. Histochem. Cytochem.* **34,** 1389–1398.

Griffiths, G., Warren, G., Quinn, P., Mathieu-Costello, O., and Hoppeler, H. (1984). Density of newly synthesized plasma membrane proteins in intracellular membranes. I. Stereological studies. *J. Cell Biol.* **98,** 2133–2141.

Gual-Arnau, X., and Cruz Orive, L. (1998). Variance prediction under systematic sampling with geometric probes. *Adv. Appl. Probab.* **30,** 1–15.

Gundersen, H. J. (1986). Stereology of arbitrary particles. A review of unbiased number and size estimators and the presentation of some new ones, in memory of William R. Thompson. *J. Microsc.* **143,** 3–45.

Gundersen, H. J., Bagger, P., Bendtsen, T. F., Evans, S. M., Korbo, L., Marcussen, N., Moller, A., Nielsen, K., Nyengaard, J. R., and Pakkenberg, B. (1988). The new stereological tools: Disector, fractionator, nucleator and point sampled intercepts and their use in pathological research and diagnosis. *APMIS* **96**, 857–881.

Gundersen, H. J. G. (1977). Notes on the estimation of the numerical density of arbitrary profiles: The edge effect. *J. Microsc.* **111**, 219–223.

Gundersen, H. J. G., and Jensen, E. B. (1987). The efficiency of stereological sampling in stereology and its prediction. *J. Microsc.* **147**, 229–263.

Gundersen, H. J. G., and Osterby, R. (1981). Optimizing sampling efficiency of stereological studies in biology: Or 'do more less well!' *J. Microsc.* **121**, 65–73.

Gunning, B. E. S., and Hardham, A. R. (1977). Estimation of average section thickness in ribbons of ultrathin sections. *J. Microsc.* **109**, 337.

Henrique, R. M., Rocha, E., Reis, A., Marcos, R., Oliveira, M. H., Silva, M. W., and Monteiro, R. A. (2001). Age-related changes in rat cerebellar basket cells: A quantitative study using unbiased stereological methods. *J. Anat.* **198**, 727–736.

Howard, C. V., and Reed, M. G. (1998). "Unbiased stereology. Three dimensional measurement in microscopy." BIOS, Oxford.

Jensen, E. B., and Gundersen, H. J. G. (1993). The rotator. *J. Microsc.* **170**, 35–44.

Lucocq, J. (1992). Quantitation of gold labeling and estimation of labeling efficiency with a stereological counting method. *J. Histochem. Cytochem.* **40**, 1929–1936.

Lucocq, J. (1993). Unbiased 3-D quantitation of ultrastructure in cell biology. *Trends Cell Biol.* **3**, 345–358.

Lucocq, J. (1994). Quantitation of gold labelling and antigens in immunolabelled ultrathin sections. *J. Anat.* **184**, 1–13.

Lucocq, J. M. (2003). Electron microscopy in cell biology. *In* "Essential Cell Biology: A Practical Approach" (J. Davey, and M. Lord, eds.), pp. 53–112 (Chapter 3) Oxford University Press.

Lucocq, J. M. (2006). Quantitative EM techniques. *In* "Encyclopedia of Genetics, Genomics, Proteomics and Bioinformatics" (L. B. Jorde, P. F. R. Little, M. J. Dunn, and S. Subramaniam, eds.) Wiley, Chichester.

Lucocq, J. M., Berger, E. G., and Warren, G. (1989). Mitotic Golgi fragments in HeLa cells and their role in the reassembly pathway. *J. Cell Biol.* **109**, 463–474.

Lucocq, J. M., Habermann, A., Watt, S., Backer, J. M., Mayhew, T. M., and Griffiths, G. (2004). A rapid method for assessing the distribution of gold labeling on thin sections. *J. Histochem. Cytochem.* **52**, 991–1000.

Mayhew, T. M., and Reith, A. (1988). Practical ways to correct cytomembrane surface densities for the loss of membrane images that results from oblique sectioning, in *Stereology and Morphometry in Electron Microscopy. Problems and Solutions* (Reith, A. and Mayhew, T. M., eds.), Hemisphere Publ. Co., New York/Washington/Philadelphia/London, pp. 99–110.

Mayhew, T., Griffiths, G., Habermann, A., Lucocq, J., Emre, N., and Webster, P. (2003). A simpler way of comparing the labelling densities of cellular compartments illustrated using data from VPARP and LAMP-1 immunogold labelling experiments. *Histochem. Cell Biol.* **119**, 333–341.

Mayhew, T. M. (1999). Second-order stereology and ultrastructural examination of the spatial arrangements of tissue compartments within glomeruli of normal and diabetic kidneys. *J. Microsc.* **195**, 87–95.

Mayhew, T. M., Griffiths, G., and Lucocq, J. M. (2004). Applications of an efficient method for comparing immunogold labelling patterns in the same sets of compartments in different groups of cells. *Histochem. Cell Biol.* **122**, 171–177.

Mayhew, T. M., and Lucocq, J. M. (2008). Quantifying immunogold labelling patterns of cellular compartments when they comprise mixtures of membranes (surface-occupying) and organelles (volume-occupying). *Histochem. Cell Biol.* **129**, 367–378.

Mayhew, T. M., and Gundersen, H. J. (1996). 'If you assume, you can make an ass out of u and me': A decade of the disector for stereological counting of particles in 3D space. *J. Anat.* **188**, 1–15.

Mayhew, T. M., Lucocq, J. M., and Griffiths, G. (2002). Relative labelling index: A novel stereological approach to test for non-random immunogold labelling of organelles and membranes on transmission electron microscopy thin sections. *J. Microsc.* **F205,** 153–164.

McCullough, S., and Lucocq, J. (2005). Endoplasmic reticulum positioning and partitioning in mitotic HeLa cells. *J. Anat.* **206,** 415–425.

Michel, R. P., and Cruz-Orive, L. M. (1988). Application of the Cavalieri principle and vertical sections method to lung: Estimation of volume and pleural surface area. *J. Microsc.* **150,** 117–136.

Mironov, A. A., Jr, and Mironov, A. A. (1998). Estimation of subcellular organelle volume from ultrathin sections through centrioles with a discretized version of the vertical rotator. *J. Microsc.* **192,** 29–36.

Nyengaard, J. R., and Gundersen, H. J. G. (1992). The isector: A simple and direct method for generating isotropic, uniform random sections from small specimens. *J. Microsc.* **165,** 427–431.

Posthuma, G., Slot, J. W., and Geuze, H. J. (1986). A quantitative immuno-electronmicroscopic study of amylase and chymotrypsinogen in peri- and tele-insular cells of the rat exocrine pancreas. *J. Histochem. Cytochem.* **34,** 203–207.

Posthuma, G., Slot, J. W., and Geuze, H. J. (1987). Usefulness of the immunogold technique in quantitation of a soluble protein in ultra-thin sections. *J. Histochem. Cytochem.* **35**(4), 405–410.

Posthuma, G., Slot, J. W., Veenendaal, T., and Geuze, H. J. (1988). Immunogold determination of amylase concentrations in pancreatic subcellular compartments. *Eur. J. Cell Biol.* **46,** 327–335.

Prior, I. A., Muncke, C., Parton, R. G., and Hancock, J. F. (2003). Direct visualization of Ras proteins in spatially distinct cell surface microdomains. *J. Cell Biol.* **160,** 165–170.

Reed, M. G., and Howard, C. V. (1999). Stereological estimation of covariance using linear dipole probes. *J. Microsc.* **195,** 96–103.

Ripley, B. D. (1981). "Spatial statistics." Wiley, New York.

Slot, J. W., Posthuma, G., Chang, L. Y., Crapo, J. D., and Geuze, H. J. (1989). Quantitative aspects of immunogold labeling in embedded and in nonembedded sections. *Am. J. Anat.* **185,** 271–281.

Small, J. V. (1968). In "Measurement of section thickness." Abstracts Fourth European Regional Conference on Electron Microscopy, Rome, 1, 609–610.

Smythe, E., Pypaert, M., Lucocq, J., and Warren, G. (1989). Formation of coated vesicles from coated pits in broken A431 cells. *J. Cell Biol.* **108,** 843–853.

Sterio, D. C. (1984). The unbiased estimation of number and sizes of arbitrary particles using the disector. *J. Microsc.* **134,** 127–136.

Stringer, B. M. J., Wynford-Thomas, D., and Williams, E. D. (1982). Physical randomization of tissue architecture: An alternative to systematic sampling. *J. Microsc.* **126,** 179–182.

Thompson, D. W. (1917). On Growth and Form. Cambridge University Press. (Second edition, reprinted 1963).

Watt, S. A., Kular, G., Fleming, I. N., Downes, C. P., and Lucocq, J. M. (2002). Subcellular localisation of phosphatidylinositol (4,5) bisphosphate using the PH domain of phospholipase C δ_1. *Biochem. J.* **363,** 657–666.

Watt, S. A., Wendy, A., Kimber, Fleming, I. N., Leslie, N. R., Downes, C. P., and Lucocq, J. M. (2004). Detection of novel intracellular agonist responsive pools of phosphatidylinositol 3,4 bisphosphate using the TAPP1 pleckstrin homology domain in immunoelectron microscopy. *Biochem. J.* **377,** 653–663.

Weibel, E. R. (1979). "Stereological Methods Vol 1 Practical methods for biological Morphometry." Academic Press, London.

Weibel, E. R., Hsia, C. C., and Ochs, M. (2007). How much is there really? Why stereology is essential in lung morphometry. *J. Appl. Physiol.* **102**(1), 459–467. Epub 2006 Sep 14.

Weibel, E. R., and Paumgartner, D. (1978). Integrated stereological and biochemical studies on hepatocytic membranes. II. Correction of section thickness effect on volume and surface density estimates. *J. Cell Biol.* **77,** 584–597.

William, G. C. (1977). "Sampling Techniques," 3rd ed. Wiley, New York.

Witgen, B. M., Grady, M. S., Nyengaard, J. R., and Gundersen, H. J. (2006). A new fractionator principle with varying sampling fractions: Exemplified by estimation of synapse number using electron microscopy. *J. Microsc.* **222**(Pt 3), 251–255.

CHAPTER 5

Combined Video Fluorescence and 3D Electron Microscopy

Alexander A. Mironov, Roman S. Polishchuk, and Galina V. Beznoussenko

Department of Cell Biology and Oncology
Consorzio Mario Negri Sud
66030 Santa Maria Imbaro
Chieti, Italy

Abstract
I. Introduction
II. Rationale
III. Method Steps
    A. Observation and Fixation of Living Cells
    B. Immunolabeling for EM
IV. Immunolabeling for EM with NANOGOLD
V. Immunolabeling for EM with HRP
    A. Embedding
    B. Locating the Cell on the Resin Block
    C. Cutting
    D. Picking up the Serial Sections
    E. EM Analysis
References

Abstract

Three-dimensional structure of cells and organelles examined with the power of resolution of electron microscopy (EM) including EM tomography represents the average view of cell processes. More precise and detailed analysis is limited by the significant variations in the structure of cells in temporal dynamics of cellular events. Therefore EM cannot identify rare and fast events that could be extremely important for understanding molecular mechanisms underlying these cellular processes. Observation of living cells under EM is still impossible. In contrast, observations of cellular dynamics with the help of laser scanning confocal microscopes or light digitalized microscopes became an indispensable tool of cell biology. However, the resolution of even confocal microscope is limited to the half of the light wave. Therefore, in studies of dynamic cellular processes, it would be ideal to be able to combine the capability of *in vivo* fluorescence video microscopy with the EM. This chapter describes this technique with details and useful tricks, including the way to localize the same cell after its transfection with a protein fused with a fluorescent tag, examination under the microscope in living condition, fixation, immunolabeling, embedding, serial sectioning from the first section, and observation under EM. We also illustrate here the kinds of questions that the CVLEM approach was designed to address, as well as the particular know-how that is important for the successful application of this technique.

I. Introduction

Many cellular functions crucially depend on rapid translocations and/or shape changes of specific intracellular organelles, including intracellular traffic, cytokinesis, and cell migration. To understand how such functions are organized and executed *in vivo*, it would be important to be able to use the degree of spatial resolution afforded by electron microscopy (EM) to observe dynamic structures in real time in living cells, such as a budding transport carrier, an elongating microtubule or a developing mitotic spindle. The most suitable methodology to achieve this is conceptually simple, yet powerful—combination dynamic observations obtainable from GFP video microscopy in living cells with EM. We refer this approach as correlative video-light EM (CVLEM, Polishchuk *et al.*, 2000), by which observations of the *in vivo* dynamics and ultrastructure of intracellular objects can indeed be combined to achieve this result.

II. Rationale

CVLEM procedure includes several stages: (1) observation of the structures labeled with green fluorescent protein (i.e., GFP) in living cells, (2) immunolabeling and embedding for EM, (3) identification of the cell on the resin block and cutting of thin or thick serial sections, (4) EM analysis and structure identification, (5) digital 3D reconstruction of the structure of interest.

During the first step, the cells are transfected with the cDNA encoding the GFP fusion protein of choice and the fluorescence of the associated structures in living cells is followed (Lippincott-Schwartz and Smith, 1997). In this way, it is possible to gain information about the dynamic properties of these structures (i.e., motility speed and direction, changes in size and shape, etc.). At the end of this stage, the cells are killed by the addition of fixative, capturing the fluorescent object at the moment of interest. As GFP is not visible under EM, immunostaining allows for the identification of the GFP-labeled structure at the EM level. The immunogold and immunoperoxidase protocols to perform this staining for EM are described later. Usually, the immunogold protocol (Burry et al., 1992) is suitable for labeling most antigens, while immunoperoxidase allows the labeling of antigens that reside only within small membrane-enclosed compartments, because the electron dense product of the peroxidase reaction tends to diffuse from the actual location of the antibody binding (Brown and Farquhar, 1989; Deerinck et al., 1994). Once stained, the cells must be prepared for EM by traditional epoxy embedding, and the cell and structure of interest must be identified in sections under EM. The finding of individual subcellular structures in single thin sections can be complex, and sometimes even impossible, simply because most of the cellular organelles are bigger than the thickness of the section and lie along a plane that is different from that of the section. So an analysis of serial sections from the whole cell is required for the identification of the structure(s) observed previously *in vivo*. Finally, the EM analysis of serial sections can be supported by high voltage EM tomography and/or digital 3D reconstruction.

III. Method Steps

A. Observation and Fixation of Living Cells

Materials:

- Cells of interest, DNA, and transfection reagents;
- MatTek Petri dishes with CELLocate cover slip (MatTek Corporation, Ashland, MA);
- HEPES buffer (0.2 M). Dissolve 4.77 g HEPES in 100 ml distilled water and add 1 N HCl to provide a pH of ~7.2–7.4;
- Fixative (0.1% glutaraldehyde-8% paraformaldehyde). Dissolve 8 g paraformaldehyde powder in 50 ml HEPES buffer, stirring and heating the solution to 60 °C. Add drops of 1 N NaOH to clarify the solution. Add 1.25 ml 8% glutaraldehyde and 50 ml HEPES buffer;
- 4% paraformaldehyde. Dissolve 4 g paraformaldehyde powder in 100 ml HEPES buffer, stirring and heating the solution to 60 °C. Add drops of 1 N NaOH to clarify the solution.

The cells are plated for CVLEM on MatTek Petri dishes that have CELLocate cover slips attached to their base. The CELLocate cover slips have etched grids with coordinates that allow the cells of interest to be found easily through all of the steps of the procedure. Transfect the cells with the cDNA of the GFP fusion protein of choice using any method available in your laboratory. After a transfected cell has been chosen and located on the CELLocate grid, its position is drawn on the map of the CELLocate grid (available from MatTek, see the example of the map in Fig. 1A). This living cell can then be observed for the GFP-labeled structures using confocal or light microscopy, which allows the grabbing of a time-lapse series of images by a computer. At the moment of interest, the fixative is added to the cell culture medium while still grabbing images (fixative: medium volume ratio of 1:1). The fixation usually induces the fast fading of the GFP fluorescence and blocks the motion of the labeled structures in the cells. Particular attention must also be paid to the temperature of the fixative, as the addition of a colder fixative to the cell medium will induce a shift in the focal plane during the time-lapse observations. The grabbing of the time-lapse images is then stopped, and the cells are kept in the fixative for 10 min. Wash once with 4% paraformaldehyde and leave the cells in 4% paraformaldehyde for 30 min.

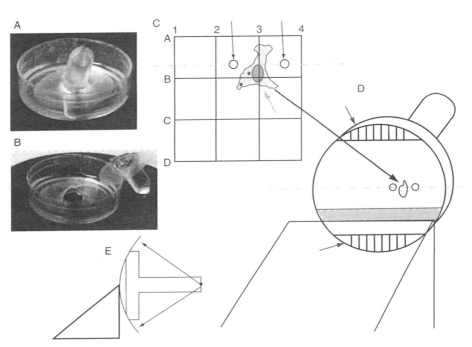

Fig. 1 Orientation of the sample. (A) The sample for CVLEM after embedding. (B) Detachment of cells from the cover slip. (C) The sample map. Red arrow shows the cell of interest, black arrows show marks that are made by a needle. (D) Orientation of the sample along the vertical direction. Black arrow shows the cells of interest and two marks formed by the needle. Blue arrows show the ends of the blocks equally cut by the knife if the position of the block is vertical. The blue area shows the shadow that indicates the distance between the knife-edge and the surface of the sample. (E) If the sample is oriented vertically, the central zone of the sample where the cells is situated will not be trimmed. Only ends will be equally cut by the knife (triangle). Arc shows the trajectory of the knife related to the sample. Arrows indicate the movement of the holder. Note that there is no danger for our cell to be damaged. (See Plate no. 10 in the Color Plate Section.)

B. Immunolabeling for EM

The immunolabeling of fixed cells for EM during the CLEM procedure can be performed using both immunogold and immunoperoxidase protocols (see Introduction). However, preliminary experiments should be done to determine whether the antibodies selected for labeling of the GFP-fusion protein work with the immuno-EM protocol. Many antibodies that give perfect results for immunofluorescence do not work for immuno-EM staining. This happens because the glutaraldehyde used in most EM fixatives tends to cross link the amino groups of the antigen epitopes, and therefore decrease the antigenicity of the target protein. However, the decreasing or removing of the glutaraldehyde in a fixative can result in poor preservation of the ultrastructure of the intracellular organelles. So, if there are problems with immuno-EM labeling, it is possible to optimize the concentration of glutaraldehyde in the fixative or to use a periodate-lysine-paraformaldehyde fixative (Brown and Farquhar, 1989). After this, it is important to select the immunoperoxidase or immunogold protocol to label the structure of interest. We would advise the use of only immunogold protocol to label epitopes of GFP-fusion proteins located in the cytosol (see Strategy), while for other epitopes, horseradish peroxidase (HRP) labeling is suitable.

Materials:

- Blocking solution. Dissolve 0.50 g BSA, 0.10 g saponin, 0.27 g NH_4Cl in 100 ml of PBS;
- Fab fragments of the secondary antibodies conjugated with HRP (Rockland, Gilbertsville, PA);
- NANOGOLD conjugated Fab fragments of the secondary antibodies (Nanoprobes Inc., Yaphank, NY);
- TRIS-HCl buffer (0.1 M). Dissolve 1.21 g TRIZMA base in 100 ml distilled water and add 1 N HCl to provide a pH of ~7.2–7.4;
- Diaminobenzidine (DAB) solution. Dissolve 0.01 g DAB in 20 ml TRIS-HCl buffer. Add 13.3 µl 30% H_2O_2 solution just before use;
- Gold-enhance mixture. Use a gold-enhance kit from Nanoprobes. Using equal amounts of the four components (Solutions A, B, C, and D), prepare about 200 l reagent per Petri dish (a convenient method is to use an equal number of drops from each bottle). First mix Solution A (enhancer; green cap) and Solution B (activator; yellow cap). Wait for 5 min, and then add Solution C (initiator; purple cap), and finally Solution D (buffer; white cap). Mix well.

IV. Immunolabeling for EM with NANOGOLD

Wash the fixed cells for 3 × 5 min with PBS. Incubate the cells with the blocking solution for 30 min, and then with the primary antibodies diluted in blocking solution, overnight. Wash the cells for 6 × 2 min with PBS. Dilute the

NANOGOLD-conjugated Fab fragments of the secondary antibodies ~50 times in the blocking solution and add this to the cells; incubate for 2 h. Wash the cells again for 6 × 2 min with PBS. Fix the cells with 1% glutaraldehyde in 0.2 M HEPES buffer for 5 min. Wash the cells for 3 × 5 min with PBS, and then for 3 × 5 min in distilled water. Incubate the cells with the gold-enhancement mixture for 6–10 min. The cells will become violet-grey in color if the gold enhancement is successful. Finally, wash cells for 3 × 5 min with distilled water, and proceed with the embedding.

V. Immunolabeling for EM with HRP

Wash the fixed cells for 3 × 2 min with PBS. Incubate the cells with the blocking solution for 30 min, and then with the primary antibodies diluted in blocking solution, overnight. Wash the cells for 6 × 2 min with PBS. Incubate the cells with the HRP-conjugated Fab fragments of the secondary antibody for 2 h. Wash the cells again for 6 × 2 min with PBS. Fix the cells with 1% glutaraldehyde in 0.2 M HEPES buffer for 5 min. Wash the cells 3 × 5 min with PBS, and then incubate them with the DAB solution. A successful peroxidase reaction results in a slightly brown staining of the cells. Finally, wash the cells for 3 × 2 min with PBS.

A. Embedding

Materials:

- Cacodylate buffer (0.2 M). Dissolve 2.12 g sodium cacodylate in 100 ml distilled water add 1 N HCl to provide a pH of ~7.2–7.4;
- OsO_4 (Electron Microscopy Sciences, Fort Washington, PA);
- Potassium ferrocyanide;
- EPON. Put 20.0 g EPON, 13.0 g Dodecenyl succinic anhydride (DDSA) and 11.5 g Methyl nadic anhydride (MNA) into the same test tube. Heat the tube in the oven for 2–3 min at 60 °C and then vortex it well. Add 0.9 g tri-Dimethylaminomethyl phenol (DMP-30, all from Electron Microscopy Sciences, Fort Washington, PA) and immediately vortex the tube again. It is possible to freeze the EPON in aliquots and to store it for a long time at −20 °C before use.

Wash the cells 5×/6× with distilled water. Be careful, because the residual phosphate may precipitate when it is mixed with OsO_4. Incubate the cells in the 1:1 mixture of 2% OsO_4 and 3% potassium ferrocyanide in 0.2 M cacodylate buffer for 1 h. Wash the cells once with distilled water, and then incubate them for 10 min each with the following ethanol solutions: 50% (once), 70% (once), 90% (once), 100% (3×), for the dehydration of the specimens. Keep the cells in a mixture of EPON and 100% ethanol (1:1; v/v) for 1–2 h. Keep the cells in EPON for 1–2 h at room temperature, and then leave them in an oven at 60 °C overnight.

B. Locating the Cell on the Resin Block

Materials:

- Hydrofluoric acid (HF)
- EPON
- Stereomicroscope

After 12 h of polymerization of the EPON, place a small droplet of a fresh resin on the site where the examined cell is located, and insert a resin cylinder (prepared before by polymerization of the resin in a cylindrical mold) with a flat lower surface; leave the samples for an additional 18 h in the oven at 60 °C. Carefully pick up the resin from the Petri dish and glass; this is easy to do by gentle bending of the resin cylinder to and fro. The resin block and the empty MatTek Petri dish after block detachment are shown in Fig. 1A and B. If the cover glass with a coordinated grid cannot be detached from the cells included into the resin, the latter should be placed into commercially available HF (do not use glassware for this) for 30–60 min. Control the completeness of the glass dissolution under a stereomicroscope. Wash the samples in water after the complete removal of the glass. Leave the samples in 0.1 M HEPES buffer (pH 7.3) for 60 min to neutralize the HF. Wash the samples in water and allow them to dry. The final resin block is shown in Fig. 1A.

C. Cutting

Materials and equipment:

- Glass and diamond knives;
- Ultratome

Find the cell of interest among the cells within the sample according to the coordinated grid, and put the resin block into the holder of an ultratome. Using a steel needle and rotating the sample in the holder, make two small cavities in such a way that they (thin arrows in Fig. 1A) form a horizontal line (broken red line in Fig. 1C and D) with the cell appearing in the center of the sample. Introduce the holder into the ultratome in such a way that the segment arc of the ultratome is in the vertical position and the two cavities form a horizontal line. By rotating the glass knife stage, align the bottom edge of the pyramid parallel to the knife-edge. Using the segment arc, orient the plane of the sample vertically. Bring the sample as close as possible towards the glass knife. Adjust the gap (which is visible as a bright band if all three of the lamps of an ultratome are switched on) between the knife-edge and the surface of the sample. The gap has to be identical in width between the most upper and lower edges of the sample during the up and down movement of the resin block. This ascertains that every point of the sample surface containing the cell of interest is at the same distance from the knife-edge. Slowly moving the sample up and down, continue its approach until the knife begins to cut one of the edges of the sample. The sectioning begins from either the upper or the lower part of the sample, the middle part of the sample where the cell of interest is situated will be unaffected because the length of the radius passing through the cell is shorter than the radii passing through the upper and the lower edges of the sample. If the sectioning is to begin from the upper part of the sample, tilt the segment arc to approach the lower edge towards the knife. If the sectioning is to begin from the lower edge of the sample, tilt the segment arc and approach the upper edge of the sample towards the knife. A vertically oriented sample should produce equal sections from both the upper and lower edges of the samples (Fig. 1D and E). Note down precisely all of parameters relating to the position of the sample in the ultratome, i.e. the degree of rotation of the sample in the holder, the degree of tilting of the segment arc, and the degree of rotation of the knife in its stage. Take the sample and trim it to provide a narrow horizontal pyramid of about 0.2×0.9 mm in size with the cell of interest at its centre.

Do not take the sample from the holder and do not rotate the sample inside the holder. The pyramid should be as narrow as possible (no wider than 200 m), and the cell of interest should be at the centre of the pyramid. An experienced person can trim a pyramid directly with a razor blade. Introduce the sample back into the ultratome, and lock it in exactly the same position as before (preserving all of the parameters of sample positioning; this is very important). Replace the glass knife with the diamond one, and position the latter towards the pyramid. If the sample is not parallel to the knife, adjust the angle of the diamond knife by rotating the knife stage to make its edge parallel to the plane of the pyramid. Do not change any other parameters of the sample position. Approach the sample towards the edge of the knife until the gap is extremely narrow. Using a 200-nm approaching step, begin the sectioning. Take serial 200-nm sections according to the instructions with the ultratome. It is enough to take only 10 sections to pass 2 µm from the height of the cell. Remember the position of the organelle of interest according to the Z-stacking, and select those thick sections that should correspond to this position. For instance, if the organelle of interest is situated at 500 nm from the bottom of the cell, it is enough to collect only the first four 200-nm serial sections. If the position is at 1 µm in height, it will be necessary to collect from the fourth to the eighth serial sections.

During the cutting of the specimen, the thickness of the serial sections needs to be selected. This should be about 80 nm for routine work, 50 nm (or less) for very precise 3D reconstruction, or 250 nm for EM tomography.

D. Picking up the Serial Sections

1. Picking up the Sections with the Empty Slot Grid

   Materials and equipment:
   - Pick-up loop (Agar, Cambridge, UK);

5. Video Fluorescence and 3D Electron Microscopy

- Empty slot grids and slot grids covered with carbon-formvar supporting film (Agar, Cambridge, England or Electron Microscopy Sciences, Fort Washington, PA);
- self closing tweezers (Agar, Cambridge, UK)

For this you need one empty (not covered) but cleaned slot grid (the donor transfer grid) (Fig. 2A) and one slot grid covered with formvar-carbon film (the acceptor grid, Fig. 2B). The former (Fig. 2A) should have a holder for the tweezer (thick arrow in Fig. 2A). This holder could be done by bending just the small part of the grid from horizontal position. For this procedure you need to have serial sections with the width smaller that 900 m and the height of about 100 μm. In this case you could place on the slot 18 serial sections (Fig. 3A).

Take the slot grid with supporting film from a container by the self-closing tweezer (Fig. 2D) and place it film-down on parallel holders covered with scotch (Fig. 2C, arrows; E).

Take the empty slot grid and place over sections not touching the water. Touch water with the donor grid in such a way that sections are inside the slot and move the slot grid away (Fig. 3B and C). The droplet of water and sections will be inside the slot (Fig. 4A; Fig. 5 arrowheads).

Using stereomicroscope put on the donor grid with serial sections on acceptor grid as it is shown in Fig. 4B and C. To avoid dirt on the sections (Fig. 5A, arrow) it

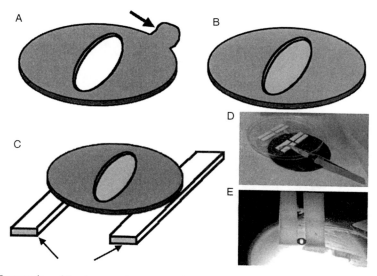

Fig. 2 Preparation of the donor and acceptor grids. (A) The empty (not covered) but cleaned slot grid (the donor transfer grid, arrow shows the holder for the tweezers). (B) The slot grid covered with formvar-carbon film (the acceptor grid, the slot is covered with the supporting film). (C) Position of the acceptor grid on parallel holders covered with scotch (arrows). (D) Placement of the acceptor grid on the scotch holders. (E) Final position of the grid. (See Plate no. 11 in the Color Plate Section.)

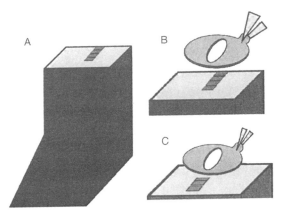

Fig. 3 Picking up of serial sections with the donor slot grid from the water. (A) Serial sections in the glass bath. (B) Approaching of the donor grid. (C) Final position of the donor grid before its contact with the water surface. (See Plate no. 12 in the Color Plate Section.)

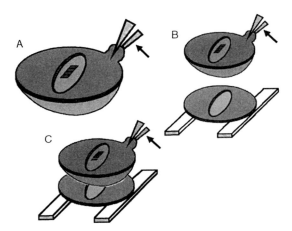

Fig. 4 Transfer of serial sections with the donor grid. (A) After touching of water in such a way that sections were inside the slot the serial section are inside the slot. The droplet of water with sections inside the slot are formed. Arrows show the tweezers. (B) Position of the donor slot grid over the acceptor grid. (C) Approaching of the donor grid towards the acceptor grid. (See Plate no. 13 in the Color Plate Section.)

is better to place a small droplet of glass handled distilled water on the acceptor slot grid (Fig. 5A).

Using sharp and very narrow filter paper (Fig. 5C and D) orient slot in parallel to each other and eliminate the excess of water from the space between slot grids (Fig. 5E). Dry grids during at least 20 min (Fig. 5E). Carefully eliminate the donor grid (Fig. 5F). Take the dried acceptor grid and check whether sections are in correct position (Fig. 5G).

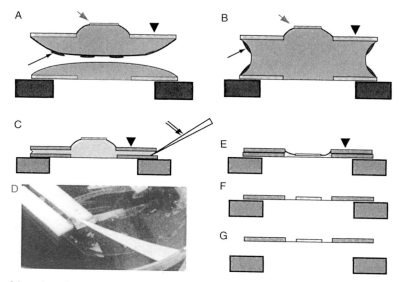

Fig. 5 Mounting of serial sections on the acceptor grid. (A, B) Touching of the acceptor grid with the donor grid containing serial sections and formation of the common droplet. Green arrows show serial sections. (C, D) Elimination of water between sections and supporting films with the thin filter paper. (E, F) Drying of water and elimination of the dried donor grid (F). (G) Removal of the acceptor grid with serial sections on the supporting film from the scotch holders. (See Plate no. 14 in the Color Plate Section.)

2. Picking up with the Pick-Up (perfect) Loop

Materials and equipment:

- Pick-up loop (Agar, Cambridge, UK);
- Slot grids covered with carbon-formvar supporting film (Agar, Cambridge, England or Electron Microscopy Sciences, Fort Washington, PA)

For the picking-up of the sections according to this method, stop the motor, and divide the band of the sections into pieces of a suitable size for collection with the pick-up loop, using two eyelashes. Touch the surface of the water with the band of sections with the pick-up loop in such a way that the band is completely inside the inner circle of the loop, without touching it (Fig. 6A). Raise the loop with the droplet of water with the sections on it, and place the loop inside the tripod near the microscope. The loop should be visible under the stereomicroscope of the ultratome. Take the slot grid coated with the formvar (or preferably butvar)/ carbon supporting film and gently touch sections on the water (do not touch the loop) with the carbon-coated surface of the grid (Fig. 6B). Very slowly, move the slot grid away from the loop. If the movement is slow enough, the water is eliminated from the surface of the supporting film, and only a very small droplet of water remains on the grid, which will not represent an obstacle for the placement of the grid directly into the grid container (Fig. 6C).

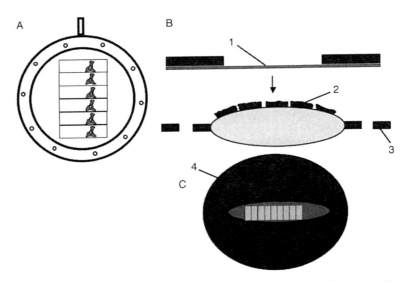

Fig. 6 Picking up of serial sections with the perfect loop. (A) Perfect loop with sections. Upper view. (B) Position of the slot grid with the supporting film (1) over the perfect loop (3) with the droplet of water and serial sections (2). (C) Position of serial sections on the slot grid (4). (See Plate no. 15 in the Color Plate Section.)

E. EM Analysis

Place the slot grid under the electron microscope, and using the traces of the coordinated grid filled with the resin on the first few sections, or the central position of the cell of interest within the pyramid, identify the cell on the sections. Take consecutive photographs (or grab the images with a computer using a video camera) of the serial sections until the organelle of interest (just observed under the LSCM) is no longer seen. If EM tomography is to be used, take a tilting series of the organelle of interest and produce the electron tomogram according to the instruction for the IMOD software (available at the following web site: http://bio3d.colorado.edu/imod/). Using the software for the 3D reconstruction, align the images and then make a 3D model according to the instructions with the software.

1. Critical Parameters and Troubleshooting

One limitation, and at the same time, attraction, of CVLEM, is its complexity. However, if researchers will follow instructions described here, they will be able to have good results. Taken together, all of the steps for CVLEM represent quite a long procedure, and they require significant effort of the experimenter. Therefore, it would be particularly disappointing to loose such *tour de force* experiments because of small problems with specimen handling. To apply

CVLEM successfully, several important parameters should always be taken into account by the experimenter.

First, it is extremely important to be able to find the cell of interest at all of the steps of the CVLEM procedure. So, only cells located on the grid of the MatTek Petri dish should be selected for time-lapse observations. The position of the cell of interest on the grid must be noted; otherwise it would be difficult to find it again. The low magnification images showing the field surrounding the cell of interest can help greatly in order to be able to trim the resin block around the right cells, and to find them later under the electron microscope. In this case, neighboring cells can be used as landmarks to identify the cell of interest. For this reason, the cells for experimentation should be plated with a lower confluence (50–60%) than usual. During the analysis of the serial sections under the electron microscope, it is useful to have the fluorescent and phase-contrast images of the target cell because particular structures (microvilli, pseudopodia, inclusions, etc.) can help greatly to find both the cell and the structure of interest.

Second, the main demand for CVLEM is to produce as many images as possible and to perform as many steps as possible (better all manipulations) under the stereomicroscope. Another possible problem could be tweezers that often lost EM grids. Therefore, we recommend the use of self closing tweezers although it could initially seem to be very inconvenient.

Acknowledgments

We thank Dr. C.P. Berrie for critical reading of the manuscript and Telethon Italia for financial support.

References

Brown, W. J., and Farquhar, M. G. (1989). Immunoperoxidase methods for the localization of antigens in cultured cells and tissue sections by electron microscopy. *Methods Cell Biol.* **31**, 553–569.

Burry, R. W., Vandre, D. D., and Hayes, D. M. (1992). Silver enhancement of gold antibody probes in pre-embedding electron microscopic immunocytochemistry. *J. Histochem. Cytochem.* **40**, 1849–1856.

Deerinck, T. J., Martone, M. E., Lev-Ram, V., Green, D. P., Tsien, R. Y., Spector, D. L., Huang, S., and Ellisman, M. H. (1994). Fluorescence photooxidation with eosin: A method for high resolution immunolocalization and *in situ* hybridization detection for light and electron microscopy. *J. Cell Biol.* **126**, 901–910.

Lippincott-Schwartz, J., and Smith, C. L. (1997). Insights into secretory and endocytic membrane traffic using green fluorescent protein chimeras. *Curr. Opin. Neurobiol.* **7**, 631–639.

Polishchuk, R. S., Polishchuk, E. V., Marra, P., Buccione, R., Alberti, S., Luini, A., and Mironov, A. A. (2000). GFP-based correlative light-electron microscopy reveals the saccular-tubular ultrastructure of carriers in transit from the Golgi apparatus to the plasma membrane. *J. Cell Biol.* **148**, 45–58.

CHAPTER 6

From Live-Cell Imaging to Scanning Electron Microscopy (SEM): The Use of Green Fluorescent Protein (GFP) as a Common Label

Sheona P. Drummond* and Terence D. Allen

Department of Structural Cell Biology
Paterson Institute for Cancer Research
University of Manchester
Manchester M20 4BX
United Kingdom

*Present address: Post-transcriptional Research Group
Manchester Interdisciplinary Biocentre
University of Manchester
Manchester M1 7ND
United Kingdom

Abstract
I. Introduction
II. Rationale
III. Methods
  A. Cell Culture and Transfection
  B. Preparation of Cells on Silicon Chips
  C. Sample Preparation for Immuno-SEM
IV. Summary
  A. Cell Culture and Transfection
  B. Processing for Real-Time Light Microscopy
  C. Processing for Scanning Electron Microscopy
V. Concluding Remarks
References

Abstract

The identification and characterization of many biological substructures at high resolution requires the use of electron microscopy (EM) technologies. Scanning electron microscopy (SEM) allows the resolution of cellular structures to approximately 3 nm and has facilitated the direct visualization of macromolecular structures, such as nuclear pore complexes (NPCs), which are essential for nucleocytoplasmic molecular trafficking. However, SEM generates only static images of fixed samples and therefore cannot give unambiguous information about protein dynamics. The investigation of active processes and analysis of protein dynamics has greatly benefited from the development of molecular biology techniques whereby vectors can be generated and transfected into tissue culture cells for the expression of specific proteins tagged with a fluorescent moiety for real-time light microscopy visualization. As light microscopy is limited in its powers of resolution relative to electron microscopy, it has been important to adapt a protocol for the processing of samples for real-time imaging by conventional light microscopy with protein labels that can also be identified by SEM. This allows correlation of dynamic events with high resolution molecular and structural identification. This method describes the use of GFP for tracking the dynamic distribution of NPC components in real-time throughout the cell cycle and for high resolution immuno-SEM labeling to determine localization at the nanometer level.

I. Introduction

Fluorescence microscopy, involving the transfection of tissue-culture cells with vector constructs for the expression of fluorescein-tagged proteins, has proved to be a powerful tool in the analysis of protein localization and dynamics. Although substantial progress and improvement has been made in the micrometer-to-nanometer resolution capacity of fluorescence microscopes, this discipline, by definition, is constrained by the wavelength of visible light (approximately 400–700 nm).

In contrast electron microscopy, both scanning and transmission (SEM and TEM respectively), allows for the direct visualization of biological structures to single nanometre resolution.

Over the last decade a burgeoning field of exploration has been the elucidation of the fine structure of the nuclear pore complex (NPC). NPCs are modular protein assemblies, embedded within the nuclear envelope (NE), that facilitate the bidirectional trafficking of molecules between the nucleoplasm and cytoplasm (reviewed in Drummond *et al.*, 2006; Lim and Fahrenkrog, 2006; Schwartz, 2005). Ultrastructural studies of the NPC involves visualization of nuclear envelope (NE) associated structures that are both annular and filamentous in arrangement and possess sizes of only single digits to tens of nanometers (Allen *et al.*, 1997; Bagley *et al.*, 2000; Goldberg *et al.*, 1997; Kiseleva *et al.*, 2004). Recent studies have shown that several nucleoporins (nups; the protein constituents of the NPC) relocate to kinetochores upon entry into open-mitosis and are then recruited back to the periphery of decondensing chromatin as the cell begins to exit mitosis (Blegareh et al., 2001; Joseph et al., 2002; Loiodice et al., 2004; Rasala et al., 2006; Salina et al., 2003). The nup107/160 complex is composed of nine nups, namely nup107, nup160, nup133, nup96, nup75, nup43, nup37, Seh1, and Sec13, which remain associated throughout mitosis (Loiodice et al., 2004). This nucleoporin subcomplex is an essential requirement for NPC assembly upon exit from mitosis (Boehmer et al., 2003; Harel et al., 2003; Walther et al., 2003) and has been shown to play an important role in the correct function of kinetochores (Orjalo et al., 2006; Zuccolo et al., 2007). The methods described here use reagents generated for the investigation of nup37 and nup43, the most recently identified members of the nup107/160 complex. Both of these proteins contain multiple tryptophan-aspartate (WD) repeat motifs. Because of resolution of the tertiary structure of other WD-repeat containing proteins it is thought that both nup37 and nup43 adopt a similar characteristic propeller structure which may play a role in initiating or maintaining pore-membrane curvature.

This protocol focuses on the location of nucleoporins at the kinetochore plates with respect to a well characterized kinetochore structural component, namely CENP-F. The methods described here show the use of GFP-labeled nup37 or nup43 to illustrate the migration, and gross associations, of nucleoporins with the NE during interphase and kinetochores during mitosis.

II. Rationale

Correlation of data relating to protein dynamics, generated by fluorescence microscopy, and that generated using scanning electron microscopy (SEM) is often complicated by the lack of probes that are effective in both techniques. Therefore the protocol detailed herein allows the identification of the location and distribution of specific proteins and protein complexes and, observation of the dynamics of these proteins within the same cell population. This method describes the use of green fluorescent protein (GFP) as a common protein label that can be directly visualized in a live cell by fluorescent microscopy and then the same cell populations can be processed for SEM and the GFP labeled with appropriate antibodies. This allows for the direct comparison of the dynamics and high resolution localization of GFP-labeled proteins.

III. Methods

A. Cell Culture and Transfection

1. Thaw a 1 ml aliquot of HeLa cells (that has been stored in cell-culture media containing 10% glycerol at −80 °C or under liquid nitrogen) in a preheated water bath at 37 °C.

2. Transfer the cells to a 75 cm² tissue culture flask containing 10 ml of L15 Media containing 10% fetal calf serum (FCS) and grow cells in an incubator at 37 °C.
3. Once the cells are confluent remove the media and incubate with Trypsin-EDTA until the cells have rounded and are easily detached by shaking from the surface of the flask (Do not incubate with Trypsin-EDTA for longer than 5 min) and transfer an aliquot of cells to a glass-bottomed dish compatible with subsequent fluorescence microscope visualization or transfer an aliquot of the trypsinized cells to a fresh flask containing 10 ml of fresh media, for subsequent processing for immunogold SEM.

The following steps should be followed for both the cells in the glass-bottomed dish and those in the tissue-culture flask

When cells are approximately 50–60% confluent, cells can be transfected for the expression of a GFP-labeled protein.

4. Fugene 6 (Roche Applied Science, Indianapolis, IN) was used as the transfection reagent of choice owing to high efficiency and reproducibility of transfection and was used according to manufacturer's instructions.
5. In these studies cells were transfected with nup37-EGFP or nup43-EGFP, in both case together with pHA-puro (confers resistance to puromycin for antibiotic selection of positively transfected cells – often it is possible to include an antibiotic resistance cassette within the GFP vector, this is preferable).
6. Cells are incubated with serum-containing media and the transfection mixture for 24 h at 37 °C and then media is changed to media containing 3 µg/ml puromycin.
7. Media is changed every 24 h for the following two days.

For cells grown in the 75 cm² tissue culture flask:

Cells are collected by 'manual mitotic shake-off.' This involves a sharp tap to the underside of the tissue culture flask sufficient to dislodge the mitotic cells from the surface of the flask and subsequent removal of media, which will contain the mitotic cells in suspension.

B. Preparation of Cells on Silicon Chips

1. Mitotic cells collected by mitotic shake off are gently collected by centrifugation at 800g at room temperature, excess buffer removed (avoiding drying of the loosely packed cell pellet), and resuspended in 1 ml of media.
2. Silicon chips (see Fig. 1) are prepared by first numbering with a diamond pen then washed with 100% acetone, gently wiped dry with a lint-free tissue and then autoclaved at a minimum of 160 °C for 12 h.

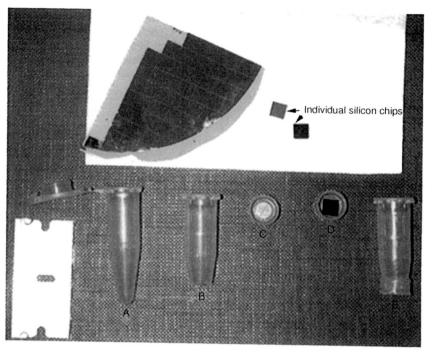

Fig. 1 Adaptation of microtubes for centrifugation of cells onto silicon chips. Remove the lid from a 1.5 ml microtube (A) using a razor blade or scissors. Trim away any remaining lid attachment from the top of the tube. Cut off the bottom of the tube to a diameter which pushes into the lid providing a tightly fitting leak proof seal (B). Retain the lid (C). When required, a 5 × 5 mm silicon chip should be placed inside the inverted lid (D) and the tube pushed on to fit tightly without damaging the chip (E). When filled with liquid, the chamber should not leak.

3. Chips then incubated with poly-L-lysine (10–20 μl pipetted onto the upper surface of the chip) for at least 1 h at room temperature and then washed briefly in water.
4. This protocol has only been attempted in adherent tissue culture cell lines, however it is likely that suspension culture cells can be arrested in mitosis, or synchronized and processed upon entry into mitosis and processed similarly. In the case of nonadherent cells it would be especially important to pre-incubate the silicon chips with poly-L-lysine to ensure cell attachment and retention during processing.
5. An aliquot of media, containing mitotic cells, was then pipetted into a silicon chip contained within a chamber constructed from a 1.5 ml Eppendorf tube (see Fig. 1) and centrifuged at $1000g$ for 10 min at 4 °C.
6. Chips are then released from the chamber by removal of the 'lid' portion of the tube, excess buffer removed, and the chips are immediately transferred to fixative (see C below), ensure that samples do not dry out at any stage.

C. Sample Preparation for Immuno-SEM

1. Cells are fixed by incubation with 2% paraformaldehyde and 0.01% glutaraldehyde in PBS for 10 s at room temperature. Each chip is held individually with forceps during the brief immersion in fixative.
2. Immediately transfer chips to 0.1 M glycine in PBS for a minimum of 10 min to quench any unreacted aldehyde.
3. Rinse chips briefly in PBS.
4. Incubate cells with 0.5% Triton X-100 in PBS for 30 min. As this protocol requires the removal of cytoskeletal elements a stringent incubation with detergent is necessary however, if the GFP-tagged protein of interest is known to be present at the plasma membrane this step can be omitted.
5. Rinse in PBS.
6. Incubate in 1% Bovine Serum Albumin (BSA) in PBS for a minimum of 20 min.

Immunolabeling:

7. Prepare a wet-chamber using a 90 mm Petri dish containing wet filter paper and Parafilm (see Fig. 2).
8. Dilute primary antibodies in PBS containing 1% BSA. In this case, the anti-GFP antibodies and kinetochore-specific antibodies can be diluted together and incubated with the sample for 2 h at room temperature in the wet chamber.
9. Wash the chips 2 × 10 min in PBS.
10. Block the samples in 1% fish skin gelatin (FSG) in PBS for 10 min.
11. Incubate the samples with different diameters of gold conjugated to appropriate secondary antibodies to facilitate discrimination between GFP and kinetochore labeling. For example, samples were incubated for 1 h at room temperature with 10 nm gold conjugated to anti-mouse IgG to identify GFP-tagged proteins and 5 nm gold conjugated to anti-rabbit IgG to visualize kinetochore-specific proteins (e.g., CENP-F).

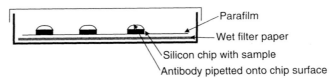

Fig. 2 Preparation of a 'wet chamber' for immunolabeling. Filter paper is placed in the bottom of a plastic Petri dish and saturated with water. The filter paper is then partially covered with a layer of Parafilm and silicon chips are placed upon the Parafilm layer, with sample facing upwards. Then a maximum of 20 μl of antibody solution of appropriate concentration is carefully pipetted onto the surface of the silicon chip.

12. Wash samples 3 × 10 min in PBS.
13. Fix samples in 3% glutaraldehyde in PBS for 1 h at room temperature or over-night at 4 °C.

Postfixation, critical point drying, cell fracture, and sample coating:

14. Wash chips briefly in Sorensen's Buffer (Prepare 0.15 M solutions of KH_2PO_4 and Na_2HPO_4 and mix in the ratio of 19:81 respectively to give a buffer of pH 7.4)
15. Postfix samples in 1% osmium tetroxide in Sorensen's buffer for 10 min. (Osmium preserves lipids and enhances contrast when visualizing the specimen).
16. Wash the samples briefly in distilled water.
17. Dehydrate samples through a graded ethanol series (30%, 50%, 70%, 95% X2 and 100% X3), incubating for 5 min at each step.
18. Transfer the chips to critical point drying apparatus containing ethanol as the transitional solvent and critical point dry from high purity liquid CO_2 (high purity =< 5ppm H_2O).
19. Cells are dry fractured after completion of critical point drying by adhering the chips to standard laboratory adhesive tape (sample directly contacting the adherent side of the tape), gently tapping the chip and removing the chip by gently pulling away from the tape avoid lateral movement.
20. Samples are then coated with 3 nm of chromium and stored under vacuum until visualized by SEM.

IV. Summary

The methods described here can be divided into three main sections:

A. Cell Culture and Transfection

This involves culturing cells to the appropriate density (approximately 50–60% confluency) for optimal transfection efficiency, either in a tissue culture flask for subsequent 'mitotic shake-off' or in glass bottomed Petri dishes compatible with an inverted microscope for live-cell imaging. In both cases, cells are then transfected using Fugene 6, or a preferred transfection agent, for incorporation of a vector for the expression of a GFP-tagged protein of interest. Any other fluorescent label can be used and then identified using the appropriate antibody in scanning electron microscopy.

B. Processing for Real-Time Light Microscopy

Approximately 48–72 h post-transfection cells should be expressing fluorescently labeled protein and will be ready for visualization; however, optimal times for data collection should be verified for each expression construct and cell-line used.

In addition, cells can be incubated with Hoechst 33342, a cell-permeable lipophilic DNA dye, for visualization of chromosome structure and condensation state. Here a Deltavision multidimensional restorative microscope, based around an Olympus BX71 microscope, fitted with an isolated environmental chamber to maintain the temperature at 37 °C, was used for live-cell image acquisition (Fig. 3). Images were collected every 15 min for a minimum of 24 h and movies were edited to show the interphase-mitosis transition, throughout mitosis and then re-entry into interphase.

C. Processing for Scanning Electron Microscopy

The visualization of chromosomes and associated intracellular structures by field-emission scanning electron microscopy (FESEM) is necessary and informative in determining the presence and function of nuclear structures and proteins (Fig. 4). Cells are permeablized and then incubated with appropriate primary antibodies, washed and then incubated with collodial gold conjugated to secondary antibodies. After thorough washing chromosome structure is revealed by subjecting the cells to a fracturing technique with an adhesive at room temperature (16–24 °C) in which the fracture planes can pass at various planes through detergent-extracted cells. In this case the sample is coated with 3 nm of chromium, before examination in the scanning electron microscope using secondary and backscatter imaging. Surface topography is apparent from the conventional secondary imaging, and backscatter imaging produces an unequivocal distribution of the gold particles (Fig. 5). Both images are acquired simultaneously, and can be superimposed to show the location of the gold.

V. Concluding Remarks

This chapter details the protocol by which one and the same tissue culture cell population can be processed for both live-cell imaging and immunogold SEM, where green (or an alternative) fluorescent protein has been used to generate a specifically tagged protein using a standard transfection protocol. Once visualized in the living cell at various stages of division and the cell cycle, the resolution of the labeling as seen by light microscopy can be increased approximately 200-fold by the use of antibodies to GFP bound to colloidal gold and visualized by SEM. This allows for correlation of the dynamic localization of specific proteins and their structural location at high resolution. Another advantage in the SEM is that the macromolecular surroundings of the labeling sites are also available in the secondary image, in contrast to fluorescence, where the signal is always a bright dot against a black background. Thus in this particular example, the kinetochore region of the chromosome appears to reside in a depression in the chromation surface, and the relative distributions of two kinetochore proteins are apparent. Here the SEM labeling has been extended to a second protein, using specific

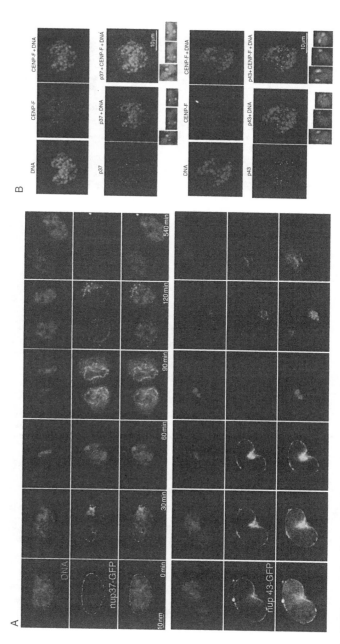

Fig. 3 Still images taken from a real-time movie of HeLa cells transfected with nup37-EGFP or nup43-EGFP. (A) For live-cell imaging of cells containing GFP-tagged nucleoporins (nup37 in the upper three rows and nup43 in the lower three rows) DNA is visualized with the cell-permeable DNA dye Hoechst 33258, imaged in blue (upper panel), the GFP-tagged nucleoporin is shown in green (middle panel) and the merged DNA and GFP-fluorescence images are shown in the bottom panel. During interphase (t = 0), the DNA is decondensed within the nucleus, and nucleoporins display a characteristic punctuate staining at the nuclear periphery as they are present within the nuclear pore complexes (NPCs), which periodically perforate the nuclear envelope (NE). Upon entry into mitosis (30 min), the DNA begins to condense and concomitantly the NPCs dissociate into specific subcomplexes and then the NE begins to disassemble. During metaphase (∼60–70 min) the DNA is fully condensed into visibly distinct chromosomes each of which aligns at the metaphase plate and the bulk of the nucleoporins are present as a diffuse cytoplasmic pool. As the daughter chromosomes migrate to opposing poles of the dividing cells, the NE and NPCs reassociate and bind to the chromatin periphery and form new NEs around each decondensing chromosome population containing NPCs functional for bidirectional nucleocytoplasmic trafficking. (B) Colocalization of Nup37 and Nup43 with CENP-F During Mitotsis. CENP-F is a well-characterized component of the mammalian kinetochore that localizes to these chromosomal sites during mitosis. In each montage, DNA is stained with Hoechst 33258 (blue), CENP-F is identified using a monoclonal antibody conjugated to Cy3 (red), the GFP-tagged nup37 or nup43 are shown in green, and the images on the right show the indicated merged images. In both the case of nup37 and nup43 it is clear that these proteins relocate to distinct sites upon condensed chromosomes (small panels below the GFP image reveals individual chromosomes, each with two clear regions of GFP-nup localization). When the CENP-F, GFP-nup, and DNA images are merged (right) it is clear that the kinetochore-specific CENP-F colocalizes with both nup37 and nup43. This localization is displayed at higher resolution in the small panels below the merged image and shows that CENP-F and each nup is present on each kinetochore pair. (See Plate no. 16 in the Color Plate Section.)

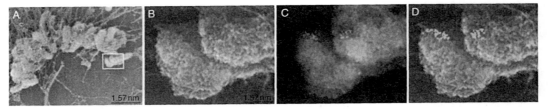

Fig. 4 S.E.M. of mitotic HeLa cells revealing kinetochores and CENP-F localization. (A) A mitotic HeLa cell prepared by detergent extraction and manual fracture for visualization by SEM shows condensed chromosome pairs aligning at the metaphase plate. The inset box is magnified in (B), (C), and (D). (A) and (B) are secondary electron images showing the surface morphology of the mitotic chromosomes and (B) reveals paired concavities at the surface of the chromosomes. This image was then visualized by backscatter imaging (C) to reveal 10 nm diameter gold particles (white dots) bound to an antibody specific for CENP-F in and around the concavities. Images (B) and (C) are captured simultaneously, allowing exact superimposition, as shown in (D). In (D) gold particles are pseudocolored yellow for clarity. (See Plate no. 17 in the Color Plate Section.)

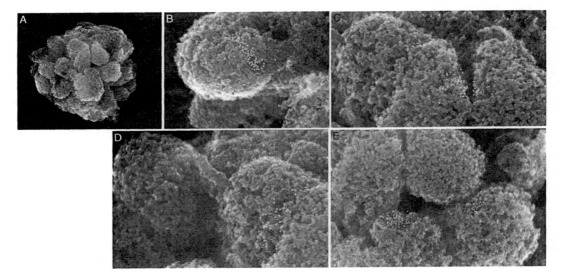

Fig. 5 Double immunogold labeling of mitotic chromosomes with antibodies specific for CENP-F and GFP-tagged Nup37. These images are collected digitally in the same manner briefly described in Fig. 4. Similarly, (A) mitotic chromosome profile of a single HeLa cell. (B–E) show magnified regions of the same chromosome complement in which the secondary and backscatter images have been merged. In this case, two different sizes of gold particles have been used. 10 nm diameter gold specific for a GFP antibody (raised in mouse) to bind to GFP-tagged nup37 (pseudocolored yellow) and 5 nm diameter gold particles specific for a CENP-F antibody (raised in rabbit), which is pseudocolored green. The distribution of the labeling indicates that nup37 is localized slightly more centrally within the kinetochore plate compared with CENP-F. (See Plate no. 18 in the Color Plate Section.)

antibodies bound to a different sized gold colloid, but this could be extended to a third antibody, using 5, 10, and 15 nm gold colloids for SEM visualization of three different colors in the fluorescent microscope.

In this protocol the manner in which each cell fractures and consequently the structures that are exposed will vary between different cells. However, with approximately 10,000 cells per chip, the population offers a sufficiently high density to generate statistically significant information. One of the adavantages of the SEM in this situation is that relatively rare events such as specific stages of division in an unsynchronized population can be searched for and localized with similar effort to that required for light microscopy.

Acknowledgments

The author thank Sandra Rutherford (Department of Structural Cell Biology, Paterson Institute for Cancer Research, University of Manchester, UK) for considerable technical input and advice at all stages of protocol optimization, Steve Murray for Fig. 2 (Paterson Institute for Cancer Research, University of Manchester, U.K.) and Steve Bagley (Advanced Imaging Facility, Paterson Institute for Cancer Research, University of Manchester, UK) for technical assistance with live-cell imaging.

References

Allen, T. D., Bennion, G. R., Rutherford, S. A., Reipert, S., Ramalho, A., Kiseleva, E., and Goldberg, M. W. (1997). Macromolecular substructure in nuclear pore complexes by in-lens field-emission scanning electron microscopy. *Scanning* **19**(6), 403–410.

Bagley, S., Goldberg, M. W., Cronshaw, J. M., Rutherford, S., and Allen, T. D. (2000). The nuclear pore complex. *J. Cell. Sci.* **113**, 3885–3886.

Blegareh, N., Rabut, G., Bai, S. W., van Overbeen, M., Beaudouin, J., Daigle, N., Zatsepina, O. V., Pasteau, F., Labas, V., Fromont-Racine, M., Ellenberg, J., and Doye, V. (2001). An evolutionarily conserved NPC subcomplex, which redistributes in part to kinetochores in mammalian cells. *J. Cell Biol.* **154**(6), 1147–1160.

Boehmer, T., Enninga, J., Dales, S., Blobel, G., and Zhong, H. (2003). Depletion of a single nucleoporin, Nup107, prevents the assembly of a subset of nucleoporins into the nuclear pore complex. *Proc. Natl. Acad. Sci. USA* **100**(3), 981–985.

Drummond, S. P., Rutherford, S. A., Sanderson, H. S., and Allen, T. D. (2006). High resolution analysis of mammalian nuclear structure throughout the cell cycle: Implications for nuclear pore complex assembly during interphase and mitosis. *Can. J. Physiol. Pharmacol.* **84**(3–4), 423–430.

Goldberg, M. W., Wiese, C., Allen, T. D., and Wilson, K. L. (1997). Dimples, pores, star-rings, and thin rings on growing nuclear envelopes: Evidence for structural intermediates in nuclear pore complex assembly. *J. Cell Sci.* **110**, 409–420.

Harel, A., Orjalo, A. V., Vincent, T., Lachish-Zalait, A., Vasu, S., Shah, S., Zimmerman, E., Elbaum, M., and Forbes, D. J. (2003). Removal of a single pore subcomplex results in vertebrate nuclei devoid of nuclear pores. *Mol. Cell.* **11**(4), 853–864.

Joseph, J., Tan, S. H., Karpova, T. S., McNally, J., and Dasso, M. (2002). SUMO-1 targets RanGAP1 to kinetochores and mitotic spindles. *J. Cell Biol.* **156**(4), 595–602.

Kiseleva, E., Allen, T. D., Rutherford, S., Bucci, M., Wente, S. R., and Goldberg, M. W. (2004). Yeast nuclear pore complexes have a cytoplasmic ring and internal filaments. *J. Struct. Biol.* **145**(3), 272–288.

Lim, R. Y., and Fahrenkrog, B. (2006). The nuclear pore complex up close. *Curr. Opin. Cell Biol.* **18**(3), 342–347.

Loiodice, I., Alves, A., Rabut, G., van Overbeek, M., Ellenberg, J., Sibarita, J. B., and Doye, V. (2004). The entire Nup107-160 complex, including three new members, is targeted as one entity to kinetochores in mitosis. *Mol. Biol. Cell.* **15**(7), 3333–3344.

Orjalo, A. V., Arnaoutov, A., Shen, Z., Boyarchuk, Y., Zeitlin, S. G., Fontoura, B., Briggs, S., Dasso, M., and Forbes, D. J. (2006). The Nup107-160 nucleoporin complex is required for correct bipolar spindle assembly. *Mol. Biol. Cell.* **17**(9), 3806–3818.

Rasala, B. A., Orjalo, A. V., Shen, Z., Briggs, S., and Forbes, D. J. (2006). ELYS is a dual nucleoporin/kinetochore protein required for nuclear pore assembly and proper cell division. *Proc. Natl. Acad. Sci. USA* **103**(47), 17801–17806.

Salina, D., Enarson, P., Rattner, J. B., and Burke, B. (2003). Nup358 integrates nuclear envelope breakdown with kinetochore assembly. *J. Cell Biol.* **162**(6), 991–1001.

Schwartz, T. U. (2005). Modularity within the architecture of the nuclear pore complex. *Curr. Opin. Struct. Biol.* **15**(2), 221–226.

Walther, T. C., Alves, A., Pickersgill, H., Loiodice, I., Hetzer, M., Galy, V., Hulsmann, B. B., Kocher, T., Wilm, M., Allen, T., Mattaj, I. W., and Doye, V. (2003). The conserved Nup107-160 complex is critical for nuclear pore complex assembly. *Cell.* **113**(2), 195–206.

Zuccolo, M., Alves, A., Galy, V., Bolhy, S., Formstecher, E., Racine, V., Sibarita, J. B., Fukagawa, T., Shiekhattar, R., Yen, T., and Doye, V. (2007). The human Nup107-160 nuclear pore subcomplex contributes to proper kinetochore functions. *EMBO J.* **26**(7), 1853–1864.

CHAPTER 7

Immunolabeling for Scanning Electron Microscopy (SEM) and Field Emission SEM

Martin W. Goldberg

Department of Biological and Biomedical Sciences
University of Durham
Durham DH1 3LE
United Kingdom

Abstract
I. Introduction
   A. Forming an Image
   B. Preparing Samples
II. Methods
   A. Materials
   B. Equipment
III. Procedure
   A. Attach Sample to Silicon Chip
   B. Fixation
   C. Antibody Labeling
   D. Final Fixation
   E. Processing and Critical Point Drying
   F. Coating
   G. Imaging
   H. Image Processing
Conclusions
References

Abstract

Scanning electron microscopy (SEM) is a high resolution surface imaging technique. Many biological process and structures occur at surfaces and if antibodies are available, their components can be located within the surface structure. This is usually done in a similar way to immuno-fluorescence, using an unconjugated primary antibody followed by a tagged secondary antibody against the primary. In this case the tag is usually a colloidal gold particle instead of a fluorophore. Therefore it is quite straightforward to adapt an immuno-fluorescence procedure for SEM, as long as certain precautions are followed, as discussed here. Progressing from immuno-fluorescence, which essentially only indicates the position of a protein within the volume of a cell, to immuno-SEM, puts the labeling into the context of cellular structures. The principles and practices of sample preparation, labeling and imaging are described here.

I. Introduction

Scanning electron microscopes (SEMs) are used to look at surfaces, which can include cell or tissue surfaces, intracellular surfaces, isolated organelles or isolated molecular structures, such as viruses, protein, or nucleoprotein complexes, individual proteins and nucleic acids. With the increased availability of high-resolution SEMs with field emission electron sources (feSEMs), SEMs can now be used to image details of molecular organization at resolutions down to 0.4 nm, which is approaching the resolution of the transmission electron microscope (TEM) and depending on the sample and how it is prepared, the level of *information* can exceed, or more importantly, complement, the information from TEM.

One application of SEM in cell biology is the imaging of the external side of the cell which can be accessed and imaged using straightforward methods. Internal surfaces can be accessed by methods such as cell fractionation (e.g. Goldberg and Allen, 1992; Kluck *et al.*, 1999), detergent extraction (Long *et al.*, 2006), or physical fracturing (e.g., Arhel *et al.*, 2007; de Souza *et al.*, 2006).

To localize proteins on a membrane or other surface, immuno-gold labeling can be used. This is similar to immuno-gold labeling for TEM, and can be considered as a modification of the immuno-fluorescence that many cell biologists are familiar with (introductions to immuno-fluorescence and immuno-TEM can be found in McNamara (2006) and Ochs (2006) respectively). Antibodies still need to be characterized and labeling conditions determined. Sometimes these can be translated directly from immuno-fluorescence studies, but often modifications are required. In particular, as the structures themselves are being viewed, and not just the position of a label (as in immuno-fluorescence), it is essential that the structures as well as their antigenicity are well preserved. Therefore time and resources do need to be set aside for this process.

Successful immuno-SEM therefore depends on a good understanding of how to prepare samples in a way that exposes the surface, preserves its structure, and maximizes the accessibility and antigenicity of the target proteins. Although SEMs are straightforward to use, there are certain variables and choices to be made in setting up the microscope which affect the ability to visualize small gold particles and the labeled structures. For this reason this chapter starts with a short discussion on the principles of image formation in the SEM and how this affects decisions on the selection of imaging parameters. It then discusses methods for labeling and preparing samples.

A. Forming an Image

1. Generating the Signal

In the TEM, a "flood" beam of electrons is accelerated through a vacuum onto the specimen. The image is formed from electrons that are scattered as they pass through the sample. High resolution information is contained in electrons that do not transfer energy to the sample. These are known as elastically scattered electrons. Inelastically scattered electrons do transfer energy, change in wavelength and hence contribute to chromatic aberrations, which degrade the image. Because of the short mean free path of an electron in matter (the distance it travels between interactions), the likelihood of inelastic scattering is increased with increasing specimen thickness. Therefore in the TEM, specimen thickness is limited by the short distance an electron can travel before being deflected and by the energy it could loose. In a 120 kV TEM, specimen thickness is limited to around 100 nm. The mean free path can be increased by using higher accelerating voltages of the primary beam. Also aberrations correctors (Walther *et al.*, 2006) and electron energy filters (Stoffler *et al.*, 2003) can be applied but high-resolution information is still only possible in specimens of a few hundred nanometers thickness.

Cells range in size from about 1 µm (bacteria) up to over 1 mm (e.g., amphibian eggs), so generally the only way a cell biologist can take advantage of the high resolution afforded by the TEM is to cut the cell into sections or to isolate cellular components and study them separately. The cross-sectional views of TEM ultrathin sections are highly informative but can be difficult and time consuming to translate into 3D. Stereology methods have been developed to quantify three-dimensional aspects of cell structure from serial sections (Griffiths *et al.*, 2001). EM tomography is also emerging as a powerful method for determining 3D cell structure from thicker sections, small cells, and isolated organelles (e.g. Beck *et al.*, 2004), although specimen thickness is still limiting and it is yet to become a routine method. Another useful technique to determine the 3D structure of cellular surfaces is based on making thin metal replicas of the surface before removing the thick biological material (Heuser, 1980). However this method is time consuming and difficult, requiring specialist equipment and dedicated experts.

An alternative to replicas is the SEM, where surfaces can be imaged directly. There are two major differences between SEM and TEM imaging. First, instead of the flood beam of electrons used in the TEM, in the SEM the electrons are focused to a very fine diameter beam. This beam is then scanned across the sample in a rectangular raster much like the laser beam of a confocal microscope. As the electron beam interacts with each part of the sample, electrons are emitted and detected. The number of electrons emitted from a point that is interacting with the primary beam depends on the physical properties and topography of that point. As the beam is scanned across the sample and electrons are emitted, a picture is built up of the area being scanned. To increase the magnification in the SEM you simply reduce the size of the area being scanned. The resolution is dependent on the diameter of the beam used to scan the sample: the smaller the diameter, the higher the resolution. If the beam has a wide diameter it will potentially interact with more close objects at the same time than if the beam is narrow (Fig. 1). One of the advantages of the feSEM over conventional SEMs with tungsten filaments is that the field emission gun generates an intense electron beam (up to 1000 times brighter) that can be focused to a very fine point and still have enough primary electrons to generate a good signal from the sample.

The second difference between TEM and SEM imaging is the type of electrons that are collected to form the image. Regardless of the type of microscope, when high energy electrons emitted from the electron gun interact with the sample, they may be elastically or inelastically scattered as discussed above. They may also be elastically reflected, or backscattered (Fig. 2). This type of interaction forms an essential basis for most immuno-SEM and will be discussed here. However, for

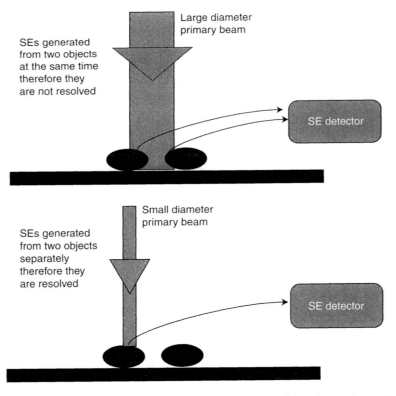

Fig. 1 Resolution in SEM is directly dependent on the diameter of the primary electron beam.

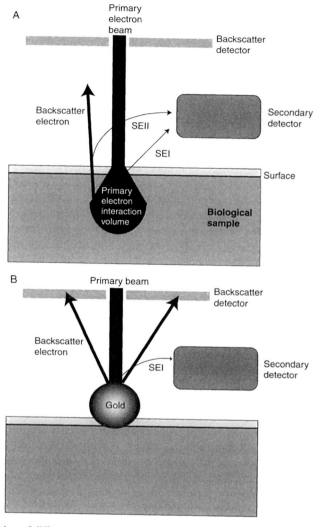

Fig. 2 Generation of different types of electrons in the SEM. (A) Due to scattering and energy loss, primary electrons exist in a "teardrop" shaped volume when they enter a thick sample. SEIs, which are generated as the primary electrons first enter the specimen surface, contain high resolution information. Backscatter electrons are primary electrons that re-emerge from the surface and generate further secondary electrons (SEIIs). (B) Biological samples do not generate many backscatter electrons, but dense material such as gold does, so that backscatter electron detection can be used to detect gold labeled antibodies.

imaging in the SEM, the most important interaction is the absorption of electron energy by the sample which results in the ejection of low energy electrons from the sample surface (SEIs, Fig. 2A). These low-energy electrons are then selectively attracted by a low voltage to the secondary electron detector. This low voltage does not attract the high-energy elastically backscattered electrons, very few of which will interact with the secondary detector, and therefore do not contribute to the secondary image.

2. Accelerating Voltage (kV)

Backscatter electrons do indirectly affect the secondary electron image which leads to an important consideration to be made for SEM and immuno-SEM: that of the accelerating voltage of the primary beam (kV). High energy primary electrons interact with a certain volume of the sample as they are scattered while passing through it (Fig. 2A). A higher accelerating voltage produces higher energy electrons that interact deeper into the sample. As the energy of the primary electrons is directly related to the energy of the backscatter electrons that are reemerging from the specimen surface, the depth from which they can come is related to the accelerating voltage. At high accelerating voltages (i.e., 10–30 kV), backscatter electrons are generated from relatively deep in the sample, whereas at low voltages (0.5–3 kV), they are only generated near the surface. Backscatter electrons also transfer energy to the sample and cause the ejection of secondary electrons. If this occurs near the surface, these are emitted and detected by the SE detector. However, the information in these SEII electrons is not primarily from the surface but from subsurface structures. This can be useful information but generally it confuses the interpretation of the image, adds to noise and reduces resolution. The reduction in resolution is because the backscatter electrons can emerge some distance from the primary beam effectively increasing the diameter that is emitting secondary electrons at any one time (see Fig. 2A). At high kVs cell membranes appear transparent and details are difficult to see, whereas at low kVs they appear as solid surfaces. It should be noted however that for some samples this does not fully hold. For instance thin specimens such as isolated membranes (Goldberg and Allen, 1992) (especially if they are on thin supports like carbon films) do not generate large amounts of backscattered electrons, and hence few SEIIs. Therefore high signal to noise ratios are achieved using high kVs, basically because most elastically and inelastically scatter primary electrons pass straight through the specimen and there is no large interaction volume in which to generate SEIIs, which would appear as noise. Therefore high kVs are generally preferable for generating high-resolution images if the specimen is thin.

3. High kVs are Required for Immuno-SEM

These considerations have led some to advocate only the use of low kVs for imaging biological samples (Pawley and Erlandsen, 1989). However, particularly for immuno-SEM, this is not always possible or desirable.

a. SEMs Resolve Better When High Accelerating Voltages are Used

First, instrument resolution is reduced at low kVs. How important this is depends on the microscope. Some modern feSEMs are designed to operate with high resolution at low kVs (in the order of 2–4 nm at 1 kV). This must be weighed against the loss of resolution at high kV due to beam interaction, which is highly specimen dependent. This stresses the need to understand how your sample is likely to interact with the electron beam so that you can select an appropriate voltage. Alternatively, simply experiment with different accelerating voltages.

b. Immuno-Gold Particles are Detected Using High Accelerating Voltages

Second, colloidal gold particles used to tag antibodies are detected using the backscatter detectors which only function at high kVs (Fig. 2B). Primary electrons are more likely to be backscattered by dense objects with high atomic numbers (Z number). Hence colloidal gold particles backscatter strongly, unlike the surrounding biological material. Therefore a backscatter image of an immuno-gold labeled sample will appear as a weak, noisy image of the sample with bright white dots where the gold particles are (Fig. 4B). A secondary image is then also obtained of the same area (Fig. 4A) and then the two images are combined to determine the position of the gold particles in relation to the structures. The backscatter image should give an unequivocal identification of the gold particles because it can differentiate objects of different atomic numbers and the gold particles have a much higher atomic number than most of the surrounding material. Although it is sometimes possible to identify immuno-gold particles in the secondary image, there is no real "Z-contrast," so gold particles can be mistaken for other similar sized particles or visa versa. Therefore backscatter detection is the method of choice for locating immuno-gold particles. However, backscatter detectors work by excluding low energy (usually secondary) electrons from the detector. Therefore if the primary beam is of low energy (low kV), low energy backscatter electrons will be produced which cannot be detected. Some backscatter detectors can work down to about 5 kV, but sensitivity is poor so only large gold particles can be detected. However most work optimally in the 10–20 kV range and even 5 kV is not really "low kV."

4. Charging

Another consideration is charging. The sample is bombarded by electrons. In the TEM, most of these pass through the sample, but in thick SEM samples that are nonconductive they cause severe imaging artifacts. Fixation methods that introduce heavy metals into the sample, such as OsO_4 make it more conductive and allow the excess electrons to escape to earth. This effect can be increased by using chemicals such as tannic acid (see Hayat, 2000) and thiocarbohydrazide (Harrison *et al.*, 1982), which increase the binding of osmium, but it is important to consider the effect that such chemicals might have on the structures of interest. For instance we have found that although tannic acid is beneficial for preserving protein filaments, it can cause the vesiculation of endoplasmic reticulum membrane tubules.

The substrate on which samples are deposited is important and must be conductive to prevent charging but it also must not generate significant secondary or backscatter electrons which will contribute to background noise. For the former reason glass coverslips are not suitable. Our preferred substrate is the silicon chip which is electrically conductive and on which cells can usually be grown in exactly the same way as on glass, but with the limitation that they are not transparent.

a. Metal Coating—Reduction of Charging and Enhancement of Secondary Signal

The most effective way to reduce charging is to coat with a thin metal film, which also generates secondary electrons and increases the signal. Metal coatings, such as gold, however, mask the signal from the colloidal gold particles attached to the antibody. Metal coating is generally done in a high vacuum by magnetron sputtering (Allen *et al.*, 1998), although other sputtering methods are also effective (Walther *et al.*, 1995). Conventional SEM coatings of sputtered gold are unsuitable both for high resolution SEM and for immuno-gold labeling. Sputtered gold has a large particle size that is resolvable even by fairly modest modern SEMs and therefore obscures surface details. Platinum has a much finer grain size and is quite effective at reducing charging and generating secondary electrons. It is often considered a metal of choice, especially for low kV work. However, platinum (and of course gold) have a very similar atomic number to colloidal gold and therefore there is little Z-contrast between the metal coating and the immuno-gold particles. This makes it difficult or impossible to pick out gold particles in the backscatter image of a sample coated with a high atomic number metal.

b. Chromium—An Effective Coating for Immuno-SEM

An important development in this respect was the use of chromium (Allen *et al.*, 1998; Peters, 1986). If sputtered in the right conditions, chromium is conductive (to reduce charging) and has a small grain size (so detail is not obscured). It also has a relatively low atomic number of 24 compared to 79 for gold, so there is good Z-contrast in the backscatter detector between the colloidal gold and the coated surface being imaged. Chromium is therefore currently the metal of choice for immuno-SEM.

Sadly, this does not come without a cost. Because of the low atomic number the secondary signal generated is less than for platinum. It is also not such a good conductor. Therefore it is less effective at reducing charging, especially on bulky specimens. Chromium is ideal for thin specimens such as isolated macromolecules, organelles, and even whole cells, but multicellular tissues and more bulky specimens can be difficult to image with chromium coats. At this point compromises are almost inevitable, as is often the case in SEM. With bulky specimens, charging can be reduced by using lower accelerating voltages, but as discussed above, this is limited by the backscatter detection in immuno-SEM studies. If charging affects are not too severe, you can simply live with them and take them into account when interpreting the images.

Further, related limitations of chromium are associated with its oxidation producing a poor coating. This leads to some important practical considerations. Chromium coating must be done in a high vacuum system, initially pumped to at least 5×10^{-7} mbar, with all oxygen and water eliminated, before introducing ultra pure argon gas. As the surface of the target immediately oxidizes on contact with air, this oxide layer must be first removed by sputtering onto a shutter before exposing the specimen. Finally, as soon as the coated specimens are removed from the vacuum chamber they will start to oxidize and so must be looked at immediately and generally cannot be stored for long periods (a few hours is OK, a few weeks is not, generally they deteriorate over a day or two). It is possible to recoat specimens at a later stage. This will increase the secondary electron signal of old specimens but increase the coating thickness so that resolution of detail is decreased.

B. Preparing Samples

There are several requirements for successful immuno-SEM:

1. The surface of interest must be exposed.
2. The structure must be well preserved.
3. Epitopes must be accessible to the antibodies.
4. Epitopes must maintain antigenicity.

How the surface of interest is exposed is dependent on the specimen and the question being asked. Therefore it is beyond the scope and competence of this chapter to go beyond general advice. The preservation of structure for EM is also a complex subject and the reader is directed to books such as Hayat (2000) for discussion on the chemistry of fixation.

1. Fixation

Fixation is one point where the transition from a light microscopy project to the SEM can be frustrating. First, it is more-or-less essential to use aldehyde fixation for SEM. A large proportion of a cell is water which needs to be removed for EM (with the exception of cryo-EM which is not covered here). In TEM thin sections samples are chemically fixed and then the water is replaced by an embedding resin. For SEM there is no embedding, so samples need to be made particularly tough in order to dry them without unduly altering the structure. Therefore fixation needs to be strong. This generally means using high concentrations of glutaraldehyde (e.g. 2%) and possibly other stabilizing chemicals such as tannic acid to preserve protein structures and OsO_4 to preserve membranes. Such a protocol is incompatible with maintaining antigenicity for almost all antibodies. Even glutaraldehyde fixation prevents labeling with the majority of antibodies. It is therefore necessary to stabilize the specimen with a relatively mild fix just to maintain it through the labeling procedure, before fixing again with stronger fixes. This generally means fixing with 2–6% formaldehyde in a suitable buffer. This should be made fresh from paraformaldehyde powder. Do not use premade commercial formaldehyde or formalin solutions which often contain methanol and could cause structural artifacts and will extract membranes. Even preparations claiming to be for EM can be unsuitable. Especially do not use organic solvent fixation methods employing chemicals such as methanol or acetone, which are commonly used for light microscopy. These may maintain antigenicity in some cases where it is destroyed by aldehyde fixation but there will be little point in using an electron microscope to look at the resulting structure. Generally if aldehyde prefixation cannot be used with a particular antibody it will be difficult to do immuno-SEM. It may also be possible to do the labeling on unfixed material then fix afterwards, but only if it is known that the antibody will not disrupt the structure or organization. Therefore careful controls are required.

2. Labeling

Having prefixed the sample, it is important to block any unreacted aldehyde groups which might bind to the antibodies. This is done by incubation with glycine or NH_4Cl.

The next steps are generally similar to immuno-fluorescence protocols except antibody-gold conjugates are used instead of fluorescent ones and usually antibodies, especially primary antibodies, are used at higher concentrations. As with any immunolabeling method it is usually necessary to find empirically the most appropriate antibody concentrations and blocking agents.

a. Gold Particle Size

Another important consideration is to choose the most appropriate gold label attached to the secondary antibody. It is generally considered that for maximum sensitivity you should use the smallest gold particles. Also smaller gold particles will obscure less of your sample and be more closely located to the epitope. However, even if it is possible to resolve a 5-nm gold particle in a particular SEM, it may not be the best choice, because magnifications of around $100,000\times$ are required just to see it. Therefore an overview of the distribution on the surface is not obtained. It will only show its localization on fine structural details. Therefore it is important to consider the question being addressed. It may be worth sacrificing some sensitivity for visibility and convenience.

For feSEM immuno-gold labeling at high resolution, we start with a high-quality 10 nm colloidal gold tagged secondary antibody (such as from Amersham which we find very reliable but other manufacturers also produce good products). This can be observed at moderate magnification ($20-30,000\times$) and gives a good overview of the level of labeling and distribution. For some conventional SEMs were it is difficult to resolve a 10 nm gold particle it may be necessary to use larger (15–20 nm) gold, but this will reduce the sensitivity. Therefore it may be better still to add a step where extra layers of silver or gold are added to the surface of a small gold particle, after labeling, to enhance its size. Such an approach may also be appropriate for feSEM studies where 1 nm colloidal gold can be used and "enhanced" (to 2–10 nm) to make it visible. Companies such as Nanoprobes Inc. sell convenient kits for this. Another useful reagent is Nanogold (Nanoprobes Inc.) which is a 1.4 nm gold compound which can be covalently bound to the antibody. Colloidal gold particles are attached to antibodies by absorption and charge interactions, whereas Nanogold is more stably covalently attached. The small size increases antibody penetration of the sample and therefore can increase the level of labeling. The particle size can then be increased by silver or gold enhancement to a size appropriate for your instrument and application. It has been possible to directly image 1 nm gold particles in a feSEM (Hermann *et al.*, 1996), but generally it is very difficult on any samples with much bulk or topography and so usually it is necessary to silver/gold enhance.

Best still, Nanogold or colloidal particles can be attached directly to a primary antibody for direct labeling. The advantage of indirect labeling (a two step procedure using a secondary antibody that recognizes the primary) in immuno-fluorescence microscopy is that the signal is enhanced because more than one secondary antibody can bind to each primary. This is, however, a disadvantage in immuno-SEM where the extra antibody molecules and gold particles can obscure the object of interest and add nothing to the detection limit. Direct labeling is often, however, impractical as relatively large quantities of antibody are required for the conjugation reactions. The highest precision labeling will though, in principle, be achieved by direct labeling using fab fragments attached to Nanogold or 1 nm gold particles.

In some circumstances, where the specimen is particularly well defined and its structure consistent and well understood, it may be possible to label with just the antibody or fab fragment and no gold tag (Hermann *et al.*, 1996). This is because with a high-resolution feSEM the antibody protein, itself, can be directly recognized. However, in complex specimens this may not be practical or even possible.

3. Sample Processing for SEM

After labeling samples are fixed more strongly, generally with glutaraldehyde and osmium tetroxide and possibly other fixatives to optimally preserve the structure and fix the antibodies in place. They are then dehydrated using an ethanol series and critical point dried (CPD). Ethanol (100%) used in the CPD should be dried with molecular sieve. The CO_2 should be high quality and dry. Some labs dry the CO_2 by passing it through a filter (Tousimis LCO2 filter), but this may not be essential. However, if preservation is poor it may be worth trying this as it allows for contaminated CO_2. Most commercial CPDs are usable but the bench CPDs such as the Bal-Tec CPD 030, where the temperature and flow of gas in and out of the unit can be controlled are particularly suitable. In particular, letting out the gas at the end should be done slowly and with great care (this should take 10–15 min). A gas flow meter will allow this to be controlled and standardized. As immuno-SEM is a multistep procedure, the more that is standardized the better, allowing effective troubleshooting.

4. Metal Coating

As discussed above magnetron sputtered chromium is the coating of choice for high-resolution immuno-SEM. Several commercial systems have been successfully used for chromium coating (e.g. Edwards, Cressington, and Denton). Crucial to this seems to be pulling a vacuum of at least 5×10^{-7} mbar and the elimination of oxygen and water from the system. The use of cryo-pumps and liquid nitrogen cooled cold traps may be effective for this. The system should be kept scrupulously clean, only using gloves to handle internal components and specimens. It should also be remembered that the chromium target will oxidize immediately upon exposure to air. Therefore it is essential to presputter onto a shutter before exposing the specimen.

The thickness of the coating is important. Thicker coatings will give more signal and reduce charging more, but will cover up detail (hence reduce resolution) and can obscure gold particles. For immuno-SEM it is necessary to find the correct thickness that allows you to see the gold particles as well as the specimen. To visualize 10 nm colloidal gold particles by feSEM a thickness of about 1–2 nm is usually suitable. Any thicker and the gold becomes difficult to detect. For reasons discussed above, larger gold particles may necessarily be used, in which case thicker coatings may be used, but this will be at the expense of resolution of structural detail. Because chromium oxidizes quickly and dry specimens can re-hydrate, they should be examined as quickly as possible. Storage in a vacuum can help prolong their useful life, but not perfectly.

5. Imaging

This is highly dependent on the sample and microscope being used. Accelerating voltage should be selected as discussed above. Ten kilovolts is used to visualize 5–10 nm gold on the Hitachi S5200 feSEM (Cotter et al., 2007), but a voltage of 25–30 kV seems more appropriate on the Topcon DS130F (Rutherford et al., 1997). The type of electron source (tungsten filament vs. field emission gun) and the type of backscatter detector are important considerations here.

Some SEMs allow acquisition of secondary and backscatter images simultaneously. This is a distinct advantage as the area scanned will then be identical for each type of image. If images have to be obtained consecutively, any drift or other image distortions can make aligning the two images difficult. Image alignment is done easily in commonly available image software such as Adobe Photoshop as described below.

II. Methods

A. Materials

1. Fixation

Stock solutions

1. 1 M sodium cacodylate, pH 7.4—note highly toxic, wear double gloves
2. 1 M Pipes, pH 6.8

3. 1 M NaCl
4. 1 M KCl
5. 1 M $MgCl_2$
6. 25 or 50% glutaraldehyde (EM grade e.g. Agar or Ted Pellar)
7. 10% paraformaldehyde

A stock solution of 10% paraformaldehyde must be made fresh each day from paraformaldehyde powder. This should be weighed out in a fume hood because it is toxic and very light, so tends to become airborne easily. Weigh 1 g paraformaldehyde into a sealable tube, make up to 10 ml and add one drop of 1 N NaOH. Close lid and place in a 60 °C water bath. Agitate every now and then until the powder dissolves.

8. Double distilled or distilled/deionised water
9. Ethanol stored with molecular sieve
10. 1% osmium

Osmium tetroxide is a highly toxic heavy metal solid salt that also forms a vapor at room temperature. Extreme caution should therefore be taken when handling this material and a full risk assessment carried out. All procedures must be carried out in a fume hood and gloves worn at all times. Vials (0.1 g) of osmium tetroxide are broken open and the whole vial placed in a 100 ml glass reagent bottle. Ten milliliters of 0.2 M sodium cacodylate is added. The lid is put on and sealed with parafilm. Gently shake the bottle to expel any air from the vial, then either sonicate until the osmium tetroxide crystals dissolve or leave in the fume hood overnight and agitate in the morning to check that the crystals have completely dissolved.

2. Buffers

1. 5× membrane fix buffer: 400 mM Pipes, pH 6.8, 5 mM $MgCl_2$, 750 mM sucrose
2. 0.2 M sodium cacodylate, pH 7.4
3. PBS: tablets obtainable from Sigma are a convenient source or make up the following: 137 mM NaCl, 2.7 mM KCl, 10 mM Na_2PO_4, 2 mM KH_2PO_4

3. PreFixes

1. PBS prefix (for cells and tissue):

Add 4 ml of 10% paraformaldehyde to PBS powder or tablets before making up to 10 ml

2. Membrane prefix (for intracellular structure):

1 ml 5× membrane fix buffer
1 ml 10% paraformaldehyde
3 ml water

4. Preservation Fixes

   3. Karnovski's fix (for cells and tissue): 0.2 M sodium cacodylate pH 7.4, 2% paraformaldehyde, 2.5% glutaraldehyde

   To make 5 ml add the following stock solutions:

   1 ml 1 M sodium cacodylate, pH 7.4
   1 ml 10% paraformaldehyde
   250 μl of 50% glutaraldehyde (500 μl of 25%)
   Water to 5 ml

   4. Membrane fix (for intracellular structures):

   1 ml 5× membrane fix buffer
   1 ml 10% paraformaldehyde
   25 μl glutaraldehyde (from 50% stock)
   2.975 ml water

   5 Tannic acid fix (for protein filaments):

   Always make fresh.
   1M Hepes pH 7.4, 0.2% tannic acid (Sigma-Aldrich), 2% glutaraldehyde

5. Post Fixes

   6. 1% osmium tetroxide in 0.2 M sodium cacodylate (after Karnovski or membrane fix)
   7. 0.1% osmium in 0.2 M sodium cacodylate (after tannic acid fix)

B. Equipment

1. Critical point drier
2. High vacuum sputter coater with turbomolecular pump and cold trap or cryo-pump, chromium target, shutter between target and specimen, and rotating specimen table.
3. Fine tweezers
4. Petri dishes

III. Procedure

A. Attach Sample to Silicon Chip

Cells can be grown on silicon chips (Agar Scientific Ltd.), which have similar chemical properties to glass coverslips. Therefore for adherent culture cells clean, sterilized chips can simply be placed into culture dishes and the cells will usually grow on them. The chips are not transparent; therefore to indirectly monitor the density of cells you can put a glass coverslip next to the chips.

Nonadherent cells or samples such as isolated organelles can usually be attached to the silicon chip by centrifugation. If your sample will stick to a coverslip it will usually adhere to a silicon chip. Modified Eppendorf tubes can be used as chambers for spinning samples onto 5 mm square chips (see Fig. 3 for construction). Samples are diluted into 1 ml of suitable buffer, pippetted into the chamber which is placed in a swing out rotor and spun at a suitable speed (just fast enough to pellet the object without distorting it—this may have to be determined empirically). The tube is then broken open and the chip removed and placed into a Petri dish containing fix.

The choice of dilution buffer will depend on the specimen. Cells can be spun in PBS. Organelles will require a suitable intracellular buffer. For SEM it is important to consider that since you are just looking at the surface, it is vital that the surface of interest is "clean." For instance, when examining nuclei present in cell extracts containing cytosol, other organelles and cytoskeletal components, samples are spun through a 10% sucrose cushion in an intracellular buffer in order to prevent other cellular components from adhering to the nuclear surface (Goldberg *et al.*, 1997).

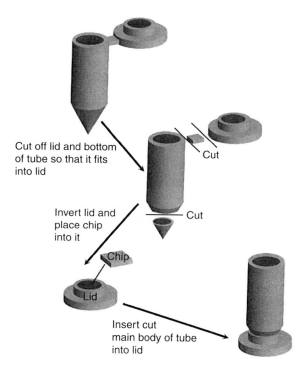

Fig. 3 Preparation of chamber from a 1.5 ml microfuge tube for centrifuging specimens onto silicon chips.

Adherence can be increased by treating the chip with poly-L-lysine (Sigma). To do this 20 µl of a 1 mg/ml solution is place on the chip for 30 min, then washed off by rinsing twice in water. Note however that placing unfixed samples, such as culture cells, onto poly-L-lysine can lead to distortion.

B. Fixation

With the sample attached, the chip is then placed in a Petri dish containing a prefix. Our usual prefix is 2–4% paraformaldehyde in a suitable buffer.

For culture cells, we usually fix in PBS containing 2–4% paraformaldehyde. Small amounts of glutaraldehyde (0.01–0.5%) can also be added if it is compatible with your antibody.

Make up PBS from components or tablets (Sigma), including 2–4% paraformaldehyde and pour into Petri dish. Chips with cells either spun onto or grown on their surface are rinsed with PBS then placed in the fix for 10 min.

Organelles and other cellular components, such as protein complexes, cytoskeletal filaments, etc., should be fixed in a buffer formulated to best preserve the structure of interest and containing paraformaldehyde.

C. Antibody Labeling

Washes and buffer incubations are done in small Petri dishes by transferring chips with fine tweezers. Antibody incubations are done in a simple "wet chamber" to prevent evaporation of the small antibody volume. This is a Petri dish lid with damp filter paper placed in it. A glass slide or Parafilm is placed on the filter paper and chips are placed on this. The Petri dish base is then used as a lid.

1. Wash in buffer (e.g. PBS or appropriate buffer) 2×2 min in Petri dishes.
2. Place into 100 mM glycine in the same buffer 10 min. This blocks any unreacted aldehyde.
3. Wash in buffer 2×2 min.
4. Place in blocking solution for 1 h.

This helps block nonspecific antibody binding. 1% fish skin gelatine (Sigma) is routinely used for blocking, but other blocking solutions are sometimes more appropriate for some antibodies and may have to be tested empirically or using information from previous experiments (such as immuno-fluorescence). Normal serum from the species used to generate the primary antibody is often used, as is bovine serum albumin (BSA). Occasionally, for particularly weak antibodies, the blocking step is omitted. Then, of course, careful controls become especially important.

5. Rinse in buffer.
6. Transfer to a dry piece of filter paper to dry the back of the chip.

It is important that the back of the chip is dry, but that the front (with the sample) does not dry out at all. Any air drying of the sample will destroy the structure.

7. Place chip on slide in wet chamber.
8. Pipette 10 µl of primary antibody onto the sample, place the lid on the wet chamber and incubate at room temperature for 1 h.

Ensure that the antibody solution remains as a drop on top of the chip and does not run down the sides onto the slide. If it does, the back of the chip was not completely dry.

The antibody dilution generally needs to be determined empirically but previous immuno-fluorescence studies can be a guide. Usually the antibody needs to be 2–10× more concentrated for SEM than for fluorescence.

It is good practice to dilute the antibody in blocking solution, but this is not always necessary if the proper controls are done and often reduces the level of labeling. If the signal is weak it can help to extend the incubation time. Usually this is done overnight in the fridge.

9. Wash chips 3 × 5 min, then 1 × 15 min in buffer.
10. Dry the back of the chip on filter paper as above and place in the wet chamber.
11. Pipette on 10 µl secondary gold tagged antibody and incubate for 1 h.

This can be diluted in blocking solution but straight buffer is usually OK. The optimum antibody concentration may again need to be determined. We usually start with dilutions of 1:50 to 1:100 for commercial preparations, but this is only a rough guide.

12. Wash chips 3 × 5 min, then 1 × 15 min in buffer.

D. Final Fixation

Place into Petri dish containing final fix.

For culture cells or tissue this can be standard EM fixes such as Karnovsky's (see Fixes for recipe). This may be suitable for some organelles and macromolecular complexes but not always. For instance, high concentrations of glutaraldehyde can cause artifactual fragmentation of endoplasmic reticulum membranes, hence the use of Membrane Fix (see Fixes). Tannic acid, on the other hand, is useful for stabilizing protein filaments, but is not ideal for membrane preservation.

E. Processing and Critical Point Drying

1. Dehydrate samples through an ethanol series of 50, 70, 90, 95, 100, 100% 2–10 min each depending on the specimen thickness. This is done in glass Petri dishes, transferring the samples individually with tweezers.

Cell monolayers and isolated structures only require 2 min or less, whereas thicker tissue samples require longer.

2. Fill CPD with ethanol (dried with molecular sieve, VWR).
3. Transfer samples quickly into CPD chamber. Do not allow any air drying to occur.
4. Cool chamber.
5. Exchange ethanol for liquid CO_2 under pressure (at least 10 exchanges).
6. Leave for 30 min then repeat exchange.
7. Heat to 40 °C during which CO_2 passes through its critical point.
8. Slowly release gaseous CO_2.

Samples can be stored at this stage for a few days in a vacuum. It is not advisable to store chromium coated samples which should be coated just prior to microscopy.

F. Coating

Place samples in the coating unit immediately and pump a vacuum of at least 5×10^{-7} mbar using an additional liquid nitrogen cooled cold trap. Introduce pure argon gas (99.999% purity) and sputter onto shutter. Open shutter and deposit 1–5 nm chromium.

G. Imaging

1. Find the object to be imaged using secondary electron imaging.

In order to detect small gold particles it is essential that all alignments and astigmatism adjustments are done perfectly.

2. Insert backscatter detector.
3. Use accelerating voltages of 10–30 kV so that backscatter electrons have enough energy to be detected. Experiment with different kVs to find the optimum for the SEM being used, as well as for the sample and gold size.
4. Use a high beam current (i.e. 20 μA), and the spot size should not be too small as you want to maximize the number of electrons hitting the gold particles in order to detect them. Although some resolution is sacrificed such compromises are often necessary. Well fixed samples that have been coated with chromium are quite tough at this stage and can generally resist quite high electron doses. Sometimes however less resistant samples can become damaged and "contaminated" as the electron beam scans an area of interest. This has the appearance of "melted cheese" as the detail disappears and fills in. It can be seen most obviously by focusing on an area at high magnification then going to a lower magnification when you will see a rectangular patch. If this happens, lower beam currents and smaller spot sizes can help, but the best advice is to take the picture as quickly as

possible and move to a new area. Alignments, stigmation, and focusing can be done on adjacent areas and/or at low magnifications, then you zoom in to your area of interest, or a randomly selected area, and quickly press the acquisition button. This may lead to many failed images, but could get you a few good images from a difficult sample.

5. If possible, acquire secondary and backscatter images simultaneously.

H. Image Processing

In order to see the position of the gold particles on the structure, it is best to superimpose the backscatter image of the gold particles onto the secondary image. However, backscatter signal from other objects than the gold particles can degrade the secondary image. In the following procedure this is circumvented by using the backscatter image as a template to mark the position of the gold particles. Fig. 4 shows an example.

1. Open both secondary and backscatter images from the same area in Adobe Photoshop.

2. If the images were acquired simultaneously,

3. Select the whole backscatter image (Click on the image and press Ctrl+A).

4. Copy the image to the clipboard (Ctrl+C).

5. Click the secondary image and paste the backscatter image as a new layer on top (Ctrl+V).

6. Open a third new layer on top (Layer → New → Layer → click OK).

7. Use the Paintbrush tool to mark the position of each gold particle with a dot in the third layer. Ideally gold particles are clearly defined dots of a uniform size. In practice this is not always the case. As shown in Fig. 4C some particles are small and ill defined (small arrows). These may be gold particles that are beneath the surface being visualized which reduces the signal, but it is difficult to be sure and these are usually disregarded (Fig. 4D). The large particle (Fig. 4C, large arrow) also may be an aggregate of gold particles, but as it is uncertain, this too is disregarded.

8. Finally delete the backscatter image layer leaving just the position of the gold particles marked on the secondary image.

If the images were obtained one after the other they may need to be aligned before marking the position of the gold (due to drift). This can also be done in Photoshop. Repeat (Steps 1–5) above and with the backscatter image layer selected, click Layer → Layer Style → Blending Options. Then in the General Blending box select Overlay from the drop down menu. This will give you a transparent backscatter image overlaid onto the secondary image where the gold particles can be seen as bright dots. Select the Move Tool from the Tools window and drag the backscatter layer around until the bright gold dots align with the corresponding dots in the secondary image where they can be seen. Then do Steps 6–7.

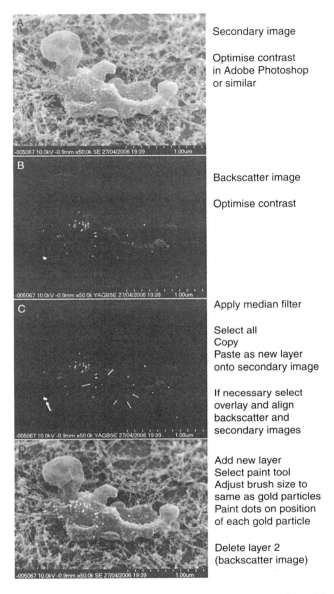

Fig. 4 Process of superimposing information from backscatter detector (gold position) onto secondary information (high resolution structures). See figure and text for details. (A) Secondary image. (B) Backscatter image. (C) Filtered backscatter image. (D) Secondary image with gold positions marked and backscatter image removed.

Conclusions

SEM is a powerful method to determine many levels of tissue, cellular and molecular structure. Modern SEMs are capable of resolving fine details of any biological surface that can be exposed and then using the methods described here constituent proteins can be located within a structure, often generating surprising and novel information.

References

Allen, T. D., Rutherford, S. A., Bennion, G. R., Wiese, C., Riepert, S., Kiseleva, E., and Goldberg, M. W. (1998). Three-dimensional surface structure analysis of the nucleus. *Methods Cell Biol.* **53,** 125–138.

Arhel, N. J., Souquere-Besse, S., Munier, S., Souque, P., Guadagnini, S., Rutherford, S., Prevost, M. C., Allen, T. D., and Charneau, P. (2007). HIV-1 DNA Flap formation promotes uncoating of the preintegration complex at the nuclear pore. *EMBO J.* **26,** 3025–3037.

Beck, M., Forster, F., Ecke, M., Plitzko, J. M., Melchior, F., Gerisch, G., Baumeister, W., and Medalia, O. (2004). Nuclear pore complex structure and dynamics revealed by cryoelectron tomography. *Science* **306,** 1387–1390.

Cotter, L., Allen, T. D., Kiseleva, E., and Goldberg, M. W. (2007). Nuclear membrane disassembly and rupture. *J. Mol. Biol.* **369,** 683–695.

de Souza, W., Campanati, L., and Attias, M. (2006). Strategies and results of field emission scanning electron microscopy (FE-SEM) in the study of parasitic protozoa. *Micron.* **39,** 77-87.

Goldberg, M. W., and Allen, T. D. (1992). High resolution scanning electron microscopy of the nuclear envelope: Demonstration of a new, regular, fibrous lattice attached to the baskets of the nucleoplasmic face of the nuclear pores. *J. Cell Biol.* **119,** 1429–1440.

Goldberg, M. W., Wiese, C. W., Allen, T. D., and Wilson, K. L. (1997). Dimples, pores, star rings and thin rings on growing nuclear envelopes: Evidence for intermediates in nuclear pore complex assembly. *J. Cell Sci.* **110,** 409–420.

Griffiths, G., Lucocq, J. M., and Mayhew, T. M. (2001). Electron microscopy applications for quantitative cellular microbiology. *Cell Microbiol.* **3,** 659–668.

Harrison, C. J., Allen, T. D., Britch, M., and Harris, R. (1982). High-resolution scanning electron microscopy of human metaphase chromosomes. *J. Cell Sci.* **56,** 409–422.

Hayat, M. A. (2000). "Principles and Techniques of Electron Microscopy: Biological Applications." Cambridge University Press, UK.

Hermann, R., Walther, P., and Muller, M. (1996). Immunogold labeling in scanning electron microscopy. *Histochem. Cell Biol.* **106,** 31–39.

Heuser, J (1980). Three-dimensional visualization of coated vesicle formation in fibroblasts. *J. Cell Biol.* **84,** 560–583.

Kluck, R. M., Esposti, M. D., Perkins, G., Renken, C., Kuwana, T., Bossy-Wetzel, E., Goldberg, M., Allen, T., Barber, M. J., Green, D. R., and Newmeyer, D. D. (1999). The pro-apoptotic proteins, Bid and Bax, cause a limited permeabilization of the mitochondrial outer membrane that is enhanced by cytosol. *J. Cell Biol.* **147,** 809–822.

Long, H. A., Boczonadi, V., McInroy, L., Goldberg, M., and Maatta, A. (2006). Periplakin-dependent re-organisation of keratin cytoskeleton and loss of collective migration in keratin-8-downregulated epithelial sheets. *J. Cell Sci.* **119,** 5147–5159.

McNamara, G. (2006). Introduction to immunofluorescence microscopy. *In* "Basic Methods in Microscopy" (D. L. Spector, and R. D. Goldman, eds.), pp. 343–358. Cold Spring Harbor Laboratory Press.

Ochs, R. (2006). Immunoelectron microscopy. *In* "Basic Methods in Microscopy" (D. L. Spector, and R. D. Goldman, eds.), pp. 145–154. Cold Spring Harbor Laboratory Press.

Pawley, J. B., and Erlandsen, S. L. (1989). The case for low voltage high resolution scanning electron microscopy of biological samples. *Scanning Microsc. Suppl.* **3,** 163–178.

Peters, K. R. (1986). Rationale for the application of thin, continuous metal films in high magnification electron microscopy. *J. Microsc.* **142,** 25–34.

Rutherford, S. A., Goldberg, M. W., and Allen, T. D. (1997). Three-dimensional visualisation of the route of protein import: The role of nuclear pore complex structures. *Exp. Cell Res.* **232,** 146–160.

Stoffler, D., Feja, B., Fahrenkrog, B., Walz, J., Typke, D., and Aebi, U. (2003). Cryo-electron tomography provides novel insights into nuclear pore architecture: Implications for nucleocytoplasmic transport. *J. Mol. Biol.* **328,** 119–130.

Walther, P., Wehrli, E., Hermann, R., and Muller, M. (1995). Double-layer coating for high-resolution low-temperature scanning electron microscopy. *J. Microsc.* **179**(Pt 3), 229–237.

Walther, T., Quandt, E., Stegmann, H., Thesen, A., and Benner, G. (2006). First experimental test of a new monochromated and aberration-corrected 200 kV field-emission scanning transmission electron microscope. *Ultramicroscopy* **106,** 963–969.

CHAPTER 8

Immunogold Labeling of Thawed Cryosections

Peter J. Peters and Jason Pierson

Department Tumor Biology: H4
The Netherlands Cancer Institute
Amsterdam, The Netherlands 1066 CX

    I. Introduction
   II. Preparation of Carbon- and Formvar-Coated Copper Girds
  III. Aldehyde Fixation of Cells for Cryoimmuno Labeling
  IV. Embedding Samples for Cryoimmuno Labeling
   V. Cryosectioning for Cryoimmuno Labeling
  VI. Immunogold Labeling
 VII. Reagents and Solutions
      A. BSA Stock Solution, 10% (W/V)
      B. Formaldehyde (FA) Fixative, 2×
      C. Formaldehyde (FA) Stock Solution, 16% (W/V)
      D. Formaldehyde/Glutaraldehyde (FA/GA) Fixative, 2×
      E. Paraformaldehyde (PFA) Fixative, 2×
      F. Gelatin, 1%, 2%, 5%, and 12% (W/V)
      G. Methyl Cellulose/Uranyl Acetate
      H. PHEM Buffer
      I. Section Retrieval Solution
      J. Storage Solution
      K. Uranyl Acetate, 4% (V/V)
      L. Uranyl Oxalate Solution
VIII. Future Outlooks
     References

I. Introduction

Immunogold labeling is an established approach to localize gene products in their cellular environment. An antibody, coupled to an electron-dense marker, is linked to an antigen present on the protein of interest within the cryosection. In 1959, Singer devised a way to couple an electron-dense protein, ferritin, to an antibody (Singer, 1959). The recognizable marker could be seen with little ambiguity using the electron microscope. Enzyme-labeled tags were later developed by Nakane and Pierce (1967), and Sternberger subsequently developed a technique that involved the peroxidase–antiperoxidase tag to improve sensitivity and resolution with enzyme tags (Sternberger *et al.*, 1970). Later, the use of colloidal gold markers became a popular method (Faulk and Taylor, 1971). Using small gold markers, many ultrastructural components of the cell can be labeled with little or no uncertainty.

In this chapter, we will focus on our current protocols for cryoimmunogold labeling in order to localize subcellular proteins of interest. Section II describes the types of grids that are most often used and how they can be prepared for labeling. Next, Sections III and IV address specimen preparation that is a prerequisite for efficient labeling. Following specimen preparation, Section V outlines a method for ultrathin sectioning. Subsequently, a method for immunolabeling, which enables the localization of specific proteins and cellular components, will be described in Section VI, reagents and solutions will follow. Finally, we provide an outlook for the future of cryosectioning and the application of this technique for tomography. Because of the nature of this chapter, the majority of the text will be focused on basic principles and variations in order to maximize labeling efficiency.

II. Preparation of Carbon- and Formvar-Coated Copper Girds

The electron microscopic grid is analogous to a light microscopic glass cover slide, both provide support for the materials to be viewed. Electron microscopic (EM) grids are made from metals to conduct heat and charge away from the sample. Carbon, Formvar, or a combination of the two films can be applied to the metal to provide support for the cryosection.

Materials

- 1.7 mol/l of ammoniumhydroxide solution (6%, w/w)
- Acetone c.z.
- Formvar
- Chloroform p.a.
- 100 mesh copper grids or slot grids (mesh size 0.5 mm × 2 mm)
- 100-ml glass-stoppered Erlenmeyer flask

- Glass test tube
- Glass microscope slides
- Coplin jar
- White address label – Adhesive paper
- Formvar/carbon coating setup (BOC Edwards or equivalent) (see Fig. 1)

Preparation of the Grids

1. Place the copper grids in a glass test tube that contains 1 ml of ammonium hydroxide solution (6%, w/w). Vortex until the color of the ammonium hydroxide solution appears slightly green against a white surface. Pour off the supernatant and replace with water. Repeat in water at least 10 times. Finally, remove the water and rinse in acetone. Pour off the acetone supernatant and place the glass test tube upside down on a piece of filter paper to dry. The grids will fall from the bottom of the glass test tube on the filter paper.
2. Dissolve 1.1 g of Formvar in 100 ml chloroform, while stirring, in a Erlenmeyer flask fit with a glass stopper for 30–60 min. After the Formvar is dissolved let the solution stand overnight to mature.

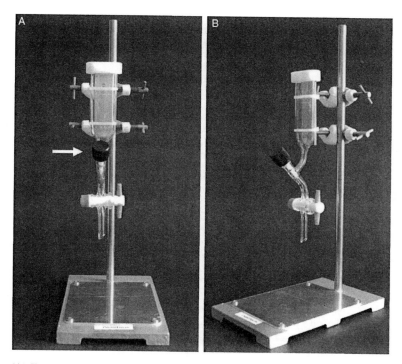

Fig. 1 (A) Formvar coating setup. The white arrow indicates the valve, which is opened to allow the Formvar to drain. (B) Side view of the Formvar coating setup. (See Plate no. 19 in the Color Plate Section.)

3. Place a microscopic slide into the Formvar coating setup (see Fig. 1) and mark the outside of the apparatus at 70% of the slide. Remove the cover slide and gently add the Formvar solution to the Formvar coating device.
4. Place the slide into the Formvar coating setup (the fluid level should be at 70% of the glass slide). Leave the slide in the Formvar solution for 1 min.
5. Open the bottom valve to drain the Formvar solution. The drain speed can be adjusted using the black valve indicated by the white arrow in Fig. 1.

The thickness of the Formvar is determined by the rate at which the Formvar flows out of the tunnel, ∼7 s. for an optimal thickness of ∼50 nm.

6. Remove the Formvar coated slide from the setup and score the edges of the glass slide with a razor blade to detach the Formvar at the edges of the slide. Cut a thin strip from the bottom side to split the film.
7. Breathe on the slide and immediately insert the slide slowly, and perpendicular, into a container filled with water. The Formvar film will float from both sides of the slide on the water surface. A grayish interference color is optimal and usually indicates an approximately 50-nm thick Formvar layer.

Coat the Girds with Formvar

8. Using fine forceps, place clean EM grids onto the floating Formvar film; the rough surface of the grid should be placed down facing the film.
9. Remove the Formvar film supporting the grids from the water surface, using an ethanol cleaned glass slide covered with a white address label on one side, by placing the slide on the Formvar film and pushing it into the water so that the grids adhere to the paper on the slide.
10. Dry the slide and evaluate the quality of the film using an electron microscope. The film should not have any significant tears or holes.

Carbon Coat the Grid

11. Coat the girds 1 day later with a thin layer of evaporated carbon at high vacuum.

Steps 1–10 (see Video 11 at http://www.currentprotocols.com).

III. Aldehyde Fixation of Cells for Cryoimmuno Labeling

Glutaraldehyde and paraformaldehyde cross-link neighboring proteins through highly reactive aldehyde groups that create interlocking structures. To minimize the denaturing effects of aldehyde fixatives, a low concentration of aldehyde solution is used for optimal fixation. The absence of organic solvents, which are routinely used in freeze substitution/plastic-embedding, results in better ultrastructural preservation and antigenicity.

Materials—Cells

- Cultures of adherent cells (in a 10-cm dish) or cells in suspension
- Culture medium appropriate to the cell type
- 2 times fixative, prewarmed to room temperature
- Storage solution (see Reagents and Solutions)
- PBS containing 0.15 mol/l glycine at room temperature
- PBS containing (1% w/w) gelatin at room temperature
- Cell scraper

1. All fixative should be prepared freshly the day of the experiment. Cells should be grown to ~75% confluency at the time of fixation.
2. For the fixation procedure mix a 1:1 volume ratio of fixative and cellular suspension at 37 °C (*Note:* The cells must be fixed directly after removing from the incubator). Incubate the cells in the fixative for 24 h in formaldehyde (FA) or for 2 h in formaldehyde/gluteraldehyde (FA/GA) at room temperature. Do not shake or vortex the cells during this fixation period.

Warning: Handle aldehyde solutions carefully and avoid exposure to skin and/or breathing of the aldehyde solutions. All steps involving aldehyde solutions should be carried out in a well-ventilated fume hood and solutions containing aldehydes should be disposed of in the proper manner.

3. For adherent cells, remove the fixative and store it for later use. The fixative should be replaced by, by washing 2 times with a 0.15 mol/l glycine solution in PBS. Remove the glycine solution and replace with 5 ml of a 37 °C, 1% (w/w) gelatin solution in PBS and leave for 30 min. at 37 °C.

   a. Remove the viscous cell suspension from the Petri dish using a cell scraper and transfer it using a glass Pasteur pipet to a 10-ml Falcon tube and centrifuge the cells for 3 min at 1000g at room temperature (centrifuge duration may be altered depending on cell type).
   b. Remove the supernatant and resuspend the cell pellet in 1ml of PBS and transfer to a 1.5-ml Eppendorf tube.
   c. Centrifuge the saved fixative, remove supernatant and resuspend the pellet in 1 ml of PBS. Combine this fraction with the resuspended fraction in 3b.

Certain (pro-apoptotic) cells can be detached before or during fixation. In order to representative information it is important to study both populations.

4. For cells in suspension, transfer suspension of fixed cells to a 15 ml tube and centrifuge for 3 min at 1000g, at room temperature.

   d. Resuspend the cells in 1 ml PBS/0.15 M glycine and transfer to a 1.5 ml microcentrifuge tube.

5. If cells must be stored for a long period of time, transfer cells to a storage buffer. Shipment can take place at room temperature. Avoid freezing of the sample.

Steps 2 and 3 (see Video 4 at http://www.currentprotocols.com).

Materials—Tissue

- Tissue sample
- 1× fixative: 1× FA, or 1× FA/GA
- Storage solution
- Razor blades, acetone cleaned
- Glass vials with screw caps
- Fine forceps

1. Place one freshly dissected tissue piece immediately to 2 ml of 1× FA fixative and another tissue piece to 2 ml of 1× FA/GA fixative both in glass vials with screw caps.
2. Cut tissue into 3-mm thick pieces using clean acetone-rinsed, air-dried razor blades and fine forceps.
3. Fix FA-fixed tissue at room temperature for 24 h and then GA/FA-fixed tissue for 2 h.
4. After fixation, transfer tissue to storage solution in glass vials. Store at 4 °C in an airtight container. Shipment can take place at room temperature, but always avoid freezing of the sample.

IV. Embedding Samples for Cryoimmuno Labeling

Fixed cells and tissues are embedded in gelatin to enable solidification of the sample. A small gelatine cube is produced which can subsequently be plunge frozen into liquid nitrogen and sectioned.

Materials

- Fixed cells or tissue in storage solution
- PBS
- PBS containing 0.15 M glycine
- 12% (w/v) gelatin in 0.1 M sodium phosphate buffer, 37 °C for tissue samples.
- 2.3 M sucrose in 0.1 M sodium phosphate buffer, pH 7.4
- Parafilm M
- Razor blades, acetone-rinsed and air-dried
- Dissecting microscope with cold-light optics

- 1 ml plastic vials
- 10 ml vials
- End-over-end rotator (~10 s per rotation)
- Fine forceps
- Greiner analyzer cup 14/24 mm, 1.5 ml, and caps (Greiner Bio-One)
- Aluminum specimen holders, or steel pins, roughened with sandpaper and acetone-cleaned, dust-free
- 15 ml aluminum cryotubes (Sanbio) with two holes punched in the top

Cells

1. Centrifuge fixed cells 3 min at 1000g, room temperature. Remove storage solution. Resuspend the cells in PBS/0.15 M glycine (same volume as storage solution) and centrifuge. Repeat washing step one more time (wash with PBS/0.15 M glycine and centrifuge).

All free aldehyde groups should be quenched, otherwise the cells and gelatine may cross-link to the free aldehyde groups.

2. Resuspend cells in 1 ml of 12%, 37 °C gelatin, then incubate 5 min at 37 °C.
3. Centrifuge cells for 3 min at 11,180g, room temperature. Let stand on ice for 10 min to allow the gelatin to solidify.
4. Cut microcentrifuge tubes just above the cell pellet and cut halves. Transfer to a 10 ml vial containing precooled PBS, and keep on ice.
5. Transfer the pellet halves to a droplet of cold PBS using forceps, but without adding pressure on the cells. Cut the cells in gelatin into slices of 0.5 mm. Cut the slices into 0.5 mm cubic blocks with a clean (acetone-rinsed, air-dried) razor blade, under a dissecting microscope. Then proceed to Step 6.

Perform steps 4–9 in a cold room to avoid melting of gelatine and drying of the sample.

Tissue

1. Wash tissue twice, each time for 5 min, in PBS/0.15 M glycine
2. Incubate 30 min each in 2%, 5% gelatin, and twice, 30 min each, in 12% gelatin in an end-over-end rotator (13 rotations/min) at 37 °C.
3. Place the tissue in a small drop of gelatin on top of a piece of Parafilm M and remove excess gelatin by pushing the piece of tissue, with a small Petri dish, along the Parafilm surface when the gelatin is still liquid. Put the Parafilm on ice so the remaining gelatin covering the tissue solidifies. Cover the gelatin-infiltrated piece of tissue with a drop of cold PBS. Then, proceed to Step 6.

4. Transfer the cubic blocks to a Grener cup with a cap that contains 2.3 M sucrose.
5. Rotate sample in sucrose for 4 h in an end-over-end rotator in the cold room.

Steps 1–7 (for cells) (see Video 5 at http://www.currentprotocols.com).

6. Transfer the sample, along with a minimal amount of sucrose from the vial to a clean aluminum specimen holder that has been roughened with sandpaper or scratched with a sharp steel pin.
7. Remove excess sucrose using a clean filter paper, and transfer holders to small aluminum cryocontainers filled with liquid nitrogen and store.

Steps 8 and 9 (see Video 6 at http://www.currentprotocols.com).

V. Cryosectioning for Cryoimmuno Labeling

Ultrathin cryosectioning has been around since the early 1970s. The idea of a device able to cut thin sections at cryogen temperatures was pioneered by Christensen (1971). The devise was called a cryobowl and could be maintained at temperatures from -30 to $-160\,°C$ by flushing the bowl with cold nitrogen gas. Tokuyasu followed and used a sucrose solution to prevent ice-crystal formation in frozen samples to be cryosectioned (Tokuyasu, 1973). A high-molar sucrose solution was used to penetrate the membrane of the specimen and also to reduce ice-crystal formation, resulting in better preservation of cellular ultra-structure.

The term cryosection is referred to a thawed section that has been produced from an aldehyde fixed, sucrose infiltrated sample and subsequently sectioned at low temperature. Contrasted to a vitreous section or frozen hydrated section that has not been subjected to fixation or stain prior to sectioning.

Materials

- Sample blocks that have been stored in liquid nitrogen
- Section retrieval solution (see reagents and solution)
- Ultramicrotome (Leica) with cryochamber and antistatic devices
- Diamond knife (Diatome, cryoimmuno 35°)
- Trimming knife (Diatome, cryotrim 45° or 20°)
- Eyelash, mounted on a wooden stick

(see Video 7 at http://www.currentprotocols.com)

- 3 mm diameter loop made of twisted 0.3 mm romanium wire (Winkelstroeter Dentaurum) on a 15 cm wooden stick

(see Video 1 at http://www.currentprotocols.com)

- Carbon- and formvar-coated copper grids

8. Immunogold Labeling of Thawed Cryosections

Procedure

1. Cool the microtome to −80 °C. Once microtome is cooled insert trimming knife, and knife for ultra-thin sectioning, once the knifes are cooled insert and secure the specimen and the knifes.
2. Face the front of the specimen block with the trimming knife at a 200 nm feed with a speed of 100 mm/sec.

An antistatic device facilitates sectioning. Make sure that the ionizer is cleaned and positioned in the microtome correctly, and operating when sectioning.

3. Trim the specimen block in the shape of a rectangle with dimensions of 250 μm by 350 μm.
   a. Trim one edge of the block at a 100 mm/s feed, 100–120 μm deep.
   b. Trim other edge, ∼100 μm from first edge, with the same cutting speed and depth.
   c. Rotate the sample at 90° and trim along each side the same as (a and b).
   d. Rotate the sample back to the original position.
4. It is advisable to check whether the trimmed block is suitable for further sectioning. This can be done by cutting a few ∼0.25 μm thin cryosections. The sections can be stained with Methylene blue and evaluated in the light microscope.

Ultrathin Sectioning

5. Set and maintain the temperature of the cryochamber, knife holder, and specimen holder of the microtome at −120 °C.
6. For quality sectioning, the feed should be 50–70 nm (gold to yellow interference color), at a speed of 1–2 mm/s.
7. After a ribbon of sections is made, carefully lay the ribbon on the platform of the knife.
8. Retrieve the ribbon sections from the cryochamber using a droplet of pick-up solution created on a 3 mm stainless steel loop attached to a 15 cm wooden stick. Place the almost frozen droplet very close to the face of the ribbon of sections until attached and quickly remove the loop carrying the sections from the cryo chamber. *The tube that contains the pick-up solution (methylcellulose/sucrose) should remain on ice throughout the sectioning procedure; a dry film may appear on the surface of the solution.*
9. Once the sections are thawed, immediately place them onto a formvar/carbon coated grid and store sections until immunogold labeling. Grids can be stored for weeks in the cold room.

Steps 8 and 9 (see Video 8 at http://www.currentprotocols.com).

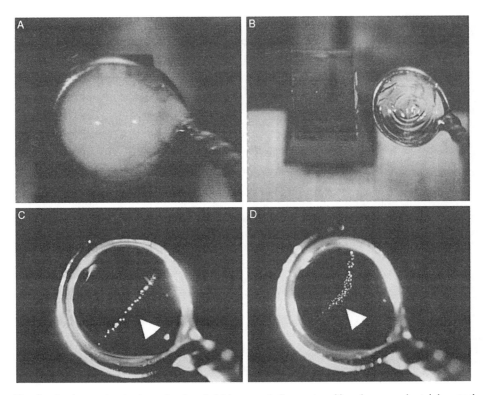

Fig. 2 Section retrieval using a droplet of picking up solution captured by a home-made stainless steel loop attached to a wooden stick. The droplet is placed onto the ribbon of cryosections and subsequently removed from the microtome chamber, thawed, and placed onto an EM grid for observation. (A) Shows a droplet of sucrose/methylcellulose solution that is beginning to freeze at liquid nitrogen temperatures. (B) Show a droplet of pure sucrose solution that was used to pick up the ribbon of cryosections. Note the different freezing properties from the sucrose/methylcellulose mixture and the pure sucrose droplet. (C) Shows the droplet from (A) warmed at room temperature, the arrowhead shows the ribbon of cryosections. (D) The droplet from (B) warmed at room temperature, the arrowhead denotes the ribbon of cryosections. Note the difference in the quality between the two solutions used for retrieving the ribbon of cryosections. In (D) the ribbon looks warped and wrinkled, but in (C) the ribbon looks straight. (Courtesy Erik Bos, NKI). (See Plate no. 20 in the Color Plate Section.)

10. When the sectioning session is finished clean knives (see Handling and Use Manual at http://www.emsdiasum.com/diatome/diamond5Fknives).

Ultrathin cryosectioning (see Video 8 at http://www.currentprotocols.com).

VI. Immunogold Labeling

This technique utilizes the antigen-antibody reaction, whereby antibodies applied to the surface of a cryosection localize an antigen. Protein A gold is then attached to the antibody. The sample is then contrasted with uranyl methylcellulose mixture.

8. Immunogold Labeling of Thawed Cryosections

Materials

- Cryosections on a grid
- 2% (w/v) gelatin-coated 2.5-cm Petri Dishes
- PBS (phosphate buffer solution) containing 0.15 M glycine
- PBS containing 1% (w/v) BSA
- Primary antibody diluted in PBS/1% BSA
- PBS/0.1% (w/v) BSA
- Secondary antibody in PBS/1% BSA
- 10 nm protein A gold particles at OD 0.1 (Cell Biology, Medical School, Utrecht University) in PBS/1%BSA
- PBS containing 1% (v/v) glutaraldehyde
- Uranyl oxalate solution
- Methyl cellulose/uranyl acetate solution
- Parafilm M
- Whatman no. 5 filter paper
- 37 °C Incubator
- Forceps
- Stainless steel loop slightly larger than grid attached to P100 (yellow, 100 µl) pipette tip.
- Stainless steel loops larger than grid attached to P1000 (Blue, 1000 µl) pipette tips.

Preparation of the Sample

1. Place all grids that will be labeled on cold (solid) 2% gelatin plates and leave for 30 min at 37 °C.

The 2% gelatin will melt and the grids will float on the surface of the gelatin liquid. The 12% gelatin used to embed the sample is melted away from the cryosections into the 2% gelatin.

2. Apply a sufficient amount of Parafilm M to the bench that will act as a surface to place the droplets of solution. (Prewetting the surface may help to stick the parafilm to the bench.)
3. Transfer the grid with fine tip forceps to a drop (100 µl) of PBS/0.15 M glycine.

The upper part of the grid should remain dry at all times. The bottom part of the grid should always remain wet during subsequent steps.

4. Block 1 time in 100 µl droplet of PBS/1% BSA for 3min.

Primary Antibody

5. Incubate grid for the appropriate time at room temperature on a 5 µl droplet of primary antibody diluted to an appropriate concentration in PBS/1% BSA.

The appropriate dilution and incubation times vary with the antibody used. The incubation times range from 30 to 100 min. The antibody must be correctly tittered to find the best specific signal-to-background signal ratio. Useful references are Aurion (2006) and Raposo et al. (1997).

6. Wash the grid five times, each time floating for 3 min on 100 µl droplets of PBS/0.15 M glycine. The final blocking step should be on PBS/0.1% BSA.

7. *Optional:* If a primary antibody is used that does not react with protein A gold, incubate with a secondary (bridging) antibody that will bind to protein A gold. The antibody must be diluted in PBS/1% BSA. The grid is exposed for 20 min and then rinsed 5 times on droplets of PBS/0.15 M glycine.

Protein A Gold

8. Incubate the grid for 20 min on 7 µl droplets of 10-nm protein A gold particles in PBS/1% BSA. The OD of the gold is 0.1.

9. Wash grid 3 times with rapid changes in 100 µl droplets of PBS.

10. Wash grid 6 times with 3 min intervals in 100 µl droplets of PBS.

Fix the Sample

11. Incubate grid for 5 min on 100 µl droplets of PBS/1% glutaraldehyde in a chemical hood.

12. Wash grid 10 times, each in 100 µl droplets of distilled water at 2 min intervals.

13. *Optional:* Incubate grid for 5 min on a 100 µl droplet of uranyl oxalate solutions and rinse by floating for 1 min on a 100 µl droplet of water.

Embed the Sample

14. Quickly wash grid in a 100 µl droplet of methylcellulose/uranyl acetate solution for 1 min, then in fresh 100 µl droplet of methylcellulose/uranyl acetate solution for 5 min *on ice*.

Steps 1–5, and 14 (see Video 2 at http://www.currentprotocols.com).

15. Retrieve the grid with a stainless steel loop that has a diameter slightly larger than the grid, glued into a P1000 pipette tip.

16. Drain excess methylcellulose/uranyl acetate solution by touching the loop at an angle of 90° to a Whatmann no. 5 filter paper. Slide the loop from left to right across the filter paper for ∼2 cm. Lift loop from the paper just before all of the volume is absorbed.

17. Air-dry grid at room temperature by placing the pipette tip upside down on a tube drying rack.

A thin film of methylcellulose/uranyl acetate is left on the grid after drying. The thickness of the film is important for optimal contrast and to preserve the integrity of membrane structure. The thickness can be assessed by the interference colors (viewed under a stereomicroscope). A gold colored film is optimal for 50–70-nm thick sections.

Steps 14–17 (see Video 3 at http://www.currentprotocols.com).

VII. Reagents and Solutions

A. BSA Stock Solution, 10% (W/V)

Prepare 10% (w/v) bovine serum albumin (BSA, fraction V; Sigma) in 0.1 M sodium phosphate buffer containing 0.02% (w/v) sodium azide. Stir slowly at 4 °C until BSA dissolves. Adjust pH to 7.4 and centrifuge 1 h at 10,000g, 4 °C. Retain supernatant and store in 1 ml aliquots at 4 °C in closed tubes.

B. Formaldehyde (FA) Fixative, 2×

Thaw a 5 ml aliquot of 16% (w/v) formaldehyde (FA) stock solution in 60 °C water. When the solution is dissolved completely, mix the following:

10 ml 0.4 M PHEM buffer (0.2 M final)
5 ml 16% FA stock (4%, w/v/final)
5 ml H_2O

This solution must be prepared fresh before the experiment.

C. Formaldehyde (FA) Stock Solution, 16% (W/V)

Dissolve 32 g of paraformaldehyde in 200 ml of water (to assist with dissolving, 3 drops of 10 M NaOH can be added and the solution can be stirred on a hotplate until completely dissolved—*do not heat above* 60 °C. Divide formaldehyde solution into aliquots in 10 ml tubes. This solution can be stored for years at −20 °C in a sealed box.

D. Formaldehyde/Glutaraldehyde (FA/GA) Fixative, 2×

Thaw a 5 ml aliquot of 16% formaldehyde stock solution in 60 °C water. When the solution is completely dissolved mix with the following:

10 ml 0.4 M PHEM buffer (0.2 M final)
5 ml 16% (w/v) FA stock solution (4%, w/v final)
1 ml 8% (v/v) glutaraldehyde (0.4%, v/v final)
4 ml H_2O

This solution must be prepared fresh before the experiment.

E. Paraformaldehyde (PFA) Fixative, 2×

Prepare immediately before use.

10 ml 0.4 M PHEM buffer
5 ml 16% PFA stock solution
5 ml dist. H_2O

Alternative is to purchase premade 16% paraformaldhyde from EMS (Electron Microscopy Sciences).

F. Gelatin, 1%, 2%, 5%, and 12% (W/V)

Prepare gelatin (Merck) solutions in 0.1 M sodium phosphate buffer, pH 7.4, containing 0.02% (w/v) azide. Store 2% gelatin to be used for labeling in small Petri dishes in a closed box at 4 °C to avoid drying.

G. Methyl Cellulose/Uranyl Acetate

For the methyl cellulose stock solution, add 2 g methyl cellulose (Sigma, 25 cP) to 98 ml prewarmed (90 °C) distilled water and stir. Place solution on ice to cool until the temperature has dropped to 10 °C. Stir overnight at low speed in the cold room. Let the solution stand for 3 days in the cold room. Centrifuge for 95 min at 97,000g, 4 °C, then divide the supernatant into 10 ml aliquots and store for up to 2 months at 4 °C.

For methyl cellulose/uranyl acetate, mix 1 ml of 4% uranyl acetate and 9 ml of methyl cellulose stock. The solution can be stored for a couple weeks at 4 °C in a dark container.

H. PHEM Buffer

Dissolve 72.5 g PIPES (240 mM final) in 600 mL 0.375 M NaOH. Gradually add 15.2 g EGTA (40 mM final) while stirring. Dissolve 23.8 g HEPES (100 mM final) in 200 ml water and add this to the other mixture. Add 1.63 g $MgCl_2$ (8 mM final). Adjust the pH to 6.9 using 1 M NaOH and adjust the volume to 1 l. The solution can be stored for years at −20 °C.

I. Section Retrieval Solution

Prepare a 1:1 mixture of 2.3 M sucrose in 0.1 M sodium phosphate buffer, pH 7.4, and 2% (v/v) methyl cellulose in distilled water. Mix on an end-over-end rotator in the cold for 20 min. This solution must be prepared fresh before the experiment.

J. Storage Solution

Thaw 5 ml aliquot of 16% formaldehyde stock solution in 60 °C water. When the solution is clear, add the following:

5 ml 0.4 M PHEM buffer

0.6 ml 16% FA stock solution, thawed and clear

14 ml H_2O

The solution can be stored for a month at 4 °C.

K. Uranyl Acetate, 4% (V/V)

Prepare a 4% stock of uranyl acetate in distilled water and adjust pH to 4 with 0.1 N HCl. This solution can be stored in the dark for up to 1 month at 4 °C. Before using the solution must be filtered through a 0.22-μm Millipore filter.

L. Uranyl Oxalate Solution

Mix one volume of 4% uranyl acetate and one volume of 0.15 M oxalic acid. Adjust pH to 7 with 25% (v/v) ammonium hydroxide, stirring well after adding each drop. This solution can be stored for several months in the dark at 4 °C.

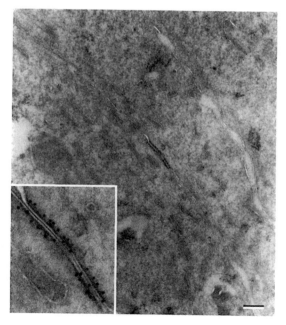

Fig. 3 An electron microscope micrograph of an ultrathin cryosection (50nm) of an A543 cell. The cryosection was immunolabeled with an antibody against α connexin 43 that was linked to 10 nm protein A gold. Inset depicts the specific labeling of the gap junction. Bar = 300 nm. (Courtesy Hans Janssen, NKI).

VIII. Future Outlooks

Microscopy enables observation of small cells and organelles, unrecognizable to the human eye. While light microscopy can provide valuable information about whole cells, the spatial resolution is limited to the wavelength of light (~0.4–0.7 μm). Therefore, in order to resolve cellular structures greater than 0.5 μm, electrons are used which can provide ~100% better resolution than imaging with light. The immunogold labeling techniques provide reliable localization of proteins in their cellular environment (van der Wel *et al.*, 2005), i.e. α connexin 43 localized in the gap junction (Fig. 4). Proper ultrastructural preservation and efficient labeling techniques afford quantification of labeling within the cryosection (Mironov *et al.*, 2003). The opportunity to quantify the distribution of a protein within the cellular environment is an important tool to understand the labeling efficiency and the proper location of the protein.

Although single image projections provide a starting point for EM imaging, the cell and cellular components are naturally three-dimensional. In order to capture the three-dimensionality of cells and cellular components, many projection images are needed at different tilting orientations, called electron tomography. Tomography involves recording images at different tilt increments by automatically tilting the stage of the microscope, the images are subsequently complied and aligned to a common origin and reconstructed into a 3D map (Koster *et al.*, 1992).

Tomography of immunolabeled cryosections can provide better spatial information about immunolabeled proteins, i.e. *Escherichia coli* tsr receptor, COPII-coated membranes and Golgi (Zhang *et al.*, 2004; Zeuschner *et al.*, 2006; Ladinsky *et al.*, 2007), within the 3D architecture of the cell. Figure 4 shows slices from a reconstructed tomogram of the gap junction labeled with an antibody that recognizes α connexin 43. For further reading on tomography of immunolabeled samples see Ladinsky and Howell (2007).

Although the EM of immunolabeled cryosections enables visualization of cellular details and organelles, individual proteins and macromolecular protein interactions are currently unreachable using conventional preparation techniques (fixation, dehydration, or staining).

For the next generation, the cell and cellular components must be imaged in their most native state. Because of the nature of the fixation process, cells are never completely free of fixation artifacts, for example aldehydes cross-link proteins and dehydration reduces the integrity of cellular macromolecules rendering the interactions nonnative. In hopes to alleviate these problems, a novel preparation technique emerged that involves rapid freezing to immobilize cells in an amorphous layer of water (Dubochet, 1983). At such high rates of freezing, water or biological material is absent of crystallization, which may damage ultrastructure of the cell. Cryoelectron tomography has revealed many novel insights into macromolecular architecture and subsequent function of complex biological machines, i.e. axoneme of *Chlamydomonas* and Sea Urchin sperm (Nicastro *et al.*, 2006).

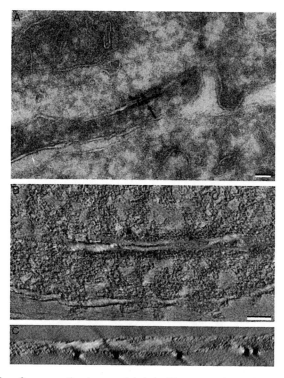

Fig. 4 Tomography of a cryosection immunolabeled with α connexin 43. A) A slice from the original tilt series acquired with an Tecnai F20. The black arrow indicates 10 nm gold that is localized to a gap junction within an A543 cell. (Similar to the area shown in Fig. 4.) Bar = 100 nm. B) A slice from the middle of a reconstructed tomogram. C) A slice from the tomogram that has been rotated perpendicular to the section plane. Note the gold label remains on the surface of the section. The tilt series was acquired using a Tecnai F20 (FEI, Eindhoven) and reconstructed using the IMOD software (Kremer et al., 1996). Bar = 50 nm.

Mammalian cells, tissue, and larger bacteria cells are too large to be viewed *in toto* by plunge freezing. Therefore, to freeze such large volumes of cells high-pressure is applied during the process of freezing (HPF). Cells larger than 0.5 μm and even small pieces of tissue can be vitrified and then sectioned using HPF technique (Studer *et al.*, 1989). Frozen thin sectioning was demonstrated in 1971 by Christensen and colleagues (Christensen, 1971), and later some of the pioneers demonstrated and developed the technique of frozen hydrated sectioning (McDowall *et al.*, 1983; Dubochet *et al.*, 1988; Fredrik *et al.*, 1982). The technique has been applied to a number of interesting biological structural studies (Al-Amoudi *et al.*, 2004; Bouchet-Marquis *et al.*, 2007; Zhang *et al.*, 2004). Although there are still problems associated with the sectioning process and imaging of the vitreous sections (Al-Amoudi *et al.*, 2005; Hsieh *et al.*, 2006), cryotomography of vitreous sections is currently the only technique to visualize cells and tissues larger than 0.5 μm close to their native state (Marko *et al.*, 2007). Cryotomography of vitreous sections may hold the key to unraveling some of the most exciting macromolecular interactions, in turn revealing the true function of supramolecular complexes *in situ*.

Acknowledgments

The authors thank Erik Bos and Hans Janssen for providing figures. Diane Huben, Matthijn Vos, and Mikkiae van for critical reading of the chapter, and Montserrat Bárcena and Abraham J. Koster, Leiden University Medical Center, for the electron tomography.

References

Al-Amoudi, A., Chang, J. J., Leforestier, A., McDowall, A., Salamin, L.-A., Norlén, L. P. O., Richter, K., Sartori Blanc, N., Studer, D., and Dubochet, J. (2004). Cryo-electron microscopy of vitreous sections. *EMBO J.* **23**, 3583–3588.

Al-Amoudi, A., Studer, D., and Dubochet, J. (2005). Cutting artifacts and cutting process in vitreous sections for cryo-electron microscopy. *J. Struct. Biol.* **150**, 109–121.

Aurion (2006). Optimised immuno labeling using Aurion Blocking Solutions and Aurion BSA-c™. *Technical Support: Newsletters & Newsflyers*, Newsletter 1.

Bouchet-Marquis, C., Zuber, B., Glynn, A. M., Eltsov, M., Grabenbauer, M., Goldie, K. N., Thomas, D., Frangakis, A. S., Dubochet, J., and Chrétien, D. (2007). Visualization of cell microtubules in their native state. *Biol. Cell* **99**.

Christensen, A. K. (1971). Frozen thin sections of fresh tissue for electron microscopy, with a description of pancreas and liver. *J. Cell Biol.* **51**, 772–804.

Dubochet, J., Adrian, M., Chang, J. J., Homo, J. C., Lepault, J., McDowall, A. W., and Schultz, P. (1988). Cryo-electron microscopy of vitrified specimens. *Q Rev Biophys* **21**, 129–228.

Dubochet, J., McDowall, A. W., Menge, B., Schmid, E. N., and Lickfeld, K. G. (1983). Electron microscopy of frozen-hydrated bacteria. *J. Bacteriol.* **155**, 381–390.

Faulk, W. P., and Taylor, G. M. (1971). An immunocolloid method for the electron microscope. *Immunochemistry.* **8**(11), 1081–1083.

Frederik, P. M., Busing, W. M., and Hax, W. M. (1982). Frozen-hydrated and drying thin cryo-sections observed in STEM. *J Microsc* **126**(Pt 1), RP1–RP2.

Hsieh, C. E., Leith, A., Mannella, C. A., Frank, J., and Marko, M. (2006). Towards high-resolution three-dimensional imaging of native mammalian tissue: Electron tomography of frozen-hydrated rat liver sections. *J. Struct. Biol.* **153**, 1–13.

Koster, A. J., Chen, H., Sedat, J. W., and Agard, D. A. (1992). Automated microscopy for electron tomography. *Ultramicroscopy* **46**, 207–227.

Kremer, J. R., Mastronarde, D. N., and McIntosh, J. R. (1996). Computer visualization of three-dimensional image data using IMOD. *J. Struct. Biol.* **116**, 71–76.

Ladinsky, M. S., and Howell, K. E. (2007). Electron Tomography of Immunolabeled Cryosections. *In* "Cellular Electron Microscopy" (J. R. McIntosh, ed.), Vol. 79, pp. 543–558. Elsevier, California.

Marko, M., Hsieh, C. H., and Mannella, C. A. (2007). Electron Tomography of Frozen-hydrated Sections of Cells and Tissues. *In* "Electron Tomography: Methods for Three-Dimensional Visualization of Structures in the Cell" (J. Frank, ed.), 2nd edn., pp. 49–81. Springer, New York.

McDowall, A. W., Chang, J. J., Freeman, R., Lepault, J., Walter, C. A., and Dubochet, J. (1983). Electron microscopy of frozen hydrated sections of vitreous ice and vitrified biological samples. *J. Microscopy.* **131**, 1–9.

Mironov, A., Latawiec, D., Wille, H., Bouzamondo-Bernstein, E., Legname, G., Williamson, R. A., Burton, D., DeArmond, S. J., Prusiner, S. B., and Peters, P. J. (2003). Cytosolic Prion Proteins in Neurons. *J. Neurosci.* **23**, 7183–7193.

Nakane, P. K., and Pierce, G. B. (1967). Enzyme-labeled antibodies for light and electron microscopic localization of tissue antigens. *J. Cell Biol* **33**, 308–318.

Nicastro, D., Schwartz, C., Pierson, J., Gaudette, R., Porter, M. E., and McIntosh, J. R. (2006). Cryotomography of axonemes: Insights into the regulation of dynein activity and microtubule stability. *Science* **313**, 944–948.

Raposo, G., Kleijmeer, M. J., Posthuma, G., Slot, J. W., and Geuze, H. J. (1997). Immunogold labeling of ultrathin cryosections: Application in immunology. *In* "Handbook of Experimental Immunology" (M. A. Cambridge, L. A. Herzenberg, D. Weir, L. A. Herzenberg, and C. Blackwell, eds.), 5th edn., Vol. 4, pp. 208, 1–11. Blackwell, Oxford.

Singer, S. J. (1959). Preparation of an electron-dense antibody conjugate. *Nature* **183**, 1523–1525.

Sternberger, L. A., Hardy, P. H. Jr., Cuculis, J. J., and Meyer, H. G. (1970). The unlabeled antibody enzyme method of immunohistochemistry. Preparation and properties of soluble antibody–antigen complex (horseradish peroxidase–antihorseradish peroxidase) and its use in identification of spirochetes. *J Histochem. Cytochem* **18**, 315–333.

Studer, D., Michel, M., and Muller, M. (1989). High pressure freezing comes of age. *Scanning Microsc.* **53**, 253–269.

Tokuyasu, K. T. (1973). A technique for ultracryotomy of cell suspensions and tissues. *J. Cell Biol.* **57**, 551–565.

Tokuyasu, K. T. (1986). Application of cryoultramicrotomy to immunocytochemistry. *J. Microsc.* **143**, 139–149.

van der Wel, N. N., Fluitsma, D. M., Dascher, C. C., Brenner, M. B., and Peters, P. J. (2005). Subcellular localization of mycobacteria in tissues and detection of lipid antigens in organelles using cryo-techniques for light and electron microscopy. *Curr. Opin. Microsc.* **8**, 323–330.

Zeuschner, D., Geerts, W. J. C., Donselaar, E. V., Humbel, B. M., Slot, J. W., Koster, A. J., and Klumperman, J. (2006). Immuno-electron tomography of ER exit sites reveals the existence of free COPII-coated transport carriers. *Nat. Cell Biol.* **8**, 377–383.

Zhang, P., Bos, E., Heymann, J., Gnaegi, H., Kessel, M., Peters, P. J., and Subramaniam, S. (2004). Direct visualization of receptor arrays in frozen-hydrated sections and plunge-frozen specimens of *E. coli* engineered to overproduce the chemotaxis receptor Tsr. *J. Microsc.* **216**, 76–83.

CHAPTER 9

Close-to-Native Ultrastructural Preservation by High Pressure Freezing

Dimitri Vanhecke, Werner Graber, and Daniel Studer

Institute of Anatomy
University of Bern
3000 Bern 9
Switzerland

Abstract
I. Introduction
   A. General Aspects Concerning Sample Preparation for TEM
   B. General Aspects of Cryofixation
   C. How Undesired Ice Is Formed
   D. Vitrification
II. Rational
III. Material
IV. Methods
V. High Pressure Frozen Samples
VI. Discussion
   References

Abstract

The objective of modern transmission electron microscopy (TEM) in life science is to observe biological structures in a state as close as possible to the living organism. TEM samples have to be thin and to be examined in vacuum; therefore only solid samples can be investigated. The most common and popular way to prepare samples for TEM is to subject them to chemical fixation, staining, dehydration, and embedding in a resin (all of these steps introduce considerable artifacts) before investigation. An alternative is to immobilize samples by cooling. High pressure freezing is so far the only approach to vitrify (water solidification without ice crystal formation) bulk biological samples of about 200 micrometer thick. This method leads to an improved ultrastructural preservation. After high pressure freezing, samples have to be subjected to follow-up procedure, such as freeze-substitution and embedding. The samples can also be sectioned into frozen hydrated sections and analyzed in a cryo-TEM. Also for immunocytochemistry, high pressure freezing is a good and practicable way.

I. Introduction

A. General Aspects Concerning Sample Preparation for TEM

In transmission electron microscopy (TEM) only solids can be investigated. Biological samples consist of about 80% water or more and this has to be either removed or solidified. The most common approach is chemical fixation, followed by dehydration and embedding in a resin (Luft, 1961). These steps introduce artifacts. Soybean root nodules cannot be well preserved by chemical fixation; however, high pressure freezing followed by freeze-substitution and Epon embedding provides an excellent ultrastructure (Studer *et al.*, 1992). The same holds for cartilage (Studer *et al.*, 1995) and other biological samples. High pressure freezing introduced by Moor in the 1970s (Moor, 1987; Moor *et al.*, 1980; Moor and Hoechli, 1970; Riehle and Hoechli, 1973) is mainly restricted by sample size. The samples have to be small (diameter: 1–2 mm, maximum thickness: 200 μm). The transition of a biological sample into a solid must occur fast. This, in fact, is possible only by cryofixation (fixation time is less than 50 ms).

B. General Aspects of Cryofixation

Cryofixation greatly increases the viscosity of the solvent (water) in a biological sample. At extremely high viscosity, all cell constituents, including water, are completely immobilized. The formation of crystalline water (ice), however, will destroy the cell's ultrastructure since only water molecules are incorporated in the ice crystal and therefore all solutes are concentrated on the surface of the ice crystal ramifications. This phase segregation leads inevitably to solute segregation, destroying the ultrastructure, and is manifested as segregation patterns (netlike structures) in the sections. Therefore ice formation must be prevented at all costs. Successful cryofixation passes a temperature window in which water usually cools that fast that no crystallization of water occurs (Fig. 1). Such state is called the vitreous state, and the solid gained amorphous ice.

C. How Undesired Ice Is Formed

The crystallization and melting of water are not the reverse transformation of each other. Melting is a single step process that occurs at the melting point when solid water is heated, whereas freezing of liquid water involves ice crystal nucleation and crystal growth. Liquid water, however, is easily supercooled to about

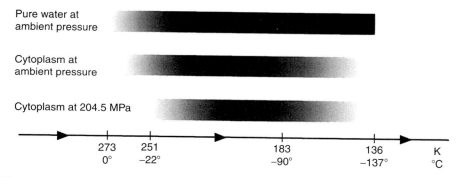

Fig. 1 The influence of solute concentration and pressure on the crystallization window for cryofixation. Upon slow cooling of pure water, ice crystals can form from 273 K (0 °C), but the effect of supercooling can postpone this process up to the nucleation temperature of 231 K (−42 °C). However, the increasing probability that ice crystals are formed at decreasing temperatures is represented by the grayscale gradient. Below 231 K (−42 °C), ice crystals will form (black region). The ice crystallization process is halted at temperatures below the recrystallization temperature of 136 K (−137 °C). The typical solute concentration of the cell's cytoplasm will lower the melting point (concentration depending) and will elevate the recrystallization temperature, as can be morphologically derived after freeze substitution. Applying a pressure of 204.5 MPa will further lower the melting point and the nucleation temperature. As a result, the temperature range suitable for ice crystal formation is decreased. The arrows indicate the direction of cooling.

248 K (−25 °C) and in 5 μm diameter tiny droplets down to about 231 K (−42 °C) under normal atmospheric pressure (Fig. 2). Liquid water may be supercooled to about 181 K (−92 °C) at 205 MPa. The extent of supercooling also increases with increasing solute concentration (Kanno *et al.*, 2004). However, ice formation also requires a minimum temperature (energy). This lower temperature limit is known as recrystallization temperature, which is about 136 K (−137 °C) for pure water. At lower temperatures, the energy is not sufficient for the formation of new crystals. Together, the nucleation temperature and the recrystallization temperature define a temperature window wherein water crystallizes (Fig. 1). Successful cryofixation passes this temperature window that fast that no ice crystals can be formed. This happens when heat is extracted faster from the system than it can be produced by the crystallization process.

D. Vitrification

Whether or not an aqueous solution will form crystalline ice depends on its ice nucleation frequency and the rate of heat removal from the system. Vitrification circumvents ice formation, and is depending on the thermal diffusivity (= thermal conductivity divided by density and thermal capacity) and solute concentration of the sample, the sample thickness, and the pressure applied. Successful vitrification is mainly influenced by two factors: the cooling rate and cryoprotective aspects. Under atmospheric conditions, only cooling rates of 500 000 K/s and higher can extract heat fast enough to prevent the formation of ice crystals in biological

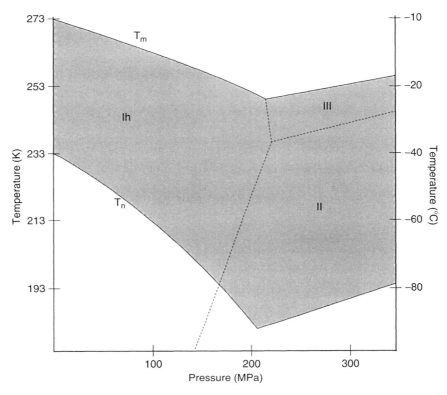

Fig. 2 Part of the phase diagram of water showing the minimal melting temperature (T_m) and the minimal nucleation temperature (T_n) around 200 MPa. The gray area marks where supercooled water can exist. At about 200 MPa, the melting point of water shows a minimum. At the same time, the possibility for supercooling is maximized. This sets out the best situation for high pressure freezing (redrawn from Kanno et al., 1975).

samples. However, heat can only be removed efficiently from the surface, and the physical properties of the sample itself controls the heat extraction in the sample (Studer et al., 1995; Shimoni and Müller, 1998). Thermal diffusivity of water and thickness of the sample will define the cooling rate in the sample. Bad thermal diffusivity of water prevents efficient heat extraction over a large distance. Even at a theoretically infinitely high cooling rate (>500,000 K/s) at the surface of a 600 μm thick aqueous sample, the cooling rate drops below 1000 K/s within 150 μm from the surface. Hence, improvement of technology will not help to vitrify thicker samples. At ambient pressure, water films of about 100 nm thickness can be vitrified (Dubochet et al., 1988).

For an aqueous solution, the time required to form an ice crystal increases with increasing solute concentration (Angell and Choi, 1986). The nucleation temperature decreases and the recrystallization temperature increases, narrowing down the crystallization temperature window of water. This implies that the needed cooling rate is reduced and heat can be extracted from the centre of thicker samples fast enough to avoid ice crystal formation. For a hypothetical biological sample with a 'typical' solute distribution and concentration, about 10–20 μm thick samples can be vitrified. Almost infinitely large samples can be vitrified by applying the Tokuaysu protocol (Tokuyasu, 1973). Such samples were mildly fixed (chemically) and then cryoprotected with 2.3 M sucrose. In these samples, ice nucleation is this rare that ice crystals almost never form. The last parameter is pressure. At 204.5 MPa, the maximum melting point depression of pure water is reached at 251 K (−22 °C; Kanno *et al.*, 1975; Fig. 2). Reducing or increasing the pressure at this temperature will induce crystallization. In hyperbaric environments, the effect of supercooling is increased to lower temperatures: the nucleation temperature has a minimum of 181 K (−92 °C) at 210 MPa (Kanno *et al.*, 1975; Fig. 2). These properties reduce the crystallization window and vitrification is achieved in the centre of the sample at lower cooling rates than at ambient pressure (Fig. 1). Practically, about 100 times smaller cooling rates are required than at ambient pressure. A cooling rate of about 5000 K/s is sufficient to vitrify at 205 MPa. As a result, vitrification at 205 MPa is achieved in approximately 10 times thicker samples than at ambient pressure. The increase in pressure to 205 MPa during freezing has a similar effect as an increase in solute concentration of the sample. Practically, this means that ice crystal formation can be prevented in 100–200 μm thick samples with a typical biological solute concentration and distribution (Sartori *et al.*, 1993; Shimoni *et al.*, 1998; Studer *et al.*, 1995). The effect of high pressure on biological samples cannot be neglected. Whenever it is possible to vitrify a sample at ambient pressure, high pressure should be avoided; however, only very thin samples will be vitreous (10–20 μm in thickness). During high pressure fixation, living cells or entire organisms are submitted to a pressure of 205 MPa for about 25–100 ms prior to freezing (Fig. 3). Experiments with Staphylococcus aureus at 200 MPa showed a survival rate of 99.99% after 25 ms of pressurization (Butz *et al.*, 1990). Around 90% survival rates of *Euglena gracilis* algae were noted (Riehle and Hoechli, 1973) after treatment with 200 MPa for considerable longer periods (100 ms). High pressures do not directly impact physical damage to living cells (Suppes *et al.*, 2003). Although exposure to pressure for short time is not lethal for most cells, pressure has an effect on the biological structure. 200 MPa will solidify the lipid bilayers, and leads to a change in biological membrane structure (Semmler *et al.*, 1998). Moreover, high pressure failed to retain the structure of cholesteric liquid crystal phase DNA (Leforestier *et al.*, 1996). This crystal structure is preserved in samples frozen at ambient pressure.

II. Rational

The main goal in electron microscopy is to show structural features reflecting the native biology. This goal can only be achieved with improved sample preparation methods. So far, high pressure freezing is the best option for small bulk biological samples.

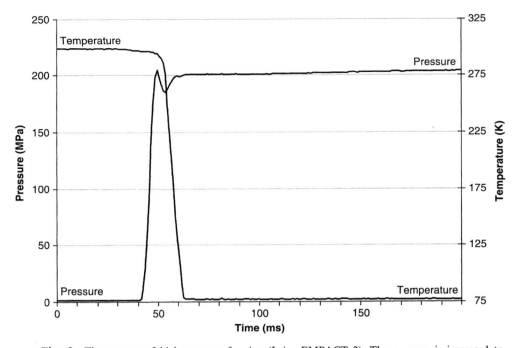

Fig. 3 The process of high pressure freezing (Leica EMPACT 2). The pressure is increased to 200 MPa within 10 ms. Once the required pressure is achieved, the temperature (starting at 37 °C) drops. Ice crystals can be formed from the moment the temperature falls below the nucleation temperature (T_m = 253 K or −20 °C) until the temperature is lower than about 173 K (Tr = −100 °C). This temperature drop lasts about 30 ms in the center of a 200 micrometer thick sample. The pressure is maintained for in total 600 ms (only 200 ms shown in the graph) at an average of 204.5 MPa.

III. Material

Three high pressure freezing machines are commercially available (Leica, Baltec, Wohlwend). We focus on the Leica EMPACT (Fig. 4; Studer *et al.*, 2001; Leica-microsystems, Vienna, Austria). Whatever machine is used, the biological sample is transferred into a specific holder (the sample holder). A range of sample holders are available, and each type of holder is tuned for optimal use of distinct types of samples: tubes for suspensions, ring holders for *in vitro* grown cell layers on sapphire disks, and microbiopsy holders for a specially designed microbiopsy system (Fig. 5). Consumables and additional tools can be obtained from the respective manufacturers.

The nature of the sample determines which type of sample holder should be used. Suspensions such as bacteria, yeast, and Trypanosoma can be easily loaded in the copper tube sample holder of the Leica EMPACT (Fig. 5, Grunfelder *et al.*, 2002; Matias *et al.*, 2003). Even for larger, nonunicellular organisms, such as the Nematode model organism *Caenorhabditis elegans* (Claeys *et al.*, 2004) and roots

9. Ultrastructural Preservation and High Pressure Freezing

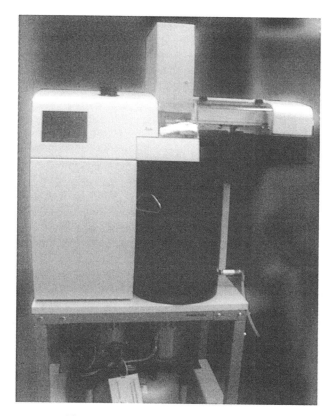

Fig. 4 The Leica EMPACT 2 (on the left).

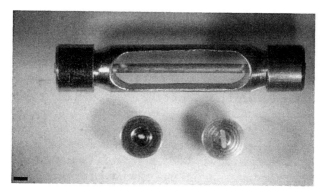

Fig. 5 Three sample holders for the Leica EMPACT. On top the tube holder, placed in its envelope prior to freezing. Below are two platelet systems: the flat platelet (left) can hold a range of samples in its cavity (200 μm deep, 1.5 mm in diameter). On its right is a specialized platelet for biopsies depicted. Bar = 1 mm.

of the plant model organism *Arabidopsis thaliana* (Van Belleghem *et al.*, 2007), this type of sample holder is suitable. The copper tubes would also be the holder of choice for fluids such as milk or blood. Additionally, the copper tubes present an efficient solution for the generation of frozen hydrated sections by CEMOVIS (Al-Amoudi *et al.*, 2004a,b; Dubochet *et al.*, 1988).

Different platelet holders (Fig. 5, discs with a central cavity of 200 μm in depth and 1.5 mm diameter) are suitable for a large variety of samples, including plant leaves (Pfeiffer and Krupinska, 2005), *Xenopus* pituary gland (Wang *et al.*, 2005), *Axolotl* embryos (Epperlein *et al.*, 2000) and *Drosophila* gestation organs (Shanbhag *et al.*, 2001), to adress but a few.

On occasion, the samples have to be trimmed down to fit the small dimensions of the holder. Often, the use of microbiopsy systems (Hohenberg *et al.*, 1996; Vanhecke *et al.*, 2003 (Leica-microsystems, Vienna)) is useful in situations where the excision is particularly difficult, has to be precise (e.g. human tissue) or fast (Hsieh *et al.*, 2006; Liu *et al.*, 2006; Muhlfeld and Richter, 2006).

IV. Methods

Suspensions are concentrated by centrifugation and the pellet is transferred with a pipette onto parafilm. Small drops (~10 μl) easily form on its hydrophobic surface. Then the cell suspension is aspirated into a copper tube by means of a piston wire, in which the pistons diameter fits the bore of the tube (~0.3 mm). Tissues are trimmed to the correct size with scalpels and punches prior to transfer to the platelets. Microbiopsy systems are specialized tools to avoid trimming of the sample and facilitate transfer to the high pressure freezing machine. Once the samples are introduced into the sample holders, they are fixed in a so-called pod and high pressure frozen in the EMPACT. The microbiopsy system allows for a handling time (from excision of the sample until the sample is frozen) of maximum 30 s.

High pressure freezing as stated earlier is limited to small samples. There is also another restriction: samples cannot contain gaseous compartments. If possible these compartments have to be filled with a liquid prior to freezing. Studer *et al.* (1989) proposed 1-hexadecene for this purpose. Gaseous compartments will collapse completely during the application of 204.5 MPa and destroy the structures.

Once the samples are frozen, follow-up procedures, such as freeze substitution, freeze drying, freeze-fracturing, or cryosectioning, are applied. The most popular one is freeze-substitution (Van Harreveld *et al.*, 1965). With the exception of yeast, all images shown originate from samples freeze-substituted at 183 K (−90 °C) in acetone containing 2% of osmiumtetroxide. Substitution is followed by embedding in Epon, curing, ultrathin sectioning, and investigation in the electron microscope. Frozen-hydrated sections are obtained by cryosectioning after high pressure freezing and investigated on a cryostage at 100 K in the electron microscope (CEMOVIS, Al-Amoudi *et al.*, 2004b).

Several fluorescent dyes are shown to survive the treatment of high pressure freezing (Biel *et al.*, 2003), thus creating possibilities for correlative confocal laser scanning microscopy and TEM.

V. High Pressure Frozen Samples

Figures 6–9 show some well preserved high pressure frozen samples. Characteristics are straight membranes and no 'empty' organelles. The overview (Fig. 6A) shows how well the ultrastructure of an ivy leaf is preserved when impregnated with 1-hexadecene. The detail shows two adjacent cells. Brain tissue sample is shown in Fig. 7. The image depicts a pyramidal cell in an organotypic tissue culture of mouse hippocampus. Cell organelles as well as the adjacent neuropil are well preserved. Fig. 8 shows Langerhans cells grown on a sapphire disc. The cells were sectioned perpendicular to the plane of the disc. The cell on the top is in the state of apoptosis. The remaining two cells are healthy. Frozen hydrated yeast cells are shown in Fig. 9. The samples are not stained with heavy metal solutions; just the vitreous biological matter is depicted by phase contrast.

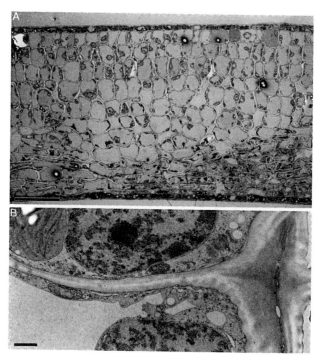

Fig. 6 A high pressure frozen and freeze-substituted ivy leaf is shown. The leaf is beautifully preserved—the overall leaf architecture (Fig 6A, bar = 50 μm) as well as the cell organelles (Fig. 6B, bar = 1 μm).

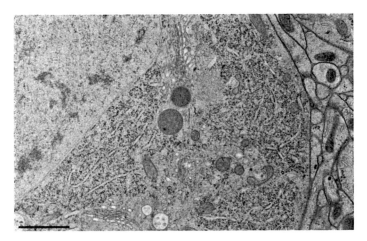

Fig. 7 A pyramidal cell high pressure frozen and freeze-substituted is depicted. The cell derives from an organotypic brain tissue culture of the hippocampus of mouse. The cell as well as the adjacent neuropil show an excellent structural preservation. Bar = 1 μm.

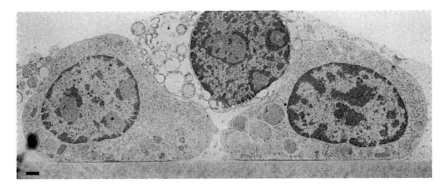

Fig. 8 Langerhans blood cells high pressure frozen and freeze-substituted are shown. The two cells adsorbed on a sapphire disc are healthy cells, whereas the top cell is apoptotic showing condensed DNA in the nucleus and a lot of empty vesicles in the cytoplasm. The section was taken perpendicular to the disc surface. Organelles of the two healthy cells are perfectly preserved. Bar: 1 μm.

VI. Discussion

It was shown on a wide variety of samples that the preservation of the ultrastructure is improved when cryofixation is applied as an alternative to chemical fixation (Kaneko and Walther, 1995; Kellenberger *et al.*, 1992; Mony and Larras-Regard, 1997; Morason *et al.*, 1994; Murk *et al.*, 2003; Royer and Kinnamon, 1996;

9. Ultrastructural Preservation and High Pressure Freezing

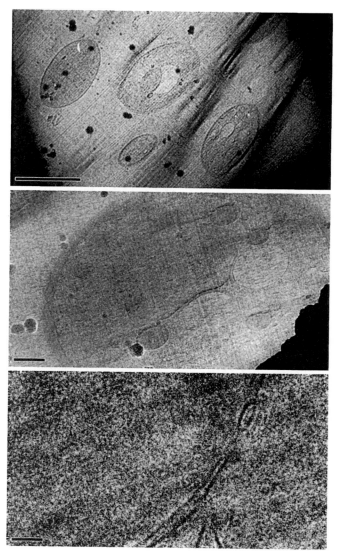

Fig. 9 Yeast cells investigated on a cryostage in the electron microscope after cryosectioning of a high pressure frozen sample. The frozen biological matter is visible because of phase contrast. This approach in electron microscopy shows biological details closest to their native state. The overview shows some single yeast cells. The image of the middle shows a single yeast cell. The cell wall is well depicted. The largest organelle is the nucleus. The bright organelles represent vacuoles and the dark ones mitochondria. The bottom image shows the nuclear membrane with pores. The nucleus is located left of the membranes. Bar: 10 μm (top), 1 μm (middle), and 100 nm (bottom).

Studer et al., 1992 and 1995; Vonschack and Fakan, 1993). Often, once-thought biological features could be debunked to aldehyde-induced artifacts (Van Harreveld et al., 1965; Wagner and Andrews, 1985). Also immunolabelling methods profit from these improvements: epitopes do not react with fixatives and therefore survive the embedding better, which leads to an increased signal (Lawrence et al., 1995).

As shown earlier, the sample thickness is limited by the physics of cooling. Thanks to the introduction of high pressure, typical biological samples can be vitrified up to 200 µm thickness and a diameter of 1–3 mm. For studying large samples, this is not sufficient. However, whenever high pressure freezing is possible, it increases the structural preservation enormously. Up to date there is no alternative available for a better preserving of small samples of a thickness of 200 µm.

References

Al-Amoudi, A., Chang, J. J., Leforestier, A., McDowall, A., Salamin, L. M., Norlen, L. P. O., Richter, K., Blanc, N. S., Studer, D., and Dubochet, J. (2004a). Cryo-electron microscopy of vitreous sections. *EMBO J.* **23**(18), 3583–3588.

Al-Amoudi, A., Norlen, L. P. O., and Dubochet, J. (2004b). Cryo-electron microscopy of vitreous sections of native biological cells and tissues. *J. Struct. Biol.* **148**(1), 131–135.

Angell, C. A., and Choi, Y. (1986). Crystallization and vitrification in aqueous systems. *J. Microsc.* **141**(3), 251–261.

Biel, S. S., Kawaschinski, K., Wittern, K. P., Hintze, U., and Wepf, R. (2003). From tissue to cellular ultrastructure: Closing the gap between micro- and nanostructural imaging. *J. Microsc.* **212**(1), 91–99.

Butz, P., Ries, J., Traugott, U., Weber, H., and Ludwig, H. (1990). The high-pressure inactivation of bacteria and bacterial-spores. *Pharmazeutische Industrie* **52**(4), 487–491.

Claeys, M., Vanhecke, D., Couvreur, M., Tytgat, T., Coomans, A., and Borgonie, G. (2004). High pressure freezing and freeze substitution of gravid Caenorhabditis elegans (Nematoda: Rhabditida) for transmission electron microscopy. *Nematology* **6**(3), 319–327.

Dubochet, J., Adrian, M., Chang, J. J., Homo, J. C., Lepault, J., McDowall, A., and Schultz, P. (1988). Cryo-electron microscopy of vitrified specimen. *Quart. rev. biophys.* **21**(2), 129–228.

Epperlein, H. H., Radomski, N., Wonka, F., Walther, P., Wilsch, M., Muller, M., and Schwarz, H. (2000). Immunohistochemical demonstration of hyaluronan and its possible involvement in axolotl neural crest cell migration. *J. Struct. Biol.* **132**(1), 19–32.

Grunfelder, C. G., Engstler, M., Weise, F., Schwarz, H., Stierhof, Y. D., Boshart, M., and Overath, P. (2002). Accumulation of a GPI-anchored protein at the cell surface requires sorting at multiple intracellular levels. *Traffic* **3**(8), 547–559.

Hohenberg, H., Tobler, M., and Muller, M. (1996). High-pressure freezing of tissue obtained by fine-needle biopsy. *J. Microsc.* **183**(2), 133–139.

Hsieh, C. E., Leith, A., Mannella, C. A., Frank, J., and Marko, M. (2006). Towards high-resolution three-dimensional imaging of native mammalian tissue: Electron tomography of frozen-hydrated rat liver sections. *J. Struct. Biol.* **153**(1), 1–13.

Kaneko, Y., and Walther, P. (1995). Comparison of ultrastructure of germinating pea leaves prepared by high-pressure freezing-freeze substitution and conventional chemical fixation. *J. Electron Microsc.* **44**(2), 104–109.

Kanno, H., Miyata, K., Tomizawa, K., and Tanaka, H. (2004). Additivity rule holds in supercooling of aqueous solutions. *J. Phys. Chem. A* **108**(28), 6079–6082.

Kanno, H., Speedy, R. J., and Angell, C. A. (1975). Supercooling of water to $-92\,°C$ under pressure. *Science* **189**, 880–881.

Kellenberger, E., Johansen, R., Maeder, M., Bohrmann, B., Stauffer, E., and Villiger, W. (1992). Artifacts and morphological-changes during chemical fixation. *J. Microsc.* **168**(2), 181–201.

Lawrence, P. A., Young, R. D., Duance, V. C., and Monaghan, P. (1995). High pressure cryofixation for immuno-electron microscopy of human cartilage. *Biochem. Soc. Trans.* **23**(4), S508.

Leforestier, A., Richter, K., Livolant, F., and Dubochet, J. (1996). Comparison of slam-freezing and high-pressure freezing effects on the DNA cholesteric liquid crystalline structure. *J. Microsc.* **184**(1), 4–13.

Liu, Y. P., Hidaka, E., Kaneko, Y., Akamatsu, T., and Ota, H. (2006). Ultrastructure of Helicobacter pylori in human gastric mucosa and *H. pylori*-infected human gastric mucosa using transmission electron microscopy and the high-pressure freezing-freeze substitution technique. *J. Gastroenterol.* **41**(6), 569–574.

Luft, J. H. (1961). Improvements in epoxy resin embedding methods. *J. Biophys. Biochem. Cytol.* **9**(2), 409–414.

Matias, V. R. F., Al-Amoudi, A., Dubochet, J., and Beveridge, T. J. (2003). Cryo-transmission electron Microscopy of frozen-hydrated sections of *Escherichia coli* and Pseudomonas aeruginosa. *J. Bacteriol.* **185**(20), 6112–6118.

Mony, M. C., and Larras-Regard, E. (1997). Imaging of subcellular structures by scanning ion microscopy and mass spectrometry. Advantage of cryofixation and freeze substitution procedure over chemical preparation. *Biol. Cell* **89**(3), 199–210.

Moor, H. (1987). Theory and practice of high pressure freezing. *In* "Cryothechniques in Biological Electron Microscopy" (R. A. S. Steinbrecht, and K. Zierold, eds.), pp. 175–191. Springer-Verlag, Berlin.

Moor, H., Bellin, G., Sandri, C., and Akert, K. (1980). The influence of high-pressure freezing on mammalian nerve-tissue. *Cell Tissue Res.* **209**(2), 201–216.

Moor, H., and Hoechli, M. (1970). The influence of high pressure freezing on living cells. *In* "Proceedings of the 7th international congress on electron microscopy" (P. Favard, ed.), pp. 445–446.

Morason, R. T., Allenspach, A. L., and Lee, R. E. (1994). Comparative ultrastructure of fat-body cells of freeze-susceptible and freeze-tolerant eurosta-solidaginis larvae after chemical fixation and high-pressure freezing. *J. Insect Physiol.* **40**(2), 155–164.

Muhlfeld, C., and Richter, J. (2006). High-pressure freezing and freeze substitution of rat myocardium for immunogold labeling of connexin 43. *Anatom. Rec. A* **288A**(10), 1059–1067.

Murk, J. L. A. N., Posthuma, G., Koster, A. J., Geuze, H. J., Verkleij, A. J., Kleijmeer, M. J., and Humbel, B. M. (2003). Influence of aldehyde fixation on the morphology of endosomes and lysosomes: Quantitative analysis and electron tomography. *J. Microsc.* **212**(1), 81–90.

Pfeiffer, S., and Krupinska, K. (2005). Chloroplast ultrastructure in leaves of *Urtica dioica* L. analyzed after high-pressure freezing and freeze-substitution and compared with conventional fixation followed by room temperature dehydration. *Microsc. Res. Tech.* **68**(6), 368–376.

Riehle, U., and Hoechli, M. (1973). The theory and technique of high-pressure freezing. *In* "Freeze-Etching Technique and Applications" (E. C. Benedetti, and P. Favard, eds.), pp. 31–61. Societe Francaise de Microscopie Electronic, Paris.

Royer, S. M., and Kinnamon, J. C. (1996). Comparison of high-pressure freezing/freeze substitution and chemical fixation of catfish barbel taste buds. *Microsc. Res. Tech.* **35**(5), 385–412.

Sartori, N., Richter, K., and Dubochet, J. (1993). Vitrification depth can be increased more than 10-fold by high-pressure freezing. *J. Microsc.* **172**(1), 55–61.

Semmler, K., Wunderlich, J., Richter, W., and Meyer, H. W. (1998). High-pressure freezing causes structural alterations in phospholipid model membranes. *J. Micros.* **190**, 317–327.

Shanbhag, S. R., Park, S. K., Pikielny, C. W., and Steinbrecht, R. A. (2001). Gustatory organs of Drosophila melanogaster: Fine structure and expression of the putative odorant-binding protein PBPRP2. *Cell Tissue Res.* **304**(3), 423–437.

Studer, D., Graber, W., Al-Amoudi, A., and Eggli, P. (2001). A new approach for cryofixation by high pressure freezing. *J. Microsc.* **203**, 285–294.

Studer, D., Hennecke, H., and Muller, M. (1992). High-pressure freezing of soybean nodules leads to an improved preservation of ultrastructure. *Planta* **188**(2), 155–163.

Studer, D., Michel, M., and Müller, M. (1989). High-pressure freezing comes of age. *Scanning Microsc.* (suppl. 3), 253–269.

Studer, D., Michel, M., Wohlwend, M., Hunziker, E. B., and Buschmann, M. D. (1995). Vitrification of articular-cartilage by high-pressure freezing. *J. Microsc.* **179**(3), 321–332.

Shimoni, E., and Muller, M. (1998). On optimizing high-pressure freezing: From heat transfer theory to a new microbiopsy device. *J. Microsc.* **192**(3), 236–247.

Suppes, G. J., Egan, S., Casillan, A. J., Chan, K. W., and Seckar, B. (2003). Impact of high pressure freezing on DH5 alpha *Escherichia coli* and red blood cells. *Cryobiology* **47**(2), 93–101.

Tokuyasu, K. T. (1973). A technique for ultracryotomy of cell suspensions and tissues. *J. Cell Biol.* **57**, 551–565.

Van Belleghem, F., Cuypers, A., Semane, B., Smeets, K., Vangronsveld, J., d'Haen, J., and Valcke, R. (2007). Subcellular localization of cadmium in roots and leaves of *Arabidopsis thaliana*. *New Phytologist* **173**(3), 495–508.

Van Harreveld, A., Crowell, J., and Malhotra, S. K. (1965). A study of extracellular space in central nervous tissue by freeze-substitution. *J. Cell Biol.* **25**, 117–137.

Vanhecke, D., Graber, W., Herrmann, G., Al-Amoudi, A., Eggli, P., and Studer, D. (2003). A rapid microbiopsy system to improve the preservation of biological samples prior to high-pressure freezing. *J. Microsc.* **212**(1), 3–12.

Vonschack, M. L., and Fakan, S. (1993). The study of the cell-nucleus using cryofixation and cryosubstitution. *Micron* **24**(5), 507–519.

Wagner, R. C., and Andrews, S. B. (1985). Cryofixation and chemical fixation of capillary endothelium – glutaraldehyde increases vesicle numbers. *Microvasc. Res.* **29**(2), 257–258.

Wang, L. C., Humbel, B. M., and Roubos, E. W. (2005). High-pressure freezing followed by cryosubstitution as a tool for preserving high-quality ultrastructure and immunoreactivity in the *Xenopus laevis* pituitary gland. *Brain Res. Protocols* **15**(3), 155–163.

CHAPTER 10

High-Pressure Freezing and Low-Temperature Fixation of Cell Monolayers Grown on Sapphire Coverslips

Siegfried Reipert and Gerhard Wiche

Department of Molecular Cell Biology
Max F. Perutz Laboratories
University of Vienna
Dr. Bohr-Gasse 9
A-1030 Vienna
Austria

I. Introduction
II. Materials and Instrumentation
III. Procedures
    A. Cell Growth on 1.4-mm Sapphire Coverslips
    B. HPF of Cells on 1.4-mm Sapphire Coverslips
    C. Freeze Substitution of Cell Monolayers
IV. Comments and Pitfalls
    A. Safety
    B. Sapphire Glass
    C. Handling of Sapphire Discs and Flat Sample Holders
    D. HPF with the EMPACT
    E. Ice Crystals
    F. Lowicryl Polymerization
    G. Detachment of the Resin from Sapphire Glass
References

I. Introduction

High-pressure freezing (HPF) in combination with substitution of the vitrified ice by organic solvent at low temperature, so-called freeze substitution, has widely been accepted as a method that is superior to conventional chemical fixation of the living state (for review, Moor, 1987; Schwarz *et al.*, 1993). Once embedded in resins and sectioned for transmission electron microscopy, such cryoprocessed tissues and cells show less extraction of biological material and less aggregation artifacts. In addition to these preparative aspects, the immobilization of the living state within milliseconds and its chemical fixation at low temperature offer bright prospects for the correlation of light- and electron microscopy on cell monolayers. The sequence of dynamic fine structural alterations, recorded by video microscopy, could be reconstructed by electron microscopic "snap shots."

Recently, sapphire glass was found to be a suitable tissue culture substratum for cell monolayer freezing at both ambient and high pressures (Hess *et al.*, 2000; Neuhaus *et al.*, 1998; Reipert *et al.*, 2003, 2004a; Schwarb, 1990; Wild *et al.*, 2001). The interest in the latter is understandable, as HPF achieves depths of ice-crystal free freezing that exceed the thickness of cells. Therefore, routinely reproducible results can be expected by this freezing technique. Apart from this, sapphire substrata are appreciated for their optical and mechanical properties.

Currently, different technical solutions for HPF are commercially available. Although initially designed for bulky tissue samples, they also allow cryofixation of cell monolayers on sapphire coverslips. While the stationary BAL-TEC (Lichtenstein) high-pressure freezer offers the possibility to freeze cells on sapphire with a large diameter (3 mm), LEICA Microsystems (Austria) follows the concept for mobile HPF that requires sapphire sizes of 1.4 mm in diameter. In its new version, the LEICA high-pressure freezer can be coupled with a rapid transfer station for sapphire coverslips coming directly from the light microscope.

Once ice-crystal damage-free freezing by one technique or the other is accomplished, success largely depends on reliable procedures for cryosample handling and on optimized freeze substitution/embedding protocols. The protocols presented here were adapted to the special needs for handling of tiny 1.4 mm sapphire coverslips. However, they consider also more general principles in cryopreparation that could be of common interest. Furthermore, a protocol is provided for best fine structural preservation in epoxy resin, comprising low temperature fixation and osmification of cell monolayers in the substitution medium acetone (Reipert *et al.*, 2004a; Zenner *et al.*, 2007). Finally, since osmification and epoxy resin embedding are not promising options for postembedding immunogold labeling, technical solutions are offered for low-temperature fixation and embedding of cell monolayers in the metacrylic resin Lowicryl HM20.

II. Materials and Instrumentation

HPF was performed by using the LEICA EMPACT high-pressure freezer with a flat specimen holder system that includes a loading device (LEICA Microsystems, Austria) for handling of flat gold-plated specimen carriers (0.5 mm thick, 1.5 mm in diameter, 200 μm deep; LEICA Cat. #16706898). Sapphire coverslips (1.4 mm diameter, 0.05 mm thick) were purchased from LEICA Microsystems (Cat. #16706849) and Wohlwend GmbH (Sennwald, Switzerland).

Freeze substitution was performed using an automatic freeze substitution unit (AFS; LEICA Microsystems). The spider cover with eight positions (LEICA Cat. #16702743), stem holder for spider cover (LEICA Cat. #16702744), and universal containers for plastic capsules (LEICA Cat. #16702735) were used in the way described below. Multiply-Pro cups, 0.2 ml, for PCR (Sarstedt, Germany; Cat. #72.737.002) were found to be particularly useful for economical use of Lowicryl and its polymerization under UV light. Attention was paid to low water content in the substitution medium (use of dried acetone, SeccoSolv with max. 0.01% H_2O, Merck, Germany; dehydration of uranyl acetate by calcium chloride). Low temperature measurements were done with an instrument, Testo 735–2 (Testo AG, Lenzkirch, Germany), equipped with an immersion probe TC Type K.

Selection of the right tweezers for sample handling was found to be of particular importance. The following models were used: tweezers, super fine points, short, Style #4; tweezers, thick point, short, Style #5SA; and wafer tweezers, Style #39S2.

Other chemicals and consumables mentioned in this article are commonly available from electron microscopy supply companies.

III. Procedures

A. Cell Growth on 1.4-mm Sapphire Coverslips

1. Clean the sapphire coverslips as follows:
   - Treat them with concentrated HCl for 2 h.
   - Rinse them under running tap water.
   - Subsequently, wash them in double distilled water.
   - Place them in a plastic Petri dish containing 70% ethanol.
2. Transfer the coverslips (one per well) into a 24-well plate with culture medium or matrix protein solution. Incubate them at 36 °C in culture medium overnight. If cells, e.g., keratinocytes, require collagen coating, place them in collagen solution and incubate them for 3 h at 37 °C or in the fridge over night.

3. Replace the content of the wells by 0.5 ml fresh medium without disturbing the coverslips at the bottom. Use a sterile Gilson pipette for this purpose.
4. Seed cells onto coverslips by seeding 1 ml cell suspension, obtained by tryzinisation of a 90% subconfluent cell layer.
5. Incubate cells (murine keratinocytes in keratinocyte full medium without Ca^{++}; PtK2 rat kangaroo cells in DMEM) for 2 days at 37 °C and 5% CO_2.

B. HPF of Cells on 1.4-mm Sapphire Coverslips

For transfer of the cells under well-defined conditions it is necessary to position the tissue culture incubator and high-pressure freezer next to each other. The mobile LEICA EMPACT HPF machine and its successor device allow temporary placement in tissue culture labs. For freezing, sapphire coverslips with a diameter of 1.4 mm have to be transferred into flat sample carriers with an inner diameter of 1.5 mm. Ideally this should be done under a stereo microscope with long distance range. Transfer and fitting of the sample carrier into a specimen pod, supported by a loading station, take about 10–15 s. Further 10–15 s are required for mounting of the specimen pods onto the loading device and starting of the freezing process via a tough-sensitive screen.

Steps

1. Keep a flat sample carrier and a sample pod ready in the loading station.
2. Take the sapphire coverslip out of the well plate (tweezers #5SA) and dip it shortly into a 20% BSA solution in phosphate buffered saline (PBS) before putting it with the cells facing up into the sample holder. Make sure that no air bubbles are trapped and the holder is not overloaded with BSA solution.
3. Shift the lever with sample carrier vertically into the flat specimen pod and seal the pod by tightening it with a torque wrench. As a result the cell layer is sandwiched between a black diamond and the sample carrier itself. The cell-free back side of the sapphire coverslip faces a small hole at the bottom of the specimen carrier through which the high pressure would be inserted (Fig. 1).
4. Mount the specimen pod onto the loading device, insert it into the HP-freezer, lock the tube connection to the sample pod and start freezing at high pressures between 1950 and 2020 bars. The high pressure can be adjusted conveniently by regulation of the internal working pressure of the machine. After freezing, the sample pod will be released automatically into a liquid N_2 bath.
5. While still in liquid N_2, separate the sample carrier from the specimen pod and transfer it to the freeze substitution unit or to a storage Dewar. Make sure to use precooled tools and tweezers for handling.
6. Since the BSA solution is sticky, always clean the sample pods after use: remove the black diamond for this purpose and clean its surface on wet filter paper. Perform ultrasonication of the pods in double distilled water. Subsequently, blow out residual water from the inside of the pod and dry it on a warm plate.

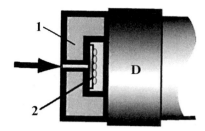

Fig. 1 Cross section through a flat sample carrier (1) with a hole at its bottom for insertion of high pressure (arrow). The cell monolayer on the sapphire coverslip (2) faces towards the opening of the carrier that is sealed by a black diamond (D). The space in between is filled with BSA solution.

C. Freeze Substitution of Cell Monolayers

Importantly, freeze substitution has to be performed while the sapphire coverslips are still in the sample carriers. This avoids the unintended loss of sample orientation during processing and handling. Despite the use of an open carrier, the coverslip will remain in it during the whole procedure.

The chamber space of a freeze substitution unit is limited. The arrangement of containers, vessels for polymerization and tools seen in Fig. 2, allows processing of a small number of samples. However, by exchange of vessels in accordance to the needs of the protocol, safe handling of up to 24 samples is possible.

1. Preparation of the AFS

1. Program the AFS for gradual warming up (35 °C/h) from the lowest temperature that can be achieved (−160 °C) to the substitution temperature of −90°. This will help to avoid cracks in the frozen sample.

2. Fill substitution medium in Sarstedt tubes, with a conical bottom (1–1.5 ml substitution medium/tube). Place these tubes into universal containers and cool them down to −160 °C. Also, insert media for exchange of solvents and screw-caped glasses for the uptake of liquid waste.

3. Put the liquid N_2 cooled samples into a small transfer container filled with liquid N_2 and transfer it into the AFS. Place one sample carrier per Sarstedt tube onto the frozen substitution medium.

4. Start the gradual warming up from −160 °C to the substitution temperature at a rate of 35 °C/h.

After 2 h, the AFS is ready to start the freeze substitution program destined for subsequent embedding either in epoxy resin or in Lowicryl HM20.

Fig. 2 Freeze substitution chamber of the AFS. Universal container containing Sarstedt tubes with samples (1); vessels for the uptake of tweezers (2), waste (3), and Lowicryl (4); 5–6 "work bench" created by two universal containers turned upside down. A plate for manipulation of the sample carriers (5) is located next to a lid with a sample in it and a mounted tube (6), which is ready to be connected with a spider cover. The spider cover (7) is placed on top of a stem holder.

2. Freeze Substitution and Epoxy Resin Embedding for Morphological and Fine Structural Studies

1. Perform freeze substitution at −90 °C for more than 48 h in a solution of 2% OsO_4 in dried acetone (made from crystalline OsO_4).
2. Activate low temperature-fixation by gradual warming up of the solvent by 2 °C/h up to −54 °C. Keep this temperature for 8 h before continuing the warming up at a rate of 3 °C/h up to −24 °C. Keep this temperature for 7 h.
3. For contrasting samples by OsO_4 more intensely switch from −24 °C to 0 °C and keep this temperature for 1 h.
4. Stop contrasting by washing the samples 3 times with acetone. For this purpose use plastic Pasteur pipettes with narrow tips. Ensure that OsO_4 vapor cannot escape (keep screw-caped bottles for uptake of OsO_4 waste inside the AFS).
5. Infiltrate with the following epoxy resin/acetone mixtures:
   - 1/3 volume Agar 100 and 2/3 volume acetone for 1 h at 10 °C.
   - 1/2 volume Agar 100 and 1/2 volume acetone for 1 h at 10 °C.
   - 2/3 volume Agar 100 and 1/3 volume acetone for 1 h at 20 °C.
6. Transfer the sapphire coverslips for infiltration with pure Agar 100:

- Place a small plate, made from polypropylene, under the stereo microscope (in a hood or in a well-ventilated room) and empty out the Sarstedt tube into it.
- Fix the sample holder with the sapphire facing up by using wafer tweezers, while lifting the coverslip with fine tweezers #4 (Fig. 3).
- Take tweezers #5SA and open them wide enough to get a grip of the opposite sides of the perimeter of the sapphire disc, transfer it into the lid of a 0.5-ml Eppendorf tube, and add a drop of pure epoxy resin.
- Subsequently, mount a tube with its bottom cut off onto the lid and fill it up with resin to the top.
- Infiltrate pure resin for 3 h, or over night.

7. Polymerize the samples in the oven at 36 °C for 36 h.
8. Detach the sapphire glass from the resin as follows:
    - Cut away the Eppendorf tube with a razor blade.
    - Gently remove resin that surrounds the perimeter of the sapphire by using a file.

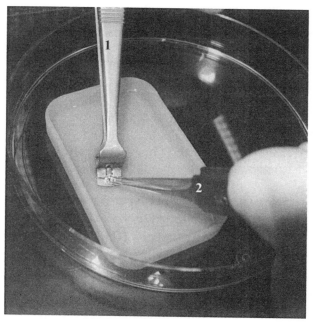

Fig. 3 Separation of the sapphire coverslip from the sample carrier. The carrier is fixed by wafer tweezers (1). While in a plate with acetone/resin, the sapphire is lifted by means of very fine tweezers #4, and subsequently transferred by using tweezers #5SA. For transfer, the tips of tweezers #5SA exert a firm grip from opposite sides of the sapphire disc perimeter (2).

- Dip the resin block with the sapphire attached to it in liquid N_2. Fix it in a bench vice. A slight lateral file stroke might be enough to detach the coverslip. If necessary, this procedure has to be repeated. Note that in almost every case the sapphire glass detaches as a whole.

9. After removal of the dust particles from the cell containing abutting face, trim a rectangular section profile with a glass knife. Subsequently, cut ultrathin sections and place them on Formvar-coated grids. After contrasting with uranyl acetate and lead citrate, they are ready for evaluation in the electron microscope.

3. Freeze Substitution and Lowicryl Embedding for Immunohistochemical Applications

As substitution media either methanol or acetone is to be chosen. To stabilize structures, uranyl acetate should be added at concentrations that are tolerated by the epitopes of the antigen (usually not above 0.5%). While uranyl salt dissolves easily in methanol, it requires dissolution in water before mixing with acetone. The water content in the uranyl/acetone solution has to be removed by drying on $CaCl_2$.

All resin/medium changes and infiltration steps will be performed safely in the AFS at −40 °C while the samples are still in the Sarstedt tubes. Importantly, the sapphire coverslips will be embedded in Lowicryl while they remain in the flat sample carrier. As a reliable option for polymerization of small amounts of Lowicryl, a system consisting of stem holder, spider cover, and PCR cups is recommended (Fig. 4).

1. Perform freeze substitution in uranyl/acetone at −90 °C for at least 2 days or in uranyl/methanol at −88 °C for at least 36 h.

2. Raise the temperature in the AFS at a rate of 2 °C/h up to −40 °C. Once the chamber is warmed up to −40 °C, perform all successive processing steps at this temperature.

3. As soon as −40 °C is reached, wash the samples 3 times with pure acetone or methanol, while they remain in the Sarstedt tubes.

4. Infiltrate your samples with Lowicryl/solvent mixtures:
   - 1/3 volume Lowicryl HM20 and 2/3 volume solvent for 30 min
   - 1/2 volume Lowicryl HM20 and 1/2 volume solvent for 1 h
   - 2/3 volume and Lowicryl HM20 1/3 volume solvent for 2 h
   - pure Lowicryl HM20 for 1 h

5. To transfer the sample carrier safely into the lids of small PCR cups create a "work bench" inside the AFS (for instance, by insertion of universal containers). Place a polypropylene plate on it for the uptake of the sample carriers (Fig. 2). Empty the sample carriers into that plate. By using precooled tweezers turn them in a way that the sapphire coverslip faces up.

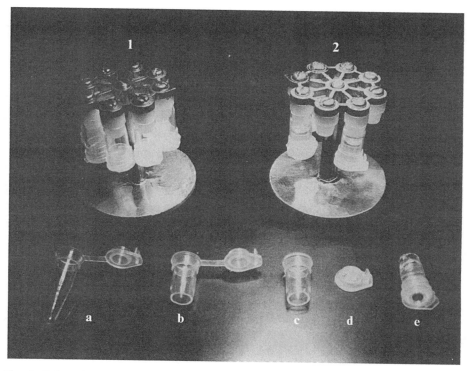

Fig. 4 Polymerization of Lowicryl in PCR cups attached to spider covers. (1) Upside down-fitting of eight cups to a spider cover placed onto a stem holder. The fitting is achieved by cutting the PCR cups (a–c). (d) Sample carrier placed into the lid. The cut cup (c) will be pressed onto the lid and filled up with resin. (2) Spider cover with already polymerized samples attached. (e) Detached polymerized sample with a carrier embedded.

6. Position the lids of the PCR cups next to the plate with the sample and fill them with a drop of fresh Lowicryl HM20.

7. Place the sample carrier into the lid while keeping its orientation with the sapphire facing up.

8. Press the tube, with its bottom cut off (Fig. 4), onto the lid and fill it up with precooled resin. Collect the sample carrier-filled PCR cups inside the AFS for mounting to a spider cover.

9. Mount them to the spider cover and place the assembly onto the stem holder (Fig. 4).

10. For UV polymerization, cover the spider with aluminum foil of the same diameter. Ensure indirect UV illumination by reflecting side walls of the AFS. Note: If the reflection from the bottom of the chamber is low, cover it with aluminum foil.

11. Place the UV lamp on top of the AFS chamber and ensure that the liquid N_2 evaporator is switched off. Polymerize for 48 h at $-40\,°C$.

12. Switch off the UV lamp and gradually warm the samples up to room temperature.

13. Subsequent curing of the resin at day light turns the color of the resin slightly pink. Now, it is safe to detach the samples from the spider cover and to release the resin block by cutting away the PCR cup with a razor blade.

14. To detach the sapphire coverslip from the Lowicryl block:
    - Gently, remove resin that surrounds the perimeter of the sample carrier by using a file.
    - Warm up the carrier by pressing the sample against a heat plate.
    - Immediately after warming, fix the sample carrier in a bench vice and gently insert force to lift the resin block. Support the lifting by a slight turning force. Usually, the sapphire glass will remain in the flat carrier while the block detaches.

15. Proceed as described for the epoxy resin embeddings to get sections on Formvar coated grids. Perform postembedding immunogold labeling, before counterstaining of sections with uranyl acetate and lead citrate.

IV. Comments and Pitfalls

HPF of monolayers provides large sample areas in equally high freezing quality (inset Fig. 5A). Their appearance in the electron microscope, however, strongly depends on the degree of extraction of biological material by solvents and possible aggregation artifacts (Fig. 5). For example, when keratin network organization in mouse keratinocytes is assessed, it becomes clear that application of methanol-based substitution in combination with Lowicryl affects fine structures in the most negative way (Fig. 6A). Structural preservation by acetone-based substitution in combination with Lowicryl embedding reduces aggregation artifacts significantly (Fig. 6B). In addition, immunogold labeling intensity might be even enhanced. However, in general, the freeze substitution protocol does not guarantee success in immunogold labeling. Therefore, one protocol or the other may prove useful for certain applications.

As evident from the keratin bundle shown in Fig. 7, embedding of 2% OsO_4/acetone-substituted samples in epoxy resin further reduces aggregation artifacts. In addition to this advantage for the preservation of cytoskeletal components, such preparations offer excellent opportunities to study membrane-bound organelles, such as mitochondria and autolytic vacuoles, in tissue culture cells (Fig. 8).

By prolongation of the time for freeze substitution, the protocols presented here are also suitable for embedding of high-pressure cryoimmobilized tissue blocks, and cell suspensions that fit in the same flat sample holders (Nowikovsky *et al.*, 2007; Reipert *et al.*, 2004b).

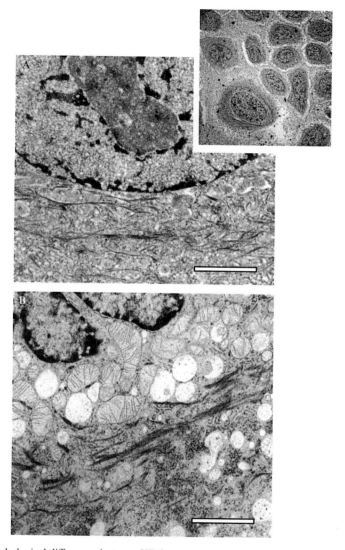

Fig. 5 Morphological differences between HP-frozen mouse keratinocytes embedded in Lowicryl or epoxy resin. (A) Freeze substitution in 0.5% uranyl acetate/methanol and subsequent embedding in Lowicryl HM20. For overview: see inset showing a confluent cell layer. Note strong aggregation of keratin bundles and the absence of membrane staining. Fine structural details of the mitochondria are disrupted. (B) Freeze substitution in 2% OsO_4/acetone and subsequent embedding in Agar 100. The cytoplasm is rich in fine structural details. Mitochondria are well preserved. Scale bars, 2 μm.

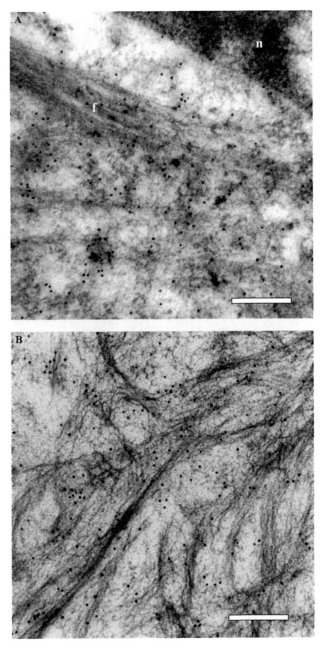

Fig. 6 Details of the keratin organization in HP-frozen mouse keratinocytes embedded in Lowicryl HM20 and immunolabeled for plectin with 10-nm colloidal gold: (A) Freeze substitution in 0.5% uranyl acetate/methanol. (B) Freeze substitution in 0.5% uranyl acetate/acetone. Note the aggregation of keratin to thick filament bundles (f) in A. n, nucleus. Scale bars, 200 nm.

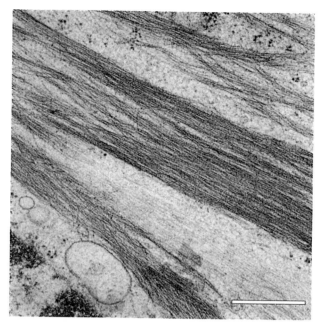

Fig. 7 Detail of the keratin organization in a HP-frozen mouse keratinocyte, freeze-substituted in 2% OsO$_4$/acetone and embedded in epoxy resin. Individual fibers of a keratin bundle are resolved. Scale bars, 200 nm.

A. Safety

For the use of the cryotechnical equipment follow the safety instructions provided by the manufacturers. Chemicals used for freeze substitution and embedding are toxic (mutagenic, allergenic, and in some cases perhaps carcinogenic) and should be handled with adequate safety precautions. Gloves have to be used that resist penetration of the resin/solvent mixtures. All preparations have to be performed in a well-ventilated hood. Exhaust gases from the AFS should be guided into a hood as well. Fine dust particles of the resins should be hoovered away immediately.

B. Sapphire Glass

Make sure that you get batches of sapphire glass that do not show scratches under the stereo microscope. Otherwise, you might put the diamond knife at risk.

C. Handling of Sapphire Discs and Flat Sample Holders

By choosing the appropriate tools, loss of samples and damage to tweezers can be avoided. Therefore, pay attention to the tweezers suggested in the protocol.

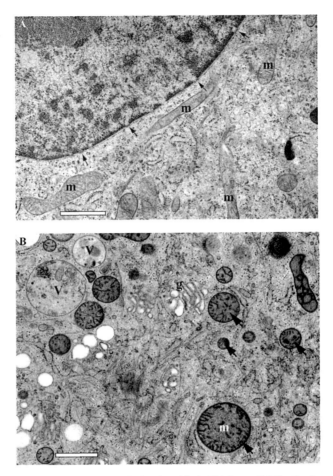

Fig. 8 Oxidative stress induced fission of mitochondria in HP-frozen PtK2 cells, freeze-substituted in 2% OsO$_4$/acetone and embedded in epoxy resin. (A) Detail of an untreated cell that contains elongated mitochondrial profiles (m). Note also well-preserved nuclear structures and nuclear pores (arrows). (With permission from Wiley–Blackwell Publishing Ltd. (Reipert *et al.*, 2004a).) (B) Detail of a hydrogen peroxide-treated cell displaying round mitochondrial profiles (m) at various sizes (arrows). Note also autolytic vacuoles (v) and parts of the Golgi complex (g). Scale bars, 1 μm.

D. HPF with the EMPACT

- Make sure that there are no air bubbles in the hydraulic liquid (methyl cyclohexan). They may cause pressure fluctuations that inflict damage to the frozen sample.
- Ensure that the working pressure required to generate the high-pressure is sufficiently high. As a precondition, the incoming pressure provided by a compressor or a laboratory air supply should be high enough.

- After each shot, inspect the flange of the sealing tube made from copper. If damages are indicated, this tube has to be exchanged.
- The freezing curves recorded by the EMPACT enable detection of gross malfunction. However, satisfactory synchronization between pressure and temperature during freezing requires fine tuning of the working pressure according to the freezing result. Not just damage of structures by ice crystals, but also fracture of otherwise well frozen cells, damage of plasma membranes and deformation of nuclei may indicate the need to alter the working pressure.

E. Ice Crystals

Small, bright spots in the cytoplasm, net-like patterns within chromatin, nucleoli, or mitochondria indicate ice crystals. They may originate from suboptimal freezing, handling errors during sample transfer into the AFS, warming up of the samples during refilling of liquid N_2 into the AFS, or freeze substitution at temperate above the re-crystallization temperature. Since there are no temperature sensors attached to the solvents it is recommended to check the substitution temperature, directly, by using a temperature measurement instrument with an immersion probe type K (Testo AG), calibrated for low temperature application.

F. Lowicryl Polymerization

Although polymerization in 0.5-ml Eppendorf tubes works well (Reipert *et al.*, 2004a), the use of smaller PCR cups harnesses all eight positions of the spider cover and saves expensive Lowicryl. For successful polymerization, do not take the samples out immediately after the resin in the tubes appears to have hardened. The inside of the sample holder with the cell monolayer polymerizes particularly slowly and, therefore, requires prolonged UV exposure. Change the UV lamp from time to time since it alters its characteristic during its life time. Alternatively, you might consider LED-UV lamps for use.

G. Detachment of the Resin from Sapphire Glass

None of the detachment procedures leads to breakage of the sapphire glass. However, since Lowicryl is brittle, during filing off the block could break at the rim of the carrier. Gentle filing off will grant a high success rate.

Acknowledgments

We thank I. Fischer, B. Wysoudil, and H. Kotisch for technical assistance. This work was supported by the Austrian Science Fund (project P19381) to S.R.

References

Hess, M. W., Müller, M., Debbage, P. L., Vetterlein, M., and Pavelka, M. (2000). Cryopreparation provides new insight into the effects of brefeldin A on the structure of the HepG2 Golgi apparatus. *J. Struct. Biol.* **130,** 63–72.

Moor, H. (1987). Theory and practice of high-pressure freezing. *In* "Cryotechniques in Biological Electron Microscopy" (R. A. Steinbrecht, and K. Zierold, eds.), pp. 175–191. Springer, Berlin.

Neuhaus, E. M., Horstmann, H., Almers, W., Maniak, M., and Soldati, T. J. (1998). Ethane-freezing/methanol-fixation of cell monolayers: A procedure for improved preservation of structure and antigenicity for light and electron microscopies. *Struct. Biol.* **121,** 326–342.

Nowikovsky, K., Reipert, S., Devenish, R. J., Schweyen, R. J. (2007). Mdm38 protein depletion causes loss of mitochondrial K(+)/H(+) exchange activity, osmotic swelling and mitophagy. *Cell Death Differ.* **14,** 1647–1656

Reipert, S., Fischer, I., and Wiche, G. (2003). Cryofixation of epithelial cells grown on sapphire coverslips by impact freezing. *J. Microsc.* **209,** 76–80.

Reipert, S., Fischer, I., and Wiche, G. (2004a). High-pressure freezing of epithelial cells on sapphire coverslips. *J. Microsc.* **213,** 81–85. Erratum in: *J. Microsc.* **215,** 313.

Reipert, S., Fischer, I., and Wiche, G. (2004b). High-pressure cryoimmobilization of murine skin reveals novel structural features and prevents extraction artifacts. *Exp. Dermatol.* **13,** 419–425.

Schwarb, P. (1990). Morphologische Grundlagen zur Zell-Zell Interaktion bei adulten Herzmuskelzellen in Kultur. Ph.D. Thesis, Swiss Federal Institute of Technology. Zurich. [Diss. ETH-Zürich, 9195].

Schwarz, H., Hohenberg, H., and Humbel, B. M. (1993). Freeze-substitution in virus research: A preview. *In* "Immunogold Electron Microscopy in Virus Diagnosis and Research" (D. Hyatt, and B. T. Eaton, eds.), pp. 349–376. CRC Press, Boca Raton, FL.

Wild, P., Schraner, E. M., Adler, H., and Humbel, B. M. (2001). Enhanced resolution of membranes in cultured cells by cryoimmobilization and freeze-substitution. *Microsc. Res. Tech.* **53,** 313–321.

Zenner, H. L., Collinson, L. M., Michaux, G., and Cutler, D. F. (2007). High-pressure freezing provides insights into Weibel-Palade body biogenesis. *J. Cell Sci.* **120,** 2117–2125.

CHAPTER 11

Freeze-Fracture Cytochemistry in Cell Biology

Nicholas J. Severs* and Horst Robenek[†]

*Imperial College London
National Heart and Lung Division
London, United Kingdom

[†]Institute for Arteriosclerosis Research
University of Münster
Münster, Germany

    Abstract
I. Introduction
II. Basic Rationale: Solving the Problems of Combining Freeze Fracture with Cytochemistry
III. Methods
    A. Freeze-Fracture Nomenclature
    B. Fracture-Label
    C. Label-Fracture
    D. Freeze-Fracture Replica Immunolabeling
IV. Discussion: Impact on Topical Questions in Cell Biology
    A. Spatial Organization of Proteins in the Plasma Membrane
    B. Lipid Droplets
V. Concluding Comment
    References

Abstract

The term *freeze-fracture cytochemistry* embraces a series of techniques which share the goal of chemical identification of the structural components viewed in freeze-fracture replicas. As one of the major features of freeze fracture is its ability to provide planar views of membranes, a major emphasis in freeze-fracture cytochemistry is to identify integral membrane proteins, study their spatial organization in the membrane plane, and examine their role in dynamic cellular processes. Effective techniques in freeze-fracture cytochemistry, of wide application in cell biology, are now available. These include *fracture-label*, *label fracture*, and the freeze-fracture replica immunolabeling technique (FRIL). In fracture-label, samples are frozen and fractured, thawed for labeling, and finally processed for viewing either by critical-point drying and platinum–carbon replication or by thin-section electron microscopy. Label-fracture involves immunogold labeling a cell suspension, processing as for standard freeze-fracture replication, and then examining the replica without removal of the cellular components. Of greatest versatility, however, is the FRIL technique, in which samples are frozen, fractured, and replicated with platinum–carbon as in standard freeze fracture, and then carefully treated with sodium dodecylsulphate (SDS) to remove all the biological material except a fine layer of molecules attached to the replica itself. Immunogold labeling of these molecules permits the distribution of identified components to be viewed superimposed upon high resolution planar views of replicated membrane structure, for both the plasma membrane and intracellular membranes in cells and tissues. Examples of how these techniques have contributed to our understanding of cardiovascular cell function in health and disease are discussed.

I. Introduction

Freeze-fracture electron microscopy was established as a major technique in ultrastructure research over 30 years ago. The success and credibility of the technique arose from its ability to provide compelling images of membranes that were entirely novel in their scientific content, set in the context of cellular features that were instantly recognizable from existing knowledge. The landmark paper by Moor and Mühlethaler (1963) presenting the first successful freeze-fracture micrographs of cells attracted wide interest, but interpretative uncertainties initially hindered full exploitation of the technique. From the early 1970s, however, with interpretative controversies resolved, the scene was set for the technique to flourish; and this it did over the following two decades, profoundly shaping our understanding of the structural organization of the plasma membrane, membrane-bound organelles and other components of the cell. With the supply of new discoveries diminishing, and molecular techniques in the ascendant, interest in freeze fracture declined in the 1990s, but a revival is currently taking place, stimulated by the development of effective approaches in freeze-fracture cytochemistry. This chapter aims to give a brief overview of the state-of-the-art in this field, focusing on those techniques that are currently contributing to new advances in cell biology.

An understanding of the principles and methodology of freeze-fracture cytochemistry requires a basic knowledge of how standard freeze fracture works. The utility of freeze fracture depends critically on the tendency of the fracture plane to follow a plane of weakness in the hydrophobic interior of frozen membranes, splitting them into half-membrane leaflets. This creates, at low magnification, spectacular three-dimensional perspectives of cellular organization in which *en face*-viewed membranes are brought to the fore. At high magnification, details of membrane structure are seen at macromolecular resolution. In particular, the distribution and organization of integral membrane proteins (seen as intramembrane particles) are viewed in the membrane plane.

There are four essential steps in making a standard freeze-fracture replica: (1) rapid freezing of the specimen, (2) fracturing it at low temperature (−100 °C or lower), (3) making a replica of the newly exposed frozen surface by vacuum-deposition of platinum and carbon, and (4) cleaning the replica to remove the biological material. The replica is then examined in the transmission electron microscope. For a detailed protocol on how to carry out freeze fracture, see Severs (2007).

In routine application, the freezing step is often preceded by pretreatment with glutaraldehyde fixation and glycerol cryoprotection. Ultrarapid freezing techniques, such as copper block impact freezing and propane jet freezing, overcome the need for cryoprotection, and provide opportunities for examining specimens frozen directly from the living state and capturing rapid, dynamic cellular events (Escaig, 1982; Heuser, 1981; Müller *et al.*, 1980). An optional step of etching, involving vacuum sublimation of ice, may be interposed between fracturing and replication to reveal the surface structure of cells and their components. One particularly effective technique for imaging surface structure is the combination of ultrarapid freezing, extended etching and low-angle rotary shadowing, a technique popularized as the "quick-freeze/deep-etch technique" (Heuser, 1981, 1989; Heuser and Salpeter, 1979).

Freeze fracture and associated techniques revolutionized the way we look at membranes. Nevertheless, without a means to identify the chemical nature of the structural components visualized, the functional role of these components all too often remained a matter of intelligent guesswork rather than firmly established fact. For this reason, the combination of cytochemistry with freeze fracture was an eagerly sought goal. The technical challenges involved in developing effective techniques in freeze-fracture cytochemistry were considerable, however, and took several decades to overcome. Ironically, with the major ultrastructural discoveries of standard freeze fracture in place, interest in freeze-fracture electron microscopy ebbed at the very time that some of the solutions to effective freeze-fracture cytochemistry were emerging. This has left something of a dearth in expertise to exploit the opportunities afforded today.

II. Basic Rationale: Solving the Problems of Combining Freeze Fracture with Cytochemistry

Originally, the combination of cytochemistry with freeze fracture was thought to be impossible because of the need to clean the replica scrupulously (e.g., with sodium hypochlorite or chromic acid) in order to remove all biological material that would otherwise obstruct the electron beam and interfere with visualization of detail in the replica. Such cleaning inevitably leads to loss of any label that might be attached to the biological material before or after freeze fracture. Thus, the only labeling procedure applied in the early days of freeze fracture involved attaching relatively nonspecific markers to the surfaces of cells in suspension, and then etching to expose these markers for visualization as shadowed particles on a margin of the true surface of the membrane, adjacent to the membrane fracture face (Pinto da Silva and Branton, 1970; Tillack and Marchesi, 1970).

During the 1980s, however, more sustained and increasingly imaginative attempts at further development of freeze-fracture cytochemistry gathered pace (see Severs, 1995a). An early attempt to solve the problem of losing preattached label along with the biological material at the cleaning stage came with "the sectioned replica technique," in which the replica was viewed en face, together with labeled biological material still in place, from within a thin section (Rash et al., 1989). Perturbation of membrane structure before freezing using specific lipid binding agents (e.g., filipin and tomatin) offered another avenue; here the structural alterations, rather than an attached label, acted as the marker (Severs and Robenek, 1983). This approach enjoyed a spell of popularity in the 1980s but, with the adoption of a more critical approach to interpretation, proved to have more restricted applications than many researchers initially anticipated (Severs, 1995b, 1997; Severs and Simons, 1983).

Significant breakthroughs came with the introduction of colloidal gold cytochemistry to electron microscopy in the early 1980s. Gold particles, because of their high electron density and small size, were quickly recognized to be ideal markers to use in conjunction with replicas, and this stimulated experimentation with new ideas for retaining label in such a way that it could be viewed superimposed on replicated structural detail. Experimental strategies for integrating gold labeling into the freeze-fracture procedure were tried at each of the key steps in the freeze-fracture procedure: (1) before the freezing step, (2) after fracturing, and (3) after replication. Notable among the techniques developed were "label-fracture" (Pinto da Silva and Kan, 1984) in category (i) and "fracture-label" (Pinto da Silva et al., 1981a,b) in category (ii). Among the techniques in category (iii) were "replica-staining label-fracture" (Andersson Forsman and Pinto da Silva, 1988), "replica-label-whole mount" (Rash et al., 1989) and the "SDS freeze-fracture replica labeling technique" otherwise known as freeze-fracture replica immunolabeling (FRIL) (Fujimoto, 1995). These various techniques differ in precisely *how* freeze fracture and cytochemistry are combined, *what* type of information is obtained, and *which* types of specimen are amenable to study. For a detailed discussion of the full range of techniques, the reader is referred to an earlier comprehensive review (Severs, 1995a). In this article, we will concentrate on three techniques that have stood the test of time, "fracture-label" (category (i)), "label-fracture" (category (ii)), and "FRIL" (category (iii)).

III. Methods

A. Freeze-Fracture Nomenclature

An understanding of these three freeze-fracture cytochemistry techniques requires a grasp of the nomenclature used for describing the aspects of the membrane viewed and accessible to label with freeze fracture and etching (Branton *et al.*, 1975). The nomenclature is best explained by envisaging the membrane as consisting of two halves—a *P half* which lies adjacent to the protoplasm, and an *E half* which lies adjacent to the extracellular, exoplasmic, or endoplasmic space (Fig. 1). The term *fracture face* is reserved for the interior views of membranes exposed by freeze fracturing, while the term *surface* is used for the true, natural surfaces of the membrane that may potentially be exposed by etching. The fracture face of the P half is thus termed the *P face*, while that of the E half is termed the *E face*. The true surfaces of the membrane are correspondingly designated the *P surface* and the *E surface*, respectively. When applying the nomenclature to intracellular membranes, the term "P" encompasses cytoplasm, nucleoplasm, the matrix of mitochondria, and the stroma of chloroplasts, while the term "E" is used to designate the spaces between inner and outer membranes of all double membrane-bound organelles (nucleus, mitochondria, and chloroplasts), and the lumina of all single-membrane organelles (Fig. 2).

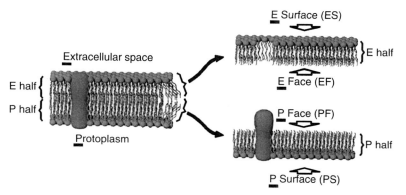

Fig. 1 Nomenclature for describing the aspects of the plasma membrane revealed by freeze fracture and etching. The membrane comprises a lipid bilayer with intercalated proteins. The half-membrane leaflet adjacent to the extracellular space is termed the E half; that adjacent to the protoplasm is termed the P half. The term *fracture face* is reserved for the interior views of membranes exposed by freeze fracturing, while the term *surface* is used for the true, natural surfaces of the membrane that may potentially be exposed by etching. The fracture face of the P half is thus termed the *P face* (or PF), while that of the E half is termed the *E face* (or EF). The true surfaces of the membrane are correspondingly designated the *P surface* and the *E surface* (PS and ES), respectively. (See Plate no. 21 in the Color Plate Section.)

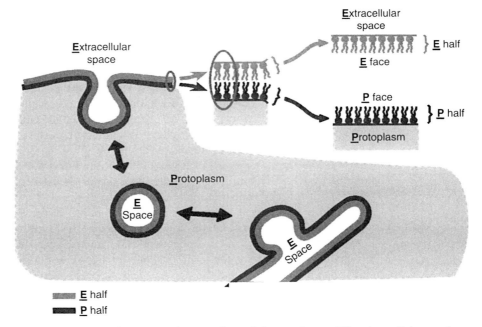

Fig. 2 Application of the nomenclature to intracellular membranes. When intracellular membrane vesicles are formed from the plasma membrane by endocytosis, the P half of the membrane forms the outer monolayer of the vesicle membrane and the E half forms the inner monolayer. The interior of the vesicle is derived from the extracellular space. Thus, convex fractures of vesicles show the E face, and concave fractures the P face of the vesicle membrane. Corresponding spatial relationships apply to exocytotic vesicles undergoing fusion with the plasma membrane, and to the numerous bidirectional membrane fusion and fission events throughout the cell. Accordingly, the term "E" designates the interior of all single-membrane organelles and the spaces between inner and outer membranes of all double membrane-bound organelles (nucleus, mitochondria, and chloroplasts). The term "P" encompasses cytoplasm, nucleoplasm, the matrix of mitochondria, and the stroma of chloroplasts. When fusion occurs between two membrane systems, P half joins with P half, and E half with E half. Correspondingly, when one membrane system pinches off from another, P half gives rise to P half, and E half to E half. Thus, during the dynamic interactions between the various compartments and membrane-bound structures of the cell, the topological relationship between extracellular, exoplasmic and endoplasmic space and that between homologous membrane-halves is preserved. (See Plate no. 22 in the Color Plate Section.)

B. Fracture-Label

In "fracture-label," cytochemical labeling is done immediately after samples have been manually freeze fractured under liquid nitrogen and thawed. This exposes for labeling the membrane halves created by freeze fracture. Fracturing the sample allows access of the label to these membrane halves, and other components, within tissue samples and cells. The labeled specimen may be examined by thin sectioning (Pinto da Silva *et al.*, 1981b) or as a replica (Pinto da Silva *et al.*, 1981a).

The basic procedure for fracture-label is straightforward (Pinto da Silva *et al.*, 1986). Tissue pieces or cell pellets (cross-linked in a gel of bovine serum albumin) are aldehyde-fixed and rapidly frozen. Fracturing under liquid nitrogen is routinely done by repeatedly crushing samples to give multiple fractures or by using a blade for directed fracturing of individual specimens. The fractured pieces are then thawed in the presence of aldehyde fixative, rinsed and blocked, and the cytochemical procedure of choice carried out.

For the replica approach, the labeled samples are critical point dried or freeze dried, and a platinum-carbon replica of the fractured surface prepared at room temperature. This step can be done using a standard vacuum evaporator if a freeze-fracture machine is not available. The platinum–carbon shadowing results in the gold marker particles being partially embedded in the replica. Because the shadowed platinum and carbon of the replica come into direct contact with the gold, the replicas are carefully cleaned using sodium hypochlorite rather than chromic acid so that the biological material is removed without dislodging the gold that is directly attached to the replica. For the thin-section approach, the samples are simply embedded by standard procedures, and sections cut at right angles to the plane of fracture.

A drawback of fracture-label is that exposure to aqueous media at the thawing stage leads to reorganization of the fractured half-membrane leaflets into a discontinuous bilayer. Underlying cytoplasmic or extracellular components are exposed for labeling in addition to epitopes of membrane proteins partitioning with a given membrane half. Aldehyde prefixation is essential to minimize these structural changes but frequently has an adverse effect on epitope preservation and hence ability to label the sample. Glutaraldehyde was extensively used in the original protocols in which lectins and other cytochemical methods not involving antibodies were applied. Owing to the deleterious effects of this fixative on epitope preservation, however, formaldehyde fixatives (freshly prepared from paraformaldehyde) are normally employed when immunogold labeling is to be undertaken. This means that ultrastructural preservation is likely to be less than ideal. A useful compromise can be to include a trace of glutaraldehyde (0.05–0.3%) with the formaldehyde fixative, but the efficacy of this approach will depend on the specific antigens to be detected and the antibodies available.

Another limitation of fracture-label is that because samples are thawed and dried before replication, the planar views of membranes have a ruffled appearance; the crisp, smoothly contoured and highly detailed faces familiar from standard freeze fracture are lost. Despite these drawbacks, fracture-label has proven to be a useful technique for analyzing the spatial organization of peripheral proteins at the cytoplasmic interface of the membrane, and has the merit that it can be done in laboratories that do not have a freeze-fracture machine.

C. Label-Fracture

The label-fracture technique provided the first really effective solution to the problem of how to retain gold markers for viewing in conjunction with high-quality freeze-fracture replica details of membranes (Kan and Pinto da Silva, 1989; Pinto da Silva, 1989; Pinto da Silva and Kan, 1984). An explanatory diagram and summary protocol is given in Fig. 3. In brief, a suspension of cells is immunogold labeled and then freeze-fractured and replicated in the standard manner. Instead of cleaning the replica, it is simply rinsed in distilled water and examined with the cells that have been fractured still attached. Where a cell is convexly fractured, the bulk of the cellular remnants completely obstructs the electron beam, preventing visualization of any detail. However, where a cell is concavely fractured, all that remains attached to the replica is the E half of the membrane together with the label attached to the E surface. A half-membrane leaflet is so thin when viewed en face that there is little interference with visibility of the replicated detail. Thus, label attached to the E surface of the half-membrane leaflet is seen superimposed upon a standard E face view of the membrane's fracture face.

The images produced by label fracture are stunning, but the limitations of the technique are that it is restricted for use with cells in suspension or the luminal portions of the membranes of tissues, and only the E face of the plasma membrane can be viewed structurally in combination only with labeling of the E surface; components of the P half of the membrane remain inaccessible to study.

Nevertheless, the realization that stabilizing fractured membrane leaflets attached to the replica could be used as a vehicle to view label in conjunction with the usual detail visible in a standard freeze-fracture replica stimulated and underpinned subsequent development in techniques. Apart from the extension of this concept in "fracture flip" (giving extended replica views of labeled membrane surfaces (Pinto da Silva et al., 1989) and dual replica techniques ("simulcast" (Ru-Long and Pinto da Silva, 1990) and "composite replica technique" (Coleman and Wade, 1989) where surface label is combined with replicas of both the surface and the membrane fracture face), it provided the foundation for FRIL (Fujimoto, 1995), which has become the preeminent technique in freeze-fracture cytochemistry today.

D. Freeze-Fracture Replica Immunolabeling

In the FRIL technique, invented by the late Kazushi Fujimoto, conventional freeze-fracture replicas are first prepared; the biological material is then dissociated using sodium dodecylsulphate (SDS) (Fujimoto, 1995, 1997). The SDS removes the bulk of the biological material, leaving a single lipid monolayer and associated integral and surface proteins adherent to the replica. As in label-fracture, this remaining layer is so thin that it does not obstruct the electron beam. The proteins (or less commonly, the lipids) are then localized by immunogold labeling. This reveals their spatial distribution superimposed upon a standard planar freeze-fracture view of the membrane interior (Fig. 4). Unlike "label-fracture" which is restricted to E-surface labeling of cells in suspension, both membrane halves are accessible to labeling. Moreover, because samples are freeze fractured prior to the cytochemistry step, target epitopes deep within tissues are rendered accessible for labeling.

1. Cells in suspension are immuno-gold labeled

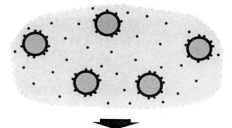

2. Labeled cells are frozen and fractured

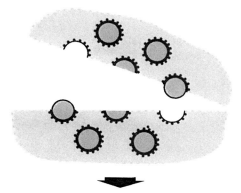

3. Platinum-carbon replica is made

4. Replica examined with fractured cell remnants still attached

Fig. 3 The key steps in label-fracture. A suspension of cells is immunogold labeled, rapidly frozen and processed for standard freeze fracture (1). The fracture may split the plasma membrane by traveling upwards and over the cell (revealing the P face), or downwards and under the cell (revealing the E face); alternatively, it may cross-fracture the cell (2). A platinum–carbon replica of the frozen, fractured surface is made, as in conventional freeze fracture (3). The replicated specimen is rinsed in distilled water and the replica, with attached cell fragments, examined in the electron microscope (4). Where cells are convexly fractured or cross fractured, the mass of cellular material attached to the replica completely obstructs the electron beam, so no structure is visible (X). However, where cells are concavely fractured, the electron beam penetrates the very thin membrane monolayer, so that the gold label on the E surface is viewed superimposed upon an E-face replica view of the membrane.

1. Frozen cell

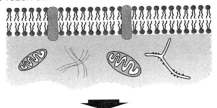

2. Freeze-fracture splits membrane

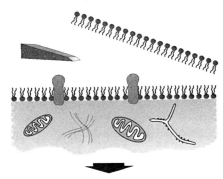

3. Replication preserves fracture face structure

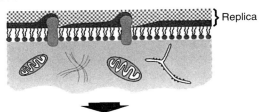

4. SDS removes cellular material leaving surface membrane components attached to the replica

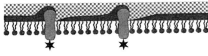

★ Epitopes available for labeling

Fig. 4 The key steps in FRIL. Tissue or cell samples are frozen and fractured (1 and 2), and then replicated with platinum–carbon (3), as in the standard freeze-fracture technique. The replicated specimen is treated judiciously with SDS to remove the cellular components apart from those attached to the replica. The cellular component of choice is then immunogold labeled. On examination in the electron microscope, the fine layer of cellular components is essentially transparent to the electron beam; the electron dense gold label is clearly visible against the replica, marking the target molecule in the membrane plane.

For the SDS to work, glutaraldehyde prefixation has to be avoided. Thus, in the original technique, samples are prepared by ultrarapid freezing. An alternative is to give a brief treatment in glycerol (with no glutaraldehyde fixation beforehand) and then use standard rapid freezing techniques. Typical preparation conditions involve treatment of the replicated specimens in Tris-buffered 5% SDS (with sucrose), pH 8.3, overnight at room temperature followed by thorough washing in PBS and blocking with 1% BSA before immunogold labeling.

More than any other technique, FRIL has meant that freeze-fracture cytochemistry has come of age. Whether applied to tissues or cell suspensions, FRIL gives superb high-resolution images in which structural and compositional information are combined. The scope of scientific information is unique to the technique, and hence its application is currently having a substantial impact in solving questions in cell biology that have hitherto been impossible to address with other ultrastructural, cell biological or molecular approaches.

IV. Discussion: Impact on Topical Questions in Cell Biology

As an illustration of how freeze-fracture cytochemistry has recently contributed to advances in cell biology, we will look at two areas of research, the spatial organization of plasma membrane proteins in cardiovascular cells, and the role of lipid droplets and their associated proteins.

A. Spatial Organization of Proteins in the Plasma Membrane

The cardiac muscle cell is uniquely specialized to contract constantly, without tiring, 3 billion times or more in an average human life-span. A detailed picture of the membrane organization of the cardiac muscle cell—to which freeze-fracture cytochemistry has made major contributions—underpins our understanding of the exquisite machinery through which individual cellular contractions are harnessed to create the heart beat.

The plasma membrane of the cardiac muscle cell is seen in conventional freeze-fracture replicas to be studded with abundant, scattered intramembrane particles, typically 3–10 nm in diameter (Fig. 5A and B). These particles represent the complement of specific channels, transporters, and receptors that endow the plasma membrane of the cardiac muscle cell with its unique electrical, transport, and signal detection/transduction properties. Upon depolarization, influx of calcium through L-type calcium channels in the plasma membrane triggers the opening of calcium release channels in the junctional sarcoplasmic reticulum (SR) membrane, resulting in a major release of calcium into the cytoplasm that triggers myofibril contraction (*calcium-induced calcium release*). Label-fracture demonstrates that L-type calcium channels are organized in discrete clusters in the plasma membrane, facing underlying junctional SR cisternae (Fig. 5C and D) (Gathercole *et al.*, 2000; Takagishi *et al.*, 1997). Such close spatial apposition of L-type calcium

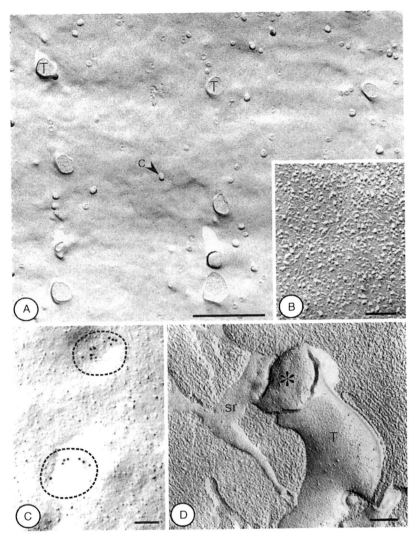

Fig. 5 Cardiac myocyte plasma membrane; structure and localization of L-type Ca channels by label fracture. (A) Planar freeze-fracture view of the plasma membrane (E-face), showing regular arrays of transverse tubule openings (T), and smaller vesicular structures, the caveolae (c). At the site of the transverse tubule openings, the plasma membrane curves upwards; before freeze-fracture, these were finger-like extensions of the plasma membrane projecting upwards, at right angles to the plane of the page. (B) High magnification freeze-fracture views disclose a heterogeneous collection of particles (3–10 nm in diameter) which represent the integral proteins of the membrane. (C) L-type calcium channels are shown by label-fracture to be aggregated in the plasma membrane (encircled). These clusters of channels lie adjacent to sacs of junctional sarcoplasmic reticulum in the cytoplasm (D), enabling the calcium-induced calcium release process that triggers contraction. The example in (D) shows a junctional sarcoplasmic reticulum sac (asterisk) apposed to transverse tubule membrane (T). The junctional SR is continuous with the free SR network (sr) that surrounds the myofibril along the length of the sarcomere. Scale bars: (A) 1 μm; (B) and (C) 100 nm; (D) 200 nm. From Severs BioEssays 22: 188–199 (2000).

channels (in both the peripheral plasma membrane and transverse tubules of the myocyte) to calcium release channels in the junctional SR facilitates optimal coupling of plasma membrane Ca^{2+} influx to SR Ca^{2+} release into the cytoplasm.

The plasma membrane of the myocyte not only serves as a vehicle to carry the electrical impulse but has to sustain the force of contraction without damage, and transmit this force both from cell to cell and laterally across the tissue. The role of fasciae adherentes junctions in cell-to-cell force transmission, and of the desmosome/intermediate filament cytoskeleton, is well established. In addition, the membrane skeleton, comprising networks of peripheral membrane proteins closely applied to the entire cytoplasmic aspect of the lateral plasma membrane, plays a key mechanical role. Freeze-fracture cytochemistry has helped unravel how these proteins are organized and interact. For example, fracture-label combined with double immunogold marking demonstrates, at the plasma membrane interface, a direct molecular interaction between the carboxyl-terminal domains of dystrophin (a peripheral membrane protein of the membrane skeleton) and β-dystroglycan (an integral membrane protein which in turn binds to laminin via α-dystroglycan on the extracellular side of the membrane) (Fig. 6A) (Stevenson et al., 1998). Dystrophin, however, is organized independently from spectrin, despite overlapping mechanical roles (Fig. 6B) (Stevenson et al., 2005).

Coordinated contraction of the cardiac chambers requires a precisely orchestrated spread of electrical excitation from cell to cell throughout the heart. The sites of electrical coupling between individual cardiac muscle cells that mediate this process are formed by gap junctions, clusters of transmembrane channels which span the closely apposed plasma membranes of neighboring cells. Gap-junctional channels are composed of connexins, a multigene family of conserved proteins.

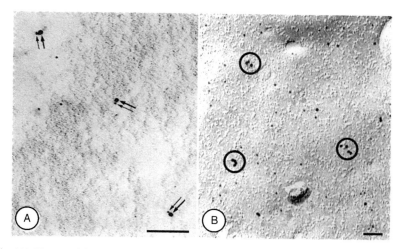

Fig. 6 (A) Fracture-label demonstration of direct molecular interaction between the carboxyl-terminal domains of dystrophin and β-dystroglycan (pairs of arrows: large gold markers, β-dystroglycan; small gold, dystrophin). (B) FRIL demonstration that spectrin (labeled with 15 nm gold) is distributed independently from dystrophin (10 nm gold). Scale bars: 100 nm.

The specific connexin type or mix of connexin types is a major determinant of the functional properties of gap junctions. Three connexin types — connexin43, connexin40, and connexin45 — are differentially expressed, in various combinations and relative quantities, in different, functionally specialized subsets of cardiac myocyte (Severs *et al.*, 2004). For example, while myocytes of the ventricle predominantly express connexin43, those of the atrioventricular node express connexin45, a connexin that forms low conductance channels, contributing to the slowing of conduction that ensures sequential contraction of atria and ventricles. A notable feature of distal Purkinje fibre myocytes, which distribute the impulse to the contractile cells of the ventricle, is the presence of high levels of connexin40, a connexin that gives high conductance channels, facilitating rapid distribution of the impulse at this stage of the cycle.

FRIL is ideally suited to exploration of the diversity of connexin expression in cardiac myocytes and other cardiovascular cells in relation to their functional properties (Severs *et al.*, 2001; Yeh *et al.*, 1998) (Fig. 7). The technique has similarly contributed to the diversity of connexin expression in neurones (Kamasawa *et al.*, 2005, 2006; Nagy *et al.*, 2004; Rash *et al.*, 2005). A protocol for FRIL, as applied to the study of gap junctions, has been published by Dunia *et al.* (2001).

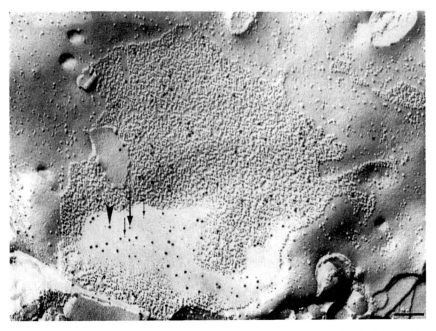

Fig. 7 FRIL demonstration of the co-assembly of three connexin types, Cx37 (5 nm gold, small arrows), Cx40 (10 nm gold, large arrow) and Cx43 (15 nm gold, large arrowhead) within the same gap-junctional plaque. This example comes from aortic endothelial cells. Scale bar: 100 nm.

B. Lipid Droplets

Until recently, the lipid droplet was arguably the most neglected organelle in cell biology, widely envisaged as little more than a relatively inactive intracellular storage depot for excess lipids. Recent research, to which FRIL has made a crucial contribution, has completely transformed this view by overturning long-held concepts on the biogenesis, structure, dynamic nature, and functions of the lipid droplet and its associated proteins.

Structurally, the lipid droplet consists of a hydrophobic neutral lipid core (containing cholesterol esters and triacyl glycerol) enveloped by a single monolayer of phospholipids. The assembly, fusion and degradation of lipid droplets, resulting in storage and release of their lipid components, is controlled by a series of proteins, in particular lipid transport proteins, acyl-CoA synthetases, caveolins and PAT family proteins (the collective term for perilipin, adipophilin and TIP47). As a storage depot, lipid droplets not only serve as a source of cellular fuel and constituents for membrane construction, but also provide precursors for lipid signaling molecules and hormones. They are thus intimately involved in the cellular influx and efflux of lipids and in the signaling and transcriptional networks central to lipid homeostasis in health and disease. Lipoatrophy (lack of mature lipid-droplet containing adipocytes) leads to diabetes and fatty liver pathology, while lipodystrophy (abnormal fat distribution), resulting from sedentary lifestyle and excess food intake, is associated with obesity and diabetes. Moreover, lipid accumulation is a critical step in the pathogenesis of atherosclerosis which, by causing coronary heart disease, is a principal cause of death and disability throughout the world.

Understanding how lipid droplets form in the cell is thus fundamental to our knowledge of these disease conditions. The mechanism of lipid droplet formation that has gained general acceptance holds that neutral lipids accumulate within the lipid bilayer of the endoplasmic reticulum (ER) membrane from where they are budded-off, enclosed by a protein-bearing phospholipid monolayer originating from the cytoplasmic monolayer of the ER membrane, to give a cytoplasmic lipid droplet This idea has the superficial attraction of explaining how the lipid droplet could acquire both its outer phospholipid monolayer and the proteins necessary for its function, but unfortunately required something of a leap of imagination in the capabilities of the principal imaging methodologies applied to support the idea (i.e., fluorescence confocal microscopy and immunogold label thin-section electron microscopy).

Results from FRIL refute the prevailing view on several counts (Robenek *et al.*, 2006b). First, freeze fracture, by permitting unique three-dimensional views of the spatial relationships of membranes and organelles, demonstrates unequivocally that at sites of close association, the lipid droplet is not situated within the ER membrane, but adjacent to it (Fig. 8). Both ER membranes clearly lie external to and follow the contour of the lipid droplet, enclosing it in a manner akin to an egg-cup (the ER) holding an egg (the lipid droplet) (Fig. 8A). Freeze-fracture cytochemistry further demonstrates that the PAT family protein adipophilin is

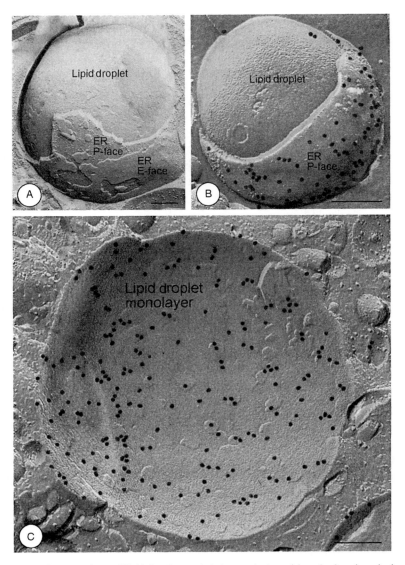

Fig. 8 Freeze-fracture views of lipid droplets and their association with endoplasmic reticulum (ER) membrane from lipid laden macrophages (i.e., macrophages fed acetylated low density lipoprotein to induce lipid droplet formation). (A) Lipid droplet situated in a cup formed from ER membranes. Both ER membranes are visible (seen in P-face and E-face view), following upwards and over the contour of the lipid droplet from below. The lipid droplet has been convexly fractured, and lies beneath (i.e., adjacent to and not within) both ER membranes exposed. (B) Similar view to (A), but with labeling for adipophilin using the FRIL technique. Abundant gold label is visible on the ER membrane (P-face) immediately adjacent to the lipid droplet. (C) Lipid droplet seen in concave fracture. FRIL demonstrates abundant labeling for adipophilin in the outer phospholipid monolayer surrounding the lipid droplet (P-face) exposed in this view. Scale bars: 200 nm.

concentrated in prominent clusters in the P-half of the ER membrane at the site of the closely apposed lipid droplet (Fig. 8B), as well as in the lipid droplet surface apposed to the ER (Fig. 8C). Adipophilin is thus strategically placed to play a role in lipid droplet growth by facilitating lipid transfer from the ER to the droplet. The evidence from these studies indicates that lipid droplets originate and develop adjacent and external to specialized domains of the ER membrane enriched in adipophilin, not within the bilayer of the ER as previously supposed.

The prevailing view is further discredited by the spatial distribution of caveolin 1, a putative mediator of intracellular lipid transport (Robenek *et al.*, 2003, 2004). As with other lipid droplet associated proteins, caveolin 1 is proposed to traffic to the lipid droplet from the ER membrane by the budding off process. As this process involves the budding off and enclosure of the lipid accumulation in the P half of the ER membrane, the model requires that lipid droplet proteins such as caveolin originate from this membrane leaflet. Contrary to this prediction, FRIL demonstrates the caveolin 1 is situated in the E-half of the ER membrane. As this membrane half does not participate in the proposed mechanism of lipid droplet production, caveolin is actually in a location in the ER membrane that would make it impossible to gain access to the forming lipid droplet.

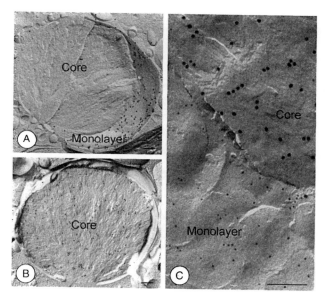

Fig. 9 FRIL images demonstrating that PAT family and other proteins are distributed not only at the lipid droplet surface but also in the cross-fractured lipid droplet core. These examples come from adipocytes and show labeling for perilipin and caveolin. (A) Example in which perilipin label is seen predominantly in the outer phospholipid monolayer (P face) of the lipid droplet. (B) Example of a lipid droplet in which perilipin label is seen throughout the core. (C) Example in which abundant caveolin label (18 nm gold) is seen in the core, and perilipin label (12 nm gold) is predominantly in the phospholipid monolayer (seen in P-face view). Scale bars: 200 nm.

The prevailing wisdom holds that caveolin and the PAT family proteins are confined exclusively to the droplet surface and that the latter are specific to lipid droplets and not present in any other organelle or membrane system of the cell. FRIL, however, demonstrates that in macrophages and adipocytes (1) PAT family proteins and caveolin are distributed not only in the surface but also throughout the lipid droplet core (Fig. 9); and (2) PAT family proteins are integral components of the plasma membrane (Fig. 10) (Robenek et al., 2005a,b). Under normal culture conditions, these proteins are dispersed in the P half of the plasma membrane (Fig. 10A). Stimulation of lipid droplet formation by incubation of the cells with acetylated low-density lipoprotein leads to clustering of the PAT family proteins in raised plasma membrane domains (Fig. 10B). Fractures penetrating beneath the

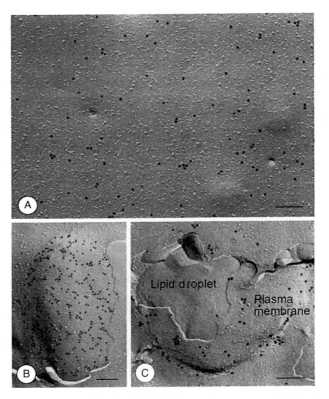

Fig. 10 FRIL images demonstrating that PAT family proteins are present in the plasma membrane, and undergo profound changes in distribution under conditions of lipid loading. (A) View of the plasma membrane (P face) of a normal cultured adipocyte after labeling for perilipin. The perilipin is widely distributed throughout the membrane. (B) Upon lipid loading, the perilipin becomes clustered in elevated domains in the plasma membrane. (C) Fractures that penetrate beneath the plasma membrane demonstrate that lipid droplets lie beneath the elevated protein-rich domains, as seen in this example from an adipophilin-labeled lipid laden macrophage. Scale bars: 200 nm.

plasma membrane demonstrate that lipid droplets are closely apposed to these domains (Fig. 10C). A similar distribution pattern of labeling in the form of linear aggregates within the clusters is apparent in the P half of the plasma membrane and the immediately adjacent outer monolayer of the lipid droplet (Robenek *et al.*, 2005b). The aggregation of the PAT family proteins into such assemblies may facilitate carrier-mediated lipid influx from the extracellular environment into the lipid droplet. The findings point to a common cellular mechanism of intracellular lipid loading in the macrophage as part of the pathogenesis of atherosclerosis and in the adipocyte during development of obesity.

A related area to which FRIL has shed new light is the mechanism of milk fat globule secretion (Robenek *et al.*, 2006a). Milk fat globule formation involves the trafficking of what is essentially a lipid droplet with an outer phospholipid monolayer (termed the "secretory granule") to the cell surface; the granule is then enveloped by a portion of plasma membrane and released from the cell as a milk fat globule. The milk fat globule thus has a lipid bilayer derived from the plasma membrane exterior to the phospholipid monolayer enclosing the neutral lipid core (Fig. 11). The molecular mechanism of the secretory process is proposed to involve

Fig. 11 Freeze-fracture view illustrating the structure of a milk fat globule secreted from a human mammary epithelial cell. The globule consists of a lipid droplet core surrounded by a phospholipid monolayer, which in turn is surrounded by a membrane bilayer derived from the plasma membrane which enwraps the droplet during secretion. These different structures are revealed as the fracture plane skips between them (bilayer seen in P-face view; phospholipid monolayer in E-face view; core cross-fractured). Scale bar: 200 nm.

formation of complexes between butyrophilin in the plasma membrane with cytosolic xanthine oxidoreductase; the resulting complexes are then believed to interact with adipophilin on the outer surface of the lipid droplet to enwrap the secretory granule in plasma membrane.

The reality demonstrated from FRIL, however, is that the topological distribution of the relevant proteins makes the proposed mechanism impossible (Figs. 12 and 13). Adipophilin is actually more abundant in the plasma membrane domains to which secretory granules are apposed in the mammary epithelial cell, and in the bilayer surrounding the secreted milk fat globule, than in the monolayer enclosing the lipid droplet (Fig. 12). Xanthine oxidoreductase is diffusely distributed in the lipid droplet monolayer. Importantly, butyrophilin in the plasma membrane is concentrated in a network of ridges that tightly appose and match the protein's

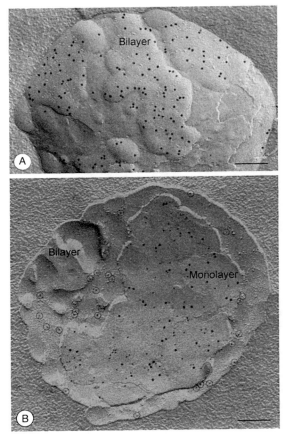

Fig. 12 FRIL images demonstrating the distribution of adipophilin and butyrophilin in the secreted milk fat globule. (A) Abundant adipophilin label is seen in the bilayer P face. (B) Double labeling reveals that adipophilin (18 nm gold) is also abundant in the phospholipid monolayer P face, while butyrophilin (12 nm gold) is present both in the bilayer E face and phospholipid monolayer P face. Scale bar: 200 nm.

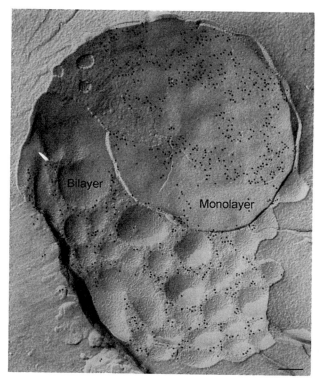

Fig. 13 FRIL image demonstrating pattern of butyrophilin labeling in concavely fractured milk fat globule. Note the network pattern of butyrophilin distribution in both the bilayer E face and phospholipid monolayer P face. This mirror distribution suggests a role for butyrophilin–butyrophilin interactions in the secretory process. Scale bar: 200 nm.

distribution in the monolayer of the lipid droplet (Fig. 13). While adipophilin-rich domains in plasma membrane may well be linked to secretory granule positioning at the cell surface, butyrophilin–butyrophilin interactions between monolayer and bilayer mediate envelopment of the granule by the plasma membrane and its release from the cell (Robenek et al., 2006a).

V. Concluding Comment

The examples discussed illustrate the recent impact of freeze-fracture cytochemistry in advancing our understanding of selected aspects of cardiovascular cell biology. The information that this approach provides is unique; without its wider application, substantial gaps in our knowledge of how membranes function will remain. Further exploitation of freeze-fracture cytochemistry may be expected as the scope and power of this technique become more widely appreciated.

Acknowledgments

We are indebted to all the members of the Severs and Robenek laboratories for their contributions to the work discussed in this review. Some of the results discussed come from studies supported by grants from British Heart Foundation (NJS) and Deutsche Forschungsgemeinschaft (HR).

References

Andersson Forsman, C., and Pinto da Silva, P. (1988). Label-fracture of cell surfaces by replica staining. *J. Histochem. Cytochem.* **36,** 1413–1418.

Branton, D., Bullivant, S., Gilula, N. B., Karnovsky, M. J., Moor, H., Muhlethaler, K., Northcote, D. H., Packer, L., Satir, B., Satir, P., Speth, V., Staehelin, L. A., *et al.* (1975). Freeze-etching nomenclature. *Science* **190,** 54–56.

Coleman, R. A., and Wade, J. B. (1989). Composite replicas: Methodologies for direct evaluation of the relationship between intramembrane and extramembrane structures. *J. Electron Microsc. Tech.* **13,** 216–227.

Dunia, I., Recouvreur, M., Nicolas, P., Kumar, N. M., Bloemendal, H., and Benedetti, E. L. (2001). Sodium dodecyl sulphate-freeze-fracture immunolabeling of gap junctions. *In* "Methods in Molecular Biology: Connexin Methods and Protocols" (R. Bruzzone, and C. Giaume, eds.), Vol. **154,** pp. 33–55. Humana Press, Totowa, NJ.

Escaig, J. (1982). New instruments which facilitate rapid freezing at 83 K and 6 K. *J. Microsc.* **126,** 221–230.

Fujimoto, K. (1995). Freeze-fracture replica electron microscopy combined with SDS digestion for cytochemical labeling of integral membrane proteins—Application to the immunogold labeling of intercellular junctional complexes. *J. Cell Sci.* **108,** 3443–3449.

Fujimoto, K. (1997). SDS-digested freeze-fracture replica labeling electron microscopy to study the two-dimensional distribution of integral membrane proteins and phospholipids in biomembrane: Practical procedure, interpretation and application. *Histochem. Cell Biol.* **107,** 87–96.

Gathercole, D. V., Colling, D. J., Takagishi, Y., Skepper, J. N., Levi, A. J., and Severs, N. J. (2000). L-type Ca^{2+} calcium channel clusters over junctional sarcoplasmic reticulum in guinea pig cardiac myocyte. *J. Mol. Cell Cardiol.* **32,** 1984–1991.

Heuser, J. (1989). Protocol for 3-D visualization of molecules on mica via the quick freeze, deep etch technique. *J. Electron Microsc. Tech.* **13,** 244–263.

Heuser, J. E. (1981). Preparing biological specimens for stereo microscopy by the quick-freeze, deep-etch, rotary-replication technique. *Methods Cell Biol.* **22,** 97–122.

Heuser, J. E., and Salpeter, S. R. (1979). Organization of acetylcholine receptors in quick-frozen, deep-etched, and rotary replicated Torpedo postsynaptic membrane. *J. Cell Biol.* **82,** 150–173.

Kamasawa, N., Furman, C. S., Davidson, K. G., Sampson, J. A., Magnie, A. R., Gebhardt, B. R., Kamasawa, M., Yasumura, T., Zumbrunnen, J. R., Pickard, G. E., Nagy, J. I., and Rash, J. E. (2006). Abundance and ultrastructural diversity of neuronal gap junctions in the OFF and ON sublaminae of the inner plexiform layer of rat and mouse retina. *Neuroscience* **142,** 1093–1117.

Kamasawa, N., Sik, A., Morita, M., Yasumura, T., Davidson, K. G., Nagy, J. I., and Rash, J. E. (2005). Connexin-47 and connexin-32 in gap junctions of oligodendrocyte somata, myelin sheaths, paranodal loops and Schmidt–Lanterman incisures: Implications for ionic homeostasis and potassium siphoning. *Neuroscience* **136,** 65–86.

Kan, F. W. K., and Pinto da Silva, P. (1989). Label-fracture cytochemistry. *In* "Colloidal Gold. Principles, Methods and Applications" (M. A. Hayat, ed.), pp. 175–201. Academic Press, Orlando, London.

Moor, H., and Mühlethaler, K. (1963). Fine structure of frozen-etched yeast cells. *J. Cell Biol.* **17,** 609–628.

Müller, M., Meister, N., and Moor, H. (1980). Freezing in a propane jet and its application in freeze-fracturing. *Mikroskopie(Wien)* **36,** 129–140.

Nagy, J. I., Dudek, F. E., and Rash, J. E. (2004). Update on connexins and gap junctions in neurons and glia in the mammalian nervous system. *Brain Res. Brain Res. Rev.* **47**, 191–215.

Pinto da Silva, P. (1989). Visual thinking of biological membranes: From freeze-etching to label-fracture. *In* "Immunogold Labeling Methods in Cell Biology" (A. Verkleij, and J. L. N. Leunissen, eds.), pp. 179–197. CRC Press, Boca Raton, FL.

Pinto da Silva, P., Andersson Forsman, C., and Fujimoto, K. (1989). Fracture-flip: Nanoanatomy and topochemistry of cell surfaces. *In* "Cells and Tissues: A Three-Dimensional Approach by Modern Techniques in Microscopy" (P. Motta, ed.), pp. 49–56. Alan R. Liss, New York.

Pinto da Silva, P., Barbosa, M. L. F., and Aguas, A. P. (1986). A guide to fracture-label: Cytochemical labeling of freeze- fractured cells. *In* "Advanced Techniques in Biological Electron Microscopy" (J. K. Koehler, ed.), **Vol. 3**, pp. 201–227. Springer, New York.

Pinto da Silva, P., and Branton, D. (1970). Membrane splitting in freeze-etching. Covalently bound ferritin as a membrane marker. *J. Cell Biol.* **45**, 598–605.

Pinto da Silva, P., Kachar, B., Torrisi, M. R., Brown, C., and Parkison, C. (1981a). Freeze-fracture cytochemistry: Replicas of critical point-dried cells and tissues after fracture-label. *Science* **213**, 230–233.

Pinto da Silva, P., Parkison, C., and Dwyer, N. (1981b). Freeze-fracture cytochemistry: Thin sections of cells and tissues after labeling of fracture faces. *J. Histochem. Cytochem.* **29**, 917–928.

Pinto da Silva, P., and Kan, F. W. K. (1984). Label-fracture: A method for high resolution labeling of cell surfaces. *J. Cell Biol.* **99**, 1156–1161.

Rash, J. E., Davidson, K. G., Kamasawa, N., Yasumura, T., Kamasawa, M., Zhang, C., Michaels, R., Restrepo, D., Ottersen, O. P., Olson, C. O., and Nagy, J. I. (2005). Ultrastructural localization of connexins (Cx36, Cx43, Cx45), glutamate receptors and aquaporin-4 in rodent olfactory mucosa, olfactory nerve and olfactory bulb. *J. Neurocytol.* **34**, 307–341.

Rash, J. E., Johnson, T. J. A., Dinchuk, J. E., and Levinson, S. R. (1989). Labeling intramembrane particles in freeze-fracture replicas. *In* "Freeze-Fracture Studies of Membranes" (S. W. Hui, ed.), pp. 41–59. CRC Press, Boca Raton, FL.

Robenek, H., Hofnagel, O., Buers, I., Lorkowski, S., Schnoor, M., Robenek, M. J., Heid, H., Troyer, D., and Severs, N. J. (2006a). Butyrophilin controls milk fat globule secretion. *Proc. Natl. Acad. Sci. USA* **103**, 10385–10390.

Robenek, H., Hofnagel, O., Buers, I., Robenek, M. J., Troyer, D., and Severs, N. J. (2006b). Adipophilin-enriched domains in the ER membrane are sites of lipid droplet biogenesis. *J. Cell Sci.* **119**, 4215–4224.

Robenek, H., Lorkowski, S., Schnoor, M., and Troyer, D. (2005a). Spatial integration of TIP47 and adipophilin in macrophage lipid bodies. *J. Biol. Chem.* **280**, 5789–5794.

Robenek, H., Robenek, M. J., Buers, I., Lorkowski, S., Hofnagel, O., Troyer, D., and Severs, N. J. (2005b). Lipid droplets gain PAT family proteins by interaction with specialized plasma membrane domains. *J. Biol. Chem.* **280**, 26330–26338.

Robenek, M. J., Schlattmann, K., Zimmer, K. P., Plenz, G., Troyer, D., and Robenek, H. (2003). Cholesterol transporter caveolin-1 transits the lipid bilayer during intracellular cycling. *FASEB J.* **17**, 1940–1942.

Robenek, M. J., Severs, N. J., Schlattmann, K., Plenz, G., Zimmer, K. P., Troyer, D., and Robenek, H. (2004). Lipids partition caveolin-1 from ER membranes into lipid droplets: Updating the model of lipid droplet biogenesis. *FASEB J.* **18**, 866–868.

Ru-Long, S., and Pinto da Silva, P. (1990). Simulcast: Contiguous views of fracture faces and membrane surfaces in a single cell. *Eur. J. Cell Biol.* **53**, 122–130.

Severs, N. J. (1995a). Freeze-fracture cytochemistry: An explanatory survey of methods. *In* "Rapid Freezing, Freeze Fracture, and Deep Etching" (N. J. Severs, and D. M. Shotton, eds.), pp. 173–208. Wiley-Liss, New York.

Severs, N. J. (1995b). Lipid localization by membrane perturbation: A cautionary tale. *In* "Rapid Freezing, Freeze Fracture and Deep Etching" (N. J. Severs, and D. M. Shotton, eds.), pp. 225–234. John Wiley & Sons, New York.

Severs, N. J. (1997). Cholesterol cytochemistry in cell biology and disease. *In* "Subcellular Biochemistry. Vol. 28. Cholesterol: Its Functions and Metabolism in Biology and Medicine" (R. Bittman, ed.), pp. 477–505. Plenum Press, London.

Severs, N. J. (2007). Freeze-fracture electron microscopy. *Nat. Protoc.* **2,** 547–576.

Severs, N. J., Coppen, S. R., Dupont, E., Yeh, H. I., Ko, Y. S., and Matsushita, T. (2004). Gap junction alterations in human cardiac disease. *Cardiovasc. Res.* **62,** 368–377.

Severs, N. J., and Robenek, H. (1983). Detection of microdomains in biomembranes—An appraisal of recent developments in freeze-fracture cytochemistry. *Biochim. Biophys. Acta (Reviews on Biomembranes)* **737,** 373–408.

Severs, N. J., Rothery, S., Dupont, E., Coppen, S. R., Yeh, H.-I., Ko, Y.-S., Matsushita, T., Kaba, R., and Halliday, D. (2001). Immunocytochemical analysis of connexin expression in the healthy and diseased cardiovascular system. *Microsc. Res. Tech.* **52,** 301–322.

Severs, N. J., and Simons, H. L. (1983). Failure of filipin to detect cholesterol-rich domains in smooth muscle plasma membrane. *Nature* **303,** 637–638.

Stevenson, S. A., Cullen, M. J., Rothery, S., Coppen, S. R., and Severs, N. J. (2005). High-resolution en-face visualization of the cardiomyocyte plasma membrane reveals distinctive distributions of spectrin and dystrophin. *Eur. J. Cell Biol.* **84,** 961–971.

Stevenson, S., Rothery, S., Cullen, M. J., and Severs, N. J. (1998). Spatial relationship of C-terminal domains of dystrophin and β-dystroglycan in cardiac muscle support a direct molecular interaction at the plasma membrane interface. *Circ. Res.* **82,** 82–93.

Takagishi, Y., Rothery, S., Issberner, J., Levi, A. J., and Severs, N. J. (1997). Spatial distribution of dihydropyridine receptors in the plasma membrane of guinea pig cardiac myocytes investigated by correlative confocal microscopy and label-fracture electron microscopy. *J. Electron Microsc.* **46,** 165–170.

Tillack, T. W., and Marchesi, V. T. (1970). Demonstration of the outer surface of freeze-etched red blood cell membranes. *J. Cell Biol.* **45,** 649–653.

Yeh, H.-I., Dupont, E., Rothery, S., Coppen, S. R., and Severs, N. J. (1998). Individual gap junction plaques contain multiple connexins in arterial endothelium. *Circ. Res.* **83,** 1248–1263.

PART II

Electron Microscopy of Specific Cellular Structure

SECTION 1

The Cell Membrane

CHAPTER 12

Three-Dimensional Molecular Architecture of the Plasma-Membrane-Associated Cytoskeleton as Reconstructed by Freeze-Etch Electron Tomography

Nobuhiro Morone,* Chieko Nakada,[†] Yasuhiro Umemura,[†] Jiro Usukura,[‡] and Akihiro Kusumi[†]

*Department of Ultrastructural Research
National Institute of Neuroscience
National Center of Neurology and Psychiatry
Kodaira 187-8502, Japan.

[†]Membrane Mechanisms Project
International Cooperative Research Project (ICORP)
Japan Science and Technology Agency
Institute for Integrated Cell-Material Sciences and Institute for Frontier Medical Sciences
Kyoto University, Shougoin
Kyoto 606-8507, Japan.

[‡]Division of Integrated Projects
EcoTopia Science Institute
Nagoya University
Furo-cho, Chikusa
Nagoya 464-8603, Japan.

I. Introduction
   A. General Introduction
   B. Introduction to Terminology: Membrane-Associated Part of the Cytoskeleton (Membrane Skeleton[MSK])
   C. Introduction to the MSK: The MSK of the Human Erythrocyte Ghost
II. Protocol for Visualization of the Three-Dimensional Structure of the MSK of the Cytoplasmic Surface of the Plasma Membrane
   A. Methods for Exposing the Cytoplasmic Surface of the Plasma Membrane
   B. Immunolabeling of the Proteins on the Cytoplasmic Surface of the Plasma Membrane
   C. Rapid-Freezing
   D. Deep-Etching and Platinum Replication
   E. Recovering the Platinum Replicas
   F. Summary of the Methods for Producing Large Plasma Membrane Fragments and Avoiding Excessive Fragmentation of Replicas
   G. Creation of Stereo Views (Anaglyphs)
   H. 3-D Reconstruction of the MSK by Electron Tomography
III. 3D Structure of the Cytoskeleton-Plasma Membrane Interface
   A. The Cytoplasmic Surface of the Plasma Membrane of Cultured Cells is Entirely Coated with the Meshwork of the Actin-Based MSK
   B. View of the MSK Using Anaglyphs
   C. Quantitative 3D Reconstruction of the Undercoat Structure on the Cytoplasmic Surface of the Plasma Membrane Using Electron Tomography
   D. Interface Structure of the MSK on the Cytoplasmic Surface of the Plasma Membrane
   E. Distribution of the MSK Mesh Size on the Plasma Membrane Determined by Electron Tomography
   F. Comparison of the MSK Mesh Size on the Plasma Membrane Determined by Electron Tomography with the Compartment Size for Membrane Molecule Diffusion
IV. Electron Tomography Clarified that Some of the Actin Filaments are Laterally Bound to the Cytoplasmic Surface of the Plasma Membrane
References

I. Introduction

A. General Introduction

The cytoskeleton and the plasma membrane are likely to carry out many functions *inter*dependently. These structures work collaboratively, and together are largely responsible for determining the movements, shapes, and shape changes of the cell. The plasma membrane dynamics, such as endocytosis, exocytosis, membrane extension, and membrane resealing after cell wounding, could be regulated by the tension exerted on the membrane by the cytoskeleton (Sheetz and Dai, 1996; Sheetz, 2001). Signal transduction by glycosylphosphatidylinositol (GPI)-anchored proteins is mediated by the binding of stimulation-induced clusters of GPI-anchored proteins to actin filaments (Suzuki et al., 2007a,b). Tension exerted on integrin clusters in the plasma membrane by the force generated by actomyosins and by binding to the extracellular matrix strengthens the linkages between the integrin clusters and the cytoskeleton (Choquet et al., 1997), possibly by way of the force-induced conformational changes of the Src family kinase substrate p130Cas at the integrin clusters (Sawada et al., 2006).

The cytoskeleton has been a target of comprehensive studies by electron microscopy (EM). The "cortical" cytoskeleton, the cytoskeleton located near the plasma membrane, has also been studied quite extensively (for example, see Hartwig et al., 1989; Svitkina et al., 2003; Yin and Hartwig, 1988), but since many of these studies employed detergents, the structures at the exact interface between the cytoskeleton and the plasma membrane were not clear. One of the best ways to observe the membrane surface is to prepare the specimens using "freeze etching", an EM technique for sample preparation. In this technique, the biological specimen is rapidly frozen, faster than the growth of the ice microcrystals, and then, after the removal of excess ice, the ice on the membrane surface is sublimed (etching) so that the membrane surface is exposed outside the ice. The surface morphology is replicated by coating this surface with platinum, and the platinum coat is observed by EM.

This technique was greatly improved and modified for its application to investigations of the plasma membrane and the cytoskeleton near the plasma membrane by Heuser and his colleagues, as well as others, in the late 1970s and early 1980s (Chandler and Heuser, 1979; Heuser and Salpeter, 1979; Heuser et al., 1979; Landis and Reese, 1981). Among them, two papers published by Heuser, Hirokawa, and their colleagues stand out in the history of the studies of both the plasma membrane and the cytoskeleton (Hirokawa and Heuser, 1981; Hirokawa et al., 1982). By using intestinal epithelial tissue, and by devising various ways of identifying a variety of intracellular structures, they clearly and impressively showed images of the cytoskeleton closely associated with the microvilli and the apical plasma membrane. Actin bundles in microvilli were shown vividly. The barbed ends (fast-growing ends) were attached to the cytoplasmic surface of the tip of the villus, while the pointed ends (slowly-growing ends) at the root of the villus were linked to a structure called the "terminal web", which was further linked to the bulk cytoskeleton consisting of actin, myosin, and other intermediate filament structures. The structures linking the actin bundles as well as those linking the actin filaments and the plasma membrane were clearly shown in their electron micrographs.

Here, we list only several of the membrane-related, representative investigations using the freeze-etching technique during the last 20 years: Fujita et al. (2007), Hartwig et al. (1989), Hanson et al. (1997, 2008), Heuser and Anderson (1989), Heuser (2005), Italiano et al. (1999), Katayama et al. (1996), Kanaseki et al. (1997, 1998), Kajimura et al. (2000), Ohno and Takasu (1989), and Rothberg et al. (1992), and Nakata and Hirokawa (1992).

We have recently adopted this technique to observe the undercoat structures of the plasma membrane. In addition, we showed that the platinum replica of the rapidly-frozen, deep-etched, plasma membrane is suitable for the three-dimensional (3D) reconstruction of the cytoplasmic surface of the plasma membrane with its membrane-associated part of the cytoskeleton, using electron tomography. Hence, the objectives of this review are:

1. To show the high potential of freeze-etch EM for studying the interface between the plasma membrane and the cytoskeleton;

2. To briefly review the structure of the plasma-membrane-associated part of the cytoskeleton;

3. To briefly summarize the protocols for preparing the plasma membrane specimen with its undercoat structures, and for the rapid-freezing, deep-etching, and platinum replication of the plasma membrane specimen, with several recent improvements; and

4. To present the 3D reconstruction data and the meshwork of the actin filaments associated with the cytoplasmic surface of the plasma membrane (within 0.85 nm from the surface), and to compare the mesh size with the size of the compartments for the diffusion of plasma-membrane molecules, detected by single-molecule tracking of phospholipids and proteins.

B. Introduction to Terminology: Membrane-Associated Part of the Cytoskeleton (Membrane Skeleton[MSK])

Distinguishing the plasma membrane-associated part of the cytoskeleton from the bulk skeleton is difficult because the membrane-associated part of the cytoskeleton is continuous with the bulk cytoskeleton. In this review, *we define the membrane-associated part of the cytoskeleton as the part of the cytoskeleton that is located within several tens of nanometers from the plasma membrane, and we call this structure the "membrane skeleton (MSK)" for convenience.* The vague distance stated here is due to the large variations in the association of the cytoskeleton with the plasma membrane among different cell types.

The reasons why the term MSK, separate from the bulk cytoskeleton, is used are as follows:

1. Since the MSK-part of the cytoskeleton often functions in close cooperation with the plasma membrane, it is often more logical and easy to consider the MSK as a part of the plasma membrane, rather than as a part of the cytoskeleton, in terms of cellular functions. As mentioned in the first paragraph of this review, the proteins in the plasma membrane constantly interact with the cytoskeleton adjacent to the plasma membrane.

2. For the interaction with the plasma membrane, the MSK is likely to involve protein species and to have structures different from those of the bulk cytoskeleton. The proteins that are more often found in the MSK, rather than the bulk cytoskeleton, include spectrin, ankyrin, band 4.1, adducin, villin, gelsolin, supervillin, filamin, dystrophin, and utrophin.

3. Since the term membrane skeleton or MSK is often used as an adjective, such as the "MSK fence model", the following term, the membrane-associated part of the cytoskeleton, which might sound more proper, is inconvenient.

C. Introduction to the MSK: The MSK of the Human Erythrocyte Ghost

EM investigations of the MSK structure have long been conducted using human erythrocytes. Since the human erythrocyte can easily be obtained, and since its plasma membrane can be readily isolated due to its lack of intracellular membrane compartments, the biochemical and structural analyses of the erythrocyte plasma membrane were much easier than those of other cell types. Therefore, it has been used as an important paradigm for studying the interaction between the MSK and the plasma membrane.

Due to the clarity of the results as well as the historical importance, first, we will describe the MSK structure, as revealed by EM (+AFM) as well as biochemical investigations. The MSK of the human erythrocyte even now provides a basic paradigm for studies of MSK structure and function. The schematic structure of the human erythrocyte MSK, obtained as a result of pioneering studies, is shown in Fig. 1 (Bennet, 1990; Byers and Branton, 1985). This amazing structure is completely different from the MSKs of other cell types, which will be described in the latter part of this review. Short actin filaments of approximately 40 nm in length, each consisting of 12–18 G-actin molecules, are bound by a tropomysin

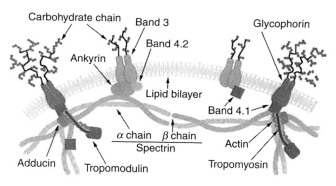

Fig. 1 Schematic model of the MSK of the human erythrocyte. (See Plate no. 23 in the Color Plate Section.)

molecule on its side (this molecule is considered to determine the length of these short actin filaments), and are linked to the plasma membrane on its barbed end (fast-polymerizing end) and to the tropomodulin at its pointed end (slow-polymerizing end). These filaments are densely distributed throughout the cytoplasmic surface of the plasma membrane. The short actin filaments are linked to the plasma membrane by way of protein complexes, called junctional complexes, consisting of a single-pass transmembrane protein, glycophorin C, band 4.1, and adducin, occurring most frequently every 78 nm (Byers and Branton, 1985). These junctional complexes are linked to each other sideways by flexible spectrin tetramers with lengths between 50–80 nm, which can be extended up to ≈200 nm. Schematically, this structure can be envisaged as a picket-fence structure, like a garden fence. In this analogy, the short actin filaments can be considered as the pickets, each of which is fixed to the ground by a glycophorin C-band 4.1-adducin complex, and these junctional complexes are linked by spectrin fences. The difference from the picket fences found in the agricultural farm is that the spectrin fences are not attached to the short actin pickets, but to the ground protein complex structures (junctional complexes + spectrin). The flexibility of individual spectrin molecules is considered to provide both the elasticity and resilience of erythrocytes in circulation (Evans, 1989; Mohandas and Chasis, 1993; Vertessy and Steck, 1989).

Observations of the erythrocyte's MSK without using detergent solubilization and negative staining of the erythrocyte membrane, which might modify the MSK structure greatly, have been carried out by several methods, including scanning EM (Hainfeld and Steck, 1977), thin-section EM (Tsukita et al., 1980), and EM after freeze-etching and platinum replication (Nermut, 1981; Ursitti et al., 1991; also see Fig. 2A in the review by Coleman et al. (1989), which was provided by Dr. J. Heuser of Washington University). These studies showed that the membrane skeleton is a dense, complex, three-dimensional network of filaments. Atomic force microscopy of the cytoplasmic surface of the erythrocyte plasma membrane, taking advantage of its high sensitivity to small height variations on the surface, were conducted using freeze-dried human erythrocyte ghosts (Takeuchi et al., 1998). The average mesh size of the spectrin network was 3000 nm^2, which is basically consistent with the number density of junctional complexes detected by the negative-staining EM.

II. Protocol for Visualization of the Three-Dimensional Structure of the MSK of the Cytoplasmic Surface of the Plasma Membrane

A. Methods for Exposing the Cytoplasmic Surface of the Plasma Membrane

Generally, two methods have been used for exposing the cytoplasmic surface of the plasma membrane of cultured cells, for immunolabeling and visualization of the cytoplasmic surface (Fig. 2). One is used to observe the upper plasma membrane (top surface in the cell culture on the coverslip) from inside the cell; and the

12. Molecular Architecture of the Plasma-Membrane-Associated Cytoskeleton

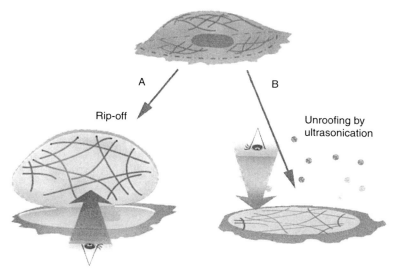

Fig. 2 The "rip-off" method (A) and the "unroofing" method (B) for exposing and observing the cytoplasmic surface of the plasma membrane of cultured cells. (See Plate no. 24 in the Color Plate Section.)

other is employed for the observation of the bottom cell membrane (the cell membrane facing the cover slip), again from inside the cell. The former method is often called the "rip-off" protocol, as it involves the placement and attachment of a coverslip from the top, and the subsequent removal of the top coverslip to rip the top membrane off from the rest of the cell. The latter method is referred to as "unroofing", because the upper plasma membrane and the majority of the cytoplasmic structures and molecules are removed by very brief, weak sonication, leaving the bottom membrane attached to the coverslip on which the cell was originally cultured. Both methods can work reasonably well, but they both have their own potential problems and limitations. Further details are given in the following subsections.

Another method frequently used to visualize the actin filament meshwork and its binding proteins near the plasma membrane is to solubilize the plasma membrane using detergents. Triton X-100 solutions, containing other stabilizing reagents such as polyethylene glycol, glycerol, and sucrose, are the most popular ones. However, these detergent solutions tend to disintegrate caveolae and clathrin-coated pits. Using this detergent-solubilization method, the development of cortical actin filaments was clearly observed (Svitkina et al., 1995, 2003).

a. The "rip-off" method for the observation of the cytoplasmic surface of the upper plasma membrane (Fig. 2A)

No.1 coverslips are cut into 5-mm square pieces. These small coverslips are cationized (coated) with alcian blue, a small-molecule reagent. The cationized coverslips are placed on cells cultured on the cell sheet at 4 °C for 10 min in order to allow the coverslips to attach to the plasma membrane (Fig. 3A). A buffer solution is then gently introduced into the gap between the coverslips and the cell sheet (Fig. 3B). The surface tension of the buffer forces the coverslips to float up, which could rip off the upper cell membrane from the rest of the cell, sometimes with very small amounts of the membrane skeleton or with contamination by whole cells (Fig. 3B). This buffer usually contains chemical fixatives such as 2% formaldehyde, so that the plasma membrane ripped off from the rest of the cell is immediately fixed, before its component proteins dissolve away into the rip-off buffer.

b. Unroofing by low-power ultrasonication for the observation of the cytoplasmic surface of the bottom plasma membrane (Fig. 2B)

A probe-type, low-power ultrasonic generator is used to remove the top membrane and the bulk cytoplasmic materials (Heuser, 2000). This method has often been plagued by the limited observation areas, the broken MSK meshwork

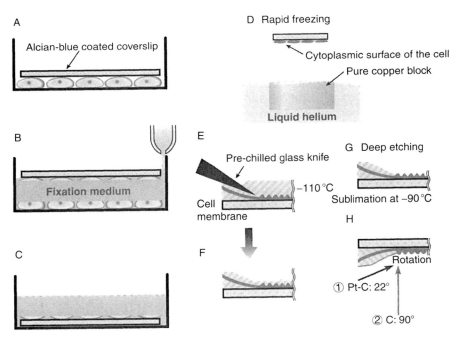

Fig. 3 Specimen preparation protocol for ripping-off the upper plasma membrane, followed by rapid-freezing, deep-etching, and platinum replication. (See Plate no. 25 in the Color Plate Section.)

structures, and the loss of small membrane invaginations, including caveolae and clathrin-coated pits. Careful adjustments of the various conditions for the sonication step are required, such as the use of smaller probes (e.g., 1 mm in diameter) and lower output powers (e.g., up to several milliwatts), for the observations of the MSK, caveolae, and clathrin-coated pits spread throughout the cytoplasmic surface of the bottom plasma membrane, just like those of the upper cell membrane.

In both cases, the plasma membrane is (further) fixed by an incubation in fresh, ice-cold 1% paraformaldehyde/0.25% glutaraldehyde for 10 min (Fig. 3C).

B. Immunolabeling of the Proteins on the Cytoplasmic Surface of the Plasma Membrane

Immunogold labeling is an excellent way to reveal the molecular identifications in the observed structures in the electron micrograph. Immunolabeling of the isolated plasma membrane on its cytoplasmic surface can be conducted in the normal manner for fixed samples. As we intend to use platinum replication of the plasma membrane specimens, colloidal gold particles are generally recommended as probes. Since each gold particle exhibits a clear dot surrounded by a halo of platinum-carbon coating, these gold probes can be easily identified in the platinum replicated specimens. Transmembrane proteins can be labeled from the extracellular surface, and these proteins with gold probes can be retained after chemical fixation and platinum replication on the replica (Fujimoto, 1995; Fujimoto et al., 1996).

C. Rapid-Freezing

The plasma membrane with the exposed cytoplasmic surface is next frozen quickly from the cytoplasmic surface, with its external surface still attached to the coverslip. One of the surfaces of a pure-copper block is polished with diamond paste to make a mirror surface, and the block is placed in liquid helium, with its mirror surface slightly exposed over the helium surface. Each coverslip is placed on the plunger tip of the rapid-freezing device, with the cytoplasmic surface of the membrane facing down (Fig. 3D). The plunger is slammed down onto the polished, pure-copper block (metal-contact method). This method takes advantage of the feature of pure copper, for which its thermal conductivity is maximal around the temperature of liquid-helium. Usually, within 20 microns from the frozen surface, the growth of ice crystals is sufficiently suppressed, due to the rapid freezing rate (10^5 °C/s, Heuser et al., 1979) so that they do not alter the cellular structures (Fig. 3D).

D. Deep-Etching and Platinum Replication

The frozen specimen is placed in liquid nitrogen, and then it is transferred into the freeze-etching-shadowing chamber, where the pressure can be lowered to approximately 10^{-6} Pa. The excess ice covering the cytoplasmic surface of the membrane is shaved off, with a prechilled glass or diamond knife, using a microtome placed in the chamber at −110 °C or below (Fig. 3E). Under optimal conditions, this shaving process leaves the structures (in ice) approximately 0.2 to 1 micron from the plasma membrane (Fig. 3F). The surface of the ice layer is then sublimed by slightly raising the temperature of the specimen to approximately −100 to −70 °C, so that the structures hidden in the ice layer are exposed (Fig. 3G). This process is called "etching (or deep-etching as, in this protocol, the etching is more extensive than other methods)", and hence this whole specimen preparation protocol is named "freeze-etching." The etched specimen surfaces are then rotary shadowed with platinum at an angle of approximately 20° from the surface, with a thickness of 1–2 nm, and then with carbon from the top (Fig. 3H). By lowering the specimen temperature and the pressure during shadowing, the platinum grains become smaller, giving images with higher resolutions. The molecules as well as the gold probes localized on the cytoplasmic surface of the cell membrane are immobilized by the deposited platinum (Fujimoto, 1995; Fujimoto et al., 1996).

E. Recovering the Platinum Replicas

The following is the procedure we learned from Drs. T. Baba and S. Ohno of Yamanashi University Medical School. Collodion is applied immediately after the platinum-carbon replicas are removed from the cold chamber, to fortify them. The platinum-carbon replica is removed from the glass coverslip by an incubation in 1% hydrofluoric acid in distilled water. After the replicas are successfully removed from the glass surface and mounted on the grid, the collodion coat is dissolved away in n-pentyl acetate. In this protocol, the sodium hypochlorite solution, which is generally used to remove the replicas from the coverslip and also to clear the membrane and the undercoat structure of the replicas, is replaced with 1% hydrofluoric acid, in order to keep the cell membrane, the undercoat structure, and the immunogold probes attached to these structures on the platinum replicas (Fujimoto, 1995; Fujimoto et al., 1996; 1% hydrofluoric acid is likely to only dissolve the glass, leaving the cell membrane molecules bound to the platinum replica). An additional advantage of using 1% hydrofluoric acid is that the platinum replicas break less often, probably because the membrane components are not removed from the replicas, leaving them rather intact.

In addition, to keep as many colloidal gold particles attached to the platinum replicas as possible, all of the solutions included 0.5–1% Kodak Photo-Flo 200, a detergent used to prevent water-drop stains on the photographic film (advice from John Heuser). After the replicas are washed with distilled water, they are mounted on 100–200 mesh copper grids coated with polyvinyl formvar, and then observed at magnifications of 10,000 ~ 70,000 with a transmission electron microscope (~80 kV).

F. Summary of the Methods for Producing Large Plasma Membrane Fragments and Avoiding Excessive Fragmentation of Replicas

The following methodological precautions and improvements will help to reproducibly produce large plasma membrane fragments and replicas without excessive fragmentation. Although individually these are minor modifications, but collectively they exert a substantial impact.

1. Employ an alcian-blue coat, rather than a poly-L-lysine coat (Rutter *et al.*, 1988; Sanan and Anderson, 1991).

2. In the protocol to observe the top membrane, before overlaying the alcian-blue-coated coverslips, remove the excess water from the specimen, leaving just enough buffer to cover the cell.

3. To cleave off the upper membrane attached to the overlaid coverslip, float the coverslip off very gently by adding cleavage medium (using the surface tension of the buffer to float the coverslip). If this is not done gently enough, the membrane will be fragmented.

4. Shave off the frozen sample with a glass or diamond knife, with the angle between the knife and the cover glass adjusted to a shallow angle (less than 6°), so that most of the excess water and the cytoplasm are removed and the cytoplasmic surface of the cell membrane could be exposed after light etching. Since replicas with too many variations in height tend to break when they are removed from the coverslip and placed on the water surface, removal of the excess cytoplasm helps to avoid replica breakage.

5. Apply collodion immediately after the replicas are removed from the cold chamber (before the replicas are removed from the coverslip on the water surface), to fortify the replica. This step helps to prevent replica breakage when the replicas are removed from the cover slip. Dissolve away the collodion coat in *n*-pentyl acetate, after the large replicas are successfully removed from the glass surface.

6. Use a solution of 1% hydrofluoric acid to slightly dissolve the glass surface, to facilitate the removal of the replicas from the cover slip.

7. To keep as many colloidal gold particles attached on the platinum replicas as possible, include 0.5–1% Kodak Photo-Flo 200 in all of the solutions used to remove the replicas from the coverslips.

G. Creation of Stereo Views (Anaglyphs)

The term "anaglyph" refers to any image that has the appearance of being raised from the surface of the paper. For the stereo views (anaglyphs), image pairs have to be obtained by tilting the stage at ±12.5° from the vertical axis. For details on the production of anaglyphs, see Heuser (2000).

H. 3-D Reconstruction of the MSK by Electron Tomography

In X-ray tomography, an increasingly popular technology for disease diagnosis, X-rays irradiate the body at various angles, and from the X-ray images taken at various rotated or tilted images, sliced images of the body are calculated for reconstructing the 3-D images of the body. In electron tomography, the specimen stage is tilted with respect to the incident electron beam in the transmission electron microscope. The mathematical formulation is basically the same, except that the tilt angle may be limited within $\pm\sim 70°$, limiting the z-resolution. In addition, since the axis of rotation slightly shifts around upon stage tilting, inducing large shifts in the location of the observed detailed structures (large as compared to the subnanometer size of the object we observe under the electron microscope), correction to compensate for this effect is needed for three-dimensional reconstruction. Usually, EM images are taken every 1°, and therefore, for the tilt angles between $\pm 70°$, 141 tilt images are taken for a single view field (Fig. 4).

Such quantitative data acquisition and three-dimensional image reconstruction in EM were developed and automated in the 1990s, resulting in the current prevalence of this method. Four key features in the development were: (1) the development of the tilting stage that allows accurate rotation; (2) the development of an electron microscope with z-axis correction and a side-entry specimen holder; (3) the development of a scientific CCD camera with high sensitivity, resolution (1024 × 1024 pixels or more, with a pixel size of 0.85 nm or better on the sample, which should be finer than the platinum grain size), and linearity; (4) the creation of software for the automatic data acquisition of a series of tilted images (Medalia et al., 2002). Recently, this method has increasingly been used to determine molecular forms as well as cellular structures (Lucic et al., 2005).

We generally use the "IMOD" software package, created by Dr. J. R. McIntosh of the University of Colorado at Boulder (Kremer et al., 1996), running on Linux,

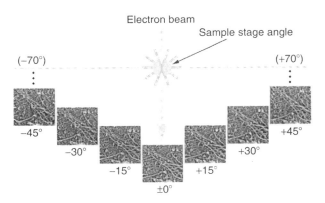

Fig. 4 EM images obtained at different tilt angles of the specimen stage with respect to the incident electron beam. (See Plate no. 26 in the Color Plate Section.)

for the generation of a series of sliced images (for example, about 100 image sections of every 0.8–1.3 nm) from a series of tilted images. Corrections for the tilt and the long-wavelength undulations of the membrane can also be achieved with the IMOD software. 3D rendering (displaying 3D images in different ways) can be carried out using the Templace Graphics AMIRA software package, operating on a Linux system.

III. 3D Structure of the Cytoskeleton-Plasma Membrane Interface

A. The Cytoplasmic Surface of the Plasma Membrane of Cultured Cells is Entirely Coated with the Meshwork of the Actin-Based MSK

A typical electron micrograph, providing a bird's-eye view of the cytoplasmic surface of a large area of the upper cell membrane of a cultured normal rat kidney (NRK) cell line, is shown in Fig. 5. A number of such EM images showing the cytoplasmic surfaces of large cell membrane fragments were obtained for NRK and fetal rat skin keratinocyte (FRSK) cells, suggesting that the entire (upper) plasma membrane, except for the places where clathrin-coated pits and caveolae exist, is coated with a filamentous, net-like structure.

The extensive filamentous, net-like structures shown in the magnified images of the cytoplasmic surface of the plasma membrane in Figs. 6A and B, which were

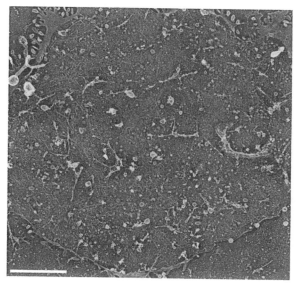

Fig. 5 EM image of the cytoplasmic surface of the upper plasma membrane (bird's-eye view). The plasma membrane fragment shown here represents about a quarter of the upper plasma membrane. Adapted from Morone *et al.* (2006). © 2003 The Rockfeller University Press.

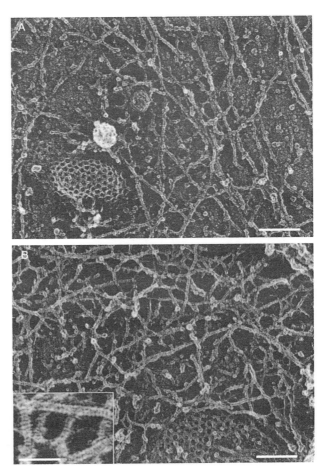

Fig. 6 EM images of the MSK of the upper plasma membrane. (A) NRK cell. (B) FRSK cell. Clathrin-coated structures (A and B) and a caveola (A) show the cytoplasmic surface. The striped banding patterns with the 5.5-nm periodicity on individual filaments are characteristic of actin filaments. These images reveal the close links of the MSK actin filaments with the clathrin-coated structures and caveolae. Bars = 100 nm. The bar in the inset in B = 50 nm. Reproduced from Morone *et al.* (2006). © 2003 The Rockefeller University Press.

obtained for an NRK cell (A) and an FRSK cell (B), respectively, are the MSK. The presence of clathrin-coated structures shows that this is indeed the cytoplasmic surface. The striped banding patterns with a 5.5-nm periodicity on individual filaments are characteristic of actin filaments, and thus indicate that these are actin filaments (Heuser and Kirschner, 1980; Katayama, 1998; Schoenenberger *et al.*, 1999). Since almost all of these filaments contain this striped pattern, it was thus concluded that *the MSK is predominantly composed of actin filaments*. This was also confirmed by immunogold staining.

The inset in Fig. 6B indicates the spatial resolution: since each band in the striped pattern with a 5.5-nm periodicity is visibly separated, the effective resolution is thought to be ≈ 2 nm (both the thickness of the platinum coating and the platinum granule size are ≤ 2 nm: Heuser and Kirschner, 1980; Heuser, 1983).

The MSK structure observed here on the upper cell membrane is similar to that on the bottom cell membrane (the part of the cell membrane facing the coverslip) observed previously (Heuser and Anderson, 1989). Based on these observations, it was concluded that the entire cytoplasmic surface of the plasma membrane is coated with the filamentous actin network (MSK), except for the places where clathrin-coated pits and caveolae are present. The notion of the complete coverage of the cytoplasmic surface of the plasma membrane by actin filaments might have existed for over 30 years in part of the EM community (Byers and Porter, 1977; see Sheetz *et al.* (2006) for a review), but the data specifically indicating that the actin filaments of the MSK may cover the entire plasma membrane had neither been presented in the literature, as done by Morone *et al.* (2006), nor shared in the cell biology community. The EM observations described here are consistent with the MSK-fence and anchored transmembrane protein picket models, in which the entire plasma membrane, except for specific membrane domains, is parceled up into apposed domains, with regard to the lateral diffusion of the molecules incorporated in the plasma membrane (Fujiwara *et al.*, 2002; Kusumi *et al.*, 2005a,b).

B. View of the MSK Using Anaglyphs

A representative anaglyph produced from images taken at ±12° is shown in Fig. 7. In these images, because of their 3D representation, it is especially clear that the MSK, which is mostly composed of actin filaments, generally spreads along the plasma membrane, covering almost the entire cytoplasmic surface of the upper plasma membrane. In addition, clathrin-coated pits and caveolae are very closely associated with the actin filaments in the MSK, as seen in this image as well as in Figs. 6A and B. These results are consistent with those reported by Fujimoto *et al.* (2000), Rothberg *et al.* (1992), and Parton (2003), but in the NRK cells studied here, 92 and 93% of clathrin-coated pits and caveolae (n = 200) are bound by the actin filaments. Furthermore, in these images, many actin filaments are associated with each clathrin-coated pit or caveola. These results are consistent with the requirement of f-actin for clathrin-coated pit internalization (cf. Merrifield *et al.*, 2002; Qualmann *et al.*, 2000).

Many short, thin filaments protrude toward the cytoplasm, mostly perpendicularly, from the membrane surface (arrows in Fig. 7; they were short probably because they were broken when the membrane was ripped off). Note that these perpendicular filaments are almost always connected to the MSK network lying on the cytoplasmic surface (see the tips of the arrows). Thus, the part of the MSK that is located on the cytoplasmic surface is connected three-dimensionally to

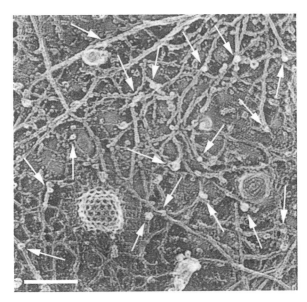

Fig. 7 Stereo electron micrograph (anaglyphs; left = red, right = green) of the plasma membrane undercoat structure generated at ±12° of the tilt angle among the 131 tilt images (acquired in the range of ±65° with 1° steps). For the 3D view, one will need red-blue viewing glasses. These glasses are widely available online or via novelty sources, such as comic book shops and toy stores, and are often attached to Journals [for example, J. Cell Biol. 2006 vol.174, No.6)] or confocal microscopes. We will send a pair upon request at singlemolecules111@frontier.kyoto-u.ac.jp. Arrows: actin filaments protruding from the membrane cytoplasmic surface toward the cytoplasm. The arrows point to the places where the protruding actin filaments intersect with the actin filaments lying horizontally on the plasma membrane. Bar = 100 nm. Adapted from Morone *et al.* (2006). © 2003 The Rockefeller University Press. (See Plate no. 27 in the Color Plate Section.)

the cytoskeleton. Together, they will provide the mechanical support for the membrane and the force for deforming the membrane.

C. Quantitative 3D Reconstruction of the Undercoat Structure on the Cytoplasmic Surface of the Plasma Membrane Using Electron Tomography

The 3D structure of the undercoat within 100–134 nm from the cytoplasmic surface of the plasma membrane, which includes clathrin-coated pits, caveolae, and the actin-based MSK, was reconstructed using electron tomography for the platinum replicated samples. Based on the 97–141 tilt images acquired in the range of ±48–70° every 1° step for a single EM view field, 100–121 sliced images of every 0.85–1.34 nm perpendicular to the z-axis (parallel to the image obtained at 0° of the tilt angle) were calculated by a computer (long-wavelength [≥∼500 nm] undulations of the cell membrane were corrected by the 3D-reconstruction software, IMOD). The 3D-image was reconstructed based on these serial thin slices.

In Fig. 8A, a typical MSK structure quantitatively analyzed in the present work is shown in an anaglyph, and its 8.5-nm thick sections (created by superimposing ten 0.85-nm sections) of the MSK of an NRK cell, starting from the cytoplasmic side toward the membrane, are shown (Fig. 8B). The actin-based MSK is visible on image sections 81 through 110. Individual actin filaments, forming a network as well as bundles, can be identified. Given the high density of the actin filament meshwork, which is much smaller than the optical resolution, conventional fluorescence microscopy cannot be used to observe the individual actin filaments, and can visualize only the bundles of actin filaments.

D. Interface Structure of the MSK on the Cytoplasmic Surface of the Plasma Membrane

The part of the actin-based MSK that contacts the cytoplasmic surface of the plasma membrane has been proposed to partition the cell membrane into 30–230 nm compartments, by the "fence and picket" effect (Edidin *et al.*, 1991; Kusumi and Sako, 1996; Kusumi *et al.*, 2005a,b), for the diffusion of membrane molecules. If these fence and picket models are correct, then the distribution of the mesh size of the MSK on the cytoplasmic surface of the plasma membrane would be practically the same as that of the compartment size determined by diffusion measurements of membrane molecules. To carry out this examination, the 3D reconstruction of the MSK by electron tomography provides a unique opportunity, because the obtained images provide quantitative data on the distance between the individual filaments and the membrane surface.

The actin filaments of the MSK that are directly associated with the cytoplasmic surface of the plasma membrane and may be involved in partitioning the plasma membrane were systematically determined. Out of the stack of 121 image slices taken every 0.85 nm from the cytoplasmic surface ($\approx$ 100-nm thick altogether), 16 consecutive image slices from the membrane surface ($\approx$ 13.6-nm thick altogether) were used for this analysis (Figs. 9A and B).

In Figs. 9A (four images on the right) and 9B (from the second to the fourth images), the square boxes in the left-most images were expanded, and the sections of every 1.7 nm (superposition of two 0.85-nm-thick slices, 330 × 330 nm) are displayed, between 0 and 11.9 nm. Using these sections, the filaments that are closely associated with the cytoplasmic surface of the cell membrane were determined. Since the width of the actin filament after platinum shadowing is between 9–11 nm (consistent with Heuser, 1983) and the thickness of the platinum replica is $\leq$ 2 nm (consistent with Heuser, 1983 and Moritz *et al.*, 2000), the height of the actin filament that is associated with the membrane will be 7–9 nm (because the height is given by the actin thickness and one replica thickness, whereas the width of the actin filament in the image is determined by the actin thickness plus two replica thicknesses), with 8 nm being a reasonable estimate.

The electron tomography sections shown in Figs. 9A and B revealed three major classes of filaments, with regard to the distance from the membrane surface.

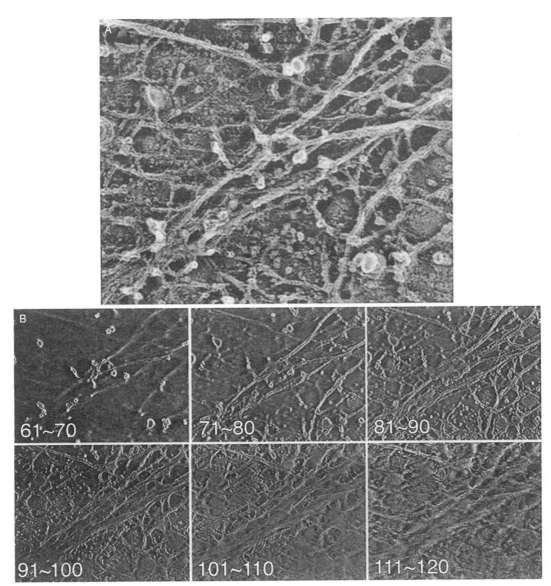

Fig. 8 A typical actin MSK structure on the cytoplasmic surface of the plasma membrane of an NRK cell. (A) An anaglyph of a typical actin MSK structure generated at $\pm 12°$ of the tilt angle among the 97 tilt images (acquired in the range of $\pm 48°$ with 1° steps). (B) A typical series of sliced images of the actin MSK. Ten consecutive sections, each 0.85-nm thick, are superimposed, and six of these superimposed images, representing 60 image sections out of 121 image sections, are shown from the cytoplasmic side toward the plasma membrane side. The numbers here indicate the number of slices counted from the cytoplasmic side. The actin-based MSK near the cytoplasmic surface of the plasma membrane is visible on images 81 through 110. Adapted from Morone *et al.* (2006). © 2003 The Rockfeller University Press. (See Plate no. 28 in the Color Plate Section.)

12. Molecular Architecture of the Plasma-Membrane-Associated Cytoskeleton

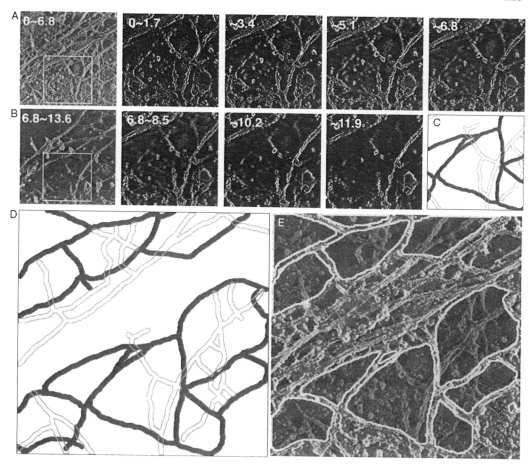

Fig. 9 The method for determining the MSK mesh on the cytoplasmic surface of the plasma membrane, which possibly delimits the compartments of the plasma membrane, using the 3D-reconstructed images of the MSK (an NRK cell). (A, B) The images on the far left are the 0 ~ 6.8 nm or 6.8 ~ 13.6 nm sections, each of which is a stack of eight 0.85-nm sections of 670 × 670 nm. These are from a series of 121 image sections (0.85-nm thick) from the cytoplasmic surface, after the tilt and the long-wavelength undulation of the cell surface were corrected. The images in the white squares in (A) and (B) (330 × 330 nm) are expanded on the right of these image stacks, with a section thickness of 1.7 nm (two 0.85-nm sections are superimposed)(330 × 330 nm for each image). (C) The outline of each actin filament adjacent to the membrane surface (green, which could not be observed above 10.2 nm) and that of each actin filament that could be observed above 10.2 nm (red). The view field and magnification are the same as those for the thinner sections shown in (A) and (B) (330 × 330 nm). See the Methods section for details. (D) The outline of actin filaments in a greater view field, which is the same as those in the thick sections (0 ~ 6.8 nm and 6.8 ~ 13.6 nm) in (A) and (B) (670 × 670 nm, expanded here). (E) The image of the 0 ~ 6.8 nm sections (670 × 670 nm), superimposed on the image of the areas surrounded by the filaments outlined in green in (D) (green areas with yellow outlines). According to the "fence" and "picket" models, these areas are likely to be the compartments where the membrane molecules are temporarily confined. Adapted from Morone *et al.* (2006). © 2003 The Rockefeller University Press. (See Plate no. 29 in the Color Plate Section.)

The actin filaments of the first class are distinct even in the first 0 ∼ 1.7 nm section (since the contrast is reversed in these micrographs, they look more lucent or white), but they fade out of the reconstructions 8–10 nm away from the membrane surface. These filaments are drawn in green in Fig. 9C. We interpreted this to mean that these filaments are in close contact with the plasma membrane, with the gap between the filament and the inner membrane surface being less than 0.85 nm, because they can be seen clearly even in the first 0.85-nm section. These filaments are likely to be the significant ones for generating membrane corrals (for more quantitative analyses and descriptions, see Morone *et al.* (2006)).

The filaments of the second class are also clearly visible in the sections very close to the membrane surface, but they do not fade out until about 14 nm away from the surface. These are probably the actin filaments that caught the platinum coating all around their surfaces because they resided slightly off the surface. The extra coating slightly exaggerated their thickness and made them look as though they were in contact with the plasma membrane, when in fact they probably were not quite in direct contact. We did *not* consider these filaments to be close enough to generate membrane corrals.

The actin filaments of the third class are not apparent in the sections closest to the plasma membrane, but they become clear some distance away from it (greater than 2–4 nm), and they also do not fade out until ∼14 nm. We interpreted this to mean that these filaments are those that definitely do not contact the plasma membrane directly, and hence should not contribute to forming corrals. The second and third classes of filaments are drawn in red in Fig. 9C.

Therefore, we considered that only the first class of filaments (those drawn in green in Figs. 9C and D) forms the MSK fences and pickets, and the area surrounded by these filaments is colored green in the 0–6.8 nm section shown in Fig. 9E. Note that there are regions that were not amenable to such an analysis. They were the areas where the bundles of actin filaments are present (e.g., the structure crossing diagonally from the lower left to the upper right in Fig. 8), the actin filaments are too crowded to be individually discerned, the actin filament is terminated in the middle of a domain (domains that contain a loose end of an actin filament) or the clathrin-coated pits, caveolae, and the smooth-surface membrane invaginations are present. They were excluded from this analysis (the white regions in Fig. 10C).

E. Distribution of the MSK Mesh Size on the Plasma Membrane Determined by Electron Tomography

A similar determination of the MSK meshwork was also made for FRSK cells. Representative meshes of the MSK are shown in Fig. 10 (for an FRSK cell, colored to aid visualization). We carried out such analyses for 10 representative stacks of image sections (1290 nm × 1290 nm plane) each for NRK cells and FRSK cells (eight different cell membrane sheets for each cell type), and identified 76 and 1300 areas bounded by the MSK meshwork, respectively. The two-dimensional area size

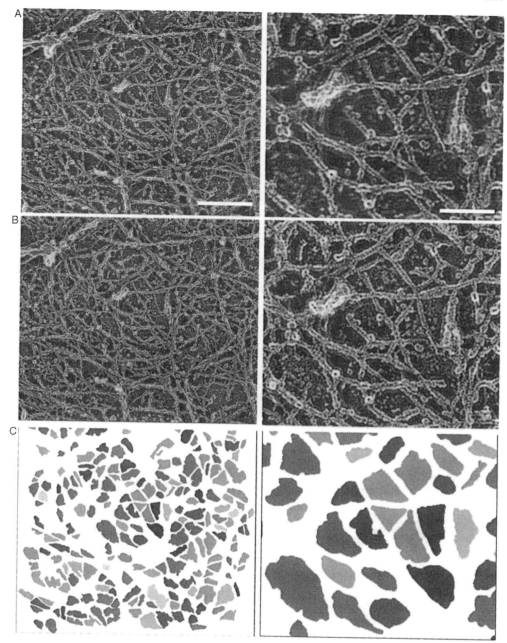

Fig. 10 The MSK meshwork directly located on the cytoplasmic surface of the plasma membrane of an FRSK cell. The central parts of the figures in the left column (bar = 300 nm) are magnified by a factor of 3, and are shown in the right column (bar = 100 nm). (Row A) Typical stereo views of the plasma membrane specimen (anaglyph; left = red, right = green). (Row B) Normal electron

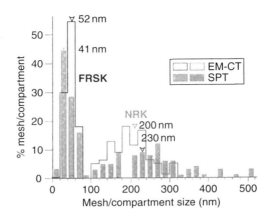

Fig. 11 Comparison of the distributions of the MSK mesh size on the cytoplasmic surface of the plasma membrane estimated by electron tomography (open bars), with that of the compartment size determined from the phospholipid diffusion data (closed bars, adapted from Fujiwara *et al.* (2002) and Murase *et al.* (2004)), for NRK (magenta) and FRSK (blue) cells. Within the same cell type, the MSK mesh size and the diffusion compartment size exhibited similar distributions (compare the open and closed bars with the same color). The actual sizes are quite different between NRK and FRSK cells. Adapted from Morone *et al.* (2006). © 2003 The Rockefeller University Press. (See Plate no. 31 in the Color Plate Section.)

for each domain was measured by the AMIRA software. The distributions of the square root of the area size (the side length, assuming a square shape for the area) for NRK (magenta open bars) and FRSK (blue open bars) cells are shown in Fig. 11. The median values of the area and its square root are 3.9×10^4 nm^2 and 200 nm, respectively, for NRK cells, and 2.7×10^3 nm^2 and 52 nm, respectively, for FRSK cells.

F. Comparison of the MSK Mesh Size on the Plasma Membrane Determined by Electron Tomography with the Compartment Size for Membrane Molecule Diffusion

Our group has proposed that a part of the MSK is directly and closely associated with the cytoplasmic surface of the plasma membrane, and that this close association of parts of the MSK meshwork induces partitioning of the plasma membrane, with regard to the translational diffusion of membrane molecules, based on high speed single-particle tracking data for membrane proteins and lipids (Fig. 12, Jacobson *et al.*, 1995; Kusumi and Sako, 1996; Kusumi *et al.*, 2005b). Namely, the entire plasma membrane is parceled up into apposed domains by the MSK meshwork associated with the plasma membrane. In the short-time regime, these

---

micrographs of the plasma membrane samples. The same view fields as those in A. (Row C) The areas delimited by the actin filaments closely apposed to the cytoplasmic surface of the cell membrane are shown. Different colors are used to aid in visualization. Adapted from Morone *et al.* (2006). © 2003 The Rockefeller University Press. (See Plate no. 30 in the Color Plate Section.)

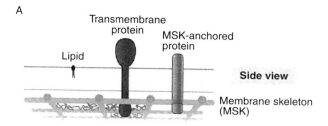
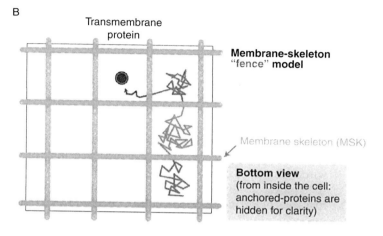
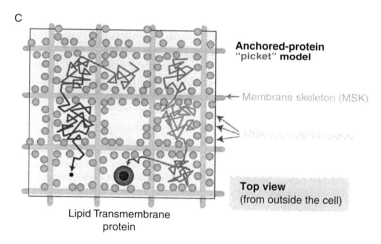

Fig. 12 Proposed mechanisms for the compartmentalization of the plasma membrane for the translational diffusion of transmembrane proteins and phospholipids (located in the *outer* leaflet) in the membrane: Corralling by the membrane-skeleton "fences" and the anchored-protein "pickets". The plasma membrane may be parceled up into closely apposed domains (compartments) for the translational diffusion of transmembrane proteins and lipids (even those located in the outer leaflet).

membrane molecules are temporarily confined within the compartments delimited by the MSK mesh, and, in the long-time regime, they undergo macroscopic diffusion by hopping between these compartments. The Singer-Nicolson model of the membrane is only suitable for a description of diffusion on the scale of ~10 nm in the plasma membrane, but the presence of such compartment boundaries must be considered for the diffusion over distances of several 10 s of nanometers.

Transmembrane proteins are temporarily confined within a compartment, due to the collision of their cytoplasmic domains with the actin filaments in the MSK mesh, closely located on the surface of the plasma membrane (MSK-"fence" model, Fig. 12; also see Sako and Kusumi, 1995; Sako et al., 1998; Suzuki et al., 2005; Tomishige et al., 1998).

Lipid molecules also undergo hop diffusion, which might be explained by the "anchored-protein picket model" (Fujiwara et al., 2002; Kusumi et al., 2005b; Murase et al., 2004). In this "picket" model, various transmembrane proteins anchored to and aligned along the actin filaments that are located right on the cytoplasmic surface of the plasma membrane might effectively act as rows of pickets against the free diffusion of all of the molecules incorporated in the cell membrane. This is due to the steric hindrance and circumferential slowing of the immobile picket proteins, anchored to and lined up along the MSK. Here, the circumferential slowing due to the hydrodynamic friction-like effect of the immobile picket proteins (Bussell et al., 1995a,b) is critical for effectively blocking the passage of membrane molecules across the picket line, because this effect will propagate quite far from the immobile protein surface. Without this effect, the picket model would not be valid. Lipid movement is affected only by pickets, whereas both pickets and fences would act on transmembrane proteins.

The size distributions of the compartments for the diffusion of membrane molecules were obtained for an unsaturated phospholipid, L-α-dioleoylphosphatidylethanolamine, by Fujiwara et al. (2002) and Murase et al. (2004), for NRK and FRSK cells, respectively. The distributions of the side lengths for NRK (magenta closed bars) and FRSK (blue closed bars) cells are shown in the histograms in Fig. 12. The median values of the compartment area and the side length are 4.3×10^4 nm^2 and 230 nm, respectively, for NRK cells, and 2.1×10^3 nm^2 and 41 nm, respectively, for FRSK cells (Murase et al., 2004).

All of the membrane constituent molecules undergo short-term confined diffusion within a compartment and long-term hop diffusion between these compartments. This may be due to corralling by two mechanisms: The membrane-skeleton "fences" and the anchored-protein "pickets". (A) Side-view schematic representation of a transmembrane protein, a phospholipid located in the outer leaflet, and an MSK-anchored protein (membrane skeleton-anchored protein, grey cylinder). The former two molecules are mobile, whereas the MSK-anchored protein is immobile. (B) The membrane-skeleton "fence" or "corral" model, showing that transmembrane proteins are confined within the mesh of the actin-based membrane skeleton, as viewed from inside the cell. Meanwhile the phospholipids located in the outer leaflet of the membrane do not directly interact with the membrane skeleton. (C) The anchored-protein "picket" model, showing the MSK-anchored proteins effectively represent immobile obstacles to the diffusion of transmembrane proteins and phospholipids, as viewed from outside the cell. (See Plate no. 32 in the Color Plate Section.)

These results indicate that in the same cell line (for both the NRK and FRSK cases), the MSK mesh size determined by electron tomography and the diffusion compartment size determined by the high-speed single-particle tracking of a phospholipid are similar to each other. However, between these two cell lines, both the MSK mesh and the diffusion compartment sizes differ greatly. The similarities between the MSK mesh sizes and the diffusion compartment sizes in cell lines that exhibit quite different distributions strongly support the MSK fence and picket models.

IV. Electron Tomography Clarified that Some of the Actin Filaments are Laterally Bound to the Cytoplasmic Surface of the Plasma Membrane

Many mammalian cells have a well-developed cytoskeleton in the bulk cytoplasm. The cytoskeleton consists of actin filaments, intermediate filaments, such as keratin filaments and neurofilaments, and microtubules. Since actin filaments are more involved in cell movement and morphological changes of the cell than the others, they have been believed to be associated with the plasma membrane, at least in the regions of leading edges and membrane ruffles. However, since the interactions of the actin filaments with the plasma membrane have traditionally been investigated in this context, in the actin literature, researchers have almost always assumed that, to understand their interactions, it is sufficient to consider that the barbed-ends of the actin filaments are bound to the plasma membrane, and that they can neglect the lateral binding of the actin filaments to the components of the plasma membrane.

In the late 1980s, actin binding and nucleation on the cytoplasmic surface of the plasma membrane were investigated (Schwartz and Luna, 1986, 1988; Tranter *et al.*, 1989), and ponticulin, a transmembrane protein in the Dictyostelium discoideum plasma membrane that laterally binds to actin filaments, was discovered (Wuestehube and Luna, 1987). Since then, biochemical analyses have revealed many more proteins that exhibit lateral binding to the actin filaments as well as binding to membranes and membrane molecules. These proteins include the Ezrin/Radixin/Moesin (ERM)-family of proteins (Hamada *et al.*, 2003; James *et al.*, 2001), the villin-gelsolin-superfamily proteins including fimbrin and supervillin (Pestonjamasp *et al.*, 1997, where the involvement of supervillin in adhesion structures is described), Epithelial protein lost in neoplasm (EPLIN), which binds to the cadherin-catenin complex (Abe and Takeichi, 2008; Maul *et al.*, 2003), filamin, which is involved in various membrane-protein functions (Stossel *et al.*, 2001; Tavano *et al.*, 2006; Uribe and Jay, 2007), dystrophin and utrophin, which, like tropomyosin, are likely to bind to actin filaments (Fig. 1; Rybakova and Ervasti, 1997; Rybakova *et al.*, 2002), and, let us not forget the myosin-family of proteins, including myosin-I. Namely, *these proteins are likely to mediate the lateral binding of actin filaments to the cytoplasmic surface of the plasma membrane*, and thus *lateral binding as well as barbed-end binding* is important for understanding the MSK functions and the interaction between the plasma membrane and the cytoskeleton.

However, very few systematic structural studies of the MSK at the interface with the plasma membrane have been carried out. The actin filaments laterally bound to the plasma membrane have not received sufficient attention, despite their possibly important functions. This is partly due to the technical difficulties in obtaining large plasma membrane specimens that allow high resolution observations, without introducing too many artifacts. We think that 3D reconstructions of the MSK, using electron tomography with frozen-etched, platinum-replicated plasma membrane specimens, may be ideal for carrying out ultrastructural observations of the interface between the MSK and the plasma membrane. We hope that this review enhances the readers' interests in the interface structures between the MSK and the plasma membrane, and that the methods described in this review will help the readers to perform electron microscopic as well as tomographic studies of the interactions of the MSK with the plasma membrane.

Acknowledgments

We would like to thank Shigeki Yuasa and John Heuser for their helpful advice and encouragement throughout this electron tomography work. This work was supported in part by World Premier International Research Center Initiative (WPI initiative) of the Ministry of Education, Culture, Sports, Science, and Technology (MEXT) of the Japanese government, and also by Health Labor Sciences Research Grant nano-001 to N. Morone, and Grants-in-Aid for Scientific Research and that on Priority Areas from the MEXT to J. Usukura and A. Kusumi.

References

Abe, K., and Takeichi, M. (2008). EPLIN mediates linkage of the cadherin catenin complex to F-actin and stabilizes the circumferential actin belt. *Proc. Natl. Acad. Sci. USA* **105,** 13–19.

Bennett, V. (1990). Spectrin-based membrane skeleton: A multipotential adaptor between plasma membrane and cytoplasm. *Physiol. Rev.* **70,** 1029–1065.

Bussell, S. J., Koch, D. L., and Hammer, D. A. (1995a). Effect of hydrodynamic interactions on the diffusion of integral membrane proteins: Tracer diffusion in organelle and reconstituted membranes. *Biophys. J.* **68,** 1828–1835.

Bussell, S. J., Koch, D. L., and Hammer, D. A. (1995b). Effect of hydrodynamic interactions on the diffusion of integral membrane proteins: Diffusion in plasma membranes. *Biophys. J.* **68,** 1836–1849.

Byers, H. R., and Porter, K. R. (1977). Transformations in the structure of the cytoplasmic ground substance in erythrophores during pigment aggregation and dispersion. I. A study using whole-cell preparations in stereo high voltage electron microscopy. *J. Cell Biol.* **75,** 541–558.

Byers, T. J., and Branton, D. (1985). Visualization of the protein associations in the erythrocyte membrane skeleton. *Proc. Natl. Acad. Sci. USA* **82,** 6153–6157.

Chandler, D. E., and Heuser, J. (1979). Membrane fusion during secretion: Cortical granule exocytosis in sex urchin eggs as studied by quick-freezing and freeze-fracture. *J. Cell Biol.* **83,** 91–108.

Choquet, D., Felsenfeld, D. P., and Sheetz, M. P. (1997). Extracellular matrix rigidity causes strengthening of integrin-cytoskeleton linkages. *Cell* **88,** 39–48.

Coleman, T. R., Fishkind, D. J., Mooseker, M. S., and Morrow, J. S. (1989). Functional diversity among spectrin isoforms. *Cell Motil. Cytoskeleton* **12,** 225–247.

Edidin, M., Kuo, S. C., and Sheetz, M. P. (1991). Lateral movements of membrane glycoproteins restricted by dynamic cytoplasmic barriers. *Science* **254**, 1379–1382.

Evans, E. A. (1989). Structure and deformation properties of red blood cells: Concepts and quantitative methods. *Meth. Enzymol.* **173**, 3–35.

Fujimoto, K. (1995). Freeze-fracture replica electron microscopy combined with SDS digestion for cytochemical labeling of integral membrane proteins. Application to the immunogold labeling of intercellular junctional complexes. *J. Cell Sci.* **108**(Pt 11), 3443–3449.

Fujimoto, K., Umeda, M., and Fujimoto, T. (1996). Transmembrane phospholipid distribution revealed by freeze-fracture replica labeling. *J. Cell Sci.* **109**(Pt 10), 2453–2460.

Fujimoto, L. M., Roth, R., Heuser, J. E., and Schmid, S. L. (2000). Actin assembly plays a variable, but not obligatory role in receptor-mediated endocytosis in mammalian cells. *Traffic* **1**, 161–171.

Fujita, A., Cheng, J., Hirakawa, M., Furukawa, K., Kusunoki, S., and Fujimoto, T. (2007). Gangliosides GM1 and GM3 in the living cell membrane form clusters susceptible to cholesterol depletion and chilling. *Mol. Biol. Cell* **18**, 2112–2122.

Fujiwara, T., Ritchie, K., Murakoshi, H., Jacobson, K., and Kusumi, A. (2002). Phospholipids undergo hop diffusion in compartmentalized cell membrane. *J. Cell Biol.* **157**, 1071–1081.

Hainfeld, J. F., and Steck, T. L. (1977). The sub-membrane reticulum of the human erythrocyte: A scanning electron microscope study. *J. Supramol. Struct.* **6**, 301–311.

Hamada, K., Shimizu, T., Yonemura, S., Tsukita, S., and Hakoshima, T. (2003). Structural basis of adhesion-molecule recognition by ERM proteins revealed by the crystal structure of the radixin-ICAM-2 complex. *EMBO J.* **22**, 502–514.

Hanson, P. I., Roth, R., Lin, Y., and Heuser, J. E. (2008). Plasma membrane deformation by circular arrays of ESCRT-III protein filaments. *J. Cell Biol.* **180**, 389–402.

Hanson, P. I., Roth, R., Morisaki, H., Jahn, R., and Heuser, J. E. (1997). Structure and conformational changes in NSF and its membrane receptor complexes visualized by quick-freeze/deep-etch electron microscopy. *Cell* **90**, 523–535.

Hartwig, J. H., Chambers, K. A., and Stossel, T. P. (1989). Association of gelsolin with actin filaments and cell membranes of macrophages and platelets. *J. Cell Biol.* **108**, 467–479.

Heuser, J. (2005). Deep-etch EM reveals that the early poxvirus envelope is a single membrane bilayer stabilized by a geodetic "honeycomb" surface coat. *J. Cell Biol.* **169**, 269–283.

Heuser, J. E. (1983). Procedure for freeze-drying molecules adsorbed to mica flakes. *J. Mol. Biol.* **169**, 155–195.

Heuser, J. E. (2000). Membrane traffic in anaglyph stereo. *Traffic* **1**, 35–37.

Heuser, J. E., and Anderson, R. G. (1989). Hypertonic media inhibit receptor-mediated endocytosis by blocking clathrin-coated pit formation. *J. Cell Biol.* **108**, 389–400.

Heuser, J. E., and Kirschner, M. W. (1980). Filament organization revealed in platinum replicas of freeze-dried cytoskeletons. *J. Cell Biol.* **86**, 212–234.

Heuser, J. E., Reese, T. S., Dennis, M. J., Jan, Y., Jan, L., and Evans, L. (1979). Synaptic vesicle exocytosis captured by quick freezing and correlated with quantal transmitter release. *J. Cell Biol.* **81**, 275–300.

Heuser, J. E., and Salpeter, S. R. (1979). Organization of acetylcholine receptors in quick-frozen, deep-etched, and rotary-replicated Torpedo postsynaptic membrane. *J. Cell Biol.* **82**, 150–173.

Hirokawa, N., and Heuser, J. E. (1981). Quick-freeze, deep-etch visualization of the cytoskeleton beneath surface differentiations of intestinal epithelial cells. *J. Cell Biol.* **91**, 399–409.

Hirokawa, N., Tilney, L. G., Fujiwara, K., and Heuser, J. E. (1982). Organization of actin, myosin, and intermediate filaments in the brush border of intestinal epithelial cells. *J. Cell Biol.* **94**, 425–443.

Italiano, J. E., Jr., Lecine, P., Shivdasani, R. A., and Hartwig, J. H. (1999). Blood platelets are assembled principally at the ends of proplatelet processes produced by differentiated megakaryocytes. *J. Cell Biol.* **147**, 1299–1312.

Jacobson, K., Sheets, E. D., and Simson, R. (1995). Revisiting the fluid mosaic model of membranes. *Science* **268**, 1441–1442.

James, M. F., Manchanda, N., Gonzalez-Agosti, C., Hartwig, J. H., and Ramesh, V. (2001). The neurofibromatosis 2 protein product merlin selectively binds F-actin but not G-actin, and stabilizes the filaments through a lateral association. *Biochem. J.* **356,** 377–386.

Kajimura, N., Harada, Y., and Usukura, J. (2000). High-resolution freeze-etching replica images of the disk and the plasma membrane surfaces in purified bovine rod outer segments. *J. Electron Microsc. (Tokyo)* **49,** 691–697.

Kanaseki, T., Ikeuchi, Y., and Tashiro, Y. (1998). Rough surfaced smooth endoplasmic reticulum in rat and mouse cerebellar Purkinje cells visualized by quick-freezing techniques. *Cell Struct. Funct.* **23,** 373–387.

Kanaseki, T., Kawasaki, K., Murata, M., Ikeuchi, Y., and Ohnishi, S. (1997). Structural features of membrane fusion between influenza virus and liposome as revealed by quick-freezing electron microscopy. *J. Cell Biol.* **137,** 1041–1056.

Katayama, E. (1998). Quick-freeze deep-etch electron microscopy of the actin-heavy meromyosin complex during the *in vitro* motility assay. *J. Mol. Biol.* **278,** 349–367.

Katayama, E., Shiraishi, T., Oosawa, K., Baba, N., and Aizawa, S. (1996). Geometry of the flagellar motor in the cytoplasmic membrane of *Salmonella typhimurium* as determined by stereophotogrammetry of quick-freeze deep-etch replica images. *J. Mol. Biol.* **255,** 458–475.

Kremer, J. R., Mastronarde, D. N., and McIntosh, J. R. (1996). Computer visualization of three-dimensional image data using IMOD. *J. Struct. Biol.* **116,** 71–76.

Kusumi, A., Ike, H., Nakada, C., Murase, K., and Fujiwara, T. (2005a). Single-molecule tracking of membrane molecules: Plasma membrane compartmentalization and dynamic assembly of raft-philic signaling molecules. *Semin. Immunol.* **17,** 3–21.

Kusumi, A., Nakada, C., Ritchie, K., Murase, K., Suzuki, K., Murakoshi, H., Kasai, R. S., Kondo, J., and Fujiwara, T. (2005b). Paradigm shift of the plasma membrane concept from the two-dimensional continuum fluid to the partitioned fluid: High-speed single-molecule tracking of membrane molecules. *Annu. Rev. Biophys. Biomol. Struct.* **34,** 351–378.

Kusumi, A., and Sako, Y. (1996). Cell surface organization by the membrane skeleton. *Curr. Opin. Cell Biol.* **8,** 566–574.

Landis, D. M., and Reese, T. S. (1981). Astrocyte membrane structure: Changes after circulatory arrest. *J. Cell Biol.* **88,** 660–663.

Lucic, V., Forster, F., and Baumeister, W. (2005). Structural studies by electron tomography: From cells to molecules. *Annu. Rev. Biochem.* **74,** 833–865.

Maul, R. S., Song, Y., Amann, K. J., Gerbin, S. C., Pollard, T. D., and Chang, D. D. (2003). EPLIN regulates actin dynamics by cross-linking and stabilizing filaments. *J. Cell Biol.* **160,** 399–407.

Medalia, O., Weber, I., Frangakis, A. S., Nicastro, D., Gerisch, G., and Baumeister, W. (2002). Macromolecular architecture in eukaryotic cells visualized by cryoelectron tomography. *Science* **298,** 1209–1213.

Merrifield, C. J., Feldman, M. E., Wan, L., and Almers, W. (2002). Imaging actin and dynamin recruitment during invagination of single clathrin-coated pits. *Nat. Cell Biol.* **4,** 691–698.

Mohandas, N., and Chasis, J. A. (1993). Red blood cell deformability, membrane material properties and shape: Regulation by transmembrane, skeletal and cytosolic proteins and lipids. *Semin. Hematol.* **30,** 171–192.

Moritz, M., Braunfeld, M. B., Guenebaut, V., Heuser, J., and Agard, D. A. (2000). Structure of the gamma-tubulin ring complex: A template for microtubule nucleation. *Nat. Cell Biol.* **2,** 365–370.

Morone, N., Fujiwara, T., Murase, K., Kasai, R. S., Ike, H., Yuasa, S., Usukura, J., and Kusumi, A. (2006). Three-dimensional reconstruction of the membrane skeleton at the plasma membrane interface by electron tomography. *J. Cell Biol.* **174,** 851–862.

Murase, K., Fujiwara, T., Umemura, Y., Suzuki, K., Iino, R., Yamashita, H., Saito, M., Murakoshi, H., Ritchie, K., and Kusumi, A. (2004). Ultrafine membrane compartments for molecular diffusion as revealed by single molecule techniques. *Biophys. J.* **86,** 4075–4093.

Nakata, T., and Hirokawa, N. (1992). Organization of cortical cytoskeleton of cultured chromaffin cells and involvement in secretion as revealed by quick-freeze, deep-etching, and double-label immunoelectron microscopy. *J. Neurosci.* **12,** 2186–2197.

Nermut, M. V. (1981). Visualization of the "membrane skeleton" in human erythrocytes by freeze-etching. *Eur. J. Cell Biol.* **25,** 265–271.

Ohno, S., and Takasu, N. (1989). Three-dimensional studies of cytoskeletal organizations in cultured thyroid cells by quick-freezing and deep-etching method. *J. Electron Microsc. (Tokyo)* **38,** 352–362.

Parton, R. G. (2003). Caveolae—from ultrastructure to molecular mechanisms. *Nat. Rev. Mol. Cell Biol.* **4,** 162–167.

Pestonjamasp, K. N., Pope, R. K., Wulfkuhle, J. D., and Luna, E. J. (1997). Supervillin (p205): A novel membrane-associated, F-actin-binding protein in the villin/gelsolin superfamily. *J. Cell Biol.* **139,** 1255–1269.

Qualmann, B., Kessels, M. M., and Kelly, R. B. (2000). Molecular links between endocytosis and the actin cytoskeleton. *J. Cell Biol.* **150,** F111–F116.

Rothberg, K. G., Heuser, J. E., Donzell, W. C., Ying, Y. S., Glenney, J. R., and Anderson, R. G. (1992). Caveolin, a protein component of caveolae membrane coats. *Cell* **68,** 673–682.

Rutter, G., Bohn, W., Hohenberg, H., and Mannweiler, K. (1988). Demonstration of antigens at both sides of plasma membranes in one coincident electron microscopic image: A double-immunogold replica study of virus-infected cells. *J. Histochem. Cytochem.* **36,** 1015–1021.

Rybakova, I. N., and Ervasti, J. M. (1997). Dystrophin-glycoprotein complex is monomeric and stabilizes actin filaments *in vitro* through a lateral association. *J. Biol. Chem.* **272,** 28771–28778.

Rybakova, I. N., Patel, J. R., Davies, K. E., Yurchenco, P. D., and Ervasti, J. M. (2002). Utrophin binds laterally along actin filaments and can couple costameric actin with sarcolemma when overexpressed in dystrophin-deficient muscle. *Mol. Biol. Cell* **13,** 1512–1521.

Sako, Y., and Kusumi, A. (1995). Barriers for lateral diffusion of transferrin receptor in the plasma membrane as characterized by receptor dragging by laser tweezers: Fence versus tether. *J. Cell Biol.* **129,** 1559–1574.

Sako, Y., Nagafuchi, A., Tsukita, S., Takeichi, M., and Kusumi, A. (1998). Cytoplasmic regulation of the movement of E-cadherin on the free cell surface as studied by optical tweezers and single particle tracking: Corralling and tethering by the membrane skeleton. *J. Cell Biol.* **140,** 1227–1240.

Sanan, D. A., and Anderson, R. G. (1991). Simultaneous visualization of LDL receptor distribution and clathrin lattices on membranes torn from the upper surface of cultured cells. *J. Histochem. Cytochem.* **39,** 1017–1024.

Sawada, Y., Tamada, M., Dubin-Thaler, B. J., Cherniavskaya, O., Sakai, R., Tanaka, S., and Sheetz, M. P. (2006). Force sensing by mechanical extension of the Src family kinase substrate p130Cas. *Cell* **127,** 1015–1026.

Schoenenberger, C. A., Steinmetz, M. O., Stoffler, D., Mandinova, A., and Aebi, U. (1999). Structure, assembly, and dynamics of actin filaments *in situ* and *in vitro*. *Microsc. Res. Tech.* **47,** 38–50.

Schwartz, M. A., and Luna, E. J. (1986). Binding and assembly of actin filaments by plasma membranes from Dictyostelium discoideum. *J. Cell Biol.* **102,** 2067–2075.

Schwartz, M. A., and Luna, E. J. (1988). How actin binds and assembles onto plasma membranes from Dictyostelium discoideum. *J. Cell Biol.* **107,** 201–209.

Sheetz, M. P. (2001). Cell control by membrane-cytoskeleton adhesion. *Nat. Rev. Mol. Cell Biol.* **2,** 392–396.

Sheetz, M. P., and Dai, J. (1996). Modulation of membrane dynamics and cell motility by membrane tension. *Trends Cell Biol.* **6,** 85–89.

Sheetz, M. P., Sable, J. E., and Dobereiner, H. G. (2006). Continuous membrane-cytoskeleton adhesion requires continuous accommodation to lipid and cytoskeleton dynamics. *Annu. Rev. Biophys. Biomol. Struct.* **35,** 417–434.

Stossel, T. P., Condeelis, J., Cooley, L., Hartwig, J. H., Noegel, A., Schleicher, M., and Shapiro, S. S. (2001). Filamins as integrators of cell mechanics and signalling. *Nat. Rev. Mol. Cell Biol.* **2,** 138–145.

Suzuki, K. G., Fujiwara, T. K., Edidin, M., and Kusumi, A. (2007a). Dynamic recruitment of phospholipase C gamma at transiently immobilized GPI-anchored receptor clusters induces IP3-Ca^{2+} signaling: Single-molecule tracking study 2. *J. Cell. Biol.* **177,** 731–742.

Suzuki, K. G., Fujiwara, T. K., Sanematsu, F., Iino, R., Edidin, M., and Kusumi, A. (2007b). GPI-anchored receptor clusters transiently recruit Lyn and G alpha for temporary cluster immobilization and Lyn activation: Single-molecule tracking study 1. *J. Cell Biol.* **177,** 717–730.

Suzuki, K., Ritchie, K., Kajikawa, E., Fujiwara, T., and Kusumi, A. (2005). Rapid hop diffusion of a G-protein-coupled receptor in the plasma membrane as revealed by single-molecule techniques. *Biophys J.* **88,** 3659–3680.

Svitkina, T. M., Bulanova, E. A., Chaga, O. Y., Vignjevic, D. M., Kojima, S., Vasiliev, J. M., and Borisy, G. G. (2003). Mechanism of filopodia initiation by reorganization of a dendritic network. *J. Cell Biol.* **160,** 409–421.

Svitkina, T. M., Verkhovsky, A. B., and Borisy, G. G. (1995). Improved procedures for electron microscopic visualization of the cytoskeleton of cultured cells. *J. Struct. Biol.* **115,** 290–303.

Takeuchi, M., Miyamoto, H., Sako, Y., Komizu, H., and Kusumi, A. (1998). Structure of the erythrocyte membrane skeleton as observed by atomic force microscopy. *Biophys. J.* **74,** 2171–2183.

Tavano, R., Contento, R. L., Baranda, S. J., Soligo, M., Tuosto, L., Manes, S., and Viola, A. (2006). CD28 interaction with filamin-A controls lipid raft accumulation at the T-cell immunological synapse. *Nat. Cell Biol.* **8,** 1270–1276.

Tomishige, M., Sako, Y., and Kusumi, A. (1998). Regulation mechanism of the lateral diffusion of band 3 in erythrocyte membranes by the membrane skeleton. *J. Cell Biol.* **142,** 989–1000.

Tranter, M. P., Sugrue, S. P., and Schwartz, M. A. (1989). Evidence for a direct, nucleotide-sensitive interaction between actin and liver cell membranes. *J. Cell Biol.* **109,** 2833–2840.

Tsukita, S., and Ishikawa, H. (1980). Cytoskeletal network underlying the human erythrocyte membrane. Thin-section electron microscopy. *J. Cell Biol.* **85,** 567–576.

Uribe, R., and Jay, D. (2007). A review of actin binding proteins: New perspectives. *Mol. Biol. Rep.*

Ursitti, J. A., Pumplin, D. W., Wade, J. B., and Bloch, R. J. (1991). Ultrastructure of the human erythrocyte cytoskeleton and its attachment to the membrane. *Cell Motil. Cytoskeleton* **19,** 227–243.

Vertessy, B. G., and Steck, T. L. (1989). Elasticity of the human red cell membrane skeleton. Effects of temperature and denaturants. *Biophys. J.* **55,** 255–262.

Wuestehube, L. J., and Luna, E. J. (1987). F-actin binds to the cytoplasmic surface of ponticulin, a 17-kD integral glycoprotein from Dictyostelium discoideum plasma membranes. *J. Cell Biol.* **105,** 1741–1751.

Yin, H. L., and Hartwig, J. H. (1988). The structure of the macrophage actin skeleton. *J. Cell Sci. Suppl.* **9,** 169–184.

CHAPTER 13

Visualization of Dynamins

Jason A. Mears and Jenny E. Hinshaw

Laboratory of Cell Biochemistry and Biology
NIDDK, NIH
Bethesda, Maryland 20892

I. Introduction
II. Methods and Materials
    A. Self-Assembly of Dynamins
    B. Dynamin–Lipid Tubes
    C. Quantifying Dynamin Oligomerization
    D. Transmission Electron Microscopy
    E. Rotary Shadowing
    F. Scanning Transmission Electron Microscopy
    G. AFM
III. Discussion
IV. Summary
    References

I. Introduction

Dynamins play a crucial role in numerous membrane remodeling events throughout eukaryotic cells and have a relatively low nucleotide affinity and high rate of GTP hydrolysis. The propensity of dynamins to self-assemble and stimulate their own GTPase activity distinguishes them from other GTPases. The founding member, dynamin, regulates vesicle scission at the plasma membrane, endosome, and trans-Golgi network during endocytosis and caveolae internalization (Hinshaw, 2000). The dynamin-related protein (Drp1/Dnm1/ADL2B) is involved in mitochondrial fission, while mitofusins (Mfn1 and Mfn2) and OPA1/Mgm1 control fusion of the outer and inner mitochondrial membranes, respectively (Hoppins *et al.*, 2007). Other dynamin family members control peroxisome (Vps1/Drp1) division as well as chloroplast division and cell wall formation in plants (ARC5/ADLs/Phragmoplastin) (Hong *et al.*, 2003; Otegui *et al.*, 2001; Praefcke and McMahon, 2004).

To achieve these varied tasks, all dynamins contain three conserved domains essential for function: a highly conserved GTPase domain, a middle domain, and a GTPase effector domain (GED) (Fig. 1). Each domain is required for self-assembly of dynamins into functional, oligomeric structures (Ingerman *et al.*, 2005; Ramachandran *et al.*, 2006; Smirnova *et al.*, 1999; Song *et al.*, 2004). In addition to these conserved motifs, dynamins contain other functional domains specific to the cellular mechanism associated with each protein (Fig. 1).

Dynamin, the family member studied most extensively, has an additional pleckstrin-homology (PH) domain and a proline-rich domain (PRD) (Fig. 1). The PH domain serves to target dynamin to negatively charged lipids (Klein *et al.*, 1998; Tuma *et al.*, 1993; Zheng *et al.*, 1996), which may concentrate dynamin at the necks of invaginating pits during endocytosis (Achiriloaie *et al.*, 1999; Artalejo *et al.*, 1997; Lee *et al.*, 1999). The PRD interacts with SH3-domain containing proteins, including endophilin, amphiphysin, intersectin, and cortactin. These dynamin partners all serve to help regulate vesicle endocytosis (Dawson *et al.*, 2006; Schmid *et al.*, 1998). Other dynamin family members contain transmembrane (TM) domains (mitofusin, Opa1/Mgm1), a mitochondrial targeting sequence (MTS; OPA1/Mgm1) and additional

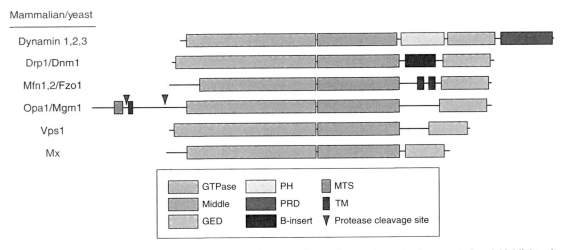

Fig. 1 A schematic alignment of mammalian and yeast dynamins is presented and highlights the domain organization for each protein. The GTPase, middle, and GED topology is conserved among all family members. GED, GTPase effector domain; PH, pleckstrin homology; PRD, proline-rich domain; MTS, mitochondria targeting sequence; TM, transmembrane. (For *Arabidopsis* dynamins see Hong *et al.*, 2003) (See Plate no. 33 in the Color Plate Section.)

inserts whose functions remain unknown (see B-insert in Dnm1/Drp1) (Fig. 1). All of these domains are tailored to the cellular function associated with the individual proteins while maintaining the conserved GTPase, middle, and GED topology. For mitofusins, the TM domains anchor the protein in opposing membranes and likely act as tethers during mitochondrial fusion (Koshiba et al., 2004) in a mechanism believed to be similar to SNARE fusion events (Choi et al., 2006). The MTS found in Mgm1/OPA1 is essential for targeting the protein to the intermembrane space in mitochondria, where it is responsible for fusion events at the inner mitochondrial membrane and regulating cristae structure (Frezza et al., 2006; Meeusen et al., 2006; Meeusen and Nunnari, 2005). Some of the smallest dynamin-related proteins are the Mx proteins, which are involved in viral resistance (Haller and Kochs, 2002). The GTPase, middle, GED topology is maintained with little added sequence and no additional domains. Of all the dynamin family members studied to date, MxA contains the minimal set of domains essential for the function of dynamins.

Large oligomers of dynamins formed upon self-assembly, are amenable to visualization using various microscopic techniques. Specifically, transmission electron microscopy (TEM), atomic force microscopy (AFM), and scanning transmission electron microscopy (STEM) have been used to examine dynamins. To quantify the assembly state of the entire sample, biochemical techniques are also an essential tool. For dynamins, sedimentation assays provide a measure of the oligomeric state, while light scattering experiments provide a measure of conformational changes in dynamin structures due to GTP hydrolysis. When combined with high-resolution imaging techniques, these methods provide a complementary representation of dynamin self-assembly and structural properties, giving a more complete interpretation.

In this chapter, we will focus on three dynamin family members: human dynamin 1, yeast Dnm1, and human MxA. Despite differences in sequence, all three proteins contain similar structural features that can be attributed to the conserved GTPase–middle–GED topology. Each protein oligomerizes in low-salt conditions or with nucleotide analogs and forms helical arrays in the presence of lipid. In the absence of lipid, both dynamin and Dnm1 assemble into spirals while MxA forms curved filaments and rings (Fig. 2). For dynamin and Dnm1, the oligomeric state is tailored to its function: dynamin forms structures with sizes comparable to the size of necks at budding vesicles ($\sim$50 nm) (Hinshaw and Schmid, 1995), while Dnm1 forms significantly larger structures required for mitochondrial fission with sizes comparable to diameters observed at mitochondrial constriction sites ($\sim$100 nm) (Ingerman et al., 2005). Furthermore, both Dnm1/Drp1 and MxA proteins have an apparent affinity for lipid despite lacking a PH domain. Therefore, the polymers of dynamins may preferentially interact with lipid bilayers due to their inherent curvature. Comparing the similarities and differences in dynamin family members using a combination of biochemical and imaging techniques provides the opportunity to understand the relationship between conserved and unique protein domains associated with varied cellular functions.

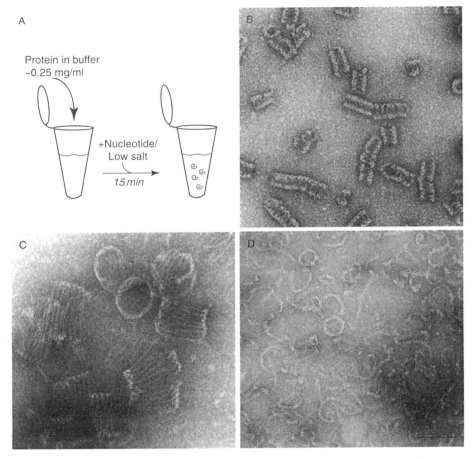

Fig. 2 Oligomeric structures of dynamins are visualized using negative stain TEM. (A) Oligomers are generated under low salt conditions or in the presence of nonhydrolyzable nucleotides. (B) Dynamin spiral structures are shown that were dialyzed in the presence of GDP/BeF. (C) Larger spiral structures are observed for Dnm1 in the presence of GMP-PCP. (D) MxA curved filaments and occasional rings are presented after incubation with GMP-PCP. All protein structures were generated in HCB100. Scale bar, 100 nm.

II. Methods and Materials

A. Self-Assembly of Dynamins

Purified dynamin in high salt exists as a tetramer/monomer (Binns *et al.*, 1999) and dilution into low salt conditions (<50 mM NaCl) forms ring and spiral structures (Hinshaw and Schmid, 1995). In addition, incubation with GDP/BeF, under physiological salt conditions, results in dynamin rings and spirals (Carr and

Hinshaw, 1997). To make spirals, dynamin (~0.2 mg/ml) in HCB100 (Hepes Column Buffer (20 mM Hepes, pH 7.2, 1 mM $MgCl_2$, 2 mM EGTA, 1 mM DTT) with 100 mM NaCl) is incubated with 1 mM GDP, 5 mM NaF, and 500 μM $BeCl_2$ for 15 min at room temperature (Fig. 2A and B). Alternatively, AlF_3 can be used in place of BeF_2 by combining 5 mM NaF and 500 μM $AlCl_2$. Dialysis of dynamin into HCB25 (25 mM NaCl) overnight at 4 °C also results in dynamin spirals, though not as consistent as with GDP/XF_x.

Dnm1 oligomerizes into curved filaments in the absence of nucleotide (HCB100 alone) and forms spirals only in the presence of GMP-PCP (Fig. 2C) (Ingerman et al., 2005). These structures are much larger than dynamin spirals (~100 nm vs. ~50 nm). The addition of GTP (1 mM) to GMP-PCP–Dnm1 spirals caused the disassembly of highly ordered rings into curved filaments that are similar to those seen in the absence of nucleotides. For GMP-PCP spirals, Dnm1(~1.0 mg/ml) is dialyzed into 1 mM GMP-PCP in HCB150 (150 mM NaCl) overnight at 4 °C.

MxA protein also oligomerizes under certain conditions (Kochs et al., 2002, 2005), but not as well as other dynamins. Upon dialysis of MxA (~1 mg/ml) in HCB25–150 and the presence of 1 mM GMP-PCP, ring structures form (Fig. 2D) (Kochs et al., 2005). Additionally, long, straight assemblies are generated by dialysis of MxA in HCB50 and the presence of 1 mM GDP, 5 mM NaF, and 500 μM $BeCl_2$ in 5% ethylene glycol for 20 h at 4 °C (Kochs et al., 2005).

B. Dynamin–Lipid Tubes

Dynamin interacts with lipid *in vitro* with a specific preference for negatively charged bilayers (Sweitzer and Hinshaw, 1998; Zheng et al., 1996). To generate dynamin oligomers on lipid bilayers, dynamin is incubated with liposomes for 1–2 h at room temperature. Liposomes are made by drying 50 μl of lipid in chloroform (100% synthetic phosphatidyl serine (PS), Avanti Polar Lipids) under nitrogen gas, keeping the lipid in vacuum overnight and resuspending to a final concentration of 2 mg/ml in buffer with physiological salt conditions (HCB100). The lipid is then extruded 11–15 times through a 1 μm polycarbonate membrane (Avanti Polar Lipids) to generate unilamellar vesicles of uniform size (Fig. 3A). Adding PS liposomes to dynamin at a final concentration of 0.2 mg/ml (protein and lipid) and incubating at room temperature for ~2 h results in the formation of long helical arrays of dynamin–lipid tubes (Fig. 3B–D). Dynamins form the best decorated tubes with 100% PS or 90% PS, 10% phosphoinositol-4,5-bisphophate ($PI_{4,5}P_2$). We have also had some success with total brain lipid (Avanti Polar Lipids), PS plus cholesterol (up to 10%), and a mixture of 70% PS and 30% galactosylceramide (GalCer). Upon addition of GTP (1 mM final), all but the GalCer tubes constrict and under certain conditions fragment (Fig. 3E–G) (Danino et al., 2004; Sweitzer and Hinshaw, 1998).

Well-ordered structures of dynamin tubes are formed with a dynamin mutant lacking the PRD (ΔPRD). ΔPRD dynamin forms tubes in the same manner as described earlier, but unlike wild-type dynamin, a constricted structure is observed

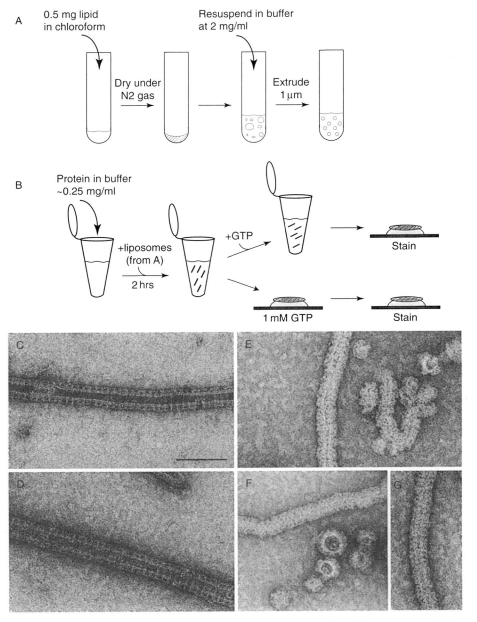

Fig. 3 Dynamin-lipid tubes generated with extruded liposomes constrict and fragment upon addition of GTP. (A) To create extruded liposomes, lipid in chloroform solvent is dried under nitrogen gas in a glass tube and stored under vacuum overnight. The lipid is then resuspended in buffer and extruded through a 1 μm polycarbonate membrane (Avanti). (B) Liposomes are added to dynamin in HCB100 and incubated for ∼2 h to generate dynamin–lipid tubes, which are observed using negative stain TEM (panels C and D). GTP is added either directly to the sample in the tube or by placing the grid with sample adhered to its surface on a drop of GTP in solution. (C, D) Negative stain EM of dynamin–lipid tubes prior to GTP addition are 50 nm in diameter. (E-G) In the presence of GTP, dynamin–lipid tubes fragment and constrict to 40 nm in diameter. Scale bar, 100 nm.

in the presence of nonhydrolysable GTP analogs (GMP-PCP, GMP-PNP, and GTPγS). To form constricted tubes, ΔPRD dynamin (0.2 mg/ml in HCB100) was preincubated with GMP-PCP (1 mM final) for 15 min at room temperature prior to the addition of liposomes (∼0.2 mg/ml) for 1–2 h at room temperature (Zhang and Hinshaw, 2001).

Dnm1 also forms helical arrays upon the addition of negatively charged liposomes despite lacking the PH domain. Unlike dynamin, a greater abundance of tubes are formed if nonextruded lipid is used in the sample, possibly due to the larger surface area required for the larger helical structures. As with dynamin, PS liposomes work well, but we also find a combination of 90% phosphatidylethanolamine (PE; Avanti Polar Lipid) and 10% $PI_{4,5}P_2$ results in tubes with a more regular diameter and length. Addition of GMP-PCP to these tubes does not constrict the lipid; however, the tubes are more ordered. GMP-PCP was added to a final concentration of 1 mM and allowed to incubate for 15 min.

MxA forms protein–lipid tubes only with nanotubes made with GalCer, a lipid that makes extended lipid tubes without protein. This suggests that MxA may not be able to deform liposomes to lipid tubes, but it is capable of binding to a lipid tube with the correct diameter. As with dynamin, incubating MxA (0.2 mg/ml) with GalCer tubes (0.2 mg/ml comprised of 70% PS and 30% GalCer synthetic lipids) at 37 °C for 1 h results in protein helical oligomerization on the lipid substrate.

C. Quantifying Dynamin Oligomerization

A common method used to quantify the amount of oligomer formation in solution is a sedimentation assay (Carr and Hinshaw, 1997; Danino *et al.*, 2004; Hinshaw and Schmid, 1995). Samples incubating at room temperature are transferred to an ice bath to prevent any additional reactions. Samples are then centrifuged at $100,000 \times g$ for 15 min at 4 °C in a Beckman ultracentrifuge (TLA 100 rotor). The supernatant and pellet fractions can then be separated and loaded onto a 4–12% SDS-PAGE gel (Invitrogen) and stained with Coomassie to determine the amounts of protein. Assembled dynamin is found in the pellet (Fig. 4A). The relative densities of dynamin in each fraction can be quantified using gel-imaging software.

Ninety degree light scattering has been used to assess conformational changes in dynamin–lipid tubes treated with GTP. Dynamin tubes are prepared as previously described and diluted 1:10 with HCB100 (dynamin at 0.02 mg/ml). A PC1 Spectrofluorometer (ISS, Champaign, IL) was used at 350 nm with a 4% screen for excitation and 355 nm for emission with excitation and emission slit widths set at 5 nm with an OD1 filter. To begin, measurements of 90° light scattering were made at 0.1-s intervals up to 15 min. After obtaining a stable background, GTP is added to a final concentration of 1 mM, and is stirred gently with a pipette tip. The scattering curves are normalized, and arbitrary units are presented. Upon GTP addition to wild-type dynamin tubes, an immediate drop in light scattering is detected (Fig. 4B), which correlates with membrane constriction and supercoiling observed by electron microscopy (Fig. 3C and D).

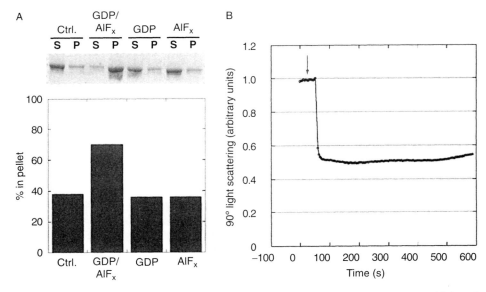

Fig. 4 Biochemical assays quantify dynamin assembly and conformational changes upon addition of GTP. (A) The sup/pellet assay quantifies dynamin assembly under different conditions by centrifuging the sample at 100,000g and running the supernatant and pellet fractions on a gel (top). Imaging software quantifies the relative amount of protein in the pellet (bottom). (B) 90° light scattering measures the relative change in tube conformation. Addition of GTP (indicated with an arrow) leads to a rapid and dramatic decrease in scattering.

D. Transmission Electron Microscopy

While sedimentation assays provide a measure of protein assembly, electron microscopy (EM) provides a means to visualize large oligomeric structures. Negative stain EM is a rapid, qualitative method to observe macromolecular structures with high contrast, although heavy atom stains introduce some structural artifacts, such as flattening. Cryo-EM overcomes many of the negative staining disadvantages by flash freezing the sample in a thin film of vitreous ice and imaging without stain. However, cryo-EM is significantly more time-consuming, and therefore, preliminary imaging of the sample by negative staining is commonly used to provide a simple, quick assessment of sample quality and structural homogeneity.

1. Negative Stain EM

To prepare a negative stain sample, the protein is adhered to carbon-coated mesh grids. To increase hydrophilicity, the grids can be glow discharged or plasma cleaned (Fischione Instruments) prior to adding sample. A small drop of sample (5–10 μl) is placed on a clean surface (i.e., parafilm) and the grid is placed, carbon-side down, onto the drop. After incubating for 0.5–2 min, the grid is washed in either buffer or 2% uranyl acetate (UA) solution in dH$_2$O and blotted with filter paper and washed again before placing the grid on a drop of UA for 2 min, blotted again and dried. The entire time needed to generate a negative stain sample is less than 5 min.

Dynamin spirals and tubes are readily seen using a TEM operated at 100 kV (Figs. 2B and 3C and D) and imaged at magnifications ranging from ∼3000× to 35,000× with a bottom mount 1K × 1K CCD camera, which increases the magnification by a factor of ∼1.3. To examine the effects of GTP, dynamin–lipid tubes are adhered to a grid and then incubated facedown on a drop of GTP in HCB100 (1–5 min) followed by subsequent staining and fixing with 2% UA. EM images of dynamin tubes before and after GTP treatment (Fig. 3C–F) show that the tubes constrict and fragment upon GTP hydrolysis.

Dnm1 spirals generated by incubation with GMP-PCP are easily visualized by negative stain EM (Fig. 2C). These structures are a great deal larger than homologous dynamin spirals. Similarly, Dnm1–lipid tubes are significantly larger than are dynamin tubes (Fig. 5A), and because of the size of these tubes, negative stain flattens the tubes. The flattening is apparent when examining the Fourier transform of these tubes (Fig. 5B), which show diffraction spots consistent with a 2D lattice as opposed to layer lines observed for well-ordered helical structures (Fig. 5D). Adding 1 mM GMP-PCP to preformed Dnm1–lipid tubes for 15 min at room temperature improves the overall order of the helical structure, but some flattening is still observed.

2. Cryo-EM

Cryo-EM eliminates the flattening effect of negative stain EM observed on Dnm1 tubes (Fig. 5C and D). In addition, cryo-EM allows for direct examination of the sample in its native state without staining. To control humidity and temperature, a Vitrobot (FEI Co.) system is used for sample vitrification. A 3–5 μl drop of sample is placed onto a holey carbon grid (Quantifoil R 3.5/1 with copper mesh) and blotted with filter paper to create a thin film of solution. The grid is immediately plunged into liquid ethane, which rapidly freezes the sample in noncrystalline, vitreous ice with the protein in its native state. The time and pressure of the blotting and subsequent freezing is computer controlled in the Vitrobot. Other manual devices work equally well with an experienced user. The advantage of the Vitrobot system is a novice can obtain good ice thickness and homogeneity. After freezing, the grid is stored under liquid nitrogen prior to examining the sample in the microscope. Using low-dose electron microscopy to prevent destruction of the sample, we are able to see ordered helical structures of the Dnm1–lipid tubes (Fig. 5C) with promising layer-line data in the Fourier transform (Fig. 5D). Cryo-electron tomography, a relatively new technique used to determine the structure of large cellular complexes, was also used to examine Dnm1 tubes. For this experiment, the vitreous sample is prepared as described earlier with the addition of gold particles applied to the grid prior to applying the sample. The sample is

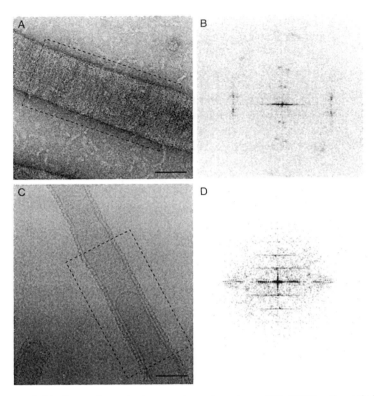

Fig. 5 Dnm1–lipid tubes are imaged using negative stain and cryo-EM. (A) Negative stain imaging of Dnm1–lipid tubes suggests flattening occurs due to stain and sample dehydration (B) Corresponding Fourier transform of image in A shows spots indicating a 2D lattice, indicative of tube flattening. (C) Cryo-EM image of Dnm1–lipid tube. (D) Fourier transform of image in (C) provides a regular helical pattern with layer lines suggesting a helical structure.

tilted to obtain a series of images at various angles (±70°) with the Serial-EM software (Mastronarde, 2005). The images are aligned and a three-dimensional reconstruction is obtained using IMOD software (Kremer *et al.*, 1996). Tomographic reconstruction of Dnm1 tubes confirms that the flattening seen with negative stain on a carbon surface is no longer observed when cryo-EM methods are used.

Cryo-EM has also been used to visualize dynamin structures in their native states (Chen *et al.*, 2004; Danino *et al.*, 2004; Sweitzer and Hinshaw, 1998; Zhang and Hinshaw, 2001). Vitreous samples of dynamin–lipid tubes can be generated as described earlier. As observed with negative stain EM, a T-shape structure is seen at the lipid interface (Fig. 6A; see insert); a common feature observed among all dynamin family members examined to date, regardless of differences in protein circumference (Ingerman *et al.*, 2005; Kochs *et al.*, 2005; Low and Lowe, 2006).

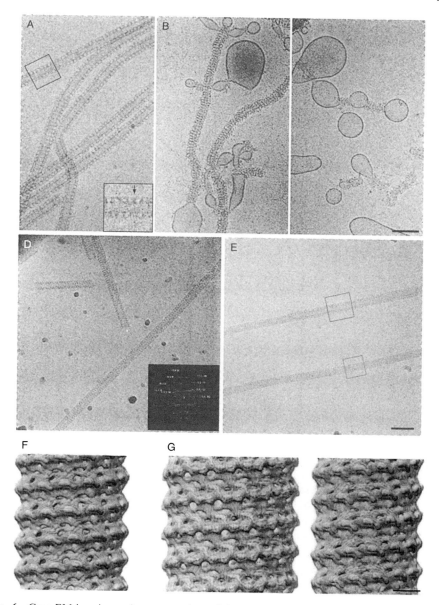

Fig. 6 Cryo-EM imaging and reconstructions of dynamin are presented in the nonconstricted and constricted states. (A) Dynamin–lipid tubes formed as in Fig. 3 and imaged using cryo-EM. (B, C) When GTP is added, the tubes constrict the lipid bilayer, leading to bulges in regions of undecorated lipid. (D) ΔPRD dynamin forms well-ordered tubes in the presence of lipid and GMP-PCP, which diffract to ~20 Å resolution (Fourier transform in inset). Such images are averaged using helical reconstruction methods to generate a three-dimensional reconstruction (panel F). (E) ΔPRD dynamin–lipid tubes in both the nonconstricted and constricted states are boxed (blue and red boxes respectively) for image processing using the IHRSR reconstruction method (Egelman, 2006). (F) 3D map of dynamin–lipid

When GTP is added to the dynamin tubes, a constriction in the helical structure occurs that changes the overall diameter of the helical array from 50 to 40 nm (Fig. 6B and C). The major difference between cryo and negative stain samples is the lack of tube fragmentation in cryo-EM. The fragmentation observed in negative stain is due to the stain and sample drying. Another advantage of cryo-EM is that it allows for shorter time points to be observed after substrate addition. After the dynamin–lipid sample is added to the holey grids in the Vitrobot apparatus, GTP is added separately to the drop of sample on the grid and allowed to incubate for a specified amount of time before blotting and plunging the sample in liquid ethane. This procedure reveals that within 5 s after addition of GTP, most of the dynamin tubes are constricted and supercoiled with undecorated lipid bulges between constricted segments (Danino et al., 2004).

Cryo-EM also provides higher resolution images with less background noise since the sample is preserved in a thin layer of ice without stain or carbon support. Conversely, the low electron dose required for preventing sample damage results in low contrast and a higher background noise. Therefore, averaging numerous images to enhance the signal-to-noise ratio is necessary for subsequent image reconstructions. For dynamin, it was determined that removing the PRD favors the formation of well-ordered tubes that are ideal for image reconstruction (Zhang and Hinshaw, 2001). Additionally, constricted ΔPRD dynamin tubes in the presence of GMP-PCP are straight with a consistent helical pitch (Fig. 6D). Using helical reconstruction techniques, a three-dimensional structure for ΔPRD dynamin tubes was obtained for the constricted (nucleotide-bound) state (Fig. 6F) (Zhang and Hinshaw, 2001). The nonconstricted ΔPRD dynamin tubes (Fig. 6E, a section is boxed in blue) are more curved and flexible with a varied pitch, and therefore the traditional helical reconstruction techniques could not be applied. Instead, the Iterative Helical Real Space Reconstruction (IHRSR) method (developed by Edward Egelman at the University of Virginia) was used to overcome these problems. The IHRSR method uses a single-particle approach to generate three-dimensional reconstructions (Egelman, 2006), and was used to generate reconstructions for ΔPRD dynamin in both the nonconstricted and constricted states (Fig. 6G) (Chen et al., 2004).

3. Docking Structures to Cryo-EM Maps

Cryo-EM with image reconstruction methods allows for visualization of protein structures at resolutions greater than ~0.5 nm. Only recently has subnanometer resolution been achieved in part due to improved optics and sample stabilization in

---

tube solved by helical reconstruction methods (Zhang & Hinshaw, 2001) represented at both high (yellow mesh) and low (blue) thresholds. (G) IHRSR image reconstructions of the nonconstricted (left) and constricted (right) ΔPRD dynamin–lipid tubes (high-yellow mesh and low–blue thresholds) (Chen et al., 2004). Scale bar for A-E, 100 nm. Scale bar for F and G, 10 nm. (See Plate no. 34 in the Color Plate Section.)

the microscope. To overcome the limited resolution, a comprehensive approach is used that combines X-ray crystallographic data with cryo-EM structures. There are numerous ways to dock X-ray structures into a cryo-EM map, including manually fitting crystal structures into cryo-EM density. Several visualization programs are available for manual fitting, including O (Jones et al., 1991), Pymol (http://pymol.sourceforge.net/), and Chimera (http://www.cgl.ucsf.edu/chimera). Using O, structures of the GTPase domain from a related dynamin family member and the PH domain from human were manually fit to the constricted ΔPRD dynamin structure determined by helical reconstruction (Zhang and Hinshaw, 2001).

The disadvantage of manual fitting is that the final model is subject to user prejudice. To prevent bias, automated fitting of X-ray structures provides a superior alternative. There are traditionally two methods used to match X-ray structures to the cryo-EM density: topology comparison and atom–voxel matching. For topology comparison, a density representation of the high-resolution structure is used to search orientations that best match the cryo-EM structure based on cross-correlation of the surfaces. For atom–voxel matching methods, the premise is to exhaustively search orientations in real or reciprocal space and find a best fit. To sample conformations with the cryo-EM density, various algorithms are available, including Monte Carlo simulations (see YAMMP/YUP (Tan et al., 1993; Tan et al., 2006) developed in group of Stephen C. Harvey). Monte Carlo simulations stochastically refine the structure(s) within the conformational space until a best fit is determined. This method requires a significant amount of user intervention but also allows a great deal of flexibility, because the force field for the simulation is user-defined. Consequently, bonds, nonbonds, volume exclusion, multidomain fitting, and other terms can be introduced along with simple rigid-body docking (Mears et al., 2006). The GTPase and PH crystal structures of dynamin were docked into the 3-dimensional density maps of dynamin using YAMMP/YUP and revealed a possible corkscrew mechanism of constriction (Mears et al., 2007). Another docking package, SITUS, uses vector quantization to fit high-resolution structures into low-resolution density (Wriggers et al., 1999). This method allows for fast and exhaustive docking of rigid-body atomic structures to the cryo-EM structure, and new methods allow for flexible fitting as well (Wriggers and Birmanns, 2001).

E. Rotary Shadowing

Rotary shadowing is commonly used to examine the shape of molecules and complexes. A thin layer of metal (usually Pt) applied at a low angle, while the sample is rotating, provides a clear replica of the top surface of the sample. This method is therefore useful for determining the hand of a helical array since only one side of the tube is imaged. With negative stain or cryo-EM the hand would be indiscernible because both the top and bottom layers of the helix are reflected equally. To determine the handedness of the ΔPRD dynamin tubes, the sample was freeze-dried and rotary-shadowed with Pt and carbon (Fig. 7A). A drop (10 μl) of ΔPRD dynamin tubes were directly added to clean mica (1 cm × 1 cm), washed

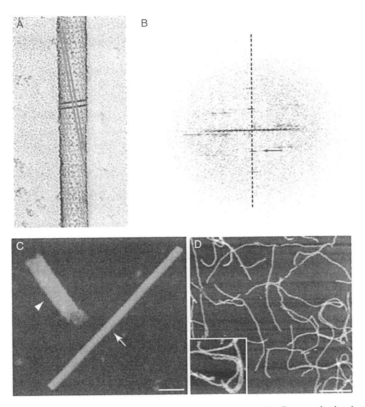

Fig. 7 Dynamin examined by several diverse image techniques. (A) Rotary shadowing of dynamin–lipid tubes reveals the surface of the tube and subsequently the hand of the helices. (B) Fourier transform of the shadowed image confirm the long (gray) pitch helix is left-handed and the short (black) pitch helix is right-handed. (C) STEM analysis of a dynamin spiral (highlighted by arrowhead) provides an accurate measure of mass over a defined length when compared with a TMV particle (arrow). Scale bar, 50 nm. (D) AFM imaging of dynamin tubes was used to examine conformational changes triggered by GTP hydrolysis. A zoomed image of dynamin tubes is shown in the inset. Scale bar, 2 μm.

with HCB0 (no NaCl) for 2 min and blotted from the back and side of the mica before freezing in liquid ethane. The sample was then stored in liquid nitrogen prior to freeze-drying and rotary shadowing. Later, the sample is placed quickly into a freeze-fracture machine (Balzers) that had been pumped down for several hours to a vacuum of 4×10^{-6} m bar and cooled to $-150\,°C$. The fracture blade from the machine is placed over the sample to act as a cold trap as the temperature is raised to $-95\,°C$. The sample is etched for $\sim$30 min, and the temperature is returned to $-150\,°C$ prior to shadowing. While the sample is rotating (speed 9) Pt was applied for 6–8 s at a 6° angle followed by $\sim$10 s of carbon. The sample is then warmed before removing from the machine. The replicas are floated off on deionized water and picked up on copper grids, and they are then examined by TEM. The hand of the long pitch and short pitch helices can be observed by eye (Fig. 7A) or determined by the Fourier transform (Fig. 7B).

F. Scanning Transmission Electron Microscopy

STEM analysis determines the mass of the specimen based on image intensity. With calibration of a specimen of known mass (usually tobacco mosaic virus (TMV); arrow in Fig. 7C), intensity in the STEM can be integrated over an isolated particle and converted to a molecular weight. STEM was used in this way to determine the number of dynamin molecules per turn of the helix (Fig. 7C). For STEM preparations, the ΔPRD dynamin spirals, made in the presence of GMP-PCP, were sent to Brookhaven National Laboratory (BNL) for imaging (Fig. 7C; arrowhead). The specimen is freeze-dried on holey carbon grids in the presence of TMV (as described on the BNL web page, http://www.biology.bnl.gov/stem/stem.html). The final concentration of dynamin spirals was 25–100 μg/ml. The images were processed using the PIC program on a VMS Dec Alpha; however, BNL currently has an updated program called PCMass25 available on the web that runs in Windows 95 or higher. The mass per length of TMV is known and acts as an internal control for the specimen. Dividing the length of the spiral by the helical pitch of 94.8 Å (determined by the diffraction pattern) and then dividing by the relative molecular mass of ΔPRD dynamin (87,100) determines the number of molecules per helical turn.

G. AFM

Dynamin tubes were examined by AFM in collaboration with Dr. Jan Hoh (Johns Hopkins University) to determine if the GTP-induced conformational change of the tubes could be observed in real time (Fig. 7D). To prepare the sample, dynamin tubes were diluted 1:1 with 20 mM NaCl prior to adding to clean mica. After 30 min, the sample was washed ten times with 20 mM NaCl and then imaged by AFM in a thin layer of liquid. Unfortunately, the conformational changes were not observed when GTP was added to tubes adhered to the mica; however, addition of GTP in solution prior to applying the sample to mica revealed supercoiled, shorter, and fewer tubes.

III. Discussion

In vitro studies of any protein require that the protein behave in a manner similar to *in vivo* preparations. For example, dynamin spirals and dynamin–lipid tubes observed *in vitro* are similar to dynamin structures observed at the necks of invaginating pits in nerve synapses (Evergren *et al.*, 2004; Koenig and Ikeda, 1989; Takei *et al.*, 1995). Also the large Dnm1 structures seen *in vitro* coincide with the mitochondrial constriction sites seen in wild-type yeast (Bleazard *et al.*,

1999; Ingerman *et al.*, 2005). In contrast, the oligomers of MxA still remain to correlate with structures seen *in vivo*; however, human MxA may also be membrane-associated as evidence suggests it interacts with the smooth endoplasmic reticulum (Accola *et al.*, 2002).

The intrinsic quality of dynamins to self-assemble makes them amenable for structural studies as illustrated in this chapter. Dynamins have a strong propensity to self-assemble because of strong interactions between the GTPase, middle and GED domains. Even in 400 mM salt, dynamin exists as a tetramer (Binns *et al.*, 1999). A decrease in salt concentration (<50 mM NaCl) leads to oligomerization of dynamin (rings/spirals) (Hinshaw and Schmid, 1995), Dnm1 (curved filaments) (Ingerman *et al.*, 2005), and MxA (curved filaments) (Kochs *et al.*, 2002). Additionally, nucleotide analogs interact with dynamins to form ring/spiral structures (Fig. 2). Dynamin forms spirals in the presence of GDP/BeF, a transition state analog, while nonhydrolysable GTP analogs have no observed affect. In contrast, Dnm1 and MxA form rings/spirals, albeit with different dimensions, in the presence of GTP analogs (GMP-PCP, GTPγS). The difference in nucleotide-dependent assembly between the dynamins may be directly related to the PRD region. ΔPRD dynamin behaves similar to Dnm1 and MxA and readily forms ring/spiral structures in the presence of GMP-PCP. In addition, ΔPRD dynamin–lipid tubes constrict when GMP-PCP is present (Fig. 6D and E). Therefore, the PRD may help regulate GTPase substrate affinity and activity.

The PH domain of dynamin is important for interactions with negatively charged lipids (Tuma *et al.*, 1993; Zheng *et al.*, 1996), and deleting this domain results in the loss of endocytosis as measured by transferrin uptake (Vallis *et al.*, 1999). Moreover, the addition of negatively charged lipid to dynamin in solution stimulates GTP hydrolysis due to increased self-assembly of dynamin on lipid bilayers (Tuma *et al.*, 1993). Dnm1 and MxA both form helical arrays in the presence of lipid, despite lacking a PH domain. The inherent curvature of the assembled oligomer may predispose the proteins to interact with membrane bilayers. However, Dnm1 favors lipid bilayers with less curvature because of the increased diameter of the oligomer.

When GTP is added, dynamin rapidly constricts and then disassociates from the lipid bilayer. The conformational change of dynamin–lipid tubes has also been examined in real time by fluorescence microscopy. In this work, Roux *et al.* observe a twisting motion resulting from GTP hydrolysis of dynamin (Roux *et al.*, 2006). If the ends of the tubes are tethered, the twisting leads to fragmentation of the tubes. The tethered dynamin tubes are similar to the negative stain results, which also show dynamin tubes fragment upon GTP hydrolysis when attached to a support. Dynamin tubes free in solution twist and supercoil as observed by cryo-EM and as untethered tubes in light microscopy (Roux *et al.*, 2006). *In vivo*, dynamin decorated necks of coated pits may be tethered by actin and the plasma membrane through numerous mediator proteins containing an SH3 domain to interact with the PRD in dynamin.

Examining the 3D structures of dynamin in the constricted and nonconstricted states show conformational changes in the "stalk" region of the T structure during lipid constriction (Chen *et al.*, 2004). The middle domain and GED, which are conserved throughout the dynamin family and are responsible for driving self-assembly, are likely contained in this region, which undergoes a dramatic change from a relatively straight pattern to a kinked, zigzag pattern when GMP-PCP is added to the ΔPRD dynamin tube. Therefore, these conserved domains not only drive assembly but also regulate conformational changes that are essential for dynamin function.

It is unclear at this time what role cofactor proteins will have on assembly and constriction of dynamins. In the case of dynamin, many SH3 domain-containing binding partners (endophilin, amphiphysin, cortactin, etc.) are involved in endocytosis (Schmid *et al.*, 1998), and many of these proteins contain lipid-binding motifs (Dawson *et al.*, 2006). Still, the PH domain effectively targets dynamin to membranes *in vitro* (Tuma *et al.*, 1993; Zheng *et al.*, 1996) and is essential for endocytosis (Achiriloaie *et al.*, 1999; Lee *et al.*, 1999). Dnm1 is largely cytoplasmic in yeast cells, and other cofactor proteins (Fis1, Mdv1, and Caf4) are responsible for recruiting Dnm1 to the outer membrane of the mitochondria (Hoppins *et al.*, 2007). Similarly, MxA is largely cytoplasmic, but some protein is associated with smooth ER (Accola *et al.*, 2002). No binding partners have been identified for MxA, although it is able to interact directly with certain viral nucleocapsid proteins (Kochs and Haller, 1999).

In the future, examining additional dynamin family members will further define the fundamental mechanism of action of the dynamins and more importantly reveal the differences that are unique to the function of each protein. For example, Opa1/Mgm1 and mitofusins are required for mitochondrial fusion and understanding how these proteins exploit dynamins' distinctive self-assembly property will provide great insight into this poorly understood mechanism. In addition, Vps1 and Mx proteins only contain the GTPase, middle, and GED domains with little additional sequence, suggesting these proteins are regulated by yet to be identified binding partners. Determining the process of how dynamins reach their functional sites is also crucial. For dynamin, the PRD has been shown to be essential for the localization of this protein to the plasma membrane while the inner membrane fusion protein (Opa1/Mgm1) contains a mitochondria targeting sequence. However, no comparable targeting sequence has been found for either Vps1 and Mx. Overall, comparison of dynamin family members provides an excellent example of modular protein domains organized around a conserved topology that dictate the function(s) for that protein.

IV. Summary

The tools presented in this chapter have been used to characterize the structural and biochemical properties of dynamins. The versatility in microscopic techniques allows for visualization of dynamins with varied shapes, including ring, spiral, and helical oligomers. Negative stain allows for structures to be examined quickly; however, larger structures may flatten, as was observed with Dnm1. Cryo-EM helps eliminate flattening, as shown with Dnm1, allows the specimen to be viewed in a more native state and freezing the sample rapidly following substrate addition (i.e., GTP) allows conformational changes to be observed in seconds. In addition, AFM, fluorescence microscopy, and light scattering assays can be used to observe structural rearrangements that occur upon GTP hydrolysis. Rotary shadowing reveals the surface structure of the sample and can be used to determine the hand of the helical array. The accuracy of STEM analysis provides a means to calculate atomic mass over a defined area or length of helix. Together, these methods provide a comprehensive approach for visualizing dynamins.

Acknowledgments

The authors thank Ye Fang and Dr. Jan Hoh (JHU) for assistance in acquiring AFM results and Dr. Blair Bowers (NHLBI/NIH) for work with rotary shadowing. We also thank Dr. Dan Sackett (NICHD/NIH) for help with light scattering and Dr. Edward Egelman (UVa) for his collaboration on image reconstruction of ΔPRD dynamin tubes using the IHRSR method.

References

Accola, M. A., Huang, B., Al Masri, A., and McNiven, M. A. (2002). The antiviral dynamin family member, MxA, tubulates lipids and localizes to the smooth endoplasmic reticulum. *J. Biol. Chem.* **277,** 21829–21835.

Achiriloaie, M., Barylko, B., and Albanesi, J. P. (1999). Essential role of the dynamin pleckstrin homology domain in receptor-mediated endocytosis. *Mol. Cell. Biol.* **19,** 1410–1415.

Artalejo, C. R., Lemmon, M. A., Schlessinger, J., and Palfrey, H. C. (1997). Specific role for the PH domain of dynamin-1 in the regulation of rapid endocytosis in adrenal chromaffin cells. *EMBO J.* **16,** 1565–1574.

Binns, D. D., Barylko, B., Grichine, N., Atkinson, M. A., Helms, M. K., Jameson, D. M., Eccleston, J. F., and Albanesi, J. P. (1999). Correlation between self-association modes and GTPase activation of dynamin. *J. Protein. Chem.* **18,** 277–290.

Bleazard, W., McCaffery, J. M., King, E. J., Bale, S., Mozdy, A., Tieu, Q., Nunnari, J., and Shaw, J. M. (1999). The dynamin-related GTPase Dnm1 regulates mitochondrial fission in yeast. *Nat. Cell Biol.* **1,** 298–304.

Carr, J. F., and Hinshaw, J. E. (1997). Dynamin assembles into spirals under physiological salt conditions upon the addition of GDP and gamma-phosphate analogues. *J. Biol. Chem.* **272,** 28030–28035.

Chen, Y. J., Zhang, P., Egelman, E. H., and Hinshaw, J. E. (2004). The stalk region of dynamin drives the constriction of dynamin tubes. *Nat. Struct. Mol. Biol.* **11,** 574–575.

Choi, S. Y., Huang, P., Jenkins, G. M., Chan, D. C., Schiller, J., and Frohman, M. A. (2006). A common lipid links Mfn-mediated mitochondrial fusion and SNARE-regulated exocytosis. *Nat. Cell. Biol.* **8,** 1255–1262.

Danino, D., Moon, K. H., and Hinshaw, J. E. (2004). Rapid constriction of lipid bilayers by the mechanochemical enzyme dynamin. *J. Struct. Biol.* **147,** 259–267.

Dawson, J. C., Legg, J. A., and Machesky, L. M. (2006). Bar domain proteins: A role in tubulation, scission and actin assembly in clathrin-mediated endocytosis. *Trends Cell Biol.* **16,** 493–498.

Egelman, E. H. (2007). The iterative helical real space reconstruction method: Surmounting the problems posed by real polymers. *J. Struct. Biol.* **157,** pp. 83–94.

Evergren, E., Tomilin, N., Vasylieva, E., Sergeeva, V., Bloom, O., Gad, H., Capani, F., and Shupliakov, O. (2004). A pre-embedding immunogold approach for detection of synaptic endocytic proteins *in situ*. *J. Neurosci. Methods* **135,** 169–174.

Frezza, C., Cipolat, S., Martins de Brito, O., Micaroni, M., Beznoussenko, G. V., Rudka, T., Bartoli, D., Polishuck, R. S., Danial, N. N., De Strooper, B., and Scorrano, L. (2006). OPA1 controls apoptotic cristae remodeling independently from mitochondrial fusion. *Cell* **126**, 177–189.

Haller, O., and Kochs, G. (2002). Interferon-induced mx proteins: Dynamin-like GTPases with antiviral activity. *Traffic* **3**, 710–717.

Hinshaw, J. E. (2000). Dynamin and its role in membrane fission. *Annu. Rev. Cell Dev. Biol.* **16**, 483–519.

Hinshaw, J. E., and Schmid, S. L. (1995). Dynamin self-assembles into rings suggesting a mechanism for coated vesicle budding. *Nature* **374**, 190–192.

Hong, Z., Bednarek, S. Y., Blumwald, E., Hwang, I., Jurgens, G., Menzel, D., Osteryoung, K. W., Raikhel, N. V., Shinozaki, K., Tsutsumi, N., and Verma, D. P. (2003). A unified nomenclature for *Arabidopsis* dynamin-related large GTPases based on homology and possible functions. *Plant. Mol. Biol.* **53**(3), 261–265.

Hoppins, S., Lackner, L., and Nunnari, J. (2007). The machines that divide and fuse mitochondria. *Annu. Rev. Biochem.* **76**, pp. 751–780.

Ingerman, E., Perkins, E. M., Marino, M., Mears, J. A., McCaffery, J. M., Hinshaw, J. E., and Nunnari, J. (2005). Dnm1 forms spirals that are structurally tailored to fit mitochondria. *J. Cell. Biol.* **170**, 1021–1027.

Jones, T. A., Zou, J. Y., Cowan, S. W., and Kjeldgaard, M. (1991). Improved methods for building protein models in electron density maps and the location of errors in these models. *Acta Crystallogr. A* **47**, 110–119.

Klein, D. E., Lee, A., Frank, D. W., Marks, M. S., and Lemmon, M. A. (1998). The pleckstrin homology domains of dynamin isoforms require oligomerization for high affinity phosphoinositide binding. *J. Biol. Chem.* **273**, 27725–27733.

Kochs, G., Haener, M., Aebi, U., and Haller, O. (2002). Self-assembly of human MxA GTPase into highly ordered dynamin-like oligomers. *J. Biol. Chem.* **277**, 14172–14176.

Kochs, G., and Haller, O. (1999). GTP-bound human MxA protein interacts with the nucleocapsids of Thogoto virus (Orthomyxoviridae). *J. Biol. Chem.* **274**, 4370–4376.

Kochs, G., Reichelt, M., Danino, D., Hinshaw, J. E., and Haller, O. (2005). Assay and functional analysis of dynamin-like Mx proteins. *Methods Enzymol.* **404**, 632–643.

Koenig, J. H., and Ikeda, K. (1989). Disappearance and reformation of synaptic vesicle membrane upon transmitter release observed under reversible blockage of membrane retrieval. *J. Neurosci.* **9**, 3844–3860.

Koshiba, T., Detmer, S. A., Kaiser, J. T., Chen, H., McCaffery, J. M., and Chan, D. C. (2004). Structural basis of mitochondrial tethering by mitofusin complexes. *Science* **305**, 858–862.

Kremer, J. R., Mastronarde, D. N., and McIntosh, J. R. (1996). Computer visualization of three-dimensional image data using IMOD. *J. Struct. Biol.* **116**, 71–76.

Lee, A., Frank, D. W., Marks, M. S., and Lemmon, M. A. (1999). Dominant-negative inhibition of receptor-mediated endocytosis by a dynamin-1 mutant with a defective pleckstrin homology domain. *Curr. Biol.* **9**, 261–264.

Low, H. H., and Lowe, J. (2006). A bacterial dynamin-like protein. *Nature* **444**, 766–769.

Mastronarde, D. N. (2005). Automated electron microscope tomography using robust prediction of specimen movements. *J. Struct. Biol.* **152**, 36–51.

Mears, J. A., Sharma, M. R., Gutell, R. R., McCook, A. S., Richardson, P. E., Caulfield, T. R., Agrawal, R. K., and Harvey, S. C. (2006). A structural model for the large subunit of the mammalian mitochondrial ribosome. *J. Mol. Biol.* **358**, 193–212.

Mears, J. A., Ray, P., and Hinshaw, J. E. (2007). A corkscrew model for dynamin constriction. *Structure* **15**, 1190–1202.

Meeusen, S., DeVay, R., Block, J., Cassidy-Stone, A., Wayson, S., McCaffery, J. M., and Nunnari, J. (2006). Mitochondrial inner-membrane fusion and crista maintenance requires the dynamin-related GTPase Mgm1. *Cell* **127**, 383–395.

Meeusen, S. L., and Nunnari, J. (2005). How mitochondria fuse. *Curr. Opin. Cell Biol.* **17**, 389–394.

Otegui, M. S., Mastronarde, D. N., Kang, B. H., Bednarek, S. Y., and Staehelin, L. A. (2001). Three-dimensional analysis of syncytial-type cell plates during endosperm cellularization visualized by high resolution electron tomography. *Plant Cell* **13**, 2033–2051.

Praefcke, G. J., and McMahon, H. T. (2004). The dynamin superfamily: Universal membrane tubulation and fission molecules? *Nat. Rev. Mol. Cell Biol.* **5**, 133–147.

Ramachandran, R., Surka, M., Chappie, J. S., Fowler, D. M., Foss, T. R., Song, B. D., and Schmid, S. L. (2006). The dynamin middle domain is critical for tetramerization and higher-order self-assembly. *EMBO J.* **26**, pp. 559–566.

Roux, A., Uyhazi, K., Frost, A., and De Camilli, P. (2007). GTP-dependent twisting of dynamin implicates constriction and tension in membrane fission. *Nature* **441**, 528–531.

Schmid, S. L., McNiven, M. A., and De Camilli, P. (1998). Dynamin and its partners: A progress report. *Curr. Opin. Cell Biol.* **10**, 504–512.

Smirnova, E., Shurland, D. L., Newman-Smith, E. D., Pishvaee, B., and van der Bliek, A. M. (1999). A model for dynamin self-assembly based on binding between three different protein domains. *J. Biol. Chem.* **274**, 14942–14947.

Song, B. D., Yarar, D., and Schmid, S. L. (2004). An assembly-incompetent mutant establishes a requirement for dynamin self-assembly in clathrin-mediated endocytosis *in vivo*. *Mol. Biol Cell* **15**, 2243–2252.

Sweitzer, S. M., and Hinshaw, J. E. (1998). Dynamin undergoes a GTP-dependent conformational change causing vesiculation. *Cell* **93**, 1021–1029.

Takei, K., McPherson, P. S., Schmid, S. L., and De Camilli, P. (1995). Tubular membrane invaginations coated by dynamin rings are induced by GTP-gamma S in nerve terminals. *Nature* **374**, 186–190.

Tan, R. K.-Z., and Harvey, S. C. (1993). Yammp: Development of a molecular mechanics program using the modular programming method. *J. Comp. Chem.* **14**, 455–470.

Tan, R. K.-Z., Petrov, A. S., and Harvey, S. C. (2006). YUP: A molecular simulation program for coarse-grained and multi-scale models. *J. Chem. Theory Comput.* **2**, 529–540.

Tuma, P. L., Stachniak, M. C., and Collins, C. A. (1993). Activation of dynamin GTPase by acidic phospholipids and endogenous rat brain vesicles. *J. Biol. Chem.* **268**, 17240–17246.

Vallis, Y., Wigge, P., Marks, B., Evans, P. R., and McMahon, H. T. (1999). Importance of the pleckstrin homology domain of dynamin in clathrin-mediated endocytosis. *Curr. Biol.* **9**, 257–260.

Wriggers, W., and Birmanns, S. (2001). Using situs for flexible and rigid-body fitting of multiresolution single-molecule data. *J. Struct. Biol.* **133**, 193–202.

Wriggers, W., Milligan, R. A., and McCammon, J. A. (1999). Situs: A package for docking crystal structures into low-resolution maps from electron microscopy. *J. Struct. Biol.* **125**, 185–195.

Zhang, P., and Hinshaw, J. E. (2001). Three-dimensional reconstruction of dynamin in the constricted state. *Nat. Cell Biol.* **3**, 922–926.

Zheng, J., Cahill, S. M., Lemmon, M. A., Fushman, D., Schlessinger, J., and Cowburn, D. (1996). Identification of the binding site for acidic phospholipids on the pH domain of dynamin: Implications for stimulation of GTPase activity. *J. Mol. Biol.* **255**, 14–21.

SECTION 2

The Cytoskeleton

CHAPTER 14

Correlated Light and Electron Microscopy of the Cytoskeleton

Sonja Auinger and J. Victor Small

Institute of Molecular Biotechnology
Dr Bohr-Gasse 3 1030
Vienna, Austria

Abstract
I. Introduction
II. Materials and Methods
    A. Materials
    B. Solutions
III. Methods
    A. Patterned Thin Films for Correlated Light Microscopy
    B. Live Cell Microscopy and Fixation
    C. Cell Relocation and Negative Staining
    D. A Note on the Fixation Procedure
IV. Results and Discussion
    A. Structural Features in Cytoskeleton Preparations
    B. Correlated Light and Electron Microscopy
    C. Evaluation of the Technique
V. Summary
References

Abstract

The cytoskeleton of cultured cells can be most easily visualized in the electron microscope by simultaneous extraction and fixation with Triton–glutaraldehyde mixtures, followed by negative staining. Actin filaments are better preserved by stabilization with phalloidin, either during or after the primary fixation step. A technique is described for the combination of this procedure with live cell microscopy. Optimal conditions for light microscopy are achieved by culturing cells on coverslips coated with formvar film. For cell relocation a gold finder grid pattern is embossed on the film by evaporation through a tailor-made mask. After video microscopy and fixation, the film is floated from the coverslip and an electron microscope grid added to the film with the central hole of the grid over the region of interest. Accurate positioning is achieved under a dissecting microscope, using forceps mounted in a micromanipulator. Examples are shown of the changes in organization of actin filaments in the lamellipodia of migrating melanoma cells resulting from changes in protrusion rate. The technique is applicable to alternative processing procedures after fixation, including cryoelectron tomography.

I. Introduction

The turnover and rearrangement of actin filaments in cells is central to morphogenetic processes. Interactions of actin filaments with the cell membrane underlie the pushing and pulling that goes on, to change shape and to move and actin filaments provide the structural scaffolding for cell–cell and cell–substrate interactions, which likewise exert a primary influence on cell form. Actin filaments were first visualized by electron microscopy in plastic sections of muscle, and their helical substructure was deduced from negatively stained images of isolated native thin filaments or from actin polymerized *in vitro* (Huxley, 1969; Steinmetz *et al.*, 1997). When attention turned to visualizing actin arrangements in nonmuscle cells by electron microscopy, the results were disappointing. It soon became evident that plastic embedding procedures were unsuited to the visualization of actin filaments in arrays other than actin bundles stabilized by interactions with proteins such as tropomyosin and myosin (Goldman and Knipe, 1972) or other cross-linkers (Tilney *et al.*, 1980). Motile regions of cells, corresponding to ruffles or lamellipodia, were either devoid of structure (Abercrombie *et al.*, 1971), or appeared in thin sections as an amorphous, "fuzzy" matrix (Wessels *et al.*, 1973). Subsequent studies (reviewed in Small, 1988) showed that actin networks are distorted by procedures that include osmium tetroxide fixation and dehydration in organic solvents. Other procedures for visualizing the cytoskeleton were therefore adopted, each with their own advantages and pitfalls (see Small, 1988; Small *et al.*, 1999). To avoid plastic embedding, present methods are so far limited to cells thin enough to be taken directly into the electron microscope after appropriate processing. Current studies are thus restricted to cells in primary culture or to cell lines. We will here focus on the technique of negative staining for contrasting, but will also discuss the applicability of the general approach to cryoelectron microscopy. Emphasis will be placed on correlating the movement of living cells in the light microscope with the organization of the actin cytoskeleton in the electron microscope. An alternative approach, described by Svitkina and coworkers (Svitkina and Borisy, 1999; Svitkina *et al.*, 1995, 2003), employs the critical point drying procedure and contrasting by rotary shadowing with platinum.

II. Materials and Methods

A. Materials

PBS
Chloroform
Formvar powder
MES, NaCl, EGTA, glucose, $MgCl_2$
Triton X-100
EM-grade glutaraldehyde
Phalloidin in MeOH (stock 1 mg/ml)
Alexa 488 or 568 Phalloidin 300U

B. Solutions

Formvar solution: 0.8–1% and 4% formvar in Chloroform; stir overnight in a closed container.

Cytoskeleton buffer (CB): 10 mM MES, 150 mM NaCl, 5 mM EGTA, 5 mM glucose, 5 mM $MgCl_2$; pH 6.1

Extraction solution: 0.25% glutaraldehyde (GA), 0.5% Triton, 1 μg/ml Phalloidin, 1:300 Alexa Phalloidin in CB; pH 6.5

Note: The ratios of glutaraldehyde and Triton can be adjusted to suit a given cell type.

Intermediate fixation solution: 2% GA, 1 μg/ml Phalloidin, 1:300 Alexa Phalloidin in CB; pH 7

Post-fixation solution: 2% GA, 10 μg/ml Phalloidin in CB; pH 7

Negative stain solutions:

4% sodium silicotungstate (SST)
2% SST
2% SST + 1% aurothioglucose
1% SST + 1% aurothioglucose

These solutions need to be pH adjusted (to around pH 7.5) and filtered (0.2 μm)! For pH adjustment it is important to avoid any intake of salts (use only NaOH) and to check pH every few days until it is stable.

III. Methods

A. Patterned Thin Films for Correlated Light Microscopy

To combine light microscopy with the electron microscopy, cells can be grown on filmed electron microscope grids and imaged live with the grid inverted in a growth chamber on the light microscope (Resch, 2006; Rinnerthaler et al., 1991). The preparation is fixed at an appropriate time and processed for electron microscopy. By using a finder grid, the cell observed by light microscopy is readily located in the electron microscope. While this technique works well (Resch et al., 2006), optimal resolution is difficult to achieve on the freely suspended film in the light microscope. We have therefore adopted a modification, originally introduced by Buckley and Porter (1967), that employs coverslips coated with formvar films. For the purpose of cell relocation, a finder grid pattern is coated onto the film. This method facilitates light microscopy under more suitable imaging conditions. The method can of course be used without a finder pattern, when identification of the same cell in the light and electron microscope is not required.

The procedure for substrate preparation is illustrated schematically in Fig. 1. A formvar film (FV) is cast on a glass slide and floated onto a water trough as usual (Steps 1–4, Fig. 1). Coverslips that fit in the incubation chamber of the light microscope are prepared beforehand: these are dipped and dried in a solution of 2.5% Triton X-100 (to ensure later release of the film) and then coated on their rim with a concentrated solution of formvar (4%), applied through a pipette tip (Step 5, Fig. 1). The latter step is necessary to facilitate later handling of the film. The coverslips are then added to the floating formvar film and retrieved with a piece of parafilm (Steps 6 and 7, Fig. 1). After drying, the coverslips are covered with copper grid masks and coated with gold in an evaporation unit (Step 8, Fig. 1).

The grid masks are custom made "negative grids" (Small, 1984) from Pyser (Edenbridge, UK) in which the grid bars are open and the squares closed (Fig. 2), so that a finder grid pattern is deposited on the film. The amount of gold deposited through the mask should be sufficient that the pattern is easily visible under a dissecting microscope. The masks are made from thin copper foil, have a total diameter of 15 mm and contain 9, separately numbered grid patterns to allow more choice in the selection of suitable cells (Fig. 2). The masks are glued to steel washers to facilitate handling and to keep them flat and can be reused many times.

The coverslip-film combination is sterilized under UV, coated with connective tissue components as required and cells plated onto the film. The cells are transfected with probes of interest expressing EGFP and RFP tags (Shaner et al., 2004) 1–2days before plating.

B. Live Cell Microscopy and Fixation

Imaging of cells can be performed in different modes (wide field, confocal, or TIRF), depending on the experimental requirements. We routinely use wide field imaging on a Zeiss Axiovert 200 inverted microscope equipped with a rear

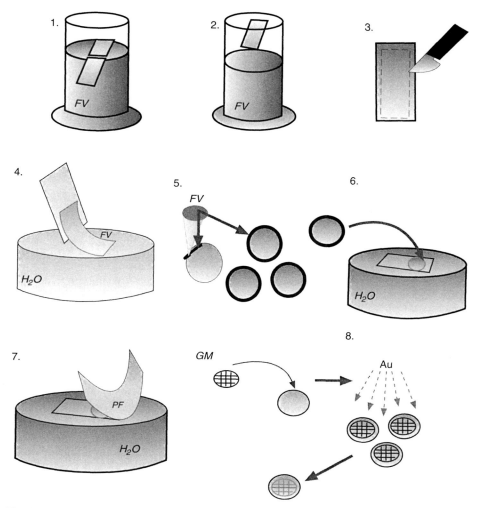

Fig. 1 Preparation of the support films for correlated light and electron microscopy. A formvar film is cast on a cleaned glass slide by dipping into a formvar solution (FV) in a measuring cylinder and drying in the cylinder above the liquid surface (Steps 1 and 2). After scoring on the edge (Step 3), the film is floated onto a water surface (Step 4). Coverslips required for light microscopy are precoated on the edge with a thick rim of formvar (4%) through a pipette tip (Step 5). The coverslips are added to the floating film and retrieved with a piece of parafilm, PF (Step 7). After drying, the coverslips are individually covered with a grid mask (GM), transferred to a vacuum evaporator and coated with a thin layer of gold (Step 8). (See Plate no. 35 in the Color Plate Section.)

illuminated, cooled CCD camera (Micromax, Roper Scientific), and filter wheel and shutter systems for alternating phase and fluorescence microscopy. Transmitted (phase contrast) and incident illumination (fluorescence) is provided by halogen or mercury lamps (with intensity control), with times between subsequent

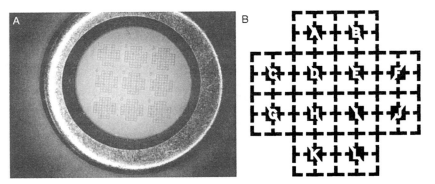

Fig. 2 The negative grid mask. (A) The copper film mask, containing nine finder patterns, mounted on a washer for easy handling. The outside diameter of the mask is 15 mm. (B) The finder pattern in which the black areas correspond to the open regions in the copper film.

frames ranging from 2 to 20 s. Imaging is performed with an oil immersion 100×, NA1.4 objective lens.

Two types of chambers are employed for imaging: an open chamber from Harvard Instruments (Nr, 64 0232) mounted on a heating platform and a homemade, flow through chamber that fits on the same platform. Similar results have been obtained with either chamber, and for simplicity, we will confine discussion to the open system that is commercially available. In the absence of a CO_2 cabinet around the microscope, we use a CO_2-independent culture medium and limit observation to normally less than 1 h. At a chosen time during imaging, the cell of interest is arrested by exchange of the growth medium for the extraction/fixation mixture, containing glutaraldehyde (0.25%), Triton (0.5%), and optionally, 1 μg/ml phalloidin in a cytoskeleton buffer (CB). After around 1 min, the fixation/extraction mix is exchanged for the intermediate fixation solution, including optionally fluorescent phalloidin, to record an actin image of the fixed cell while still on the microscope. The coverslip is then removed from the chamber and transferred to the post-fixation solution, which also includes phalloidin. Post-fixation is performed for at least 30 min at RT and is typically continued overnight at 4 °C. Cells can be kept in this solution for several days at 4 °C before further processing.

C. Cell Relocation and Negative Staining

The next step involves removal of the film from the coverslip, location of the grid square carrying the imaged cell and the application of an EM grid onto the film with the cell in the centre of the grid. This is achieved by following the steps depicted in Figs. 3 and 4.

The coverslip is placed, film side up in a 9-cm Petri dish containing CB. Using forceps, the film is carefully detached from the edge of the coverslip where it is supported by the thick formvar rim. With two pairs of forceps the film is then inverted and brought to the buffer surface so that it spreads out by surface tension,

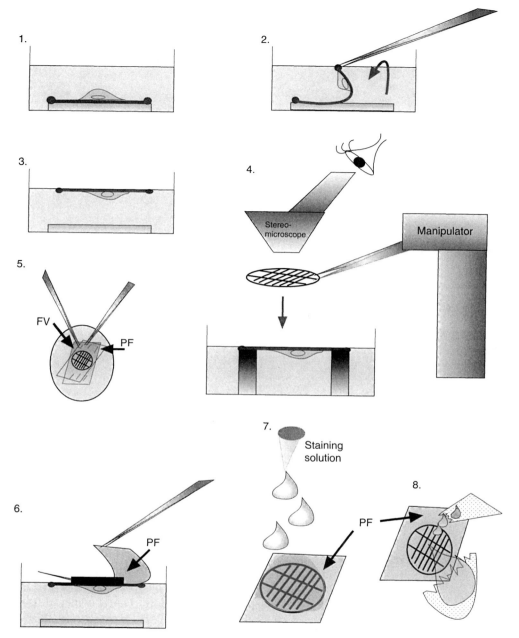

Fig. 3 Film retrieval and negative staining. With the coverslip immersed in CB in a Petri dish (Step 1), the film is loosed on the edges and flipped onto the buffer surface, with the cells down (Steps 2 and 3). Under a dissecting microscope, the film is maneuvered onto a ring and the grid placed on the film using forceps mounted in a micromanipulator (Step 4). Parafilm is layed over the floating grid and the parafilm and formvar film pressed together at the periphery with forceps (Step 5). The parafilm-grid combination is then removed, rinsed with several drops of stain and blotted on the edge with filter paper (Steps 6–8).

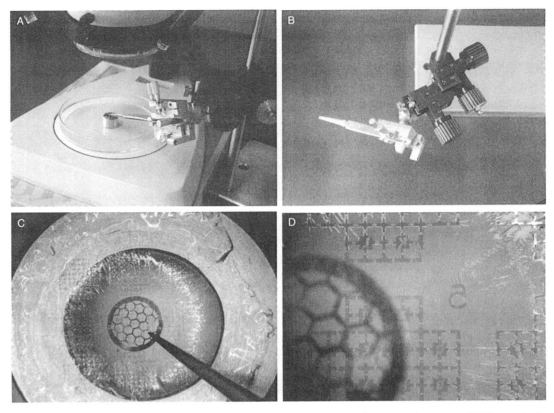

Fig. 4 Maneuvering the grid onto the film. (A) Overview of the mounting set-up. (B) Forceps in the modified dual pipette holder, mounted on the micromanipulator. (C) Application of the grid with the film immobilized on the support ring. (D) Close-up view of the 50 mesh hexagonal grid over the finder grid pattern on the film.

with the cells facing down (Steps 2 and 3; Fig. 3). Next, the film is immobilized on a stainless steel ring platform (height, 1 cm, inner and outer diameters 8 and 15 mm, respectively; Fig. 4A and C) by floating it over the ring and removing enough liquid from the dish to capture the film, with the grid pattern centered (Fig. 4D). Under a dissecting microscope, a 50 mesh hexagonal copper grid is placed onto the film so that the cell of interest lies in the central hole. This is achieved using forceps in a micromanipulator to hold the grid, with an arrangement that allows gentle release of the grid onto the film. For this operation, we use a modified dual pipette holder from Leica to hold the forceps, fixed on a manual micromanipulator from Narashiga (Fig. 4B).

The grid is retrieved and negatively stained as depicted in Steps 5–8 in Fig. 3. With the grid attached, the film is floated off the platform and covered with a piece of parafilm not much larger than the film. Forceps are used to press the floating film to the parafilm surface (Step 5) so that it remains attached in the next step. The parafilm–film combination is removed from the dish with forceps, rinsed on the cell side with several drops of negative staining solution and blotted on the sides with the torn edge of a filter paper (Steps 6–8; Fig. 3). Care should be taken that no macroscopic drops remain on the grid. After drying, the grid can be observed in the microscope. To prevent recrystallization of the stain by uptake of moisture from the atmosphere, grids should be stored in an exicator, but are preferably observed immediately.

Neutral negative stains have been found to be most suitable for cytoskeleton preparations (see also Hoglund *et al.*, 1980). We have used sodium silicotungstate, phosphotungstic acid, and mixtures of sodium silicotungstate with aurothioglucose or trehalose. The concentrations tried range from 1 to 4%, with the pH adjusted between 7 and 8. Acidic uranyl acetate gives high contrast, but the actin filaments are more distorted, suggestive of undue collapse of actin networks during drying Small, J. V., and Celis, J. E. (1978). This contrasts with the results obtained with uranyl acetate for synthetic actin filaments stabilized on a support film, as evidenced from the wealth of literature on actin filament ultrastructure (e.g., Steinmetz *et al.*, 1997). Some three dimensionality of lamellipodia networks is preserved in sodium silicotungstate, as can be illustrated with stereopairs (Small, 1981, 1988) and actin filament substructure is readily visualized in bundles of actin, in filopodia (Small, 1981). Neutral uranyl acetate, buffered with EDTA, can also give good contrast (Resch, private communication), but has not been used enough on cytoskeletons to warrant discussion of its potential here. Suffice it to say that there is room for more experimentation with new mixtures of heavy metal stains for contrasting cytoskeletons.

D. A Note on the Fixation Procedure

The fixation protocol described earlier, comprising a mixture of glutaraldehyde, Triton, and phalloidin (optional), has evolved from original efforts to preserve cell morphology by rapid fixation and at the same time extract cells sufficiently to make the cytoskeleton visible (see Small, 1988; Small *et al.*, 1999). The most suitable ratio of glutaraldehyde to Triton will depend on cell type and should be first assessed in control experiments. Phalloidin has a dual function: as a fluorescent label to control actin preservation in the fluorescence microscope and as a stabilizer of actin filaments. Without phalloidin, actin filaments in negatively stained cytoskeleton networks are often distorted, presumably because they are more susceptible to drying effects than filaments directly attached to the supporting film. Most of our earlier studies employed phalloidin after fixation. However, for correlated live cell microscopy and EM, we have found it advantageous to include phalloidin in the primary fixative. Since the basic structural details are the same, whether phalloidin is added during or after fixation, it appears that the time of cell arrest by the fixative mixture (a fraction of a second) is too short for phalloidin to modify the endogenous filament pool.

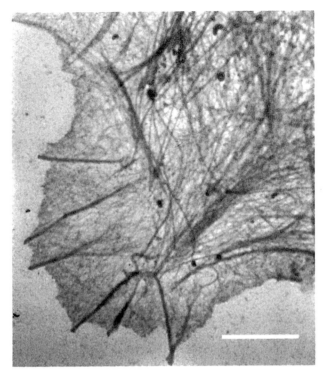

Fig. 5 Low magnification view of the peripheral region of a fish fibroblast cytoskeleton (initial fixation: 0.5%Triton; 0.25%glutaraldehyde), stained with 2% sodium silicotungstate. Bar, 2 µm.

IV. Results and Discussion

A. Structural Features in Cytoskeleton Preparations

Figures 5 and 6 show some general features of negatively stained cytoskeletons prepared by simultaneous extraction and fixation with Triton–glutaraldehyde mixtures. At low magnification, cells should display a continuous, nonfragmented cell edge and a smooth appearance of lamellipodia. The appearance of holes is diagnostic of poor preservation or poor staining. The spreading cell edges show actin networks and actin bundles (filopodia), with all transition stages in between (Fig. 6A; Small, 1988; Small et al., 1982). Behind the lamellipodium, in the so-called lamella zone, the three filament systems of the cytoskeleton can be distinguished (Fig. 6B).

B. Correlated Light and Electron Microscopy

We give two examples here of correlated light microscopy and electron microscopy of motile B16 melanoma cells. In both cases, the cells were treated with aluminum fluoride, which leads to the activation of Rac and to continuous

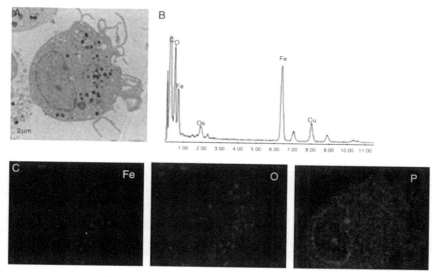

Plate 1 (Figure 2.2 on page 22 of this volume)

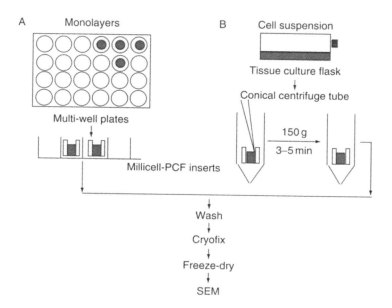

Plate 2 (Figure 2.3 on page 24 of this volume)

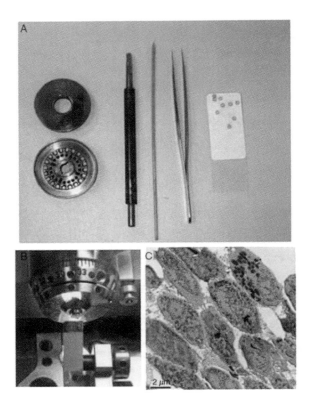

Plate 3 (Figure 2.6 on page 29 of this volume)

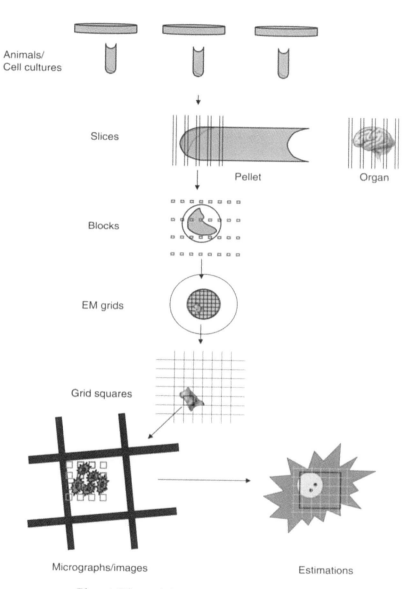

Plate 4 (Figure 4.1 on page 61 of this volume)

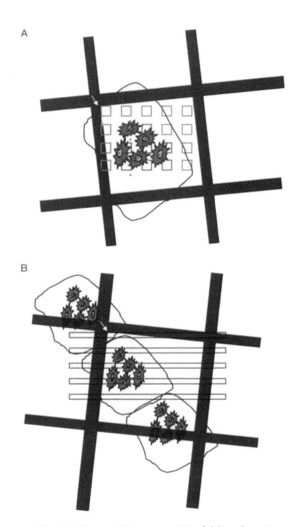

Plate 5 (Figure 4.2 on page 63 of this volume)

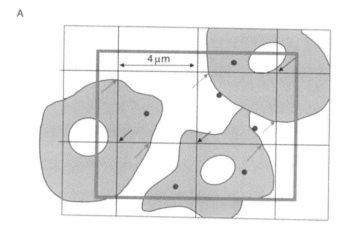
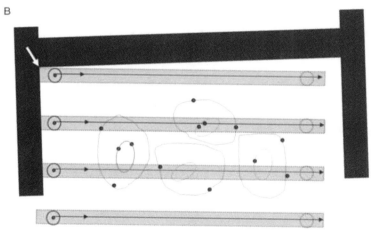

Plate 6 (Figure 4.3 on page 64 of this volume)

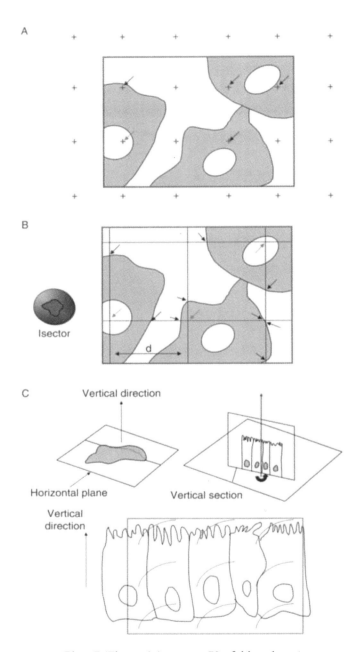

Plate 7 (Figure 4.4 on page 72 of this volume)

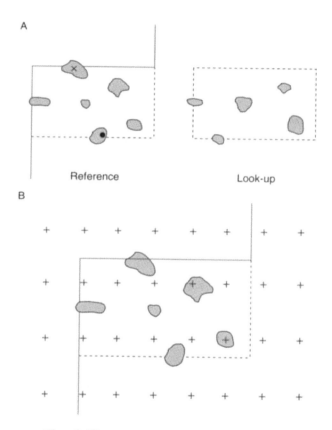

Plate 8 (Figure 4.5 on page 75 of this volume)

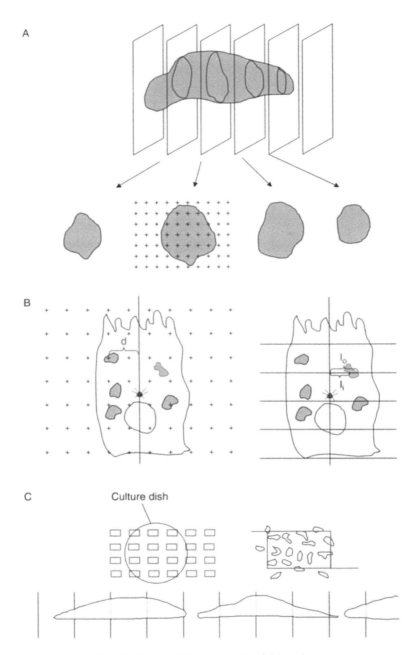

Plate 9 (Figure 4.6 on page 77 of this volume)

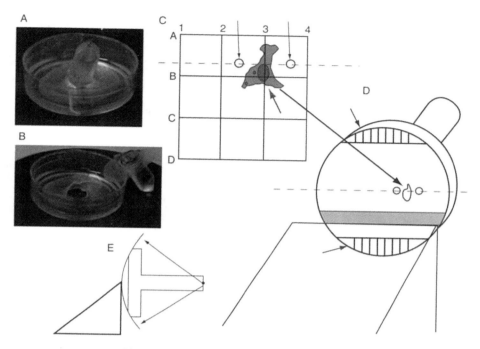

Plate 10 (Figure 5.1 on page 86 of this volume)

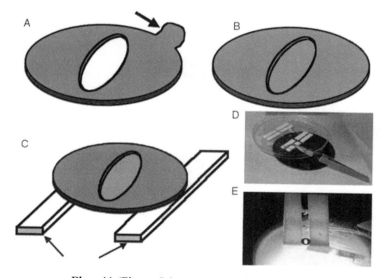

Plate 11 (Figure 5.2 on page 91 of this volume)

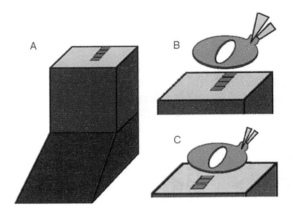

Plate 12 (Figure 5.3 on page 92 of this volume)

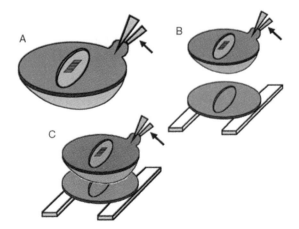

Plate 13 (Figure 5.4 on page 92 of this volume)

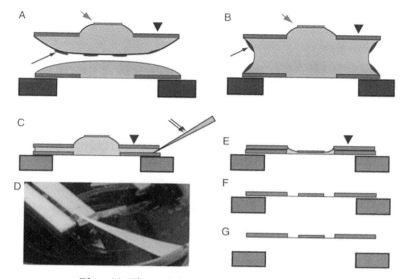

Plate 14 (Figure 5.5 on page 93 of this volume)

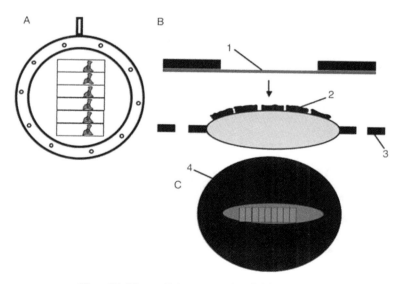

Plate 15 (Figure 5.6 on page 94 of this volume)

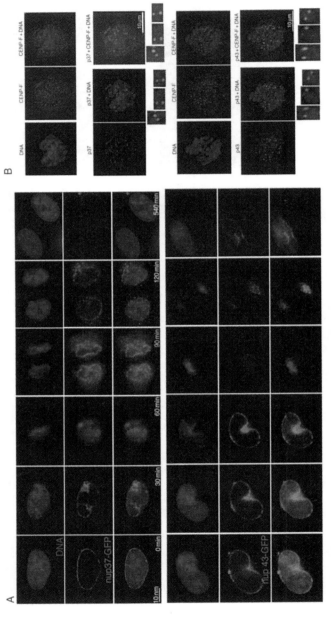

Plate 16 (Figure 6.3 on page 105 of this volume)

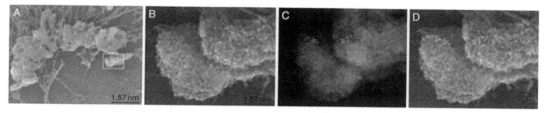

Plate 17 (Figure 6.4 on page 106 of this volume)

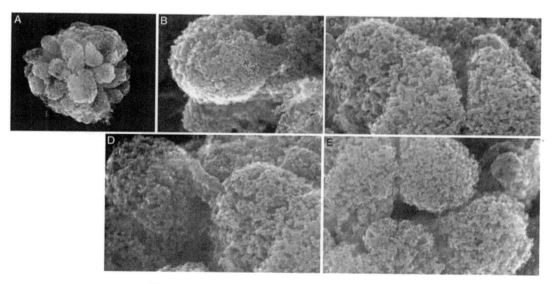

Plate 18 (Figure 6.5 on page 106 of this volume)

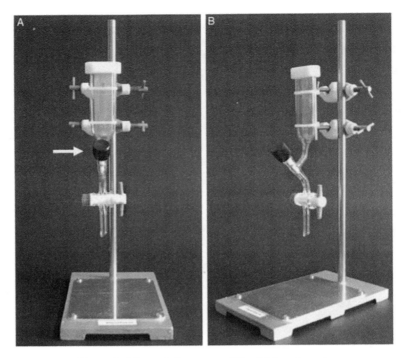

Plate 19 (Figure 8.1 on page 133 of this volume)

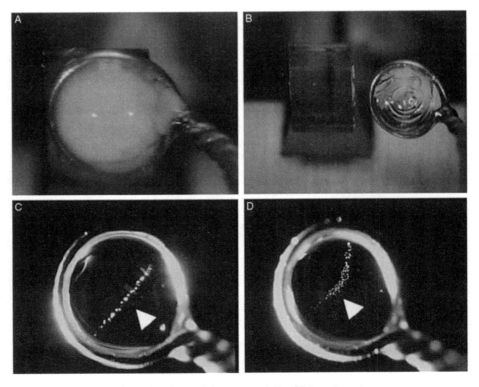

Plate 20 (Figure 8.2 on page 140 of this volume)

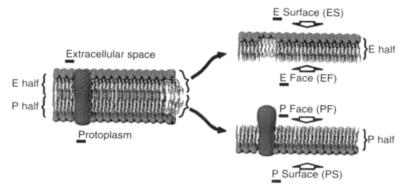

Plate 21 (Figure 11.1 on page 185 of this volume)

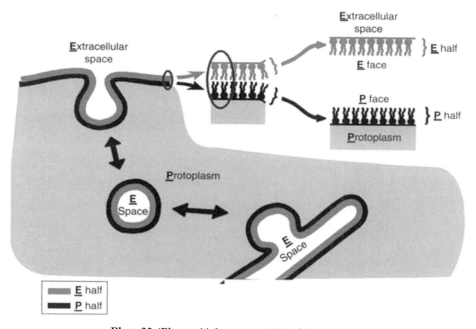

Plate 22 (Figure 11.2 on page 186 of this volume)

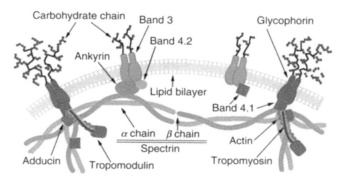

Plate 23 (Figure 12.1 on page 211 of this volume)

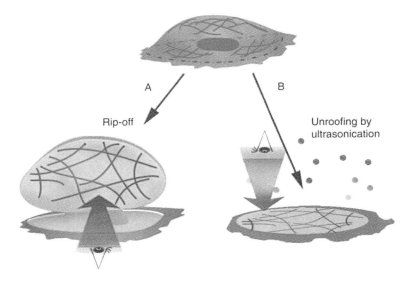

Plate 24 (Figure 12.2 on page 213 of this volume)

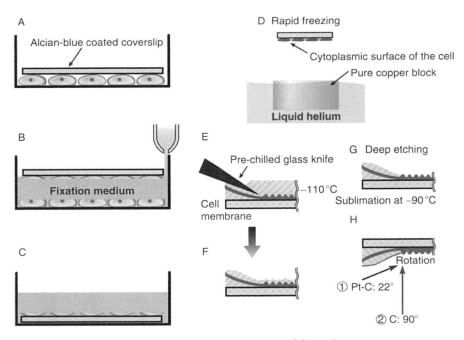

Plate 25 (Figure 12.3 on page 214 of this volume)

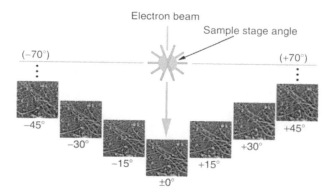

Plate 26 (Figure 12.4 on page 218 of this volume)

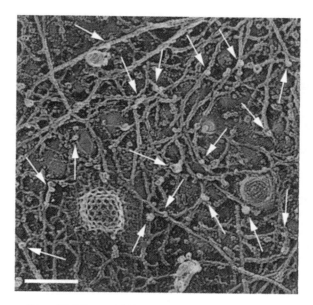

Plate 27 (Figure 12.7 on page 222 of this volume)

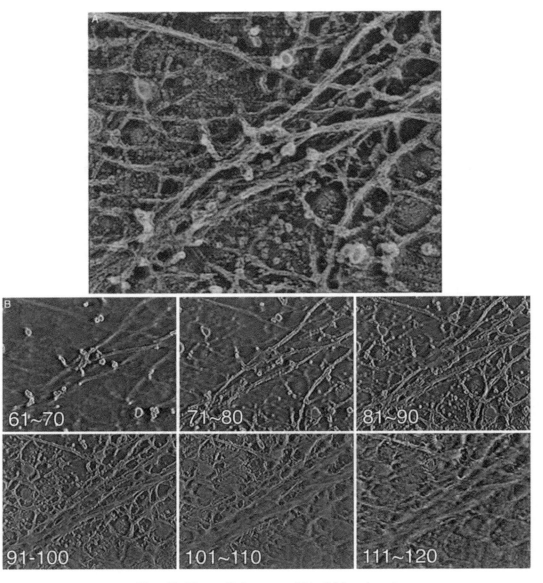

Plate 28 (Figure 12.8 on page 224 of this volume)

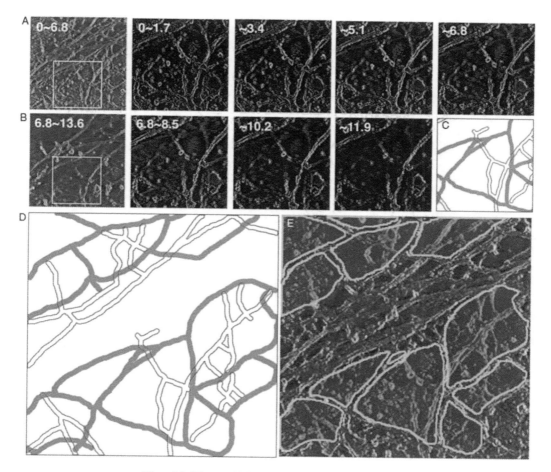

Plate 29 (Figure 12.9 on page 225 of this volume)

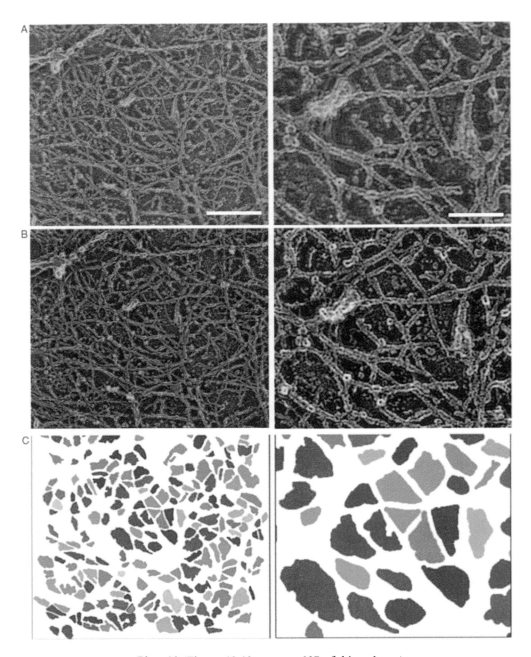

Plate 30 (Figure 12.10 on page 227 of this volume)

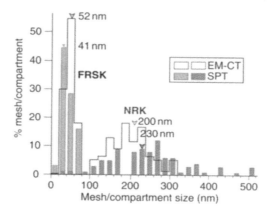

Plate 31 (Figure 12.11 on page 228 of this volume)

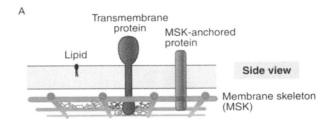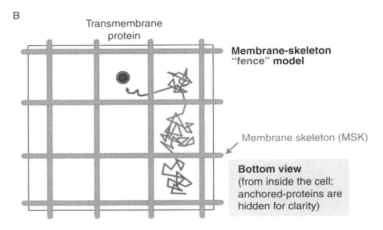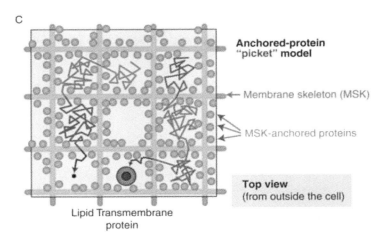

Plate 32 (Figure 12.12 on page 229 of this volume)

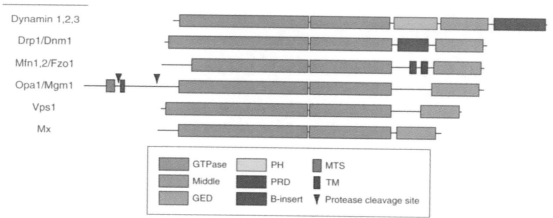

Plate 33 (Figure 13.1 on page 238 of this volume)

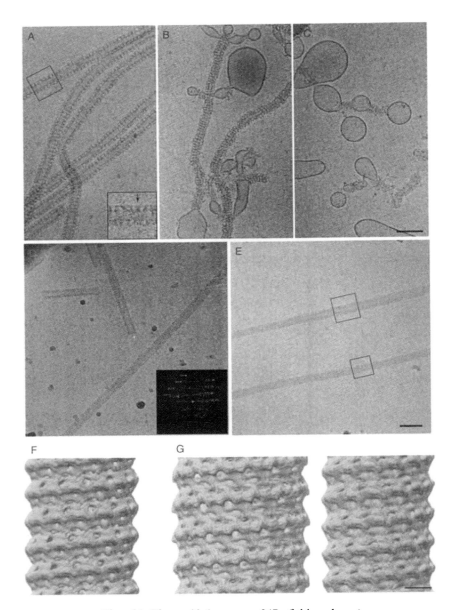

Plate 34 (Figure 13.6 on page 247 of this volume)

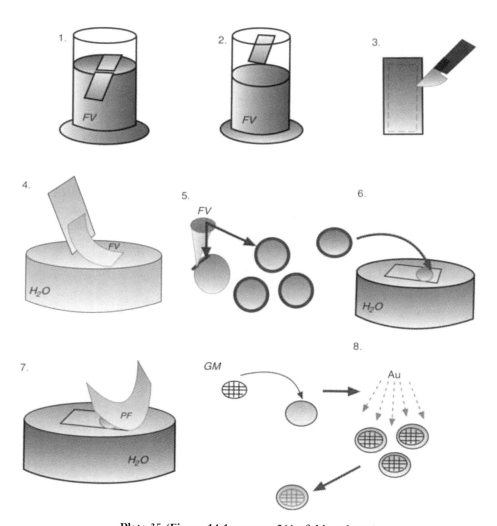

Plate 35 (Figure 14.1 on page 261 of this volume)

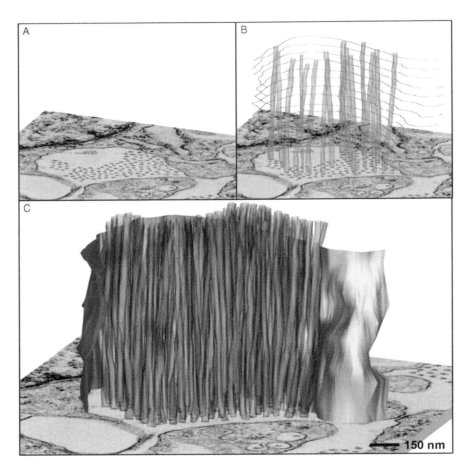

Plate 36 (Figure 17.3 on page 325 of this volume)

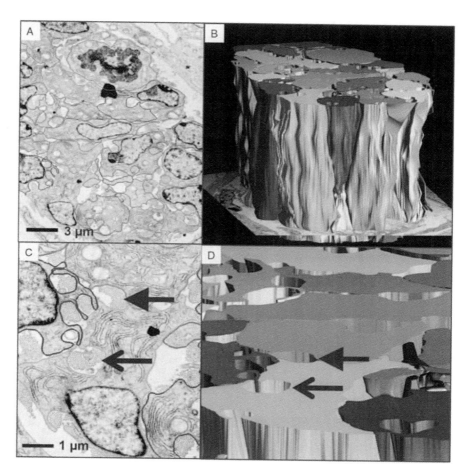

Plate 37 (Figure 17.10 on page 335 of this volume)

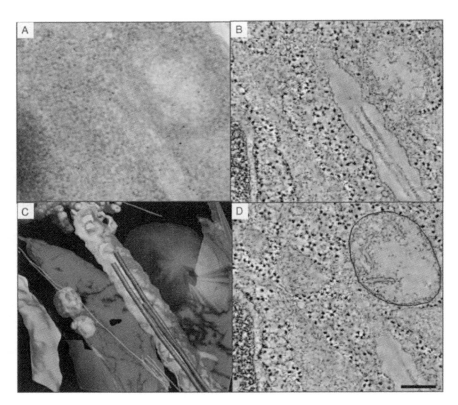

Plate 38 (Figure 17.12 on page 338 of this volume)

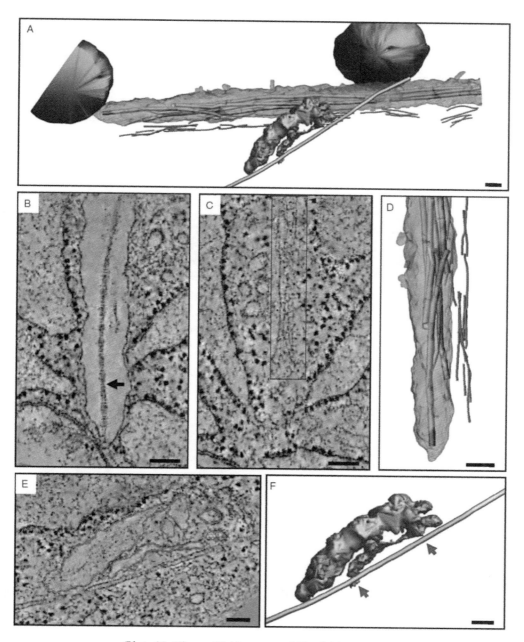

Plate 39 (Figure 17.13 on page 339 of this volume)

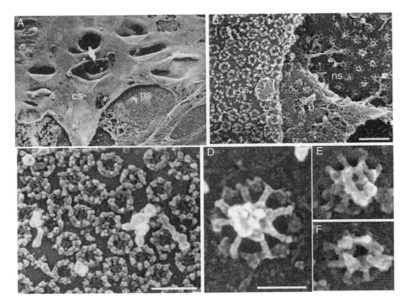

Plate 40 (Figure 20.1 on page 394 of this volume)

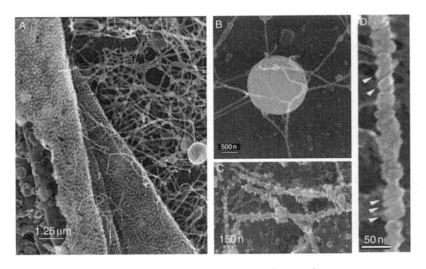

Plate 41 (Figure 20.2 on page 395 of this volume)

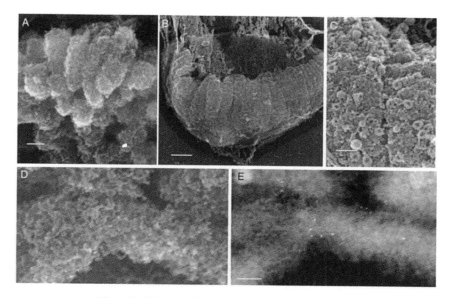

Plate 42 (Figure 20.5 on page 398 of this volume)

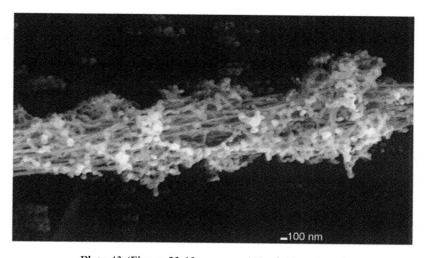

Plate 43 (Figure 23.12 on page 468 of this volume)

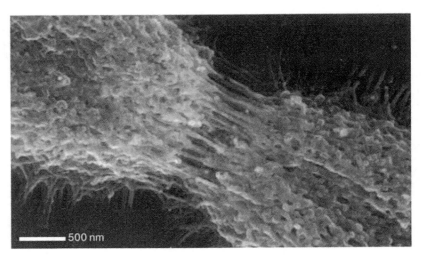

Plate 44 (Figure 23.18 on page 472 of this volume)

14. Electron Microscopy of the Cytoskeleton

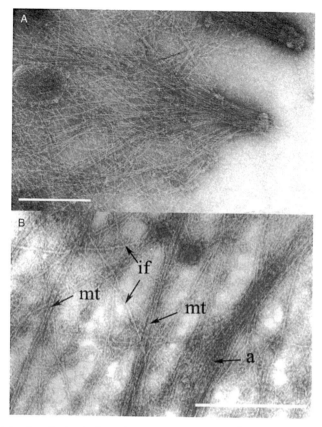

Fig. 6 (A) Details of a cell edge, in the region of a transition between the lamellipodium network and a filopodium. (B) A lamella region showing the three filaments of the cytoskeleton: actin, a; intermediate filaments, if and microtubules, mt. Bars, 0.5 μm.

protrusive activity over a period of around 30–60 min. In the first example (Fig. 7), the cell was fixed at an early stage of stimulation, when protrusion over the whole cell front was continuous at about 1 μm/min. The cell was transfected with GFP-VASP, which localizes to adhesion structures and to the tips of protruding lamellipodia (Rottner et al., 1999). The figure shows the first and last frames of the video of the cell in the GFP channel on the fluorescence microscope, up to the fixation step (Fig. 7A) and the overview of the cell in the EM (Fig. 7B). An overlay of the EM image with the last frame of the video (not shown) confirmed that the cell was rapidly arrested by the fixation step. The region boxed in the overview is depicted at high magnification in Fig. 7C and shows a distinct diagonal network of actin filaments in the protruding lamellipodium.

In the second example (Fig. 8), the cell was fixed at a later stage of aluminum fluoride stimulation, when the protrusion rate was decreasing to different degrees

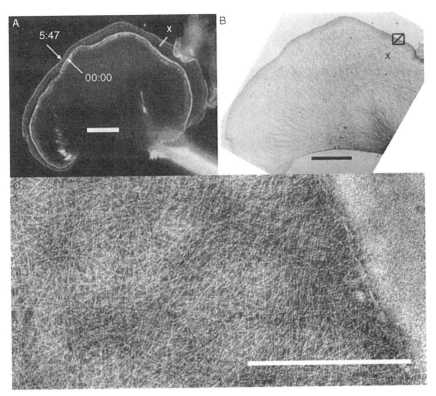

Fig. 7 Correlated light and electron microscopy of a B16 melanoma cell tagged with VASP-GFP. A. First and last video frames (5 s between frames) of the living cell, taken at the times indicated. Cell was fixed immediately after the final frame at 5 min 47 s. In the region indicated by the white line, the cell front was advancing continuously in the final 30 s at 1 μm/min. (B) EM overview; boxed region corresponds to position shown in (A) and to the close up in (C). (C) Actin filament organization in the region indicated in (A) and (B). Note more or less regular diagonal network of filaments. Bars: A and B, 10 μm; C, 0.5 μm.

along the cell edge. Again, the cell was transfected with GFP-VASP, in this case resulting in a lower expression as compared to Fig. 7. Figure 8A and B correspond to the first and last frames of the video and C to the EM overview of the cell, fixed immediately after the last frame. The regions marked "x" and "y" in the final video frame correspond to sites on the cell edge that were moving slowly, 0.8 μ/min (x) or that were stationary (y) during the final 30 s of the video. The same positions are marked on the EM overview. Note that a decrease in protrusion rate is associated with a less organized actin network (Fig. 9A) than seen in Fig. 7C. And the arrest of protrusion is associated with the development of actin arrays parallel to the cell edge (Fig. 9B), through a rearrangement of filaments in the lamellipodium (see also Rinnerthaler *et al.*, 1991; Small *et al.*, 1998).

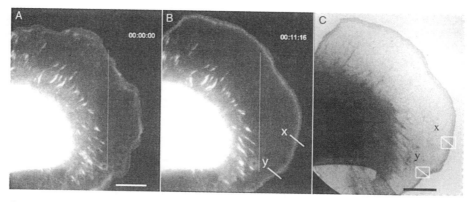

Fig. 8 (A and B) The first (0.0 min) and last video frame (11 min 16 s) of a B16 melanoma cell transfected with VASP-GFP. The vertical white line marks a reference position in the cell. (C) Overview of the cell in the EM. The positions marked "x" and "y" correspond to sites where the cell edge was stationary during the final 30 s (y) and moving forward at 0.8 μm/min(x). The same positions are marked on (C) and shown at higher magnification in Fig. 9. Bars, 10 μm.

C. Evaluation of the Technique

With the method described, the success rate of cells that make it from the light microscope to the EM is in the range of 50%. The yield of good cells with clear filament contrast is a little lower, owing to some uncontrolled variability in the depth of the negative stain. Actin filaments, intermediate filaments and microtubules are clearly resolved. In particular the rearrangements of actin at the cell front, associated with motility and adhesion are most accessible by this technique. Some variability in staining does however occur, leading to cells that are too weakly or too strongly contrasted. Whatever method may be employed, it is important to have light microscope controls of the extraction/fixation process to ensure that it faithfully arrests the cell without visibly changing morphology. The use of GFP tagged proteins is obviously advantageous as it opens up the possibility of controlling more structural parameters. Using cells tagged with actin-GFP, we have monitored the fixation process by live cell microscopy and shown that the Triton–glutaraldehyde mixtures described here preserve the gradient of actin density in the lamellipodium (Koestler et al., 2008).

In an alternative approach for correlated light and electron microscopy, Svitkina et al. (1995) have developed a procedure based on processing of cells for critical point drying and contrasting by metal shadowing. The advantage of this technique is that the cells can be grown on glass coverslips and processed on the coverslips through to the metal coating stage. The replica is then floated off and the picked up on a formvar film under a dissecting microscope. This technique has been used to great advantage in studies of the cytoskeleton (Biyasheva et al., 2004; Svitkina and Borisy, 1999; Svitkina et al., 2003). One drawback of this approach is

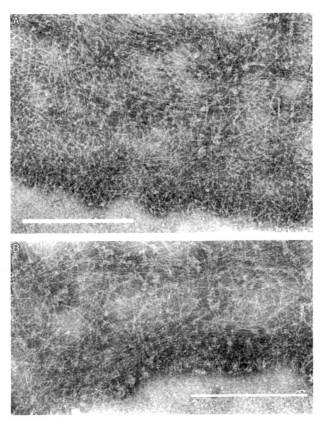

Fig. 9 (A) and (B) correspond respectively to the slowly advancing region "x" and the stationary region "y" in Fig. 8B and to the boxed regions in Fig. 8C. See text for details. Bars, 0.5 μm.

the requirement for multiple processing steps after fixation, which can potentially introduce artifacts (Resch *et al.*, 2002a).

The general technique of live cell imaging on support films described here can in principle be adapted for other procedures. Cytoskeletons prepared by extraction with detergent alone can be "dissected" to remove specific components before fixation to reveal other details, for example by the use of actin depolymerising proteins (Small *et al.*, 1982; Svitkina *et al.*, 1996; Verkhovsky *et al.*, 1995) and inhibitors. Likewise, alternative processing and imaging modes can be employed, including critical point drying (Svitkina *et al.*, 1995) and cryo-EM (Resch *et al.*, 2002b). In view of the new possibilities opened by cryo-EM tomography (Medalia *et al.*, 2002), cytoskeleton preparations for which correlated live cell imaging was performed offer interesting potential for correlating structure and function. The latter technique, combined with new methods for tagging proteins for recognition in the EM, promises to open important new avenues in cytoskeleton research.

V. Summary

The technique described here is applicable to the thin regions of cultured cells, which are well preserved after embedding in negative stain. Some degree of three-dimensionality is inevitably lost during drying, particularly in lamella regions behind the lamellipodium. The present methods have been used to advantage to relate the organization and filament density in lamellipodia to protrusion speed (Koestler et al., 2008). Further advances are to be expected from the development of new methods for high-resolution protein localization and the application of cryoelectron tomography to gain three-dimensional information of filament arrangements and interconnections.

Acknowledgements

We thank Guenter Resch for discussion and assistance with electron microscopy

References

Abercrombie, M., Heaysman, J. E., and Pegrum, S. M. (1971). The locomotion of fibroblasts in culture. IV. Electron microscopy of the leading lamella. *Exp. Cell Res.* **67**, 359–367.

Biyasheva, A., Svitkina, T., Kunda, P., Baum, B., and Borisy, G. (2004). Cascade pathway of filopodia formation downstream of SCAR. *J. Cell Sci.* **117**, 837–848.

Buckley, I. K., and Porter, K. R. (1967). Cytoplasmic fibrils in living cultured cells. A light and electron microscope study. *Protoplasma* **64**, 349–380.

Goldman, R. D., and Knipe, D. M. (1972). Functions of cytoplasmic fibers in non-muscle cell motility. *In* "Cold Spring Harbor Symposium on Quantitative Biology." Vol. XXXVII, Cold Spring Harbor, pp. 523–534.

Hoglund, A. S., Karlsson, R., Arro, E., Fredriksson, B. A., and Lindberg, U. (1980). Visualization of the peripheral weave of microfilaments in glia cells. *J. Muscle Res. Cell Motil.* **1**, 127–146.

Huxley, H. E. (1969). The mechanism of muscular contraction. *Science* **164**, 1356–1365.

Koestler, S. A., Auinger, S., Vinzenz, M., Rottner, K., and Small, J. V. (2008). Differentially oriented populations of actin filaments generated in lamellipodia collaborate in pushing and pausing at the cell front. *Nature Cell Biol.* **10**, 306–313.

Medalia, O., Weber, I., Frangakis, A. S., Nicastro, D., Gerisch, G., and Baumeister, W. (2002). Macromolecular architecture in eukaryotic cells visualized by cryoelectron tomography. *Science* **298**, 1209–1213.

Resch, G. P., Goldie, K. N., Hoenger, A., and Small, J. V. (2002a). Pure F-actin networks are distorted and branched by steps in the critical-point drying method. *J. Struct. Biol.* **137**, 305–312.

Resch, G. P., Goldie, K. N., Krebs, A., Hoenger, A., and Small, J. V. (2002b). Visualisation of the actin cytoskeleton by cryo-electron microscopy. *J. Cell Sci.* **115**, 1877–1882.

Resch, G. P., Small, J. V., and Goldie, K. N. (2006). Electron microscopy of the cytoskeleton: negative staining, cryo-EM and correlation with light microscopy. *In* "Cell Biology: A Laboratory Handbook III" (J. E. Celis, ed.), Vol. 3, pp. 267–275. Academic Press.

Rinnerthaler, G., Herzog, M., Klappacher, M., Kunka, H., and Small, J. V. (1991). Leading edge movement and ultrastructure in mouse macrophages. *J. Struct. Biol.* **106,** 1–16.

Rottner, K., Behrendt, B., Small, J. V., and Wehland, J. (1999). VASP dynamics during lamellipodia protrusion. *Nat. Cell Biol.* **1,** 321–322.

Shaner, N. C., Campbell, R. E., Steinbach, P. A., Giepmans, B. N., Palmer, A. E., and Tsien, R. Y. (2004). Improved monomeric red, orange and yellow fluorescent proteins derived from *Discosoma* sp. red fluorescent protein. *Nat. Biotechnol.* **22,** 1567–1572.

Small, J., Rottner, K., Hahne, P., and Anderson, K. I. (1999). Visualising the actin cytoskeleton. *Microsc. Res. Tech.* **47,** 3–17.

Small, J. V., and Celis, J. E. (1978). Filament arrangements in negatively stained cultured cells: The organization of actin. *Eur. J. Cell Biol.* **16,** 308–325.

Small, J. V. (1981). Organization of actin in the leading edge of cultured cells: Influence of osmium tetroxide and dehydration on the ultrastructure of actin meshworks. *J. Cell Biol.* **91,** 695–705.

Small, J. V. (1984). Simple procedures for the transfer of grid images onto glass coverslips for the rapid relocation of cultured cells. *J. Microsc* **137,** 171–175.

Small, J. V. (1988). The actin cytoskeleton. *Electron. Microsc. Rev.* **1,** 155–174.

Small, J. V., Rinnerthaler, G., and Hinssen, H. (1982). Organization of actin meshworks in cultured cells: the leading edge. *Cold Spring Harb. Symp. Quant. Biol.* **46**(Pt. 2), 599–611.

Small, J. V., Rottner, K., Kaverina, I., and Anderson, K. I. (1998). Assembling an actin cytoskeleton for cell attachment and movement. *Biochim. Biophys. Acta* **1404,** 271–281.

Steinmetz, M. O., Stoffler, D., Hoenger, A., Bremer, A., and Aebi, U. (1997). Actin: From cell biology to atomic detail. *J. Struct. Biol.* **119,** 295–320.

Svitkina, T. M., and Borisy, G. G. (1999). Arp2/3 complex and actin depolymerizing factor/cofilin in dendritic organization and treadmilling of actin filament array in lamellipodia. *J. Cell. Biol.* **145,** 1009–1026.

Svitkina, T. M., Bulanova, E. A., Chaga, O. Y., Vignjevic, D. M., Kojima, S., Vasiliev, J. M., and Borisy, G. G. (2003). Mechanism of filopodia initiation by reorganization of a dendritic network. *J. Cell. Biol.* **160,** 409–421.

Svitkina, T. M., Verkhovsky, A. B., and Borisy, G. G. (1995). Improved procedures for electron microscopic visualization of the cytoskeleton of cultured cells. *J. Struct. Biol.* **115,** 290–303.

Svitkina, T. M., Verkhovsky, A. B., and Borisy, G. G. (1996). Plectin sidearms mediate interaction of intermediate filaments with microtubules and other components of the cytoskeleton. *J. Cell. Biol.* **135,** 991–1007.

Tilney, L. G., Derosier, D. J., and Mulroy, M. J. (1980). The organization of actin filaments in the stereocilia of cochlear hair cells. *J. Cell. Biol.* **86,** 244–259.

Verkhovsky, A. B., Svitkina, T. M., and Borisy, G. G. (1995). Myosin II filament assemblies in the active lamella of fibroblasts: Their morphogenesis and role in the formation of actin filament bundles. *J. Cell. Biol.* **131,** 989–1002.

Wessels, N. K., Spooner, B. S., and Luduena, M. A. (1973). Surface movements, microfilaments and locomotion. *In* "Ciba Found. Symposium on Locomotion of Tissue Cells," Vol. 14, pp. 53–77. Elsevier, Amsterdam.

CHAPTER 15

Electron Microscopy of Intermediate Filaments: Teaming up with Atomic Force and Confocal Laser Scanning Microscopy

Laurent Kreplak,* Karsten Richter,† Ueli Aebi,* and Harald Herrmann‡

*M. E. Müller Institute for Structural Biology
Biozentrum, University of Basel
Klingelbergstrasse 70, 4056 Basel
Switzerland

†Department of Molecular Genetics
German Cancer Research Center (DKFZ)
69120 Heidelberg, Germany

‡Functional Architecture of the Cell
German Cancer Research Center (DKFZ)
69120 Heidelberg, Germany

- Abstract
- I. Introduction
- II. Rationale
- III. Visualization of Intermediate Filaments *in vitro* and in Cultured Cells
    - A. Following the Process of Vimentin IF Assembly by Electron Microscopy
    - B. Atomic Force Microscopy (AFM)
    - C. Correlative Fluorescence and Electron Microscopy of IFs in Cells
- IV. Materials
    - A. IF Proteins
    - B. Atomic Force Microscope
    - C. Substrates for AFM
- V. Conclusions and Outlook
- References

Abstract

Intermediate filaments (IFs) were originally discovered and defined by electron microscopy in myoblasts. In the following it was demonstrated and confirmed that they constitute, in addition to microtubules and microfilaments, a third independent, general filament system in the cytoplasm of most metazoan cells. In contrast to the other two systems, IFs are present in cells in two principally distinct cytoskeletal forms: (i) extended and free-running filament arrays in the cytoplasm that are integrated into the cytoskeleton by associated proteins of the plakin type; and (ii) a membrane- and chromatin-bound thin 'lamina' of a more or less regular network of interconnected filaments made from nuclear IF proteins, the lamins, which differ in several important structural aspects from cytoplasmic IF proteins. In man, more than 65 genes code for distinct IF proteins that are expressed during embryogenesis in various routes of differentiation in a tightly controlled manner. IF proteins exhibit rather limited sequence identity implying that the different types of IFs have distinct biochemical properties. Hence, to characterize the structural properties of the various IFs, *in vitro* assembly regimes have been developed in combination with different visualization methods such as transmission electron microscopy of fixed and negatively stained samples as well as methods that do not use staining such as scanning transmission electron microscopy (STEM) and cryoelectron microscopy as well as atomic force microscopy. Moreover, with the generation of both IF-type specific antibodies and chimeras of fluorescent proteins and IF proteins, it has become possible to investigate the subcellular organization of IFs by correlative fluorescence and electron microscopic methods. The combination of these powerful methods should help to further develop our understanding of nuclear architecture, in particular how nuclear subcompartments are organized and in which way lamins are involved.

I. Introduction

The discovery of intermediate filaments (IFs) in animal cells is tightly connected to advancements in electron microscopy. Buckley and Porter had early on investigated cultured rat embryo cells by a combination of light microscopy, photomicrography, and electron microscopy (Buckley and Porter, 1967). In this 'correlative' light and electron microscope study, they described 'curving filaments' of diameter 75 Å, forming elaborate arrays that were distinctly different from the filament bundles known as 'stress fibers' from phase contrast microscopy. These new filaments were associated with various cellular organelles, surrounding mitochondria 'and weaving about the segments of the endoplasmic reticulum.' Although they were measured to be of diameter 7.5 nm and therefore similar to those found in the bundles, individual filaments outside of bundles exhibited tiny dense nodes of diameter 8 nm, giving the filaments a 'faintly beaded appearance.' Considering the later characterization and designation of these wavy filament arrays in mouse 3T3 fibroblasts as vimentin, referring to the latin *vimentum* for wickerwork or brushwood (Franke *et al.*, 1978), it is quite obvious that Buckley and Porter observed IFs of the vimentin type (Bär *et al.*, 2007; Buckley and Porter, 1967).

IFs were definitely identified a year later as unique structural elements of approximately 10 nm diameter by the group of Howard Holtzer in skeletal muscle cells cultured from chick embryos (Ishikawa *et al.*, 1968). These authors noticed them as unique structures at all stages of development, because they were scattered throughout the sarcoplasm exhibiting no obvious association with myofibrils and because their diameter was clearly different from actin filaments, 5–7 nm in diameter. Furthermore, the latter were easily identified in the developing muscle cells by their association with the thick, i.e., myosin filaments. Interestingly, it was noted that the diameter of these new filaments varied between 8 and 11 nm with a prominent peak at 10 nm, for which reason they were referred to as 10 nm filaments or 'intermediate filaments,' i.e., being in diameter between actin and myosin filaments. As their cell cultures also contained fibroblasts, these authors analyzed them for the new intermediate filaments too and indeed found IFs in abundance, here measuring between 7 and 11 nm. In retrospect, it is notable that with their measurements they already elaborated the difference in diameter between desmin IFs, characteristic for muscle cells, and vimentin IFs typically found in fibroblasts (see e.g., Wickert *et al.*, 2005). In particular, after colcemid treatment of both myoblasts and fibroblasts, they were able to correlate the nuclear 'caps' observed by phase contrast with extensive perinuclear filament arrays in electron microscopy, and furthermore, metaphase cells exhibited 10 nm filaments abundantly too. Because metaphase cells from cultures of skeletal muscle did not bind fluorescein-labeled antibodies against myosin or actin, it was clear that the new filaments were biochemically not related to these muscle proteins (Okazaki and Holtzer, 1965).

In the following years, filaments of similar diameter and shape were identified by electron microscopy in other cell types and tissues of higher vertebrates, including mesenchymal cells, neurons, glia cells, oocytes, smooth muscle, epidermis, and brain (Bond and Somlyo, 1982; Franke *et al.*, 1978; Goldman and Follett, 1969, 1970; Small and Sobieszek, 1977; Steinert and Parry, 1985). Moreover, the different surface properties of microtubules, microfilaments, and IFs have been visualized by transmission electron microscopy of rotary-deposited platinum replicas of rapidly frozen, freeze-dried cytoskeletons from mouse fibroblasts (Heuser and Kirschner, 1980). Also in this study, the powerful approach by EM allowed to approve the distinct nature of IFs, demonstrated by the lack of decoration of these new 'thin' filaments by the S1 fragment of myosin as well as their specific interaction with anti-vimentin antibodies. In corresponding experiments, the quick-freeze, deep-etch and rotary replication method was employed to visualize the fine details of neurofilaments being cross-bridged to microtubules (Hirokawa, 1982) as well as those of keratin IFs and microfilaments in the terminal web from mouse intestinal epithelial cells (Hirokawa et al., 1982). The involvement of plectin in the interaction of IFs with both microtubules and microfilaments in cultured cells was eventually demonstrated by immunoelectron microscopic methods (Clubb et al., 2000; Svitkina et al., 1996).

It was noted frequently that IFs may position the nucleus within the cell and that they may also connect to the plasma membrane. This notion was particularly evident in an immunoelectron microscopic study of chicken erythrocytes, which still contain a nucleus but are mostly devoid of cytoplasmic organelles (Granger and Lazarides, 1982). Employing TEM after low-angle rotary shadowing with platinum of sonicated erythrocytes adsorbed to a glass substrate, it was demonstrated that IFs attached to both the plasma membrane and the outer nuclear membrane, thereby positioning the nucleus within the erythrocyte. In the same study, antibodies against synemin, a large IF protein originally described in muscle, were used to visualize the distribution of synemin along vimentin IFs. Synemin constitutes only a minor part of chicken erythrocyte IF proteins (1 in 50 molecules), whereas vimentin accounted for the rest, and therefore anti-IgG antibodies were employed to enhance the signal for synemin within the vimentin IFs. With this trick, a spacing of 180 nm $\pm$ 40 was determined for synemin-positive domains along the vimentin IFs in the case of adult erythrocytes. Given that vimentin IFs exhibit a structural periodicity of about 46 nm, as determined on negatively stained specimens (Steven et al., 1982), there should be one synemin dimer in every fourth unit-length filament (ULF) segment – assuming that four ULFs contain 64 vimentin dimers (Herrmann et al., 1996). Moreover, IFs are not restricted to vertebrates. Homologous filaments were also detected in invertebrates such as the giant axon of annelid worms or the somatic cells of nematodes (Bartnik et al., 1986; Krishnan et al., 1979).

IFs constitute a distinct cytoplasmic structural system that is interconnected with microtubules, microfilaments and organelles, as well as adhesion complexes at the plasma membrane through associated proteins such as the cross-bridging molecules of the plakin family. And the same type of molecule, e.g. plectin, is engaged to connect the outer nuclear membrane proteins to the cytoskeleton (Herrmann et al., 2007). Interestingly, when cells are cut such that the observed plane exhibits the cytoplasm, the nuclear envelope and the nucleoplasm, it is obvious that the cytoplasm contains a multitude of filaments, microtubules, microfilaments and IFs side by side. On occasion, bundles of IFs come very close to the outer nuclear membrane and large arrays of parallel bundles appear to emerge at that side. In contrast, within the nucleus, similar patterns of filament organization are not seen. Thus, filaments may simply be absent or they may be organized in a way to be less visible. In contrast, by electron microscopy a fine structural network has been demonstrated to tightly adhere to the inner nuclear membrane of the giant oocytes of *Xenopus laevis* (Aebi et al., 1986). At the same time, genetic cloning of the proteins involved, i.e. the lamins, provided proof that they structurally relate to *bona fide* cytoplasmic IF proteins. In the following, lamins were cloned also from lower invertebrates such as *Caenorhabditis elegans* and *Hydra* (Dodemont et al., 1994; Erber et al., 1999). Their electron microscopic characterization is, however, still restricted to pre-embedding gold-labeling immunelectron microscopy methods and no detailed structural data at a meaningful resolution are available for somatic cells (Cohen et al., 2002). Although IF-related proteins naturally populate the nucleus, the question arises why they are not as readily visible as those in the cytoplasm. To approach this question, vimentin, a cytoplasmic IF protein was engineered to contain a classical nuclear localization signal (NLS) such that it would not impede normal filament assembly. These experiments clearly revealed that vimentin filaments are easily detectable in the nucleus of cells (Herrmann et al., 1993). A similar observation has been made before with keratins, both with domain-deleted simple epithelial (Bader et al., 1991) and with ectopically expressed full-length epidermal keratins (Blessing et al., 1993). In both cases, extensive keratin bundles were found to distribute between interphase chromosomal territories. Nevertheless, a clearly defined authentic intranuclear filament system has not been demonstrated yet, except for the nuclear envelope-bound lamina of the giant oocyte of *Xenopus laevis* (Aebi et al., 1986). Despite the attempt to visualize the ultrastructure of what had been observed as a 'nuclear haze' or intranuclear lamin structures, a convincing way to visualize such elusive intranuclear lamin scaffolds has failed to reproduce convincing results. Nevertheless, microinjection of isolated lamins into cells has revealed that an internal nuclear space for lamin deposits exists (Goldman et al., 1992). Moreover, we do not want to give the impression that we ignore the many micrographs published in the last 20 years with impressive examples of what can be obtained with various EM methods (see, for instance Capco et al., 1982; Carmo-Fonseca et al., 1988; Hozak et al., 1995; Paddy et al., 1990). In some cases, immunoelectron microscopy was employed too. However, the degree of preservation of ultrastructure is in most cases questionable and in the end uncertainties persist whether the fibrillar arrays bind the gold-labeled antibodies. For this reason, we employed cDNA-transfection of cultured cells with GFP-chimeras. Thereby we were able to complement live-cell observations with ultrastructural data by straightforward processing of individual cells of interest for electron microscopy following their light-microcopic investigation.

II. Rationale

In this article we give insight into novel achievements in the treatment of IFs both for *in vitro* experiments and for the study of living cells and tissues. This includes electron microscopy as used for fast kinetic experiments, atomic force microscopy of single filaments and filament precursors, as well as correlative fluorescence and electron microscopy.

III. Visualization of Intermediate Filaments *in vitro* and in Cultured Cells

A. Following the Process of Vimentin IF Assembly by Electron Microscopy

1. Glycerol Spraying/Low-Angle Rotary Metal Shadowing of Vimentin Tetramers

Vimentin, recombinantly expressed in bacteria, is easily purified from inclusion bodies after dissolution with 8 M urea and column chromatography (Herrmann *et al.*, 2004; Strelkov *et al.*, 2004). The protein can then be stored at −80 °C in 8 M urea for several months. For the assembly test, an aliquot of vimentin is dialyzed in steps of 15 min, reducing the urea concentration by 2 M at each step, into a low ionic strength buffer, i.e., 5 mM Tris-HCl, pH 8.4, 1 mM DTT (T1-buffer). Analytical ultracentrifugation and small angle X-ray scattering studies have demonstrated that vimentin then reconstitutes mainly into a tetrameric unit containing two α-helical coiled-coil dimers (Mücke *et al.*, 2004b). These tetramers can be observed by EM using a preparation technique called glycerol-spraying/rotary-metal-shadowing (Fowler and Aebi, 1983). In this method, pure glycerol is diluted into a protein solution and sprayed on a freshly cleaved piece of mica. Then the sample is mounted in an evaporator and shadowed at a low angle with a thin layer of platinum and carbon. Carbon backing strengthens the layer, which can then be floated off onto a water surface and transferred to a copper grid for further EM inspection. The achievable resolution of about 2 nm is sufficient to reveal the overall size and shape of the vimentin tetramers, being 2 to 3 nm in diameter and 65 nm in length (Fig. 1A). The anti-parallel orientation of the two vimentin dimers within the tetramer was elucidated by chemical crosslinking (Mücke *et al.*, 2004b)

2. ULF Capture

Upon reconstitution in T1-buffer, vimentin assembles into filaments. We usually use concentrations of 0.1 mg/ml and increase the ionic strength by adding 1/9 volume of 200 mM Tris-HCl, pH 7.0, 1.6 M NaCl. Assembly of human vimentin is performed at 37 °C and stopped by simple dilution, i.e., adding an equal volume of 25 mM Tris-HCl, pH 7.5, 160 mM NaCl containing 2% glutaraldehyde. Within 1–10 s of assembly ULFs, rod-like structures 15 nm in diameter and 60–65 nm in length, form in abundance (Fig. 1B and C, and (Herrmann *et al.*, 1999). Early on, ULFs are often seen to longitudinally fuse with each others (Fig. 1B and C, arrowhead). The fusion event is best visualized with tailless vimentin. Because of the unique A_{11} orientation of tetramers, i.e., the antiparallel, half-staggered overlap of two dimers via their amino-terminal halves, these early products of lateral association have the carboxy-terminal ends on either side of the ULF, which gives them somehow tapered ends (Fig. 1C). Glycerol-spraying/low angle rotary-metal-shadowing preparations reveal a 21.5 nm banding pattern that corresponds to the overlap between the two dimers in a vimentin tetramer (Fig. 1B, see also Henderson *et al.*, 1982). Note that in some cases, fine projections are visible on

15. Electron Microscopy of Intermediate Filaments

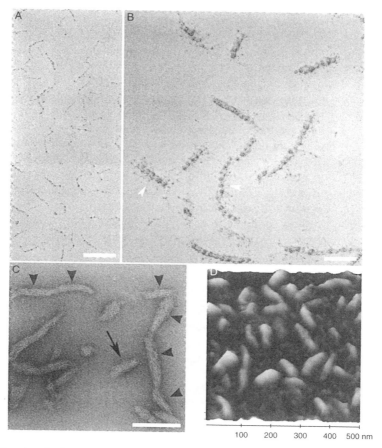

Fig. 1 Imaging the first steps of intermediate filament assembly by electron microscopy (EM) and atomic force microscopy (AFM). (A) Glycerol spraying rotary metal shadowing of vimentin tetramers in T1-buffer (5 mM Tris-HCl, pH 8.4, 1 mM DTT) and (B) Vimentin unit-length filaments (ULFs), notice the 21.5 nm beading typical of IFs prepared by rotary metal shadowing and the millipede appearance of the ULFs (arrowheads). Bar: 100 nm. (C) Negative staining electron microscopy of recombinant human tailless vimentin for 10 s. ULFs are visible (arrow) as well as longer assemblies (arrowheads; for details see Herrmann *et al.*, 1996). Bar: 100 nm. (D) Tapping mode AFM image in buffer imaging vimentin assembled for 10 s in phosphate buffer (2 mM sodium phosphate, pH 7.5, 100 mM KCl) and fixed with 0.1% glutaraldehyde. The solution was adsorbed to highly oriented pyrolitic graphite. The solid support was covered with ULFs of 10 nm in height (Mücke *et al.*, 2005).

the side of the ULFs, giving rise to a millipede appearance (Fig. 1B, arrowheads). In the case of the ULFs, the only drawback of electron microscopy is the difficulty to estimate the height of these rod-like structures. This problem can be solved using atomic force microscopy (see below). We were able to image glutaraldehyde fixed ULFs adsorbed to a surface in assembly buffer condition (Fig. 1D), yielding an average height of 10 nm (Mücke *et al.*, 2005).

Quantitative analysis of the length distribution of filaments fixed at different time points as observed by both electron microscopy and AFM revealed that end-to-end fusion is indeed the main mechanism of elongation while other potential mechanisms, like the addition of individual tetramers to a free-end, are not operative (Kirmse et al., 2007). Still another hallmark of vimentin – and desmin – assembly was revealed by electron microscopy, i.e. a radial compaction event. This means that during the assembly process vimentin filaments reduce their diameter from around 16 nm, exhibited by ULFs and short filaments, to 11 nm as measured with mature filaments (compare Fig. 1C and Fig. 2A; see also Herrmann et al., 1996). This reduction in diameter has also been observed with the assembly of recombinant desmin: At 10 s the average diameter was 17.1 nm, at one minute and 10 min it was 15.5 and 13.9 nm, respectively (Herrmann et al., 1999). Moreover, radial compaction has been observed under various ionic conditions with authentic desmin too. Here, on average a reduction of the filament diameter by 1.3 nm was observed between 5 and 60 min of assembly (14.9–13.6 nm) (Stromer et al., 1987).

3. Co-Assembly of Vimentin and Desmin

Electron microscopy is not only very useful to follow the time course of IF assembly. It can also be employed to morphologically distinguish between different types of IFs. As an example, we have analyzed the co-assembly of human vimentin with human desmin by mixing the two proteins at different stages of assembly. First, we characterized the two types of filaments separately using negative staining EM. Desmin appeared to be wider on average than vimentin filaments (Wickert et al., 2005). The larger width of desmin filaments correlates with a higher mass-per-length (MPL) as measured by scanning transmission electron microscopy

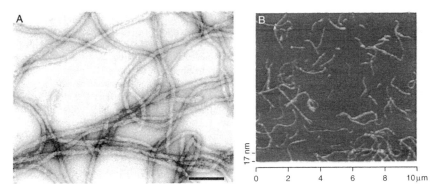

Fig. 2 (A) Mouse desmin (0.05 mg/ml) was assembled for 1 h at 37 °C in 25 mM Tris-HCl, pH 7.5, 50 mM NaCl. Negative staining electron microscopy. Bar: 100 nm. (B) Mouse desmin (0.05 mg/ml) was assembled for 1 h at 37 °C in 25 mM Tris-HCl, pH 7.5, 100 mM NaCl. The solution was adsorbed to mica for ten seconds and the surface was imaged by contact mode AFM. The filaments appeared 4.5 nm in height. Bar: 1 μm.

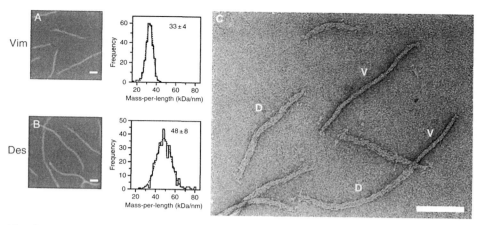

Fig. 3 Filaments with different mass-per-length (MPL) can fuse end-to-end. Dark-field Scanning Transmission Electron Microscopy (STEM) and Mass-per length (MPL) distribution (kDa/nm) of (A) vimentin and (B) desmin filaments. Bars: 50 nm. Notice that the desmin filaments have a higher MPL than the vimentin filaments (A and B adapted from Wickert et al., 2005. (C) Fusion events between vimentin and desmin filaments and ULFs as seen by negative staining electron microscopy, V stands for vimentin, and D stands for desmin. Bar in (C): 100 nm.

(STEM) (Fig. 3, A and B; see also (Wickert et al., 2005). Employing a Vacuum Generators HB-5 scanning transmission electron microscope (East Grinstead, UK), this technique can accurately determine the mass of a dried protein sample by counting the number of elastically scattered electrons (Engel, 1978a,b). Second, we mixed vimentin ULFs with desmin ULFs and let them assemble at 37 °C. Negative staining EM revealed two types of filaments with clearly different diameters that were able to fuse end-to-end (Fig. 3C). This result indicated that ULFs with different number of subunits per cross-section were able to anneal longitudinally to form a filament containing both vimentin and desmin.

4. The Diameter of Ectopically Expressed Vimentin IFs in Cultured Cells

By ultrathin sectioning of vimentin-free cultured cells cDNA-transfected with human vimentin and a tailless vimentin derivative, the properties of the assembly products were followed in living cells. In the bovine mammalian gland epithelial (BMGE) cell, vimentin and tailless vimentin accumulate near the nucleus in a structure called 'central aggregate,' as revealed by immunofluorescence microscopy (arrows in Fig. 4A and B; see also Herrmann et al., 1996). In these aggregates, loosely oriented arrays of single IFs were observed by conventional electron microscopy (Fig. 4C and D). Their diameter was determined to be on average 10.2 versus 13.4 nm for wild type and tailless vimentin IFs (Fig. 4E). This difference correlated well with the difference in number of molecules per cross-section as determined by STEM of both types of filaments assembled in vitro, i.e. 29

versus 54, which corresponds to a MPL of 37 versus 59 kDa/nm (Fig. 4F and G; and Herrmann *et al.*, 1996). However, to perform such experiments, stable cell lines are needed. Otherwise the chance to find an appropriate transfected cell is very low and would need tedious searches. To circumvent this problem, correlative fluorescence and electron microscopy has to be employed (see part C).

B. Atomic Force Microscopy (AFM)

1. Sample Preparation for AFM

In vitro assembled or tissue extracted IFs are very flexible filaments that can adsorb physically to several solid supports like glass, graphite, and muscovite mica (Mücke *et al.*, 2005). Typically 20–100 μl of filaments (protein concentration 0.001–0.01 mg/ml) are adsorbed to a 10–50 mm^2 substrate for several minutes. The surface should not be too crowded with filaments in order to limit contamination of the atomic force microscope tip by proteins detaching from the filaments (Fig. 2B). For *in vitro* assembled filaments, the imaging buffer is generally identical to the assembly buffer.

2. Contact Mode versus Tapping Mode

'Contact mode' is the most common AFM imaging method where the tip scans the surface at a constant deflection of the cantilever. For IFs, very soft types of cantilevers with spring constants below 0.05 N/m must be employed and the applied force must remain in the 100–200 pN range. Higher applied forces can lead to the mechanical disruption of the filaments. Hence 'contact mode imaging' is not well suited to obtain high-resolution pictures of IFs in fluid. However it can be used as a preliminary screening method (Fig. 2B).

To observe the molecular architecture of IFs, we generally use 'tapping mode.' In this case a cantilever with a spring constant between 0.1 and 0.5 N/m is oscillated at its resonance frequency. Because of intermittent contact, the force applied by the tip onto the filament is lower and better controlled, compared with the 'contact mode.'

3. Unraveled Filaments by EM and AFM

'Tapping mode imaging' in assembly buffer enabled us to directly observe the presence of twisted subfilaments in several IFs (Mücke *et al.*, 2005) and to confirm previous electron microscopy observations that had been made after treatment of adsorbed authentic keratin IFs with phosphate buffer (Aebi *et al.*, 1983). Correspondingly, the recombinant keratin pair K5/K14 and the *Epidermolysis Bullosa Simplex* mutant pair K5/K14 R125H form filaments that massively unravel into subfilaments when exposed to 10 mM sodium phosphate, pH 7.5 (Fig. 4A, arrows and arrowhead; Herrmann *et al.*, 2002). K5/K14 filaments adsorbed to graphite

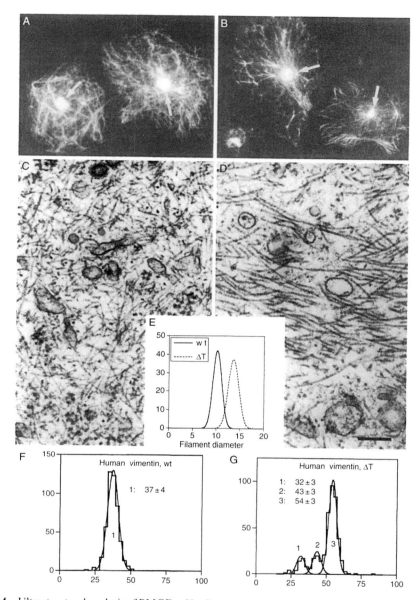

Fig. 4 Ultrastructural analysis of BMGE + H cells stably transfected with wild-type human vimentin (A, C) and tailless human vimentin (B, D) cDNAs. (A, B) Immunofluorescence images. Notice the formation of 'central aggregates' (arrows) containing vimentin-positive material close to the nucleus. (C, D) Embedding and ultrathin sectioning of both types of 'central aggregates.' Bar: 200 nm. (E) Histograms displaying the apparent diameters measured for vimentin IFs within the 'central aggregates' of cell lines transfected with wild-type human vimentin (continuous curve) and tailless human vimentin (broken curve). The histograms are represented as single Gaussian curves fitted to the measured apparent diameter values (adapted from Herrmann *et al.*, 1996). Tailless vimentin forms wider filaments

and imaged by atomic force microscopy also show a strong unraveling, probably because of the high negative charge of the surface (Fig. 4B; Kreplak et al., 2005). The AFM cantilever was furthermore used to stretch a piece of filament that appears thinner than the neighboring unstretched filaments (Fig. 4B, asterisk). Notice that subfilaments are not visible anymore within the stretched segment.

In neurofilaments, the light chain NF-L forms filaments exhibiting a strong 21.5 nm beading as observed after glycerol spraying/rotary metal shadowing (Fig. 4C, arrows) (Heins et al., 1993). However, partial unraveling after adsorption to mica was observed by AFM with neurofilaments (NFs) purified from rat spinal cords (Fig. 4D; see also (Kreplak et al., 2005). The resolution is high enough to distinguish the unraveled segments (Fig. 4D, arrowheads) from the more compact ones. Local unraveling was also observed for vimentin filaments observed by cryo-electron tomography (Fig. 4E, arrowhead, for details see Goldie et al., 2007) or adsorbed to a mica surface and imaged by AFM (Fig. 4F, arrowheads).

In summary, while performing structural investigations on IFs one should keep in mind that the architecture of the filaments is considerably sensitive to buffer change as well as adsorption to a surface or even fast freezing. Their stability against high concentrations of salt may have generated the misconception that IFs are as stable as concrete: On the contrary, most IFs are easily dissolved in distilled water and hence they have to be fixed when prepared for negative staining or STEM, as in both procedures the samples have to be washed with distilled water.

C. Correlative Fluorescence and Electron Microscopy of IFs in Cells

1. Cell Lines to Study Patterns of Recombinant Vimentin Expression

To exclude interferences with endogenous vimentin upon transfection with recombinant vimentin modifications, vimentin-free cells may be used for the investigation of processes involved with the formation of vimentin networks. Model cell systems are BMGE+H (bovine mammary gland epithelium) and SW13 (a human adrenal cortex carcinoma-derived cell line). The subclone E7/300 of SW13, used for some of the presented studies, is stably transfected for expression of a modified *Xenopus* vimentin carrying within its head domain a lamin-type NLS, i.e. *Xenopus* NLS-vimentin, (Bridger et al., 1998; Herrmann et al., 1993). The NLS-modification does not interfere with filament formation *in vitro* (Reichenzeller et al., 2000). For direct fluorescence microscopic observation of NLS-vimentin in E7/300 cells, GFP-*Xenopus* NLS-vimentin was transiently

---

than wild-type vimentin within BMGE + H cells, i.e. 10.2 ($\pm$ 1.0) versus 13.4 ($\pm$ 1.3). Ordinate, number of filaments; abscissa, filament diameter in nanometers. These diameter histograms correlate well with the mass-per-length (MPL) measurements of *in vitro* assembled wild-type human vimentin IFs (F) and tailless human vimentin (G). Ordinate, number of segments; abscissa, mass-per-length (kDa/nm). Tailless vimentin exhibits higher MPL values than wild-type vimentin (adapted from Herrmann et al., 1996).

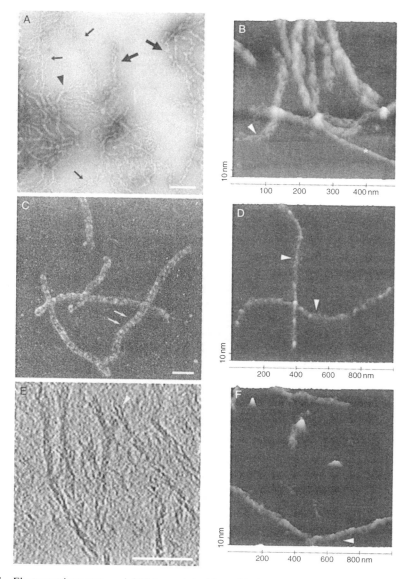

Fig. 5 Electron microscopy and AFM reveal the hierarchical organization of various IFs. The sub-filamentous nature of IFs was first revealed by negative staining EM of keratin filaments that were incubated with 10 mM sodium phosphate, pH 7.5. (A) K5/K14 R125H filaments after phosphate treatment (adapted from Herrmann *et al.*, 2002). One filament was captured during the unraveling process (arrowhead). Other filaments were either partially (large arrows) or totally (small arrows) unraveled into sub-filamentous structures. Bar: 100 nm. (B) K5/K14 filaments assembled in 10 mM Tris-HCl, pH 7.5 for 1 h at 37 °C and adsorbed to highly oriented pyrolitic graphite. 'Tapping mode' AFM imaging reveals partially unraveled filaments (arrowhead; adapted from Kreplak *et al.*, 2005). The AFM tip was used to pull on one of the filaments (asterisk in the lower right). (C) Glycerol spraying/low-

cotransfected. The GFP-cDNA was cloned into the *Xenopus* NLS-vimentin vector such that the GFP was in front of the amino-terminal end. Although this GFP-tagged *Xenopus* NLS-vimentin is incapable of forming filament on its own, copolymerization with *Xenopus* NLS-vimentin produces filament patterns indistinguishable from those of *Xenopus* NLS-vimentin alone (Herrmann *et al.*, 2003; Reichenzeller *et al.*, 2000).

2. Correlation of Confocal Laser Scanning and Electron Microscopy

For the ultrastructural investigation of transfected chimeras of IF proteins and fluorescent proteins, correlation with fluorescent light microscopy can be exploited to locate the protein of interest on the electron micrograph. In contrast to immunolocalization, this approach imposes no restriction with respect to the fixation protocol, allowing the use of glutaraldehyde and osmium tetroxide for optimized structural preservation.

Cells were grown on gridded cover slips (for this review: *Cellocate*®, Eppendorf AG Germany were employed; these cover slips are no longer available and etched grids from Bellco Biotechnology, USA, can be used instead). This allows to observe the fluorescent signal of the expressed, fluorescent-protein tagged protein either *in vivo* or post fixation (i.e. 1 h fixation with 4% freshly prepared formaldehyde plus 1% glutaraldehyde, EM-grade, plus 1 mM $MgCl_2$ in 100 mM sodium phosphate buffer at pH 6.8, rinsing with buffer followed by embedding in a water soluble mounting medium such as Vectashield, Vector Laboratories, USA). For the identification of subcellular structures, it is advantageous to use confocal laser scanning microscopy (LSM, model TCS-SP II from Leica Microsystems, Germany) to obtain 3D information on the IF-distribution patterns. Stacks of optical sections were acquired at maximum resolution (e.g., 100×/1.41 oil objective, pinhole-size 1 Airy, voxel-size $73 \times 73 \times 300$ nm^3). In addition, differential interference contrast images from a central plane of the cell of interest as well as low-power views were acquired to assist the subsequent relocation on ultrathin sections for electron microscopy.

Following the light microscopic investigation, the coverslips were further processed for electron microscopy, i.e., upon a rinse in buffer, they were postfixed

---

angle rotary metal shadowing of filaments assembled from neurofilament light chain (NF-L) protein (adapted from Heins *et al.*, 1993). Notice the 21.5 nm beading typically observed with IFs prepared for glycerol spraying/rotary metal shadowing (arrows). Bar: 100 nm. (D) Neurofilaments extracted from rat spinal cords, adsorbed to mica and imaged in 'tapping mode' in buffer containing 20 mM Tris-HCl, pH 7.0, 50 mM NaCl. Notice the subfilamentous architecture within open regions (arrowheads). (E) Vimentin filaments assembled for 1 h at 37 °C in 25 mM Tris-HCl, pH 7.5, 50 mM NaCl. Selected slice of a cryoelectron tomogram. Bar: 100 nm. Notice the unraveled ends of some of the filaments (arrowhead). (F) Vimentin filaments assembled for 1 h at 37 °C in phosphate buffer (2 mM sodium phosphate, pH 7.5, 100 mM KCl). 'Tapping mode' AFM image of filaments adsorbed to mica (courtesy of Norbert Mücke). The filaments are partially unraveled; thin strands extend out of the filaments (arrowhead).

in 1% buffered OsO$_4$ (30 min at room temperature), then carefully rinsed in buffer followed by a wash in water to eliminate phosphate as a source of precipitates. Dehydration was performed in aqueous ethanol of increasing concentration, including *en-bloc* staining with uranylacetate saturated in 70% ethanol overnight. Transfer to epoxy-resin (Epon 812 Kit, Polyscience, Germany) was done from 100% ethanol through propylene oxide. For final polymerization, resin-filled gelatin capsules were inverted on top of the gridded field of the cover slips. The cover slips detached from the polymerized resin samples under liquid nitrogen, depicting the negative imprint of the etched grid on the shiny block-surfaces. Ultrathin sections (at nominal thickness of 70 nm) were prepared from the grid region of interest, taking care not to loose the first section that includes the grid imprint for orientation and to obtain continuous ribbons in order to preserve the three-dimensional information of the section stack.

For straightforward relocation of the cells of interest, careful trimming of the block faces before ultrathin sectioning is essential. The block face should be as small as possible, and its shape should be nonsymmetrical, e.g., by breaking one of its edges, to facilitate recognition of the section's orientation in the EM. A strict protocol should be followed to mount the sections and transfer them into the EM to ensure consistent handedness of the electron micrographs. Take care to avoid confusion of handedness between light and electron microscopic images.

Ultrathin sections were mounted on formvar-coated slot-carriers and post-stained in aqueous lead citrate (2 mg/ml) for observation in a conventional transmission electron microscope at 80 kV (Philips EM410). The first two ultrathin sections depict the etched pattern of the gridded coverslip (Fig. 6). Together with the geometry of trimming and sectioning, and taking into account the information from overview DIC-images about cells in the surroundings, the cell of interest was identified. Micrographs were taken from consecutive sections and searched for structural features, which correlate with the fluorescent signals. Though fluorescence microscopy cannot resolve the structure of IFs, it is feasible to detect the fluorescence signal from a single filament. Thus, correlation with electron microscopy is a powerful method to relocate macromolecules at ultrastructural resolution (Fig. 7).

3. Ectopic Expression of Vimentin

The behavior of cytoplasmic IFs in the nucleus was studied by correlative fluorescent light and electron microscopy of ectopically expressed modifications of vimentin from human and *Xenopus* origin. The nuclear expression represents a system to study filament formation at physiological pH and osmotic pressure, but lacking the typical, cytoplasmic environment (Reichenzeller *et al.*, 2000).

The polymerization of *Xenopus* vimentin can be triggered by a temperature shift from 37 to 28 °C (Herrmann *et al.*, 1993). In contrast, polymerization of human vimentin cannot be triggered. However, this construct comes closer to human-related clinical questions. At first glance, both the frog and the human proteins

Fig. 6 Correlation of cellular ultrastructure with the fluorescent-protein reporter signal. SW13 cells were transfected with modified *Xenopus* Vimentin, linked to GFP in addition to an NLS. Cells had been incubated at 28 °C for 45 min to induce polymerization of vimentin filaments. Upper row: Cells were grown on a gridded coverslip. The cell of interest (arrow) and its environment are documented by DIC images : (A) overview of grid location 'K'; (A') enlarged view. (B) The first ultrathin section carries the imprinted relief of the coverslip-grid. The section is thicker in the darker regions. (B') For curiosity, note that the coverslip-grid is also visible on the second section, however as a negative. This is an artifact

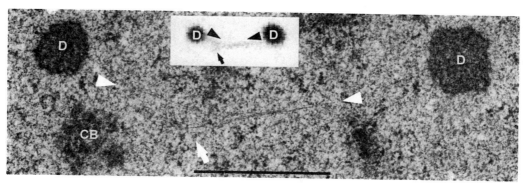

Fig. 7 *Xenopus* vimentin, ectopically expressed in the nucleus of an SW13 cell, formed body-like dots of fine-fibrillar, convoled texture, two of which are shown (D; CB: Cajal body). From these two sites, few filaments radiated (arrowheads) towards each other to meet somewhere halfway. While the right branch passed beyond (arrow), the left branch appears to reorient its slope in parallel with the right branch. This geometry is also appreciable in the fluorescence data (inset). Note, the entire filament array is caught in the single ultrathin section. Scale bar: 1 µm. Reproduced with permission from Richter *et al.*, 2005.

behave very similar. Three modifications of structured vimentin were observed: filament arrays, body-like fibrillar convolutes, and focal accumulations. However, there were also significant differences in structure. Thus, filament arrays typically associated laterally with body-like fibrillar convolutes of human vimentin (Fig. 8), in contrast to the frontal association in the case of *Xenopus* vimentin (Fig. 6). Furthermore, filaments of human vimentin appeared thicker and barbed by attached material.

Of the three structural forms of vimentin observed in the nucleus, filaments are easily recognized as vimentin IFs after ectopic expression of the NLS-vimentin. In contrast, the ultrastructural investigation of body-like convolved vimentin dots requires their unequivocal identification, e.g., by correlation with their fluorescence signal (Fig. 8). Similar to bundled filament arrays, vimentin dots are not pervaded by structural components of the nucleoplasm. Although they often associate with filament arrays, which typically run between dots, the vimentin structures within the dots distinctly differs from filaments: The fine-fibrillar, convoluted ultrastructure reveals a much higher flexibility of their structural elements.

---

due to cutting-induced tear of material from the block face, which depends on section thickness. When cutting the first section, more material is torn from the block face at the thicker parts of the section, i.e. those representing the grid relief. Consequently, the next section is thinner in these regions. (C) Part of the nucleus of the cell of interest. The expressed vimentin organized in several 'vimentin dots' (white broken rings) and started to polymerize in filaments forming arrays. A corresponding optical section (insert) correlates the dots unequivocally with fluorescence signals of the GFP-tagged vimentin (black broken rings). N, nucleoli; C, cytoplasm. (D) Enlarged view of the area boxed in (A), including the two vimentin dots (D). A bundle of vimentin filaments formed an arch to the right of the left dot (see fluorescence image), part of which is caught in the section plane (F). Scale bar: 1 µm.

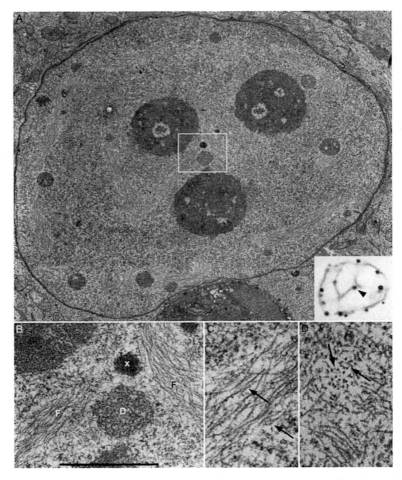

Fig. 8 Nuclear expression of human vimentin. (A) The human vimentin targeted to the nucleus of an MCF7 cell organized into arrays of filament bundles and dots. Dots typically associate laterally to bundle arrays (Inset: corresponding optical section of the fluorescence signal from the YFP-tagged protein). (B) Detailed view of the boxed area in (A). Like with *Xenopus* vimentin, the ultrastructure of dots (D) reveals a convoluted, fibrillar texture, which is very different from the comparably stiff filaments in bundle arrays (F). Note, that the dark body (x) on top of the dot contains no vimentin (see insert, arrowhead pointing to the dot). Scale bar: 1 μm. (C) Enlarged view from the lower left of part (B). Single filaments appear barbed by attached material (arrows). (D) More distant part of the same array at the same magnification. Cross-sections of filaments reveal connecting strands (arrow).

Accordingly, filaments were never seen to permeate the dots: Vimentin dots and filament arrays do not merge. A third variation of vimentin structures is represented by the focal accumulations, which are only occasionally observed (Fig. 9). Foci are built from a number of stick-shaped units (about 15–20 nm thick

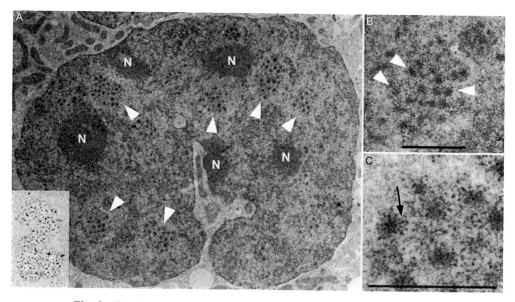

Fig. 9 Focal accumulations of *Xenopus* vimentin in an SW13 cell nucleus upon overnight incubation at 28 °C. (A) Overview micrograph of lower part from the nucleus with speckled fluorescence signal (insert). Several groups of focal accumulations are scattered throughout the nucleus (arrowheads; N: nucleoli). (B) Foci within one group often align in rows (arrowheads). (C) Single foci are about 200 nm in size, and composed of stick-shaped units in radially orientated from the centre to the periphery of foci. Scale bars: 1 μm.

50–70 nm long) that appear to associate with a central matrix, oriented with their free ends towards the periphery. Several of these foci group together and several of these groups can be found in one nucleus. This hierarchical organization is not revealed by light microscopy owing to its limited resolution.

4. Overexpression of Human YFP–Lamin A

Expression of transfected lamin A, even when transiently overexpressed, does not apparently influence nuclear architecture, except for local increases in thickness of the lamina (Fig. 10). Moreover, the transgenic YFP-lamin A does not accumulate in the nucleoplasm but instead localizes to the lamina. This possibly relates to physiological constraints of the nucleus to preserve a balanced composition for the rest of the lamina. Alternatively, it may reflect the behavior of nonprocessed forms of lamin A: In this study it had not been established to which extent the transgenic construct matures physiologically. The YFP-tagged lamin A accumulated in lens-shaped thickenings, which bulged the nuclear envelope towards the cytoplasm, whereas the inner contour of the nuclear periphery remained flat. Evidently, the nucleoplasmic surface is more restricted by tension forces than the cytoplasmic surface of the nuclear envelope. Moreover,

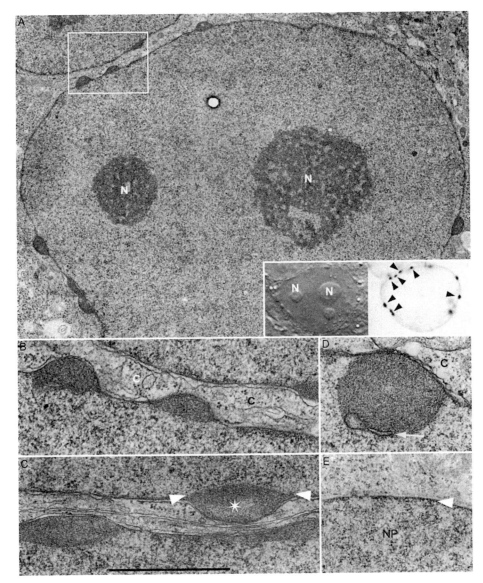

Fig. 10 Ultrastructure of a HeLa cell nucleus upon overexpression of YFP-lamin A. (A) The overexpression of YFP-tagged lamin A caused its accumulation in lens-shaped thickenings of the nuclear lamina. The fluorescence signal (lower right inset, corresponding optical section) unequivocally demonstrates their load with the YFP-tagged protein (arrowheads point to fluorescent dots seen on the electron micrograph as lens shaped thickenings; left inset: DIC image to assist identification of the nucleus; N: nucleoli). (B) Enlarged view of the boxed area in (A). The lens-shaped thickenings of the nuclear lamina protrude farther to the cytoplasm (C) than to the nuclear interior. Their content is of homogeneous, fine-fibrillar texture. (C) The interphase of the thickened lamina with the nucleoplasm is

this behavior does not indicate the existence of a system generating major filament arrays within the nucleoplasm, in sharp contrast to the ectopically expressed vimentin.

IV. Materials

A. IF Proteins

1. Vimentin and Desmin

Recombinant vimentin and desmin are expressed and purified according to a protocol that has been described elsewhere (Herrmann *et al.*, 2004). The proteins were stored at $-80\,°C$ in 8 M urea, 5 mM Tris-HCl, pH 7.5, 1 mM DTT, 1 mM EDTA, 0.1 mM EGTA, 10 mM methyl ammonium chloride. To reconstitute the protein, the stock solution was dialyzed at about 1 mg/ml into 5 mM Tris-HCl, pH 8.4, 1 mM DTT (T1-buffer) or into 2 mM sodium phosphate, pH 7.5, 1 mM DTT (T2), at room temperature by lowering the urea concentration stepwise from 6 M to 0 M. Dialysis was continued overnight at 4 °C into fresh buffer without urea. The next day, the protein was further dialyzed for 1 h at room temperature against fresh buffer.

2. Neurofilaments

Neurofilaments (NFs) containing the NF triplet proteins NF-L, NF-M, and NF-H, were purified from rat spinal cords (Leterrier *et al.*, 1996). The crude NF pellet was purified by sedimentation at 200 000g for 3 h 30 min at 4 °C onto 5 ml of 1.5 M sucrose in 0.1 M Mes, pH 6.8, 1 mM $MgCl_2$, 1 mM EGTA (RB), and 15 ml of 0.8 M sucrose in RB. The purified NFs were then dialyzed overnight at 4 °C against RB containing 90% D_2O in order to avoid aggregation over time. Dialyzed NFs were gently homogenized with a Teflon glass potter, and 20 µl aliquots were frozen in liquid nitrogen for storage at $-80\,°C$. Before use, the sample was thawed on ice and then stirred on a vortex mixer at maximal speed for a controlled time: 20 times 3 s with standing in ice for 2 s between stirring.

---

often sharply delimited (arrowheads). The packing-density of the fine-fibrillar material decreased in the center of the lens (asterisk). However, no peculiar substructural organization becomes visible. Apart from these locally restricted lens-shaped accumulations of lamin A, the nuclear periphery has a regular appearance. In particular, the thickness of the lamina did not significantly increase globally. (D) The overexpressed lamin A specifically accumulated in a region between the outer nuclear envelope and an indentation of the inner nuclear membrane (arrow). Note that the accumulation is shaped like a droplet of water between two hydrophobic surfaces. (E) Nuclear periphery of a normal, nontransfected control cell (NP: nucleoplasm). Because of reduced osmium-fixation, the nuclear membranes have weak contrast only. Thus, the lamina becomes visible as a thin, dark line of granular appearance (arrowhead). Scale bar in B–E: 1 µm.

B. Atomic Force Microscope

A commercial AFM (Nanoscope IIIa, Veeco Inc., Santa Barbara, CA) equipped with a 120 μm scanner (j-scanner) and a liquid cell is employed. Before use, the liquid cell has to be cleaned with normal dish cleaner, gently rinsed with ultrapure water (18 MΩ/cm; Branstead, Boston, MA), sonicated in ethanol (50 kHz) and sonicated in ultrapure water (50 kHz).

C. Substrates for AFM

The substrate that we routinely use is a mica sheet (Mica house, Calcutta, India), which is punched to a diameter of about 5 mm and glued onto a teflon covered steel disc with water insoluble epoxy glue (Araldit, Ciba Geigy AG, Basel, Switzerland). The steel disc is required to magnetically mount the mica disc onto the piezoelectric scanner. Notice that other solid supports can be easily glued to the Teflon disc such as glass cover slips or graphite sheets (highly oriented pyrolitic graphite); any atomic flat solid substrate can be used.

V. Conclusions and Outlook

IFs are relatively stable structures under physiological conditions, in particular when they are in association with proteins such as plectin (Foisner *et al.*, 1988; Herrmann and Wiche, 1983) or desmoplakin (Mueller and Franke, 1983). However, on occasion open filaments can be observed after reconstitution *in vitro* (Herrmann and Aebi, 1998), and furthermore in a pronounced way after treatment with physiological sodium phosphate solutions on EM grids (see, e.g. (Aebi *et al.*, 1983; Herrmann *et al.*, 2002). In addition, unraveling occurs upon binding to mica (Mücke *et al.*, 2004a) or during preparation for cryo-electron microscopy (Goldie *et al.*, 2007). These properties have drawn the attention to the fact that IFs may respond more sensitive to buffer changes as anticipated by many workers in the field. Notably, for electron microscopy in most cases IFs were fixed before being treated with distilled water. Hence, individual unmodified IFs as obtained by recombinant techniques, may significantly differ from one another, which has be left unnoted as tissue IFs are regularly isolated in buffers containing high concentrations of salt. This may then somehow reflect their differential ability for dynamic processes *in vivo* such as lateral subunit exchange.

Acknowledgements

We thank Michaela Reichenzeller for providing cells transfected with YFP-lamin A. H.H. acknowledges support from the German Research Foundation (DFG grant number HE 1853/4–3). U.A. and H.H. acknowledge the support by the European Commission (Contract LSHM-CT-2005–018690). L.K. was supported by a grant from the Swiss Society for Research on Muscular Diseases awarded to U.A.

and Sergei Strelkov. U.A. was also supported by funds from the National Centre of Competence in Research in Nanoscale Science (NCCR-Nano), an IF grant from the Swiss National Science Foundation, and funds from the Canton Basel-Stadt and the M.E. Müller Foundation of Switzerland.

References

Aebi, U., Cohn, J., Buhle, L., and Gerace, L. (1986). The nuclear lamina is a meshwork of intermediate-type filaments. *Nature* **323**, 560–564.

Aebi, U., Fowler, W. E., Rew, P., and Sun, T. T. (1983). The fibrillar substructure of keratin filaments unraveled. *J. Cell. Biol.* **97**, 1131–1143.

Bader, B. L., Magin, T. M., Freudenmann, M., Stumpp, S., and Franke, W. W. (1991). Intermediate filaments formed de novo from tail-less cytokeratins in the cytoplasm and in the nucleus. *J. Cell. Biol.* **115**, 1293–1307.

Bär, H., Goudeau, B., Wälde, S., Casteras-Simon, M., Mücke, N., Shatunov, A., Goldberg, Y. P., Clarke, C., Holton, J. L., Eymard, B., Katus, H. A., Fardeau, M., *et al.* (2007). Conspicuous involvement of desmin tail mutations in diverse cardiac and skeletal myopathies. *Hum. Mutat.* **28**, 374–386.

Bartnik, E., Osborn, M., and Weber, K. (1986). Intermediate filaments in muscle and epithelial cells of nematodes. *J. Cell. Biol.* **102**, 2033–2041.

Blessing, M., Ruther, U., and Franke, W. W. (1993). Ectopic synthesis of epidermal cytokeratins in pancreatic islet cells of transgenic mice interferes with cytoskeletal order and insulin production. *J. Cell. Biol.* **120**, 743–755.

Bond, M., and Somlyo, A. V. (1982). Dense bodies and actin polarity in vertebrate smooth muscle. *J. Cell. Biol.* **95**, 403–413.

Bridger, J. M., Herrmann, H., Munkel, C., and Lichter, P. (1998). Identification of an interchromosomal compartment by polymerization of nuclear-targeted vimentin. *J. Cell. Sci.* **111**(Pt 9), 1241–1253.

Buckley, I. K., and Porter, K. R. (1967). Cytoplasmic fibrils in living cultured cells. A light and electron microscope study. *Protoplasma* **64**, 349–380.

Capco, D. G., Wan, K. M., and Penman, S. (1982). The nuclear matrix: Three-dimensional architecture and protein composition. *Cell* **29**, 847–858.

Carmo-Fonseca, M., Cidadao, A. J., and David-Ferreira, J. F. (1988). Filamentous cross-bridges link intermediate filaments to the nuclear pore complexes. *Eur. J. Cell. Biol.* **45**, 282–290.

Clubb, B. H., Chou, Y. H., Herrmann, H., Svitkina, T. M., Borisy, G. G., and Goldman, R. D. (2000). The 300-kDa intermediate filament-associated protein (IFAP300) is a hamster plectin ortholog. *Biochem. Biophys. Res. Commun.* **273**, 183–187.

Cohen, M., Tzur, Y. B., Neufeld, E., Feinstein, N., Delannoy, M. R., Wilson, K. L., and Gruenbaum, Y. (2002). Transmission electron microscope studies of the nuclear envelope in *Caenorhabditis elegans* embryos. *J. Struct. Biol.* **140**, 232–240.

Dodemont, H., Riemer, D., Ledger, N., and Weber, K. (1994). Eight genes and alternative RNA processing pathways generate an unexpectedly large diversity of cytoplasmic intermediate filament proteins in the nematode *Caenorhabditis elegans*. *Embo. J.* **13**, 2625–2638.

Engel, A. (1978a). Molecular weight determination by scanning transmission electron microscopy. *Ultramicroscopy* **3**, 273–281.

Engel, A. (1978b). The STEM: An attractive tool for the biologist. *Ultramicroscopy* **3**, 355–7.

Erber, A., Riemer, D., Hofemeister, H., Bovenschulte, M., Stick, R., Panopoulou, G., Lehrach, H., and Weber, K. (1999). Characterization of the *Hydra* lamin and its gene: A molecular phylogeny of metazoan lamins. *J. Mol. Evol.* **49**, 260–271.

Foisner, R., Leichtfried, F. E., Herrmann, H., Small, J. V., Lawson, D., and Wiche, G. (1988). Cytoskeleton-associated plectin: in situ localization, *in vitro* reconstitution, and binding to immobilized intermediate filament proteins. *J. Cell. Biol.* **106**, 723–733.

Fowler, W. E., and Aebi, U. (1983). Preparation of single molecules and supramolecular complexes for high-resolution metal shadowing. *J. Ultrastruct. Res.* **83**, 319–334.

Franke, W. W., Schmid, E., Osborn, M., and Weber, K. (1978). Different intermediate-sized filaments distinguished by immunofluorescence microscopy. *Proc. Natl. Acad. Sci. USA* **75**, 5034–5038.

Goldie, K. N., Wedig, T., Mitra, A. K., Aebi, U., Herrmann, H., and Hoenger, A. (2007). Dissecting the 3-D structure of vimentin intermediate filaments by cryo-electron tomography. *J. Struct. Biol.* **158**, 378–385.

Goldman, A. E., Moir, R. D., Montag-Lowy, M., Stewart, M., and Goldman, R. D. (1992). Pathway of incorporation of microinjected lamin A into the nuclear envelope. *J. Cell. Biol.* **119**, 725–735.

Goldman, R. D., and Follett, E. A. (1969). The structure of the major cell processes of isolated BHK21 fibroblasts. *Exp. Cell. Res.* **57**, 263–276.

Goldman, R. D., and Follett, E. A. (1970). Birefringent filamentous organelle in BHK-21 cells and its possible role in cell spreading and motility. *Science* **169**, 286–288.

Granger, B. L., and Lazarides, E. (1982). Structural associations of synemin and vimentin filaments in avian erythrocytes revealed by immunoelectron microscopy. *Cell* **30**, 263–275.

Heins, S., Wong, P. C., Muller, S., Goldie, K., Cleveland, D. W., and Aebi, U. (1993). The rod domain of NF-L determines neurofilament architecture, whereas the end domains specify filament assembly and network formation. *J. Cell. Biol.* **123**, 1517–1533.

Henderson, D., Geisler, N., and Weber, K. (1982). A periodic ultrastructure in intermediate filaments. *J. Mol. Biol.* **155**, 173–176.

Herrmann, H., and Aebi, U. (1998). Structure, assembly, and dynamics of intermediate filaments. *Subcell. Biochem.* **31**, 319–362.

Herrmann, H., Bar, H., Kreplak, L., Strelkov, S. V., and Aebi, U. (2007). Intermediate filaments: from cell architecture to nanomechanics. *Nat. Rev. Mol. Cell. Biol.* **8**, 562–573.

Herrmann, H., Eckelt, A., Brettel, M., Grund, C., and Franke, W. W. (1993). Temperature-sensitive intermediate filament assembly. Alternative structures of *Xenopus laevis* vimentin *in vitro* and *in vivo*. *J. Mol. Biol.* **234**, 99–113.

Herrmann, H., Häner, M., Brettel, M., Ku, N. O., and Aebi, U. (1999). Characterization of distinct early assembly units of different intermediate filament proteins. *J. Mol. Biol.* **286**, 1403–1420.

Herrmann, H., Häner, M., Brettel, M., Müller, S. A., Goldie, K. N., Fedtke, B., Lustig, A., Franke, W. W., and Aebi, U. (1996). Structure and assembly properties of the intermediate filament protein vimentin: the role of its head, rod and tail domains. *J. Mol. Biol.* **264**, 933–953.

Herrmann, H., Hesse, M., Reichenzeller, M., Aebi, U., and Magin, T. M. (2003). Functional complexity of intermediate filament cytoskeletons: from structure to assembly to gene ablation. *Int. Rev. Cytol.* **223**, 83–175.

Herrmann, H., Kreplak, L., and Aebi, U. (2004). Isolation, characterization, and *in vitro* assembly of intermediate filaments. *Methods Cell. Biol.* **78**, 3–24.

Herrmann, H., Wedig, T., Porter, R. M., Lane, E. B., and Aebi, U. (2002). Characterization of early assembly intermediates of recombinant human keratins. *J. Struct. Biol.* **137**, 82–96.

Herrmann, H., and Wiche, G. (1983). Specific in situ phosphorylation of plectin in detergent-resistant cytoskeletons from cultured Chinese hamster ovary cells. *J. Biol. Chem.* **258**, 14610–14618.

Heuser, J. E., and Kirschner, M. W. (1980). Filament organization revealed in platinum replicas of freeze-dried cytoskeletons. *J. Cell. Biol.* **86**, 212–234.

Hirokawa, N. (1982). Cross-linker system between neurofilaments, microtubules, and membranous organelles in frog axons revealed by the quick-freeze, deep-etching method. *J. Cell. Biol.* **94**, 129–142.

Hirokawa, N., Tilney, L. G., Fujiwara, K., and Heuser, J. E. (1982). Organization of actin, myosin, and intermediate filaments in the brush border of intestinal epithelial cells. *J. Cell. Biol.* **94**, 425–443.

Hozak, P., Sasseville, A. M., Raymond, Y., and Cook, P. R. (1995). Lamin proteins form an internal nucleoskeleton as well as a peripheral lamina in human cells. *J. Cell. Sci.* **108**(Pt 2), 635–644.

Ishikawa, H., Bischoff, R., and Holtzer, H. (1968). Mitosis and intermediate-sized filaments in developing skeletal muscle. *J. Cell. Biol.* **38**, 538–555.

Kirmse, R., Portet, S., Mucke, N., Aebi, U., Herrmann, H., and Langowski, J. (2007). A quantitative kinetic model for the *in vitro* assembly of intermediate filaments from tetrameric vimentin. *J. Biol. Chem.* **282**, 18563–18572.

Kreplak, L., Bar, H., Leterrier, J. F., Herrmann, H., and Aebi, U. (2005). Exploring the mechanical behavior of single intermediate filaments. *J. Mol. Biol.* **354**, 569–577.

Krishnan, N., Kaiserman-Abramof, I. R., and Lasek, R. J. (1979). Helical substructure of neurofilaments isolated from Myxicola and squid giant axons. *J. Cell. Biol.* **82**, 323–335.

Leterrier, J. F., Kas, J., Hartwig, J., Vegners, R., and Janmey, P. A. (1996). Mechanical effects of neurofilament cross-bridges. Modulation by phosphorylation, lipids, and interactions with F-actin. *J. Biol. Chem.* **271**, 15687–15694.

Mücke, N., Kirmse, R., Wedig, T., Leterrier, J. F., and Kreplak, L. (2005). Investigation of the morphology of intermediate filaments adsorbed to different solid supports. *J. Struct. Biol.* **150**, 268–276.

Mücke, N., Kreplak, L., Kirmse, R., Wedig, T., Herrmann, H., Aebi, U., and Langowski, J. (2004a). Assessing the flexibility of intermediate filaments by atomic force microscopy. *J. Mol. Biol.* **335**, 1241–1250.

Mücke, N., Wedig, T., Bürer, A., Marekov, L. N., Steinert, P. M., Langowski, J., Aebi, U., and Herrmann, H. (2004b). Molecular and biophysical characterization of assembly-starter units of human vimentin. *J. Mol. Biol.* **340**, 97–114.

Mueller, H., and Franke, W. W. (1983). Biochemical and immunological characterization of desmoplakins I and II, the major polypeptides of the desmosomal plaque. *J. Mol. Biol.* **163**, 647–671.

Okazaki, K., and Holtzer, H. (1965). An analysis of myogenesis *in vitro* using fluorescein-labeled antimyosin. *J. Histochem. Cytochem.* **13**, 726–739.

Paddy, M. R., Belmont, A. S., Saumweber, H., Agard, D. A., and Sedat, J. W. (1990). Interphase nuclear envelope lamins form a discontinuous network that interacts with only a fraction of the chromatin in the nuclear periphery. *Cell* **62**, 89–106.

Reichenzeller, M., Burzlaff, A., Lichter, P., and Herrmann, H. (2000). In vivo observation of a nuclear channel-like system: evidence for a distinct interchromosomal domain compartment in interphase cells. *J. Struct. Biol.* **129**, 175–185.

Richter, K., Reichenzeller, M., Gorisch, S. M., Schmidt, U., Scheuermann, M. O., Herrmann, H., and Lichter, P. (2005). Characterization of a nuclear compartment shared by nuclear bodies applying ectopic protein expression and correlative light and electron microscopy. *Exp. Cell. Res.* **303**, 128–137.

Small, J. V., and Sobieszek, A. (1977). Studies on the function and composition of the 10-NM(100-A) filaments of vertebrate smooth muscle. *J. Cell. Sci.* **23**, 243–268.

Steinert, P. M., and Parry, D. A. (1985). Intermediate filaments: conformity and diversity of expression and structure. *Annu. Rev. Cell. Biol.* **1**, 41–65.

Steven, A. C., Wall, J., Hainfeld, J., and Steinert, P. M. (1982). Structure of fibroblastic intermediate filaments: analysis of scanning transmission electron microscopy. *Proc. Natl. Acad. Sci. USA* **79**, 3101–3105.

Strelkov, S. V., Kreplak, L., Herrmann, H., and Aebi, U. (2004). Intermediate filament protein structure determination. *Methods Cell. Biol.* **78**, 25–43.

Stromer, M. H., Ritter, M. A., Pang, Y. Y., and Robson, R. M. (1987). Effect of cations and temperature on kinetics of desmin assembly. *Biochem. J.* **246**, 75–81.

Svitkina, T. M., Verkhovsky, A. B., and Borisy, G. G. (1996). Plectin sidearms mediate interaction of intermediate filaments with microtubules and other components of the cytoskeleton. *J. Cell. Biol.* **135**, 991–1007.

Wickert, U., Mücke, N., Wedig, T., Müller, S. A., Aebi, U., and Herrmann, H. (2005). Characterization of the *in vitro* co-assembly process of the intermediate filament proteins vimentin and desmin: mixed polymers at all stages of assembly. *Eur. J. Cell. Biol.* **84**, 379–391.

CHAPTER 16

Studying Microtubules by Electron Microscopy

Carolyn Moores

School of Crystallography
Birkbeck College
London WC1E 7HX
United Kingdom

Abstract
I. Introduction
    A. Introduction to Microtubules
    B. Introduction to Electron Microscopy
II. Molecular Electron Microscopy
    A. Background
    B. Materials and Methods for Microtubule Polymerization
    C. Negative Stain Electron Microscopy
    D. Metal Shadowing for Electron Microscopy
    E. Cryo-Electron Microscopy
III. Cellular Electron Microscopy
    A. Background
    B. Electron Tomography
IV. Outlook
References

Abstract

Microtubules are one of the three components of the eukaryotic cytoskeleton and play a central role in many aspects of cell function, including cell division and cell motility. Electron microscopy, of both isolated microtubules and of microtubules in their cellular context, is an essential tool in understanding their structure and function. These studies have been particularly important because the size and heterogeneity of microtubules mean that they are not readily studied by other structural methods. Electron microscopy at different levels of detail can bridge the gap between atomic resolution structures solved by X-ray crystallography or NMR, and the dynamic but relatively coarse information that is achieved in light microscopy experiments. This chapter provides an overview of current approaches to studying microtubules by electron microscopy. Negative stain studies can be relatively quick and easy, and provide a useful overview of microtubules and their interactions prior to undertaking more complicated experiments such as high resolution structure determination by cryo-electron microscopy or reconstruction of unique cellular entities by electron tomography. As more microtubule-associated proteins are uncovered, EM studies will make a significant contribution to our understanding of how these proteins influence microtubules and where in the cytoskeleton they act.

I. Introduction

A. Introduction to Microtubules

Microtubules are involved in many essential aspects of cell life, including cell division, cell motility, defining cell architecture, and intracellular transport (see for example Guzik and Goldstein, 2004; Wittmann et al., 2001). Electron microscopy (EM) has been an essential technique in characterizing microtubules (Ledbetter and Porter, 1963) and in understanding their contributions to cellular processes. The relative ease with which microtubules can be visualised by EM, coupled with their inherent beauty, continues to inspire biologists to study and understand them, although only an overview of the different types of EM-based approaches can be discussed here.

Microtubules – hollow cylindrical polymers with a diameter of ~25 nm – are built from heterodimers of α- and β-tubulin (reviewed in Nogales, 2000). The individual monomers are folded together by dedicated chaperones so that only $\alpha\beta$-heterodimers are present in cells. α- and β-tubulin show 40% sequence identity and are structurally very similar (Nogales et al., 1998; Fig. 1); both contain a guanine nucleotide binding site but are biochemically distinct. The GTP bound to α-tubulin is buried within the tubulin dimer and is protected from hydrolysis, while GTP bound to β-tubulin can be hydrolyzed to GDP. In fact, microtubule polymerization stimulates hydrolysis of β-tubulin-bound GTP, such that tubulin acts as its own GAP (GTPase Activating Protein). GTP-bound β-tubulin enables microtubule polymerization while GDP-β-tubulin does not. Microtubule polymers are very dynamic as a result of these cycles of GTP-stimulated growth and hydrolysis-induced depolymerization, a property known as dynamic instability (reviewed in Desai and Mitchison, 1997).

Cellular microtubules interact with, and are regulated by, a plethora of microtubule-associated proteins. These proteins are even more diverse than the

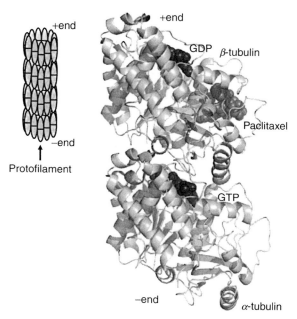

Fig. 1 Microtubule organization and structure of the αβ-tubulin heterodimer. Microtububules are built from αβ-tubulin heterodimers that join head-to-tail to form protofilaments and side-by-side to form the microtubule wall. The structure of the αβ-tubulin heterodimer was determined using electron crystallography. In this figure, the dimer is viewed as from the outside of the microtubule wall. β-tubulin is located towards the plus-end of the microtubule, while α-tubulin is towards the minus-end. α- and β-tubulin both bind guanine nucleotides (shown as space-filling models) but the α-tubulin GTP-binding site is buried at the intradimer interface and always has GTP bound, whereas the β-tubulin-bound GTP is subject to hydrolysis as polymerization proceeds. Paclitaxel, which stabilizes lateral inter-protofilament contacts and suppresses dynamic instability, is shown as a space-filling model. This figure was prepared using 1JFF.pdb (Löwe et al., 2001) and Pymol (DeLano, W.L. The PyMOL Molecular Graphics System (2002) DeLano Scientific, Palo Alto, CA, USA. http://www.pymol.org).

functions performed by the microtubules, and understanding how all these proteins impact on microtubule function *in vivo* is a significant challenge. Electron microscopists have, broadly speaking, taken two complementary approaches to understanding these microtubule-associated proteins (MAPs): (1) a divide-and-conquer approach in which microtubules are assembled *in vitro* and the interaction of individual MAPs (or recombinantly expressed fragments of MAPs) are examined, providing insight into the molecular details of their interaction (Section II); (2) a top-down approach in which microtubules are investigated in their cellular context (Section III). Complete mechanistic insight into MAP function lies at the intersection of these two approaches, in combination with many cell biology and biochemical tools.

B. Introduction to Electron Microscopy

Images are formed in an electron microscope when the electron beam interacts with the sample. The wavelength of electrons is around 0.005 nm; this is much smaller than that of visible light, so the details that can be seen by an EM are much finer. The EM operates under a vacuum, and biological samples must be fixed in some way so that they can withstand both the vacuum of the microscope and the damaging electron beam. Samples observed by EM are therefore necessarily static and thus EM is essentially a tool for studying snapshots of cellular life. The images obtained by EM are two-dimensional projections of the sample because the depth of focus of the imaging lens is large compared with the sample, and this must be taken into consideration when EM images are interpreted.

II. Molecular Electron Microscopy

A. Background

Many striking insights into microtubule biology have been made by studying microtubules in isolation. Molecular microscopy can either observe specific phenomena directly or, with computational image analysis, yield macromolecular structures. Dynamic microtubules are hard to study and for most experiments, artificially stabilized microtubules are used (see below). Tubulin also forms diverse non-microtubule oligomers and polymers, which are also amenable to study by EM and which have provided additional insight into structural aspects of microtubule dynamics (for example, Wang and Nogales, 2005). Probably the best example of this is the determination of the structure of the $\alpha\beta$-tubulin dimer at near-atomic resolution using EM images of zinc-induced two-dimensional tubulin sheets (Löwe *et al.*, 2001; Nogales *et al.*, 1998; Fig. 1).

B. Materials and Methods for Microtubule Polymerization

Tubulin can be isolated in reasonable quantities from mammalian brains using well-established protocols (Shelanski *et al.*, 1973). However, the success of these protocols depends on the availability of brain tissue from freshly slaughtered livestock (and therefore proximity to an abattoir). Small amounts of tubulin can also be obtained from cell cultures. However, since relatively little material is needed for molecular EM experiments, by far the simplest way to obtain tubulin is to buy it (http://www.cytoskeleton.com/). Microtubules can be stabilized, either using the tubulin-binding drug paclitaxel (http://www.merckbiosciences.co.uk/home.asp) or by polymerization in the presence of the nonhydrolyzable GTP analogue GMPCPP (http://www.jenabioscience.com/). The number of protofilaments in a microtubule can vary from 9 to 18 (Chrétien *et al.*, 1992) and polymerization conditions can be modified according to the microtubule architecture required (Ray *et al.*, 1993). In the first instance, this is unlikely to be a critical consideration, but since the majority of microtubules *in vivo* are built from 13 protofilaments (Tilney *et al.*, 1973), protofilament architecture may prove important for a subset of MAPs (Moores *et al.* 2004; Sandblad *et al.*, 2006).

Tubulin is labile and must be stored at −80 °C and should be rapidly thawed immediately prior to use. Typically, we mix equal volumes of tubulin with 2× polymerization buffer (80 mM Pipes, pH 6.8, 16% DMSO, 3 mM $MgCl_2$, 1 mM EGTA, 2.5 mM GTP) and incubate at 37 °C for 30–60 min. After this time, paclitaxel, dissolved in DMSO, should be added in excess (1 mM). Unstabilized microtubules are extremely temperature sensitive, so only remove the polymerizing solution from 37 °C for long enough to mix the paclitaxel and then leave it at 37 °C for another 2–3 h. If paclitaxel is present at the outset of polymerization, tubulin sheets (unclosed microtubules) will also be observed. Twenty four hours should elapse between polymerization and microtubule use in EM experiments, since some dynamic instability continues up to this point and leads to a high background of unpolymerized tubulin on the EM grids. Once they have stabilized, paclitaxel microtubules are quite robust and can be diluted into most buffers. Alternatively, GMPCPP can be added to the polymerization buffer instead of GTP and no paclitaxel will be needed. If helical microtubules (microtubules with a perfectly crystalline lattice with no discontinuities most frequently built from 15 or 16 protofilaments) are required for subsequent structure determination, polymerization should be performed at 34 °C in the presence of GTP. It is surprising and aggravating how temperature sensitive helical microtubule polymerization can be, but their formation can be slightly enhanced in the presence of higher concentrations of magnesium (Chretien *et al.*, 1992). GMPCPP overwhelmingly favors polymerization of 14 protofilament nonhelical microtubules (Hyman *et al.*, 1992).

C. Negative Stain Electron Microscopy

1. Rationale

A great deal of useful preliminary information can be obtained by negative stain EM of microtubules and their associated proteins. In fact, this is a recommended first step in the characterization of any new MAP before attempting technically challenging cryo-EM experiments. Negative stain experiments are, in principle, quick and easy and require minimal specialist equipment aside from a standard TEM. Negative stain works by surrounding the microtubule with grains of heavy metal salts and is so-called because it is the stain envelope that is imaged rather than the protein itself. The stain strongly scatters electrons and generates well-contrasted images; it also provides some protection for the biological sample in the destructive electron beam.

2. Materials and Methods

EM grids covered with continuous carbon film are standard tools for negative stain experiments. Home-made grids are easiest (and cheapest) to use but specialist equipment is required to prepare the carbon, so these grids can be purchased (http://www.grid-tech.com/). We have also found that glow-discharging grids, so that their surface is charged, gives a much better distribution of stain across the grid and improves the appearance of the microtubules (see http://www.emitech.co.uk/vacuum-coating-brief5.htm). We find that staining with aqueous 1% uranyl acetate (or uranyl formate) enables good visualization of the microtubule lattice (Fig. 2), despite the nonphysiological pH of both of these stains (uranyl acetate = pH 4.2 and uranyl formate = pH 4.5–5.2). Both these salts are weakly radioactive; specific arrangements will probably need to be made for their disposal but no particular precautions are needed during handling. However, the quality of the microtubule images obtained using uranyl salts is far superior to other stains.

0.2 mg/ml of polymerized tubulin gives a good distribution of microtubules across the grid and roughly 0.2 mg/ml of the MAP should enable its binding to be visualized, although this depends on its microtubule affinity. MAP

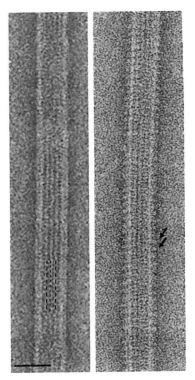

Fig. 2 Images of paclitaxel-stabilized microtubules visualized using negative stain EM. The microtubule lattice can be easily visualized by staining with uranyl acetate; protofilaments are indicated with dotted lines on the left image. The image on the left is of a microtubule-only sample, while the image on the right is of a microtubule bound by kinesin motor domain, which can be seen protruding every 8 nm along the lattice (indicated by arrows). Note that the apparent diameter of these microtubules is much larger than the expected 25 nm because they are flattened by the stain. Bar = 40 nm.

concentrations much higher than 0.2 mg/ml will give a very high background on the EM grid, making direct observation of microtubule binding harder. Typically, we apply 4 µl of polymerized tubulin to glow-discharged grids, wash with 4 µl MAP buffer, apply 4 µl of the MAP of interest, wash again and apply 4 µl of 1% uranyl acetate, wicking off nearly all of each application using filter paper before applying the next solution. This seems to give even staining over most of the grid squares.

3. Results

Visual comparison of microtubules in the presence and absence of the MAP of interest will often provide direct information on whether an interaction is occurring. In our work on kinesin motor proteins, the compact globular motor domain (~40 kDa) can be seen very easily and binds every 8 nm to each tubulin dimer (Fig. 2). Smaller binding domains or less compact proteins may be harder to see. Negative stain EM studies can also be helpful in determining if the MAP of interest has an effect on the microtubule population. In our study of the motor domain of a microtubule depolymerizing kinesin-13, negative stain images allowed us to visualize the intermediates of depolymerization in the form of curved tubulin rings, bound and bent by the attached kinesin (Moores *et al.*, 2002; Fig. 3).

4. Limitations

The most obvious limitation of negative stain studies is that the images collected are not of the microtubules themselves but are of the stain surrounding them. No information is available about the internal structure of the complex, and the detail that can be extracted from the images is limited by the granularity of the stain used. In addition, negative stain flattens the hollow cylindrical microtubule structure, which may also suffer from dehydration. It is never possible to say that such images are completely artefact-free but, as a first and easy tool for direct observation of microtubule-MAP interactions, negative stain experiments are invaluable.

D. Metal Shadowing for Electron Microscopy

1. Rationale

An alternative method for visualizing microtubules is to evaporate a fine layer of metal over the sample, either from a rotating source or by unidirectional shadowing. The resulting metal replica of the microtubule surface is then placed on an EM grid and viewed. Once again, the metal atoms strongly scatter electrons, giving images with relatively high contrast that often allow individual molecules to be distinguished.

2. Materials and Methods

Microtubules, applied to a support film, must be rapidly frozen in nitrogen to avoid ice crystal formation and to preserve their solution structure. The protein must then be freeze-dried and the ice sublimated away from the microtubule

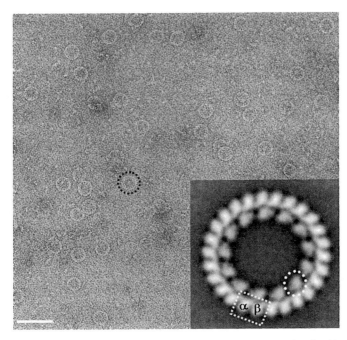

Fig. 3 Microtubule depolymerization intermediates of a kinesin-13 motor visualized by negative stain EM. In the presence of the non-hydrolyzable ATP analogue AMPPNP, a kinesin-13 motor core forms tubulin rings from paclitaxel-stabilized microtubules (a single ring is circled). The rings have a diameter of 40 nm and can be aligned and averaged together to yield a two-dimensional average that reveals the molecular details of ring formation (inset). The rings are apparently formed from the sequential binding and bending of kinesin-13 motors trapped in the ATP-bound state of their ATPase cycle. The two-dimensional average reveals kinesin-13 motor densities (white oval) on the inside of the rings; these are each bound to an $\alpha\beta$-tubulin dimer in the outer ring. The kinesin-13 motor binds to what would have been the outer surface of the tubulin protofilament and bends it away from the microtubule axis, much like unpeeling a banana (Moores et al., 2002). Bar = 100 nm.

surface by warming to $\sim-95\,°C$ under vacuum for ~ 2 h. This should leave the protein structure intact, although microtubules are usually flattened by this process. Different approaches to generating the metal replica have been taken. The most widely used technique is to rotary shadow with a mixture of platinum and carbon which, although it yields relatively coarse images, is still useful for many applications (see for example Hirokawa, 1986; Szajner et al., 2005). More recently, Andreas Hoenger and colleagues have advocated use of angular, unidirectional shadowing of a tantalum/tungsten alloy to give images with much finer detail. This alloy is very susceptible to oxidative degradation, however, and so must be transferred directly from the shadowing chamber into the microscope, a procedure that requires a specially modified EM (Gross et al., 1990).

3. Results

The high-contrast images generated by metal shadowing have been particularly useful in providing information about filament networks (Hirokawa et al., 1988; Svitkina et al., 1996; Fig. 4A). Meanwhile, the finer details observed by Hoenger and colleagues have recently yielded some striking results about the ways in which MAPs bind microtubules and that would not have been accessible using other EM methods (Krzysiak, T. C. et al., 2006; Sandblad et al., 2006; Fig. 4B).

4. Limitations

Similar concerns to those with negative stain exist for metal shadowing experiments, since the microtubules are only indirectly observed and sample preparation involves dehydration of the microtubules in a vacuum, after which they are covered with a layer of metal.

E. Cryo-Electron Microscopy

1. Rationale

The use of heavy metals, either by staining or by shadowing, provides high-contrast images of microtubules and bound MAPs. Cryo-EM, on the other hand, enables visualization of microtubules in as close to a physiological state as possible. Sample

Fig. 4 Images of microtubules obtained by metal shadowing. (A) Microtubules pelleted in the presence of the MAP tau and visualized by rotary-shadowing. Arrows indicate tau molecules projecting from the microtubules and cross-linking them (Hirokawa et al., 1988). Reproduced from Journal of Cell Biology, 1988, 107:1449–1459. Copyright 1988 The Rockefeller University Press. Bar = 50 nm. (B) Microtubule lattice bound by kinesin motor domains visualized by unidirectional shadowing. The individual motor domains can be discriminated (dotted rectangle) and the regular pattern they form on the microtubule lattice (dotted white lines) allows visualization of the microtubule seam. Bar = 20 nm. This image was kindly provided by the lab of Andreas Hoenger, Colorado University, Boulder (http://hoengerlab.colorado.edu/).

preparation for cryo-EM involves ultra-rapid freezing of microtubules in a thin layer of amorphous (vitreous) ice (reviewed in Dubochet *et al.*, 1988; Stewart and Vigers, 1986). Such samples must be maintained at liquid nitrogen temperatures while viewing them in the microscope, to preserve their vitreous state and reduce radiation damage. The resulting images arise from the direct interaction of electrons with the specimen; they have low contrast but should be free from artefact and distortion. Cryo-EM is becoming more of a standard tool in structural biology, but cannot be performed on all EMs because of essential requirements for expensive and specialized sample holders, sample stability, vacuum quality, and the ability to image the electron-sensitive cryo-EM samples under low-electron dose conditions. A coherent Field Emission Gun electron source is also a great benefit, especially when the cryo-EM images will subsequently be used for structure determination. An overview of the details of microtubule preparation for cryo-EM is given later, whereas more general aspects of cryo-EM technique have been reviewed extensively elsewhere (Kuhlbrandt and Williams, 1999; Saibil, 2000; Subramaniam and Milne, 2004; Unger, 2001).

2. Materials and Methods

Microtubules are prepared for cryo-EM by application to glow-discharged, holey carbon film, either home-made or purchased (http://www.emsdiasum.com/microscopy/products/grids/cflat.aspx). The grid is blotted to give a very thin layer of solution, 50–100 nm thick, which is then rapidly frozen by plunging it into ethane slush. A simple guillotene-like device can perform this task very well, although reproducible sample preparation and ice quality are facilitated by undertaking grid preparation in a humid environment. Higher concentrations of microtubules and MAPs are needed for cryo-EM than for negative stain grids, and typically we use solutions of 2 mg/ml polymerized tubulin. 4 µl of this solution is applied to the grid, wicked off, the grids are washed with 4 µl of buffer and the grid is blotted and rapidly frozen. To prepare MAP-microtubule complexes, we have found it easiest to apply a solution of the MAP (@3–5 mg/ml) after adsorbing microtubules to the grid because this minimizes the tendency of some MAPs to bundle microtubules. The MAP solution should be free from cryo-protectants such as glycerol because they interfere with vitrification. Working with stabilized microtubules is almost essential for cryo-EM work; as has been emphasized earlier, unstabilized microtubules are very fragile, and in particular, the changes in temperature that occur due to water evaporation from the thin layers of solution can be sufficient to induce complete microtubule depolymerization (Chretien *et al.*, 1992).

Cryo-EM images arise from direct interactions of electrons with all levels of the sample, including its internal features. This information can be extracted computationally and used for structure determination, with the goal of understanding the detailed organization of the microtubule and bound proteins. Images of the microtubule lattice contain multiple views of individual $\alpha\beta$-tubulin dimers, captured from many different angles as they wind around the microtubule axis.

Structure calculations aim to extract the information about the different views of the dimer from the noisy EM images and combine them to produce an average three-dimensional model of tubulin within the lattice. These calculations necessarily assume that all molecules that are combined in such calculations are structurally identical. Different computational strategies have been taken to achieving this goal (Hirose *et al.*, 2006; Kikkawa and Hirokawa, 2006; Li *et al.*, 2002; Rice *et al.*, 1999), depending on the underlying protofilament architecture of the microtubules used, and these approaches are described in detail elsewhere (Hoenger and Nicastro, 2007).

3. Results

Despite the technical challenges, cryo-EM is an essential tool in understanding many aspects of microtubule biology. Important experiments have included the visualization of different structural states in dynamic microtubules (Mandelkow *et al.*, 1991; Fig. 5) and observation of microtubule architectural variation (Chretien *et al.*, 1992; Ray *et al.*, 1993), all of which are most easily visualized in cryo-EM images.

Many cryo-EM studies have been aimed at understanding the structural basis of the ATP-dependent mechanisms of members of the kinesin superfamily of microtubule motors (for example, Hirose *et al.*, 2006; Kikkawa and Hirokawa, 2006; Rice *et al.*, 1999; Fig. 6A). Because the kinesin ATPase cycle is microtubule-stimulated, several important structural transitions can only occur in the presence of microtubules and are therefore only accessible by cryo-EM.

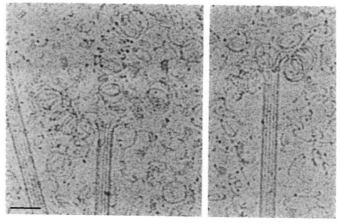

Fig. 5 Cryo-EM images of depolymerizing microtubules. Unstabilized microtubules, induced to depolymerize in the presence of divalent cations, were imaged by cryo-EM to reveal the "rams-horn"-like appearance of their unpeeling protofilaments (Mandelkow *et al.*, 1991). Bar = 50 nm. Reproduced from Journal of Cell Biology, 1991, 114:977–991. Copyright 1991 The Rockefeller University Press.

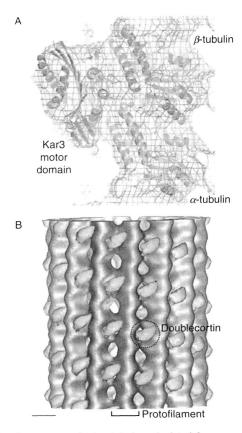

Fig. 6 Three-dimensional structures of microtubules calculated from cryo-EM images using helical reconstruction. (A) The structure of the kinesin motor domain of Kar3 bound to microtubules has been calculated at high resolution (shown here in its nucleotide-free state), enabling the crystal structures of the $\alpha\beta$-tubulin dimer and the Kar3 motor domain to be docked within the EM density map (shown as mesh). Reprinted from Hirose et al., 2006 with permission from Elsevier. (B) The neuronal MAP doublecortin binds between the protofilaments in the microtubule wall every 8 nm along the microtubule lattice (one binding site is circled). This unusual binding site confers unique microtubule-stabilizing properties on doublecortin compared to other MAPs (Moores et al., 2004). Bar = 5 nm.

Nucleotide analogues are used to capture kinesins at different points in their ATPase cycles and the resulting structures are compared, providing information about the nucleotide-dependent conformational changes of the motors. In several cases, structures have been calculated such that secondary structural elements within the motors and microtubules can be visualized, enabling very detailed analyses of conformational changes. Cryo-EM studies have also yielded information about the binding sites of a variety of nonmotor MAPs on microtubules. We are interested in the influence of microtubule architecture on the ability of certain MAPs to bind microtubules (Moores *et al.*, 2004; Fig. 6B), a feature of MAP binding that is of particular interest since most microtubules *in vivo* are built from 13-protofilaments. Atomic resolution structures from X-ray crystallography experiments can be placed inside the molecular envelopes of the resulting three-dimensional structures to yield pseudoatomic details about the cytoskeletal complexes (Fig. 6A). Data from these structural experiments have been combined with those from biophysical and biochemical experiments to provide an integrated view of the interactions.

III. Cellular Electron Microscopy

A. Background

Ultimately, we would like to understand the precise roles of MAPs and their mechanisms of regulation of the microtubule cytoskeleton within the cell. Microtubules were originally observed in sections of chemically fixed and stained cells and their dimensions make them relatively easy to identify. Such studies have contributed a good deal to our understanding of microtubules, although concerns exist about the integrity of subcellular structures in such chemically fixed samples (for example Winey *et al.*, 1995). Additionally, only a single slice of the complex cellular environment is seen, but we would like to understand the organization of the cytoskeletal network in three dimensions.

B. Electron Tomography

1. Rationale

As mentioned in Section II, three-dimensional structure calculation involves the collection and combination of multiple views of the biological entity under investigation. However, each cell is essentially a unique biological structure, which means that views of different cells cannot usefully be averaged together. Electron tomography (ET) enables multiple views to be collected and combined from the same physical object, be it a cell section, organelle, or large macromolecular complex. Because it does not rely on averaging, ET is the most versatile form of EM and the structure of virtually any object can, in theory, be calculated. However, because multiple views of the same object are extracted, sample damage by the electron beam is a major consideration.

ET is by no means a new research tool, but recent advances in EM hardware, computational capacity, and software development have meant that ET is being applied to greater numbers of previously intractable biological problems. It is best undertaken in laboratories specializing in ET and details of the methods employed have been reviewed elsewhere (Hoenger and Nicastro, 2007; Höög and Antony, 2007; Lucic *et al.*, 2005; McIntosh *et al.*, 2005).

2. Materials and Methods

Ideally, biologists would like to understand cellular organization to the finest level of detail. In practice, a number of compromises in sample preservation and visualization must be made. Accurate imaging of any sample by EM requires that each electron interact once with the sample without losing energy (elastic scattering). In some cases, the edges of intact cells are thin enough (~200 nm) for this to be true, but usually cell sections will need to be cut before they can be visualized by EM. Prior to sectioning, cells must be rapidly frozen to ensure that destructive ice crystals do not accumulate, and high pressures are needed to ensure that these thicker cellular samples freeze with sufficient rapidity (see Chapter 11 by Studer, this volume). Cells frozen in this way are often viable when thawed, attesting to the fidelity with which their structures are preserved during freezing. Direct sectioning of these frozen cells may then be attempted – this approach has the advantage that, as with molecular cryo-EM, the undistorted native cell structures are preserved. However, preparation of these sections is challenging, data must be collected in an ultra-low-dose regimen because the samples are extremely electron-sensitive and the lack of artificial contrast (for example from metal stains) also presents challenges in structure determination and interpretation. Nevertheless some early results are impressive (for example Bouchet-Marquis et al., 2006; Garvalov et al., 2006).

More common is the practice of freeze-substitution in which the frozen tissue section is gradually infiltrated over the course of days with solvents to replace the water in the sample. Fixatives such as glutaraldehyde are also used along with heavy metal stains such as osmium tetroxide or uranyl acetate. The slow infiltration is believed to be a superior method of structure preservation while still allowing the cell to be dehydrated and the sections embedded in plastic and sectioned, thereby providing additional sample preservation in the EM and providing contrast for the cellular components (Höög and Antony, 2007; Winey et al., 1995).

Data collection for ET involves incremental tilting of the sample between, at best, ±70° around two tilt axes. This maximizes the numbers of different views collected that will ultimately contribute to the structure calculation. Automated image acquisition to a digital camera greatly facilitates this otherwise time-consuming process, as does improved stage stability in more recent EMs. Digital data collection also greatly facilitates the electron dose fractionation that must occur so that structural information can still be extracted at each tilt angle while minimizing the radiation damage accumulated by the sample.

Subsequent data analysis involves the alignment of individual images from the tilt series, usually facilitated by electron-dense gold fiducial markers. Data analysis is also relatively automated and new developments are improving the ease with which reconstructions can be calculated and information extracted. Structures calculated using ET will always contain some distortion because it is not physically possible to collect data from all views, and such distortions must be considered during interpretation. The smallest features that can be reliably extracted from tomographic reconstructions will usually be around 2–5 nm in dimension. Further details can sometimes be visualized by averaging multiple tomographic volumes together (see for example Nicastro *et al.*, 2005).

3. Results

Some microtubule-containing cellular complexes can be extracted intact from tissue and their unique molecular organization can be studied by cryo-ET. A good example is the eukaryotic flagellum, a microtubule based macromolecular machine. Characteristic views of flagella and their axonemal cores are readily seen in cell sections, but to properly understand their global organization and to probe the molecular mechanism of their dynein-based contractility, three-dimensional reconstructions are needed. Flagella are large and complex and cannot be subjected to averaging procedures used in molecular EM. Instead, they are perfect specimens for cryo-ET and this approach has yielded great insight into their molecular organization and mechanism (Nicastro *et al.*, 2005, 2006; Sui and Downing, 2006; Fig. 7).

Striking views of microtubules inside cells have also been obtained by ET and have yielded insight into, for example, the organization and structural transitions that occur in the mitotic spindle (Winey *et al.*, 1995). ET studies of whole cells have also revealed unique features of microtubules *in situ*. These types of experiments are especially powerful when correlated with dynamic information from light microscopy of living cells (Garvalov *et al.*, 2006; Fig. 8).

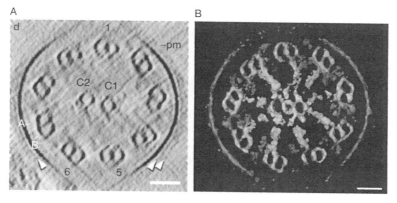

Fig. 7 Structure of eukaryotic flagella determined using cryo-ET. (A) Cross-sectional slice of cryo-ET reconstruction of sea urchin sperm flagellum. The central singlet microtubules are labelled C1 and C2, the (A) and (B) tubules are marked in each doublet, the plasma membrane is labelled pm and the extracellular material at the plasma membrane is indicated with arrow heads. Bar = 50 nm. (B) Visualisation of the three-dimensional reconstruction of the flagellum following axial averaging. Bar = 40 nm. Reprinted with permission from Nicastro *et al.*, 2005. Copyright 2005 National Academy of Sciences, USA.

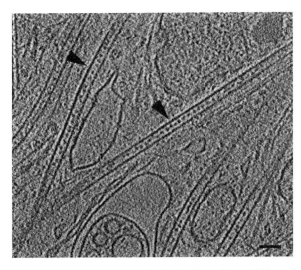

Fig. 8 Cryo-ET of neuronal processes reveals luminal particles within cellular microtubules. (A) A slice through a tomogram of a neuronal process imaged using cryo-ET reveals the presence of particles within the neuronal microtubules (arrowheads; Garvalov et al., 2006). Bar = 50 nm. Reproduced from Journal of Cell Biology, 2006, 174:759–765. Copyright 2006 The Rockefeller University Press.

IV. Outlook

The simplest EM experiments that characterize microtubules using negative stain can be undertaken without recourse to any of the latest technologies. Such experiments will always be the starting point for characterization of novel MAPs and their interaction with microtubules. In addition, all of the techniques described earlier will continue to contribute complementary information about microtubule form and function. Structure calculations that reveal very fine details will yield additional insights into the mechanistic subtleties by which microtubules are modulated by their interactions with MAPs. High resolution structure calculation will be possible through a combination of data collection on superior EMs with very stable stages, improvements in image processing software, and through the brute-force collection of large amounts of cryo-EM data such that only the very highest quality data is fed into structure calculations.

A key question, however, is how do such high resolution characteristics relate to the properties of microtubules in cells? We still have a great deal to learn about the behaviors of microtubules in specific cellular contexts, and ET is just beginning to open up new vistas of cellular architecture and environment to answer such questions. We can begin to convince ourselves that we understand many of the fundamentals of microtubule biology. However, we are just beginning to grasp the potential consequences of phenomena such as site-specific protein translation, post-translational modifications of tubulin and MAPs, and the idiosyncratic cell localization and cooperation seen for the hundreds of potential microtubule ligands. The ability to identify individual MAP molecules in a cellular context by both light and electron microscopy (using for example technologies described in Chapter 8 by Deerinck, this volume) and to visualize the effect that they have on individual microtubules will gradually enable us to understand the intricate mechanisms by which microtubules participate in cell function.

Acknowledgments

I thank the many scientists with whom I have collaborated on microtubules, especially my own group at Birkbeck College (Andy, Carsten, Christina, and Franck) and my mentor Ron Milligan (The Scripps Research Institute, La Jolla). I am also grateful to the B.B.S.R.C. for my David Phillips Fellowship.

References

Bouchet-Marquis, C., Zuber, B., Glynn, A. M., Eltsov, M., Grabenbauer, M., Goldie, K. N., Thomas, D., Frangakis, A. S., Dubochet, J., and Chretien, D. (2007). Visualisation of cell microtubules in their native state. *Biol. Cell* **99**, 45–53.

Chretien, D., Metoz, F., Verde, F., Karsenti, E., and Wade, R. H. (1992). Lattice defects in microtubules: Protofilament numbers vary within individual microtubules. *J. Cell Biol.* **117**, 1031–1040.

Desai, A., and Mitchison, T. J. (1997). Microtubule polymerisation dynamics. *Ann. Rev. Cell Dev. Biol.* **13**, 83–117.

Dubochet, J., Adrian, M., Chang, J. J., Homo, L. C., Lepault, J., McDowall, A. W., and Schultz, P. (1988). Cryo-electron microscopy of vitrified specimens. *Q. Rev. Biophys* **21**, 129–228.

Garvalov, B. K., Zuber, B., Bouchet-Marquis, C., Kudryashev, M., Gruska, M., Beck, M., Leis, A., Frischnecht, F., Bradke, F., Baumeister, W., Dubochet, J., and Cyrklaff, M. (2006). Luminal particles within cellular microtubules. *J. Cell Biol.* **174**, 759–765.

Gross, H., Krusche, K., and Tittmann, P. (1990). Recent progress in high resolution shadowing for biological samples. In "Proceedings of the XIIth International Congress for Electron Microscopy" (L. D. Peachery, and D. B. Williams, eds.), San Francisco Press, San Francisco, CA.

Guzik, B. W., and Goldstein, L. S. (2004). Microtubule-dependent transport in neurons: steps towards an understanding of regulation, function and dysfunction. *Curr. Opin. Cell Biol.* **16**, 443–450.

Hirokawa, N. (1986). Quick freeze, deep etch of the cytoskeleton. *Methods. Enzymol* **134**, 598–612.

Hirokawa, N., Shiomura, Y., and Okabe, S. (1988). Tau proteins: The molecular structure and mode of binding on microtubules. *J. Cell Biol.* **107**, 1449–1459.

Hirose, K., Akimuru, E., Akiba, T., Endow, S. A., and Amos, L. A. (2006). Large conformational changes in a kinesin motor catalysed by interaction with microtubules. *Mol. Cell* **23**, 913–923.

Hoenger, A., and Nicastro, D. (2007). Electron microscopy of microtubule-based cytoskeletal machinery. *Methods Cell Biol.* **79**, 437–462.

Höög, J. L., and Antony, C. (2007). Whole-cell investigation of microtubule cytoskeleton architecture by electron tomography. *Methods Cell Biol.* **79**, 145–167.

Hyman, A. A., Salser, S. S., Dreschel, D. N., Unwin, N., and Mitchison, T. J. (1992). Role of GTP hydrolysis in microtubule dynamics: information from a slowly hydrolysable analogue GMPCPP. *Mol. Biol. Cell* **3**, 1155–1167.

Kikkawa, M., and Hirokawa, N. (2006). High-resolution cryo-EM maps show the nucleotide binding pocket of KIF1A in open and closed conformations. *EMBO J.* **25**, 4187–4194.

Krzysiak, T. C., Wendt, T., Sproul, L. R., Tittmann, P., Gross, H., Gilbert, S. P., and Hoenger, A. (2006). *EMBO J.* **25**, 2263–2273.

Kuhlbrandt, W., and Williams, K. A. (1999). Analysis of macromolecular structure and dynamics by electron cryo-microscopy. *Curr. Opin. Chem. Biol.* **3**, 537–543.

Ledbetter, M. C., and Porter, K. R. (1963). A "microtubule" in plant cell fine structures. *J. Cell Biol.* **19**, 239–250.

Li, H., DeRosier, D. J., Nicholson, W. V., Nogales, E., and Downing, K. H. (2002). Microtubule structure at 8Å resolution. *Structure* **10**, 1317–1328.

Löwe, J., Li, H., Downing, K. H., and Nogales, E. (2001). Refined structure of alpha beta-tubulin at 3.5Å resolution. *J. Mol. Biol.* **313**, 1045–1057.

Lucic, V., Forster, F., and Baumeister, W. (2005). Structural studies by electron tomography: from cells to molecules. *Ann. Rev. Biochem* **74**, 833–865.

Mandelkow, E. M., Mandelkow, E., and Milligan, R. A. (1991). Microtubule dynamics and microtubule caps: a time-resolved cryo-electron microscopy study. *J. Cell Biol.* **114**, 977–991.

McIntosh, R., Nicastro, D., and Mastronade, D. (2005). New views of cells in 3D: an introduction to electron tomography. *Trends Cell Biol.* **15**, 43–51.

Moores, C. A., Yu, M., Guo, J., Beraud, C., Sakowicz, R., and Milligan, R. A. (2002). A mechanism for microtubule depolymerisation by KinI kinesins. *Mol. Cell* **9**, 903–909.

Moores, C. A., Perderiset, M., Francis, F., Chelly, J., Houdusse, A., and Milligan, R. A. (2004). Mechanism of microtubule stabilisation by doublecortin. *Mol. Cell* **18**, 833–839.

Nicastro, D., McIntosh, J. R., and Baumeister, W. (2005). 3D structure of eukaryotic flagella in a quiescent state revealed by cryo-electron tomography. *Proc. Natl. Acad. Sci.* **102**, 15889–15894.

Nicastro, D., Schwartz, C., Pierson, J., Gaudette, R., Porter, M. E., and McIntosh, J. R. (2006). The molecular architecture of axonemes revealed by cryoelectron tomography. *Science* **313**, 944–948.

Nogales, E., Wolf, S. G., and Downing, K. H. (1998). Structure of the alpha beta tubulin dimer by electron crystallography. *Nature* **391**, 199–203.

Nogales, E. (2000). Structural insights into microtubule function. *Ann. Rev. Biochem* **69**, 277–302.

Ray, S., Meyhofer, E., Milligan, R. A., and Howard, J. (1993). Kinesin follows the microtubule's protofilament axis. *J. Cell Biol.* **121**, 1083–1093.

Rice, S., Lin, A. W., Safer, D., Hart, C. L., Naber, N., Carragher, B. O., Cain, S. M., Pechatnikova, E., Wilson-Kubalek, E. M., Whittaker, M., Pate, E., Cooker, R., Taylor, E. W., *et al.* (1999). A structural change in the kinesin motor protein that drives motility. *Nature* **402**, 778–784.

Saibil, H. R. (2000). Conformation changes studied by cryo-electron microscopy. *Nat. Struct. Biol.* **7**, 711–714.

Sandblad, L., Busch, K. E., Tittmann, P., Gross, H., Brunner, D., and Hoenger, A. (2006). The Schizosaccharoyces pombe EB1 homolog Mal3p binds and stabilises the microtubule lattice seam. *Cell* **127**, 1415–1424.

Shelanski, M. L., Gaskin, F., and Cantor, C. R. (1973). Microtubule assembly in the absence of added nucleotides. *Proc. Natl. Acad. Sci.* **70**, 765–768.

Stewart, M., and Vigers, G. (1986). Electron microscopy of frozen-hydrated bioloigical material. *Nature* **319**, 631–636.

Subramaniam, S., and Milne, J. L. (2004). Three-dimensional electron microscopy and molecular resolution. *Ann. Rev. Biophys. Biomol. Struct* **33**, 141–155.

Sui, H., and Downing, K. H. (2006). Molecular architecture of axonemal microtubule doublets revealed by cryo-electron tomography. *Nature* **442**, 475–478.

Svitkina, T. M., Verkhovsky, A. B., and Borisy, G. G. (1996). Plectin sidearms mediate interaction of intermediate filaments with microtubules and other components of the cytoskeleton. *J. Cell Biol.* **135**, 991–1007.

Szajner, P., Weisberg, A. S., Lebowtz, J., Heuser, J., and Moss, B. (2005). External scaffold of spherical immature poxvirus particles is made of protein trimers, forming a honeycomb lattice. *J. Cell Biol.* **170**, 971–981.

Tilney, L. G., Bryan, J., Bush, D. J., Fujiwara, K., Mooseker, M. S., Murphy, D. B., and Snyder, D. H. (1973). Microtubules: Evidence for 13 protofilaments. *J. Cell Biol.* **59,** 267–275.

Unger, V. M. (2001). Electron cryomicroscopy methods. *Curr. Opin. Struct. Biol.* **11,** 548–554.

Wang, H. W., and Nogales, E. (2005). Nucleotide-dependent bending flexibility of tubulin regulates microtubule assembly. *Nature* **435,** 911–915.

Winey, M., Mamay, C. L., O'Toole, E. T., Mastronade, D. N., Giddings, T. H., Jr., McDonald, K. L., and McIntosh, J. R. (1995). Three-dimensional ultrastructural analysis of the Saccharomyces cerevisae mitotic spindle. *J. Cell Biol.* **129,** 1601–1615.

Wittmann, T., Hyman, A., and Desai, A. (2001). The spindle: A dynamic assembly of microtubules and motors. *Nat. Cell Biol.* **3,** E28–E34.

SECTION 3

Extracellular Matrix and Cell Junctions

CHAPTER 17

Electron Microscopy of Collagen Fibril Structure *In Vitro* and *In Vivo* Including Three-Dimensional Reconstruction

Tobias Starborg, Yinhui Lu, Karl E. Kadler, and David F. Holmes

Wellcome Trust Centre for Cell-Matrix Research
Faculty of Life Sciences
University of Manchester
Manchester M13 9PT
United Kingdom

Abstract
I. Introduction
   A. Extracellular Collagen Matrix
   B. Visualization of Individual Procollagen Molecules by Rotary Shadowing Electron Microscopy
   C. 3D Reconstruction from Serial Sections
   D. 3D Reconstruction from Tilt Series
II. Electron Microscopy of Isolated Collagen Fibrils
   A. Fibrils Formed from Purified Type I Collagen
   B. Axial Structure of Collagen Fibrils
   C. Fibrils Isolated from Tissue
   D. 3D Structure of Fibrils

III. Fibroblast/Fibril Interface in Developing Tendon
    A. Extracellular Channels Containing Bundles of Collagen Fibrils
    B. Fibripositors
    C. Fibripositor-Cell Tomography
  IV. Discussion
    A. Future Work
  V. Conclusions
    References

Abstract

Tissue development in multicellular animals relies on the ability of cells to synthesize an extracellular matrix (ECM) containing spatially organized collagen fibrils, whose length greatly exceeds that of individual cells. The importance of the correct regulation of fibril deposition is exemplified in diseases such as osteogenesis imperfecta (caused by mutations in collagen genes), fibrosis (caused by ectopic accumulation of collagen), and cardiovascular disease (which involves cells and macromolecules binding to collagen in the vessel wall). Much is known about the molecular biology of collagens but less is known about collagen fibril structure and how the fibrils are formed (fibrillogenesis). This is explained in part by the fact that the fibrils are noncrystalline, extensively cross-linked, and very large, which makes them refractory to study by conventional biochemical and high-resolution structure-determination techniques. Electron microscopy has become established as the method of choice for studying collagen fibril structure and assembly. This article describes the electron microscopic methods most often used in studying collagen fibril assembly and structure.

I. Introduction

A. Extracellular Collagen Matrix

The extracellular matrix (ECM) of animal tissues is essentially a complex fiber-composite material in which collagen fibrils are a major component (Goh et al., 2007). The fibrils are the major load-bearing scaffold in the ECM and their diameter, length, and spatial arrangement vary according to the type of tissue and stage of development (Parry et al., 1978). The fibrils in cornea, for example, are arranged in layers of uniformly spaced and parallel fibrils; the layers are orthogonal to the light direction and rotated 90° between adjacent layers (Bergmanson et al., 2005). In contrast, the fibrils in tendon are aligned predominantly parallel to the long-axis of the tissue (Kannus, 2000). A typical transmission electron microscope (TEM) image of a transverse section of embryonic vertebrate tendon having well-defined bundles of collagen fibrils of uniform diameter and spacing is shown in Fig. 1. Images such as these pose questions such as what is the mechanism of

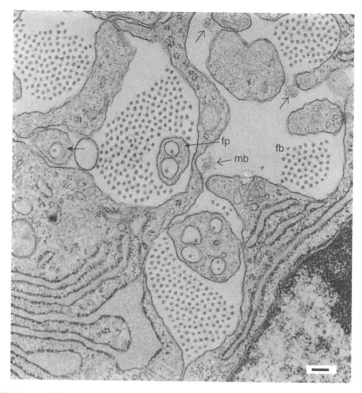

Fig. 1 TEM image of a transverse 80 nm section of 15.5 day mouse embryonic tail tendon. Collagen fibril bundles (one example is labeled fb) are prominent in the extracellular spaces. The collagen fibrils at this stage of tendon development have a uniform diameter (mean 28 nm, s.d. = 4 nm). The fibrils have a near-hexagonal arrangement in bundles with a regular centre-to-centre spacing (mean 58 nm, s.d. = 9 nm). Some isolated fibrils are seen surrounded by a plasma membrane (marked by closed arrows with one example labeled fp) and form characteristic structures that have been called "fibripositors". Bundles of fibrillin microfibrils (marked by open arrows with one labeled mb) are also apparent and one or two of these bundles are frequently found next to each collagen fibril bundle. Scale bar = 300 nm.

fibril assembly?; what is the mechanism of diameter regulation?; how is parallelism achieved and maintained?; what are the relative contributions of collagen fibril self-assembly and cellular control of fibril organization?; and how is fibrillogenesis coupled to tissue assembly during embryonic development and repair? It has been proposed that long-range spatial order of collagen fibrils in tissue might result from molecular interactions involving liquid crystal-like structure (Mosser *et al.*, 2006). Although fibrils of limited diameter can be formed from purified components in the absence of cells, the spatial arrangement of fibrils is disorganized. The current consensus is that the spatial-patterning of collagen fibrils in developing tissues is an example of cell-regulated protein self-assembly, although the cellular machinery that directs this process has not been identified.

B. Visualization of Individual Procollagen Molecules by Rotary Shadowing Electron Microscopy

Vertebrates contain at least 28 different collagens, of which types I, II, III, V, XI, XXIV, and XXVII belong to the fibril-forming subgroup. Type I collagen is the most abundant and is the focus of the methods described in this chapter. A detailed description of collagen biosynthesis is outside the scope of this article but several reviews are available (Canty and Kadler, 2005; Huxley-Jones et al., 2007; Kadler et al., 2007; Myllyharju and Kivirikko, 2004). In brief, fibrillar collagens are synthesized as procollagens comprising three polypeptide chains wound into an uninterrupted triple helix that is ~300 nm in length. The main triple helix is flanked by propeptides that are removable by procollagen N- and C-proteinases (Fig. 2). The C-propeptides are highly soluble and bulky, which makes cleavage of the C-propeptides a prerequisite for normal collagen fibril assembly (Kadler et al., 1987; Suzuki et al., 1996). Removal of the N-propeptides of type I procollagen (the precursor of type I collagen) is not an absolute requirement for fibril formation but failure to do so can result in skin and joint hyperextensibility in individuals with the Ehlers-Danlos syndrome (for review see Colige et al., 1999).

The extended structure of the procollagen molecule can be visualized using low-angle rotary-shadowing electron microscopy in which evaporated platinum is used to contrast the molecules (Fig. 2A, see Mould et al., 1985 for a detailed protocol). Under physiological conditions of buffer and temperature, the procollagen molecules are elongated semiflexible molecules containing globular propeptides at each end. Evidence from TEM suggests that the procollagen molecules can laterally aggregate in the secretory pathway (Hulmes et al., 1983; Weinstock and Leblond, 1974). Furthermore, the procollagen N- and C-proteinases are activated to their mature forms in the *trans*-Golgi network (Leighton and Kadler, 2003; Wang et al., 2003) and can cleave procollagen to collagen within the cell, prior to secretion (Canty et al., 2004, 2006). Identification of where in the secretory pathway the procollagen is cleaved has proved difficult because antibodies to the active forms of the proteinases are not available. However, progress has been made using three-dimensional (3D) reconstruction electron microscopy. Both 3D reconstruction from serial sections and electron tomographic reconstruction from tilt series of semithick sections are giving new insights into the spatial organization of proteins in the ECM. Other methods, including TEM of freeze-fracture deep-etched samples (Barge et al., 1991; Hirsch et al., 1999) as well as scanning electron microscopy (Provenzano and Vanderby, 2006; Raspanti et al., 2007), have also provided insight into the 3D structure of the ECM.

C. 3D Reconstruction from Serial Sections

A major drawback of TEM is the necessity for the sample to be thin. In an instrument with an accelerating voltage of 120 kV, the specimen should be no thicker than ~100 nm. It is possible to image thicker samples, say 500 nm, in

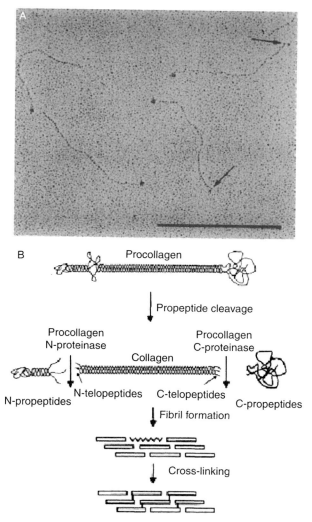

Fig. 2 (A) Transmission electron image of type I procollagen after freeze-drying on freshly cleaved mica and low-angle (5°) rotary shadowing with platinum. The globular C-propeptides are consistently visible as blobs at one end of the linear molecule. The extended N-propeptide can be folded back along the main collagen triple helix or be extended at angle; two examples of the latter conformation are marked by solid arrows. The scale bar corresponds to 300 nm. From Mould *et al.* (1985). (B) Flow chart to show the processing steps to convert procollagen to collagen by specific protease cleavage of the propeptides prior to fibril assembly. The fibril assembly occurs by multiple weak interactions and the fibrils are subsequently stabilized by intermolecular cross-link formation. From Holmes *et al.* (2001b).

higher accelerating voltage instruments (≥ 300 kV) especially with the use of energy filtration. 3D reconstruction from serial sections allows the structure of a greatly extended volume to be obtained. Essentially, a 3D image is generated by accumulation of a succession of 2D section images (Hoffpauir et al., 2007). The z-resolution is limited to twice the section thickness but the z-range can extend over hundreds of sections; a large set of serial sections could run to ~500, giving a range of 50 μmin the tissue. An important practical requirement is the need to avoid folding and distortion of the sections. Image recording on film gives a relatively large field of view. Each negative can be digitized at ~4 k × 5 k, and for a 1 nm pixel size this corresponds to 4 × 5 μm specimen area. However, the process of recording a large number of images on film, involving multiple cassette changes, and subsequent high-resolution scanning of film images can extend over several days. Direct capture of images on digital camera is a more time-efficient method. To maintain the image resolution and field of view it is necessary to capture a slightly overlapping raster of images for each section and stitch these together into a montage image (Ma et al., 2007). Typically, for a 2 k × 2 k camera, a montage of 2 × 3 images would be needed to match the information on a single micrograph on film.

The process of generating useful 3D-structural information is illustrated in Fig. 3, which shows a study of transverse sections of embryonic tendon. The first step in generating the 3D reconstruction involves alignment of the images, applying an x-y shift and rotation to bring neighboring images into alignment (assuming there is no significant change in magnification). The MIDAS routine in the IMOD software package (Kremer et al., 1996) supports this image alignment operation. The second, more time-consuming, process is the generation of a 3D model of the structures of interest. The various structures of interest have to be outlined (the segmentation process) in each serial section. At worst this involves tracing around the boundaries by hand, which is simplified by using a graphics screen tablet. This segmentation can often be semiautomated (Sandberg, 2007).

D. 3D Reconstruction from Tilt Series

3D reconstruction from tilt series (electron tomography) is possible from a wide range of specimen types (McIntosh et al., 2005) including plastic embedded sections (O'Toole et al., 2007), negatively stained fibrils/microfibrils (Baldock et al., 2001, 2002), and frozen hydrated samples (cryo-electron tomography; (Nickell et al., 2006). A set of electron images is obtained for a tilt series of the specimen using computer-controlled image acquisition (Koster et al., 1992). After alignment of the images the 3D structure, or tomogram, is computed by r-weighted back-projection or alternative methods (Lawrence et al., 2006; Tong et al., 2006). In practice the angular range is restricted to less than ±70° and the loss of high-tilt projection leads to aberrations .in the tomogram, the "missing-wedge" effect (Fig. 4). This can be reduced by the use of dual-axis orthogonal tilt series and the combination of the two resultant tomograms (Mastronarde, 1997). This procedure greatly improves the visibility of surfaces and filaments that lie in the plane of the section. Even with this procedure the resolution in the Z-direction is generally less than half of that in the x-y plane. The application of electron

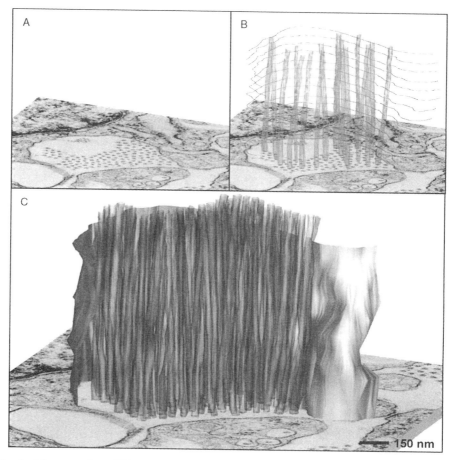

Fig. 3 The generation of 3D reconstruction of a fibril bundle and surrounding cell membrane from a set of serial sections of embryonic mouse tail tendon. (A) Individual transverse sections are imaged in the microscope. (B) By tracing the cell membrane and a few collagen fibril outlines the computer is able to generate a wire-frame model. (C) The remaining fibrils are traced and a "skin" is overlaid making it possible to view a portion of the tendon fibril bundle. The tendon runs top to bottom, with the fibrils aligned parallel to the fibril axis. Adapted from Starborg and Kadler (2005). (See Plate no. 36 in the Color Plate Section.)

tomography to isolated negatively stained collagen fibrils can be performed on a standard 120 kV instrument. The application to tissue sections of tendon however are usually on semi-thick (~300 nm) sections and requires an intermediate-voltage (300 kV) instrument. Both applications (see later sections) have employed automated tilt series acquisition software, either EM-MENU3 (TVIPS, Gauting) or SerialEM (3DEM Centre, University of Colorado at Boulder).

A typical 3D reconstruction requires hundreds of images, either tilt series, or images of serial sections. The process of acquisition of the images on the electron

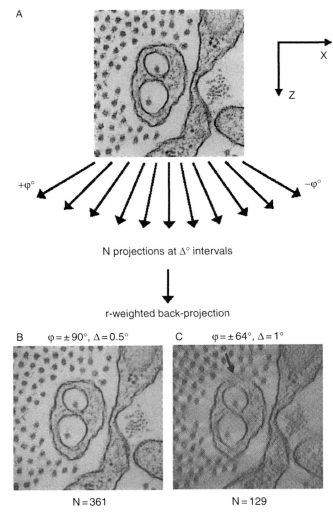

Fig. 4 Process of electron tomography by weighted back-projection from a tilt series of projections. The test image (A) is from the transverse section shown in Fig. 1, but it is used to represent the x-z plane through a longitudinal thick section. Projections have been computed from the test image and reconstructions calculated by r-weighted back-projection for a range of projection angles ($\pm\phi$) and values of angular increments (Δ). (B) A full range of projections ($\pm 90°$) at a sampling interval of $0.5°$ yields a reconstruction virtually indistinguishable from the original image. In practice both the projection range and number of projections are limited. (C) Shows a reconstruction for a typical angular range of $\pm 64°$ and angular increment of $1°$. Loss of high-tilt projections ("missing wedge" effect) leads to a blurring in the Z-direction and poor visibility of linear features running in the X-direction—see plasma membrane marked with an arrow in (C). Increasing the angular sampling interval (reducing the number of projections) leads to overall loss of resolution.

microscope takes ~1 h. Alignment of the images set takes 2–3 h. The time required to compute the 3D-reconstruction (tomogram) from the aligned images is typically ~1 h for r-weighted back-projection with 2 k × 2 k images. The final step of model generation is however the most time intensive. Manual tracing of cell membranes, other intracellular features and extracellular components from one 3D image data set can take several days. More advanced automated segmentation procedures are however being developed (Sandberg, 2007).

II. Electron Microscopy of Isolated Collagen Fibrils

A. Fibrils Formed from Purified Type I Collagen

It has long been known that native-like collagen fibrils will self-assemble from purified type I collagen molecules in warm, neutral solution (Gross and Kirk, 1958). The resultant fibrils (see Fig. 5) have a polarized arrangement of molecules, an axial periodicity of 67 nm, and a similar stain pattern to type I collagen fibrils in tissue. The diameter distribution of the fibrils in the reconstituted gel is dependent on the temperature, ionic strength, and collagen concentration (Wood and Keech, 1960). The reconstituted fibril diameters are generally greater than those observed

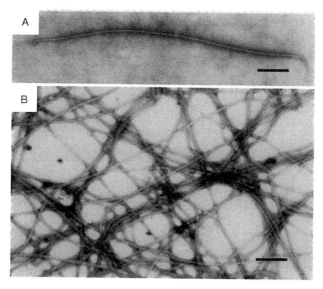

Fig. 5 Reconstitution of native-banded collagen fibrils from extracted and purified type I collagen. (A) A typical "early fibril" precursor to the final gel. These are observed in large numbers in the initial stages of gel formation. The fibrils are narrow (~25 nm in diameter), D-periodic, unipolar and have two smoothly tapered tips. (B) Part of the final gel showing a mesh-work of D-periodic fibrils. There is some local clustering of fibrils but the gel has no long-range structural order. No fibril ends are visible at this final stage of self-assembly. From Holmes and Kadler (2006a). Scale bars: 500 nm.

in embryonic vertebrate tissues and have a broader distribution. Fibril assembly is critically dependent on the extra-helical terminal regions (telopeptides) of the collagen molecule. Partial loss of these amino acids leads to a changed morphology and/or loss of molecular polarity; extensive loss can completely inhibit fibril formation (Capaldi and Chapman, 1982).

Different assembly pathways have been observed in the reconstitution of collagen fibrils. The transfer route used to raise the pH and temperature from the initial conditions (cold, acid solution) to the fibril reconstitution conditions (warm, neutral solution) is a major determinant of the assembly pathway (Holmes *et al.*, 1986). Warming before neutralization leads to abundant short "early fibrils" with tapered tips (see Fig. 5A). These fibrils have a native banding pattern and are typically ~25 nm in diameter. Similar early fibrils (alternatively called "fibril segments") are also observed in embryonic connective tissues (see Section C, below). In contrast, neutralization followed by warming leads to the early accumulation of unbanded fine filaments (Gelman *et al.*, 1979).

B. Axial Structure of Collagen Fibrils

The axial arrangement of 300 nm-long collagen molecules to produce fibrils with an axial periodicity of 67 nm, known as the quarter-stagger model (Hodge, 1989) and references therein, is shown in Fig. 6A. This arrangement generates the gap-overlap zones of each D-period, which can be identified in the negatively-stained fibril (Fig. 6B). Both the positive and negative stain patterns can be predicated in detail from the amino-acid sequence of the collagen molecules (Chapman *et al.*, 1990). This analysis has been extended to the axial structure of heterotypic fibrils

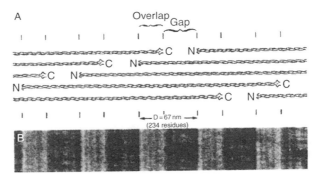

Fig. 6 Axial structure of type I collagen fibrils. (A) Schematic diagram to show the axial arrangement of collagen molecules in the native fibril. Short nontriple-helical domains ("telopeptides") are located at the end of the 300 nm-long triple-helical domain. (B) Characteristic negative stain pattern (using 1% sodium phosphotungstate, pH 7) of a native-type, D-periodic collagen fibril reconstituted from acetic-acid-soluble type I collagen. This is shown at the same magnification and aligned with the schematic axial structure in (A). The gap/overlap structure of the fibril and the fine stain excluding telopeptide regions at the gap/overlap junctions are apparent. From Holmes and Kadler (2006a).

from cartilage (containing collagen types II, IX, and XI) and yielded compositional analysis of fibrils of varying diameters (Bos *et al.*, 2001). Analysis of the negative stain pattern has also led to the identification of an N,N-bipolar form of the type I collagen fibril as described in the next section.

C. Fibrils Isolated from Tissue

Isolation of intact collagen fibrils from vertebrate tissues has been limited to embryonic tissues (Birk *et al.*, 1995; Holmes *et al.*, 1994). These fibrils have lengths typically in the range of 2–200 μm, possess smoothly tapered tips, and have a shaft region of narrow near-uniform diameter. Examination of the staining patterns of these entire fibrils has revealed the N,N-bipolar fibril (Fig. 7; Holmes *et al.*, 1994) in addition to the expected unipolar fibril. The N,N-bipolar fibril has both tips with collagen molecules oriented N-C from the fibril end inwards and a well-defined central region along the axis, the "polarity transition zone," with an antiparallel packing of the collagen molecules. These "early" collagen fibrils from tissue, both unipolar and bipolar are prevented from fusing laterally by certain small

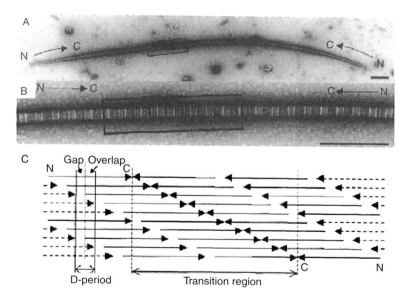

Fig. 7 N,N-bipolar collagen fibrils isolated from tissue. (A) Negatively-stained N,N-bipolar collagen fibril with two smoothly tapered tips. Inspection of the stain pattern in every D-period shows that the tips have collagen molecules oriented N-C from the fibril end inwards. There is a well-defined central region marked by the box where the stain pattern indicates an anti-parallel packing of collagen molecules. (B) Enlargement of the central part of the fibril image to show the 8D-long polarity transition region. (C) Diagram showing the axial arrangement of collagen molecules in the polarity transition region. From Holmes *et al.* (2001b).

proteoglycans bound at low concentration to specific surface sites (Graham et al., 2000). The regulation of lateral growth (diameter limitation) in these fibrils has been studied by means of mass-per-unit length measurements in the Scanning Transmission Electron Microscope (STEM; Holmes et al., 1998). Noteworthy, the collagen fibrils from mature echinoderm dermis and ligament can be isolated with lengths up to 1 mm (Trotter et al., 2000b) and these fibrils are exclusively N,N-bipolar (Thurmond and Trotter, 1994). Furthermore, the mass profiles measured on these fibrils in the length range of 14–444 µm are consistent with a growth model in which the polarity transition region serves as a surface nucleation site for lateral growth (Trotter et al., 1998, 2000a).

D. 3D Structure of Fibrils

X-ray diffraction determination of collagen fibril structure has been limited to the cases of mature rat-tail (Wess, 2005) and lamprey notochord (Eikenberry et al., 1984), which are unique in producing detailed fiber diffraction pattern from the collagen matrix. X-ray diffraction studies on rat-tail tendon over a 30-year period have yielded a variety of models for the fibril structure. A quasi-hexagonal packing model (Hulmes and Miller, 1979) has been further developed into a microfibrillar structural model (Orgel et al., 2006). Electron tomography offers an alternative approach in which the 3D structure of individual collagen fibrils from any tissue can be studied. The structure and location of noncollagen macromolecules bound to the fibril surface can also be visualized as well as the 3D-arrangement of collagen microfibrils. Here we describe the application of electron tomography to studies of the structure of collagen fibrils from the mature vertebrate corneal stroma (Holmes et al., 2001a) and to the structure of thin collagen fibrils from embryonic vertebrate cartilage (Holmes and Kadler, 2006b).

Corneal collagen fibrils are heterotypic structures composed of type I collagen molecules co-assembled along with those of type V collagen (Birk et al., 1988). In cornea, the mature form of type V collagen retains a large N-terminal propeptide domain (Linsenmayer et al., 1993), which has been implicated in fibril diameter limitation (Birk, 2001 and references therein). Fibrils were isolated from bovine cornea by crushing the tissue in liquid nitrogen and then dispersion of the powder using a Dounce homogenizer. Fibrils were then negatively stained with 5% uranyl acetate on carbon-filmed grids (Fig. 8A). By using automated low-dose technique, tilt series were recorded from 12 different fibrils over a tilt range of ±70° at 1 or 2° intervals, using a 200 kV FEG electron microscope, at an instrumental magnification of 20 K with a total accumulated dose of typically ∼600 electrons/Å^2. Automated low-dose acquisition of tilt series with standard tracking and focusing procedures was controlled by an external computer running the microscope control and image acquisition software (TVIPS, Gauting, Germany). Images were recorded on a cooled CCD camera (1024 × 1024 pixels). Reconstruction of 3D volumes from tilt series was performed by the weighted back-projection method by using the IMOD software package (Kremer et al., 1996). Fourier transforms and

17. Electron Microscopy of Collagen Fibril Structure

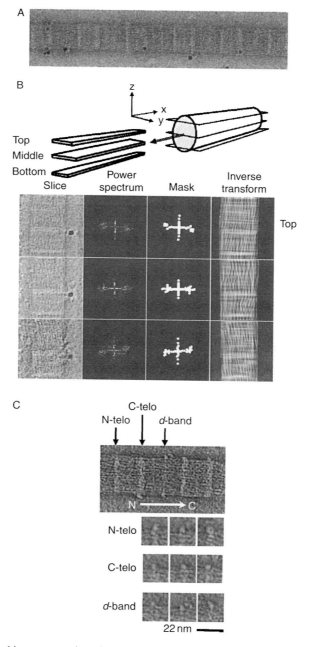

Fig. 8 Tomographic reconstruction of a corneal collagen fibril. (A). TEM image of untilted corneal collagen fibril negatively stained with 5% aqueous uranyl acetate. Gold colloid (10 nm) has been added to the grid sample to provide fiducial markers to aid the alignment of projection series prior to computing the 3D reconstruction. (B) Visualization of microfibrillar structure. Longitudinal virtual

Fourier filtering on virtual slice image data were done by using the SEMPER image analysis software (Synoptics, Cambridge, UK). For analysis of the longitudinal slices of the reconstruction, Fourier filtering was used to enhance the signal-to-noise ratio and to allow clearer visualization of microfibrillar substructure. Figure 8B, shows x–y slices through the top, middle, and bottom of a typical tomogram. Visual inspection of x–y slices through the reconstructions shows that the fibrils are constructed from narrow diameter microfibrils. Fourier transform intensities of x–y slices corresponding to the top and bottom of the fibril (i.e., high and low values on the z axis, respectively, relative to the carbon support film on which the fibril was resting) exhibit equatorial reflections that were consistent with tilted microfibrils. To display the arrangement of microfibrils, we produced a Fourier mask (Fig. 8B "mask") and inverted the filtered transform (Fig. 8B "inverse transform"). The inverted transforms show that the microfibrils are tilted, in a right-hand orientation, with respect to the fibril long axis. Fibril surface-bound particles were observed in the virtual slices of the fibril tomograms (Fig. 8C). These were located at three axial locations in the D-period, corresponding to the location of the N- and C-telopeptides and to a mid-gap location ("d-band").

Another application of 3D structure determination is in the thin heterotypic collagen fibrils that occur in cartilage and are composed of three collagen types (II, IX and XI). These fibrils, about 15 nm in diameter, can be isolated from tissue (see Fig. 9A) and have recently been studied by STEM mass-mapping, analysis of the axial negative stain pattern and electron tomography (Holmes and Kadler, 2006b). The axial stain pattern is consistent with a II:IX:XI molecular ratio of 8:1:1 and the fibril mass-per-unit length is consistent with 14×5-stranded microfibrils in cross-section. Initial attempts at 3D reconstruction using a similar electron tomography procedure as described earlier for the corneal collagen fibrils, failed to directly visualize microfibrils in the transverse section of the thin cartilage fibril. If, however, the microfibrils are arranged in an inner core and outer shell each with rotational symmetry then only a limited angular range of projections is needed for reconstruction of the transverse structure as shown in Fig. 9B–E. The reconstruction does not then suffer from the "missing wedge" effect and the microfibrils can

---

slices (x–y) through the 3D reconstruction sampling of the fibril in the top, middle, and bottom zones, as indicated schematically. The raw slice images are shown together with the power spectra masks that include the main peak intensities and the Fourier-filtered images generated by using these masks. A filamentous substructure is apparent in the original images and this is enhanced after filtering. The filaments show a predominant tilt of about $+15°$ and $-15°$ in the upper and lower zones of the fibril, respectively. The tilt direction changes rapidly in the central zone. Both tilt components can be seen in the middle slice. (C) Visualization of surface-bound macromolecules. The upper image is part of a central slice of a 3D reconstruction of a negatively stained collagen fibril, showing macromolecules bound at preferred axial locations along the fibril. The arrow shows the molecular polarity of the fibril. The gallery shows views of nine individual macromolecules that were bound to N-telopeptides (N-telo), C-telopeptides (C-telo), and the gap zone. Three macromolecules are shown for each location. Note the doughnut (ring-shaped) structure at the N-telopeptides and the tadpole-shaped molecule bound to the C-telopeptides. The macromolecules bound to the gap zone were smaller and more conspicuous than those bound to the telopeptides. Adapted from Holmes et al. (2001a).

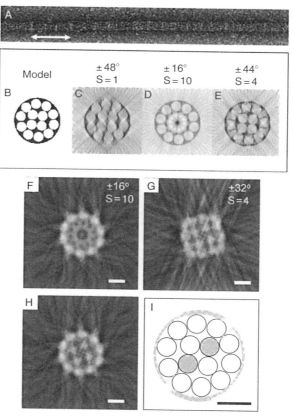

Fig. 9 (A) TEM image of a typical thin collagen cartilage fibril, about 15 nm in diameter, isolated from 14-day chick embryonic sternum and negatively stained with 4% uranyl acetate/1% trehalose. A single D-period is indicated. Light bands correspond to stain excluding regions. These heterotypic fibrils contain collagen types II, IX, and XI and the negative stain pattern differs from that of type I collagen fibrils. (B–E) Simulated r-weighted back-projection reconstructions of a test fibril containing 14 microfibrils in cross-section, consistent with measurements of mass-per-unit length. The model (B) has an outer shell with a 10-fold (S = 10) rotational symmetry and an inner core has a 4-fold (S = 4) rotational symmetry. (C) The reconstruction from a tilt series over ±48° showing "missing-wedge" artifacts. (D) Reconstruction using projections over ±16° and applying 10-fold rotational symmetry, showing the arrangement of microfibrils in the outer shell of the model. (E) Reconstruction using projections over ±44° and applying 4-fold rotational symmetry, showing the arrangement of microfibrils in the inner core of the model. (F, G, H) Experimental reconstructions of the transverse section of the negatively-stained thin cartilage fibril by applying r-weighted backprojection with rotational symmetry. (F) Reconstruction showing the arrangement of microfibrils in the outer shell. (G) Reconstruction showing the arrangement of microfibrils in the inner core. (H) Composite image combining the inner core region of (G) with the outer shell region of (F). (I) Schematic model of the transverse structure of the thin cartilage fibril. This transverse section corresponds to the axial location of the nontriple-helical component of the collagen-XI N-propeptide. The structural components are five-stranded microfibrils of collagen type II (clear circles) and collagen type XI (gray circles). Each microfibril of type XI collagen would result in one N-propeptide per D-period and the hypothetical circumferential extents of these domains are shown. The dotted lines show the projection of the tilted minor triple helix of the N-propeptides. Scale bars: 5 nm. Adapted from Holmes and Kadler (2006b).

be clearly located. Figure 9F–H show the application of this method to the negatively stained thin cartilage fibrils. The reconstructions show the arrangement of microfibrils in the transverse section of the fibril but do not however show the surface components, which include molecular strands of type IX collagen and the nontriple helical component of type XI collagen. Figure 9I is a schematic structure to show the transverse arrangement of type II and type XI microfibrils in the thin fibril. The inner location of the type XI triple helices and outer location of the type XI N-propeptide are consistent with immunolabeling studies (Keene et al., 1995). The direct visualization of the nonmicrofibrillar surface structures, including type IX collagen molecules and the extrahelical N-propeptides of type XI remain as a challenging further phase of study.

These studies demonstrate the power of automated electron tomography in the study of individual elongated protein polymers. Importantly, this technique allows "individual" collagen fibrils to be examined, which is particularly relevant to examining domains of collagen fibrils (e.g., the molecular switch region of N,N-bipolar fibrils or the tapering tips of fibrils) that would not be possible by other high-resolution structural techniques.

III. Fibroblast/Fibril Interface in Developing Tendon

A. Extracellular Channels Containing Bundles of Collagen Fibrils

A first step in examining the ultrastructual detail of cell–cell and cell–matrix interactions is to embed the tissue in plastic and take a single section. A single section oriented at 90° to the long axis of a developing tendon gives some insight into how the cells are packed together, and how they interact with the surrounding matrix (Fig. 10A). The image reveals that the developing tendon is highly cellular, showing bundles of collagen fibrils packed in gaps between the developing cells. The cell cross-sections have a convoluted plasma membrane with relatively large, almost circular, invaginations at the cell surface and some fine cellular protrusions reaching out to neighboring cells. The single section gives some indications of the complexity of the cell–cell interactions that appear to be surrounding the collagen bundles. However, the single section gives little information about the overall 3D shape of the cells in the tissue. From the information within one section it is easy to think that the cells would look the same irrespective of the section orientation, i.e., the cell might appear roughly spherical with random invaginations across its surface.

When serial sections of the same region are taken and aligned it is possible to reconstruct the original shape of the cells in 3D (Fig. 10B). The 3D view clearly shows that the cells are long and cylindrical with channels running down their sides (Fig. 10D, with the corresponding section shown in 10C). The channels are formed by a mixture of intracellular and intercellular membrane interactions. The intracellular interactions are formed when plasma-membrane lamella from a single cell form a sheath around the collagen bundle (Fig. 10D open arrow). These structures are similar to those described previously (Birk and Trelstad, 1986).

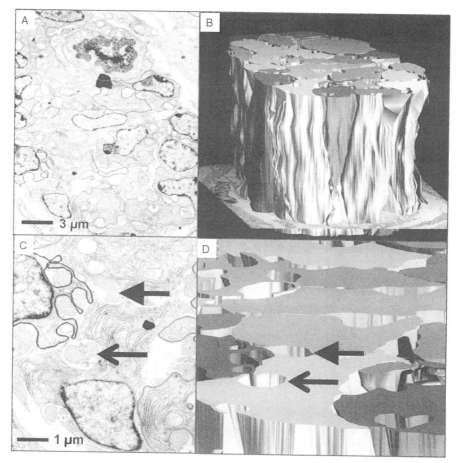

Fig. 10 3D Reconstruction of Embryonic Tendon cells. (A) Standard transmission electron micrograph showing a transverse section across an embryonic (15.5 dpc) mouse tail tendon. The image shows one of a series of 254 consecutive transverse sections through a single tendon fascicle. The plasma membranes of the cells have been traced to model the cell shape. (B) 3D reconstruction of the fascicle. A single EM image is superimposed to aid visualization. (C) Enlargement of the section shown in (A) showing collagen bundles surrounded by either a single cell (open arrow), or multiple cells (closed arrow). (D) Enlargement of the top of the reconstructed volume. The image shows channels running parallel to the tendon long axis (vertical axis). The channels can be formed by single cells (examples indicated by open arrows), or by the plasma membrane of adjacent cells (filled arrows). Adapted from data used in Richardson et al. (2007). (See Plate no. 37 in the Color Plate Section.)

The intercellular interactions occur where the fibril bundle channel is surrounded by more than one cell (Fig. 10D filled arrow). When the fibril bundles are examined in detail it can be seen that each bundle is supported by a mixture of intra and intercellular interactions that are stabilized by a series of adherens junctions that run in strips along the length of the cells (Richardson *et al.*, 2007). It is possible to follow the fibril bundles as they course through the tendon, often passing from an intracellular sheath on one cell to a similar sheath formed by the following cell, with an intercellular zone in between. The collagen fibrils frequently do not remain within the same bundle through the length of the tendon. Fibrils from one bundle often pass to another bundle via bundle branches as first described in (Birk *et al.*, 1989a). This is a brief overview of the sort of information that serial sections give that is not discernable from a single slice through the volume.

B. Fibripositors

When serial sections through a developing tendon are examined in more detail it becomes apparent that the cells contain collagen fibrils within the lumen of long, thin, membrane-bound compartments. These fibrils are aligned with the tendon long axis and with the collagen fibrils in the extracellular bundles. The fibrils can often be found exiting the cell, directly from the cell body to the matrix, as described by Birk *et al.* (1989b), or more often they exit the cell via long finger-like projections termed fibripositors (Fig. 11). The identification of the fibripositor (Canty *et al.*, 2004) was possible only through the use of serial section reconstruction. The section images were first aligned at low magnification based on the shape of the cells, before a more detailed examination of the collagen matrix was possible. The highly regular hexagonal packing of the fibrils within the collagen bundles means that it is very easy to confuse similar bundle cross-sections without the low level alignment. Once the cells are aligned it is relatively easy to follow internal collagen fibrils through multiple sections (Fig. 11). The collagen within a fibripositor has one end within the cell body, completely enclosed by a single membrane. At this point there is often more than one fibril within each lumen. As the region is followed through multiple sections (i.e., tracking along the fibril length) the fibril numbers get reduced until there is a single fibril in the lumen. In the next few sections the fibril exits the body of the cell, but it is still surrounded by a small sleeve of cellular material, which has been termed the fibripositor. The fibril can remain within the fibripositor for several micrometres, until it exits into the matrix, where it integrates into the collagen bundle.

C. Fibripositor-Cell Tomography

To examine the 3D structure of the embryonic tendon at sufficiently high resolution to reveal the cytoskeletal structures and their relationship to the collagen fibrillar matrix we have used electron tomography. In the example we show a region of a cell sectioned longitudinal to the tendon long axis (Fig. 12). The first image shows a typical electron micrograph image (Fig. 12A). The image clarity is reduced because of the section thickness (300 nm), but it is possible to make out a "dark blur" (bottom left) and a pale ovoid region (right) along with a faint linear feature running from the centre of the image to the bottom right. Following

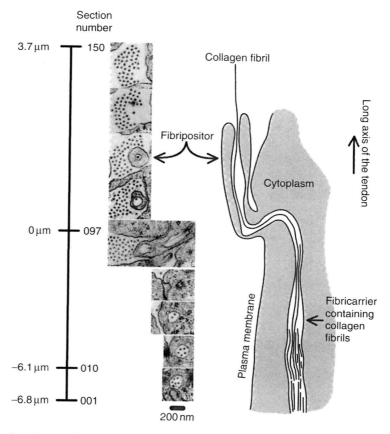

Fig. 11 Tracking a collagen fibril from the ECM to within the cell. Images from a serial section reconstruction through a collagen fibril within a fibripositor. The images follow a single fibril and show the fibril within three domains: the fibril bundle, the lumen of a finger-like projection (fibripositor), and the intracellular side of the fibripositor containing multiple fibrils. A reference bar shows the section number, the depth of the fibripositor lumen within the cell, and the length of the fibripositor. The schematic shows the longitudinal structure of the fibripositor. Adapted from Canty et al. (2004).

tomographic reconstruction from the tilt series the computer generates a representation of the entire volume of the section. It is possible to take virtual slices through the volume at any given angle. The images in Fig. 12B and C show typical virtual slices taken in the x-y plane. The original image in Fig. 12A can be seen as being the sum of all of the virtual slices from the reconstructed volume but the details are obscured by superposition of the many layers. By examining the slices in turn it is possible to identify the features that were only vaguely visible in the standard image. The "dark blur" (bottom left) can now be identified as part of the nucleus and nuclear membrane of the cell. The pale ovoid region (right) can be identified as

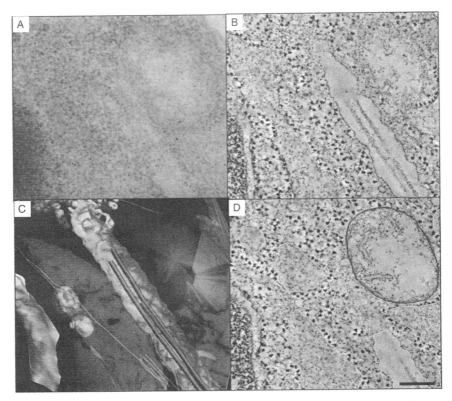

Fig. 12 Tomographic reconstruction of an internal collagen fibril. (A) STEM image of a 300 nm thick longitudinal section through an embryonic (15.5 dpc) tendon cell. (B) Virtual slice, 1 nm thick, taken from the top of the tomographic reconstruction of the region shown in (A). Collagen fibrils are clearly visible in the vesicular compartment that runs diagonally across the image. (C) Virtual slice, 1 nm thick, taken from the bottom of the tomographic reconstruction of the region shown in A. Some of the features of interest have been outlined. (D) 3D model of the intracellular compartments found within the tomographic reconstruction. Collagen fibrils (purple), fibril compartment (semi transparent cyan), mitochondria (blue), other compartments (Semi-transparent green). Scale bar = 200 nm. (See Plate no. 38 in the Color Plate Section.)

a mitochondria, and the faint linear feature is an example of a fibripositor similar to those described in the previous section. By examining all of the virtual slices and tracing individual components, in the same way as cell outlines were traced for the serial sections, it is possible to create a model of the components within the section (Fig. 12D). When serial tomograms are combined it is possible to track a feature that passes through multiple sections. In this application three tomograms were combined using the data shown in Fig. 12, along with data from two other sections. In this combined tomogram it was possible to track all the way to the base of the fibripositor (Fig. 13). Analysis of the data within the serial tomograms identified an

17. Electron Microscopy of Collagen Fibril Structure

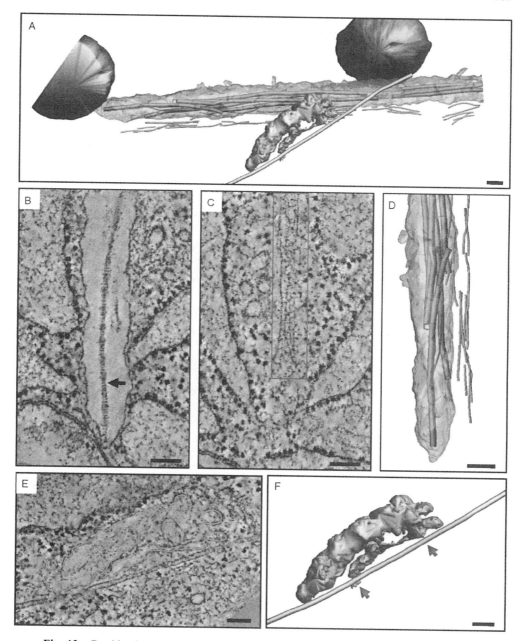

Fig. 13 Combination of tomograms from three serial sections. (A) 3D reconstruction showing several features found in a three section reconstruction of the base of a collagen fibril containing compartment. Two mitochondria are shown (blue) to indicate the dimensions of the features. (B) Virtual slice (7 nm thick) showing one of the cross-banded collagen fibrils (black arrow) present within the vesicle. (C) Virtual slice (7 nm thick) showing an actin cable (boxed) that runs alongside the collagen containing

actin bundle that runs alongside the collagen fibril-containing compartment. The visualization of an actin bundle suggests a role for actin in the stabilization and motility of these large collagen containing vesicles. Biochemical analysis confirmed that removing actin caused the fibripositor structures to disappear (Canty et al., 2006). Within the volume there was also a compartment that was large enough to carry nonfibrillar collagen molecules. This compartment was found to be closely associated with a single microtubule (Fig. 13E and F), which suggested that unprocessed collagen is secreted in a microtubule dependent manner.

IV. Discussion

There has been intensive development in methods related to 3D electron microscopy over the last 10 years. There are now several implementations of automated tomography image acquisition software that can be readily employed outside the specialist development laboratories. Similarly there are excellent software packages for 3D reconstruction from tilt series and serial sections, and for subsequent model building. These methods have now been applied to the ECM and have provided new insights into the structure and assembly of the collagen fibril matrix.

A. Future Work

Current developments at expert centers are directed at further improving the efficiency and rate of throughput of these 3D imaging operations. Acquisition time for tilt series can be reduced by precalibration and prediction of specimen movement during the tilt series (Mastronarde, 2005; Ziese et al., 2002). Fiducial-free image alignment and rapid reconstruction has led to the ability to generate 3D images reconstructions during the electron microscope session (Zheng et al., 2007). The process of volume segmentation and model generation remains the most time-consuming part of obtaining a useful 3D structure but this will become easier and more efficient with the continued development of automated segmentation methods.

High-pressure freezing combined with low-temperature embedding is now recognized to give superior cellular structure preservation and has become a preferred method to generated samples for electron tomography (McDonald and Auer, 2006). This can be combined with improved techniques of 3D labeling

---

vesicle. The actin cable is found in a different plane to that of the fibripositor and collagen fibrils. (D) 3D reconstruction showing the relative positions of the collagen containing vesicle membrane (semitransparent cyan), collagen fibrils (purple), and the actin cable (red). (E) Virtual slice (7 nm thick) showing part of the microtubule and an associated compartment (outlined) that contains fibrous material. (F) 3D reconstruction shows a number of points in the model (gray arrows) where the large compartment (orange) appears to be associated with the microtubule. Scale bars 100 nm. From Canty et al. (2006). (See Plate no. 39 in the Color Plate Section.)

(Morphew, 2007). Cryotomography (using frozen hydrated samples) has now been successfully implemented at expert centers and despite low limits on accumulated electron exposure, resolutions of 4–5 nm have been achieved in the tomograms of frozen, hydrated cells (Baumeister and Steven, 2000).

V. Conclusions

Electron microscopic studies on the ECM over several decades have revealed a highly regulated process of collagen matrix deposition. In recent years our understanding of the nature and mechanisms of this assembly process has been greatly enhanced by the application of new methods to generate 3D images. Future developments include cryopreparation methods, to improve structural preservation, and also the application of 3D protein labeling and tagging methods. The extraction of structural models from the initial 3D images is still a slow process but improved methods of computer automation will lead to a higher throughput. Ultimately it will be possible to combine structures determined by X-ray crystallography and NMR with electron tomography data to determine the dynamic molecular machinery that cells use to deposit an ECM containing highly organized collagen fibrils and associated macromolecules.

Acknowledgments

We are grateful to the Wellcome Trust for a Programme Grant to KEK. We wish to thank Sally Humphries, University of Manchester, for the model of tendon cells from serial sections. We thank Dr. Abraham Koster, University of Utrecht, for assistance with tilt-series acquisition on negatively stained collagen fibrils. We are grateful for help from Dr. David Mastronarde and Dr. Eileen O'Toole, Centre for 3DEM, University of Colorado at Boulder, for assistance with tilt series collection on 300 kV FEG, IMOD software and SerialEM.

References

Baldock, C., Gilpin, C. J., Koster, A. J., Ziese, U., Kadler, K. E., Kielty, C. M., and Holmes, D. F. (2002). Three-dimensional reconstructions of extracellular matrix polymers using automated electron tomography. *J. Struct. Biol.* **138**, 130–136.

Baldock, C., Koster, A. J., Ziese, U., Rock, M. J., Sherratt, M. J., Kadler, K. E., Shuttleworth, C. A., and Kielty, C. M. (2001). The supramolecular organization of fibrillin-rich microfibrils. *J. Cell Biol.* **152**, 1045–1056.

Barge, A., Ruggiero, F., and Garrone, R. (1991). Structure of the basement membrane of corneal epithelium: Quick-freeze, deep-etch comparative study of networks deposited in culture and during development. *Biol. Cell* **72**, 141–147.

Baumeister, W., and Steven, A. C. (2000). Macromolecular electron microscopy in the era of structural genomics. *Trends Biochem. Sci.* **25**, 624–631.

Bergmanson, J. P., Horne, J., Doughty, M. J., Garcia, M., and Gondo, M. (2005). Assessment of the number of lamellae in the central region of the normal human corneal stroma at the resolution of the transmission electron microscope. *Eye Contact Lens* **31**, 281–287.

Birk, D. E. (2001). Type V collagen: Heterotypic type I/V collagen interactions in the regulation of fibril assembly. *Micron* **32,** 223–237.

Birk, D. E., Fitch, J. M., Babiarz, J. P., and Linsenmayer, T. F. (1988). Collagen type I and type V are present in the same fibril in the avian corneal stroma. *J. Cell Biol.* **106,** 999–1008.

Birk, D. E., Nurminskaya, M. V., and Zycband, E. I. (1995). Collagen fibrillogenesis *in situ*: Fibril segments undergo post-depositional modifications resulting in linear and lateral growth during matrix development. *Dev. Dyn.* **202,** 229–243.

Birk, D. E., Southern, J. F., Zycband, E. I., Fallon, J. T., and Trelstad, R. L. (1989a). Collagen fibril bundles: A branching assembly unit in tendon morphogenesis. *Development* (Cambridge, England) **107,** 437–443.

Birk, D. E., and Trelstad, R. L. (1986). Extracellular compartments in tendon morphogenesis: Collagen fibril, bundle, and macroaggregate formation. *J. Cell Biol.* **103,** 231–240.

Birk, D. E., Zycband, E. I., Winkelmann, D. A., and Trelstad, R. L. (1989b). Collagen fibrillogenesis *in situ*: Fibril segments are intermediates in matrix assembly. *Proc. Natl. Acad. Sci. USA* **86,** 4549–4553.

Bos, K. J., Holmes, D. F., Meadows, R. S., Kadler, K. E., McLeod, D., and Bishop, P. N. (2001). Collagen fibril organisation in mammalian vitreous by freeze etch/rotary shadowing electron microscopy. *Micron* **32,** 301–306.

Canty, E. G., and Kadler, K. E. (2005). Procollagen trafficking, processing and fibrillogenesis. *J Cell Sci.* **118,** 1341–1353.

Canty, E. G., Lu, Y., Meadows, R. S., Shaw, M. K., Holmes, D. F., and Kadler, K. E. (2004). Coalignment of plasma membrane channels and protrusions (fibripositors) specifies the parallelism of tendon. *J. Cell Biol.* **165,** 553–563.

Canty, E. G., Starborg, T., Lu, Y., Humphries, S. M., Holmes, D. F., Meadows, R. S., Huffman, A., O'Toole, E. T., and Kadler, K. E. (2006). Actin filaments are required for fibripositor-mediated collagen fibril alignment in tendon. *J. Biol. Chem.* **281,** 38592–38598.

Capaldi, M. J., and Chapman, J. A. (1982). The C-terminal extrahelical peptide of type I collagen and its role in fibrillogenesis *in vitro*. *Biopolymers* **21,** 2291–2313.

Chapman, J. A., Tzaphlidou, M., Meek, K. M., and Kadler, K. E. (1990). The collagen fibril—A model system for studying the staining and fixation of a protein. *Electron Microsc. Rev.* **3,** 143–182.

Colige, A., Sieron, A. L., Li, S. W., Schwarze, U., Petty, E., Wertelecki, W., Wilcox, W., Krakow, D., Cohn, D. H., Reardon, W., Byers, P. H., and Lapière, C. M. (1999). Human Ehlers-Danlos syndrome type VII C and bovine dermatosparaxis are caused by mutations in the procollagen I N-proteinase gene. *Am. J. Hum. Genet.* **65,** 308–317.

Eikenberry, E. F., Childs, B., Sheren, S. B., Parry, D. A., Craig, A. S., and Brodsky, B. (1984). Crystalline fibril structure of type II collagen in lamprey notochord sheath. *J. Mol. Biol.* **176,** 261–277.

Gelman, R. A., Williams, B. R., and Piez, K. A. (1979). Collagen fibril formation. Evidence for a multistep process. *J. Biol. Chem.* **254,** 180–186.

Goh, K. L., Meakin, J. R., Aspden, R. M., and Hukins, D. W. (2007). Stress transfer in collagen fibrils reinforcing connective tissues: Effects of collagen fibril slenderness and relative stiffness. *J. Theor. Biol.* **245,** 305–311.

Graham, H. K., Holmes, D. F., Watson, R. B., and Kadler, K. E. (2000). Identification of collagen fibril fusion during vertebrate tendon morphogenesis. The process relies on unipolar fibrils and is regulated by collagen-proteoglycan interaction. *J. Mol. Biol.* **295,** 891–902.

Gross, J., and Kirk, D. (1958). The heat precipitation of collagen from neutral salt solutions: Some rate-regulating factors. *J. Biol. Chem.* **233,** 355–360.

Hirsch, M., Noske, W., Prenant, G., and Renard, G. (1999). Fine structure of the developing avian corneal stroma as revealed by quick-freeze, deep-etch electron microscopy. *Exp. Eye Res.* **69,** 267–277.

Hodge, A. J. (1989). Molecular models illustrating the possible distributions of "holes" in simple systematically staggered arrays of type I collagen molecules in native-type fibrils. *Connect Tissue Res.* **21,** 137–147.

Hoffpauir, B. K., Pope, B. A., and Spirou, G. A. (2007). Serial sectioning and electron microscopy of large tissue volumes for 3D analysis and reconstruction: A case study of the calyx of Held. *Nat. Protocols* **2**, 9–22.

Holmes, D. F., Capaldi, M. J., and Chapman, J. A. (1986). Reconstitution of collagen fibrils *in vitro*; the assembly process depends on the initiating procedure. *Int. J. Biol. Macromol.* **8**, 161–166.

Holmes, D. F., Gilpin, C. J., Baldock, C., Ziese, U., Koster, A. J., and Kadler, K. E. (2001a). Corneal collagen fibril structure in three dimensions: Structural insights into fibril assembly, mechanical properties, and tissue organization. *Proc. Natl. Acad. Sci. USA* **98**, 7307–7312.

Holmes, D. F., Graham, H. K., and Kadler, K. E. (1998). Collagen fibrils forming in developing tendon show an early and abrupt limitation in diameter at the growing tips. *J. Mol. Biol.* **283**, 1049–1058.

Holmes, D. F., Graham, H. K., Trotter, J. A., and Kadler, K. E. (2001b). STEM/TEM studies of collagen fibril assembly. *Micron* **32**, 273–285.

Holmes, D., and Kadler, K. (2006a). Protocol 6.44 Collagen fibril assembly *in vitro*. *In* "Cell Biology Protocols" (J. R. Harris, J. M. Graham, and D. Rickwood, eds.), pp. 375–378. Wiley, Chichester, UK.

Holmes, D. F., and Kadler, K. E. (2006b). The 10 + 4 microfibril structure of thin cartilage fibrils. *Proc. Natl. Acad. Sci. USA* **103**, 17249–17254.

Holmes, D. F., Lowe, M. P., and Chapman, J. A. (1994). Vertebrate (chick) collagen fibrils formed *in vivo* can exhibit a reversal in molecular polarity. *J. Mol. Biol.* **235**, 80–83.

Hulmes, D. J., Bruns, R. R., and Gross, J. (1983). On the state of aggregation of newly secreted procollagen. *Proc. Natl. Acad. Sci. USA* **80**, 388–392.

Hulmes, D. J., and Miller, A. (1979). Quasi-hexagonal molecular packing in collagen fibrils. *Nature* **282**, 878–880.

Huxley-Jones, J., Robertson, D. L., and Boot-Handford, R. P. (2007). On the origins of the extracellular matrix in vertebrates. *Matrix Biol.* **26**, 2–11.

Kadler, K. E., Baldock, C., Bella, J., and Boot-Handford, R. P. (2007). Collagens at a glance. *J. Cell Sci.* **120**, 1955–1958.

Kadler, K. E., Hojima, Y., and Prockop, D. J. (1987). Assembly of collagen fibrils *de novo* by cleavage of the type I pC-collagen with procollagen C-proteinase. Assay of critical concentration demonstrates that collagen self-assembly is a classical example of an entropy-driven process. *J. Biol. Chem.* **262**, 15696–15701.

Kannus, P. (2000). Structure of the tendon connective tissue. *Scand. J. Med. Sci. Sports* **10**, 312–320.

Keene, D. R., Oxford, J. T., and Morris, N. P. (1995). Ultrastructural localization of collagen types II, IX, and XI in the growth plate of human rib and fetal bovine epiphyseal cartilage: Type XI collagen is restricted to thin fibrils. *J. Histochem. Cytochem.* **43**, 967–979.

Koster, A. J., Chen, H., Sedat, J. W., and Agard, D. A. (1992). Automated microscopy for electron tomography. *Ultramicroscopy* **46**, 207–227.

Kremer, J. R., Mastronarde, D. N., and McIntosh, J. R. (1996). Computer visualization of three-dimensional image data using IMOD. *J. Struct. Biol.* **116**, 71–76.

Lawrence, A., Bouwer, J. C., Perkins, G., and Ellisman, M. H. (2006). Transform-based backprojection for volume reconstruction of large format electron microscope tilt series. *J. Struct. Biol.* **154**, 144–167.

Leighton, M., and Kadler, K. E. (2003). Paired basic/Furin-like proprotein convertase cleavage of Pro-BMP-1 in the trans-Golgi network. *J. Biol. Chem.* **278**, 18478–18484.

Linsenmayer, T. F., Gibney, E., Igoe, F., Gordon, M. K., Fitch, J. M., Fessler, L. I., and Birk, D. E. (1993). Type V collagen: Molecular structure and fibrillar organization of the chicken alpha 1(V) NH2-terminal domain, a putative regulator of corneal fibrillogenesis. *J. Cell Biol.* **121**, 1181–1189.

Ma, B., Zimmermann, T., Rohde, M., Winkelbach, S., He, F., Lindenmaier, W., and Dittmar, K. E. (2007). Use of autostitch for automatic stitching of microscope images. *Micron* **38**, 492–499.

Mastronarde, D. N. (1997). Dual-axis tomography: An approach with alignment methods that preserve resolution. *J. Struct. Biol.* **120**, 343–352.

Mastronarde, D. N. (2005). Automated electron microscope tomography using robust prediction of specimen movements. *J. Struct. Biol.* **152**, 36–51.

McDonald, K. L., and Auer, M. (2006). High-pressure freezing, cellular tomography, and structural cell biology. *Biotechniques* **41**, 137, 139, 141 passim.

McIntosh, R., Nicastro, D., and Mastronarde, D. (2005). New views of cells in 3D: An introduction to electron tomography. *Trends Cell Biol.* **15**, 43–51.

Morphew, M. K. (2007). 3D immunolocalization with plastic sections. *Methods Cell Biol.* **79**, 493–513.

Mosser, G., Anglo, A., Helary, C., Bouligand, Y., and Giraud-Guille, M. M. (2006). Dense tissue-like collagen matrices formed in cell-free conditions. *Matrix Biol.* **25**, 3–13.

Mould, A. P., Holmes, D. F., Kadler, K. E., and Chapman, J. A. (1985). Mica sandwich technique for preparing macromolecules for rotary shadowing. *J. Ultrastruct. Res.* **91**, 66–76.

Myllyharju, J., and Kivirikko, K. I. (2004). Collagens, modifying enzymes and their mutations in humans, flies and worms. *Trends Genet.* **20**, 33–43.

Nickell, S., Kofler, C., Leis, A. P., and Baumeister, W. (2006). A visual approach to proteomics. *Nat. Rev. Mol. Cell Biol.* **7**, 225–230.

O'Toole, E. T., Giddings, T. H., Jr., and Dutcher, S. K. (2007). Understanding microtubule organizing centers by comparing mutant and wild-type structures with electron tomography. *Methods Cell Biol.* **79**, 125–143.

Orgel, J. P., Irving, T. C., Miller, A., and Wess, T. J. (2006). Microfibrillar structure of type I collagen in situ. *Proc. Natl. Acad. Sci. USA* **103**, 9001–9005.

Parry, D. A., Barnes, G. R., and Craig, A. S. (1978). A comparison of the size distribution of collagen fibrils in connective tissues as a function of age and a possible relation between fibril size distribution and mechanical properties. *Proc. R. Soc. Lond. B* **203**, 305–321.

Provenzano, P. P., and Vanderby, R., Jr. (2006). Collagen fibril morphology and organization: Implications for force transmission in ligament and tendon. *Matrix Biol.* **25**, 71–84.

Raspanti, M., Viola, M., Sonaggere, M., Tira, M. E., and Tenni, R. (2007). Collagen fibril structure is affected by collagen concentration and decorin. *Biomacromolecules* **8**, 2087–2091.

Richardson, S. H., Starborg, T., Lu, Y., Humphries, S. M., Meadows, R. S., and Kadler, K. E. (2007). Tendon development requires regulation of cell condensation and cell shape via cadherin-11-mediated cell-cell junctions. *Mol. Cell Biol.* **27**, 6218–6228.

Sandberg, K. (2007). Methods for image segmentation in cellular tomography. *Methods Cell Biol.* **79**, 769–798.

Starborg, T., and Kadler, K. E. (2005). The power behind the electron microscopist. *Biol. Sci. Rev.* **18**, 16–20.

Suzuki, N., Labosky, P. A., Furuta, Y., Hargett, L., Dunn, R., Fogo, A. B., Takahara, K., Peters, D. M., Greenspan, D. S., and Hogan, B. L. (1996). Failure of ventral body wall closure in mouse embryos lacking a procollagen C-proteinase encoded by Bmp1, a mammalian gene related to Drosophila tolloid. *Development* (Cambridge, England) **122**, 3587–3595.

Thurmond, F. A., and Trotter, J. A. (1994). Native collagen fibrils from echinoderms are molecularly bipolar. *J. Mol. Biol.* **235**, 73–79.

Tong, J., Arslan, I., and Midgley, P. (2006). A novel dual-axis iterative algorithm for electron tomography. *J. Struct. Biol.* **153**, 55–63.

Trotter, J. A., Chapman, J. A., Kadler, K. E., and Holmes, D. F. (1998). Growth of sea cucumber collagen fibrils occurs at the tips and centers in a coordinated manner. *J. Mol. Biol.* **284**, 1417–1424.

Trotter, J. A., Tipper, J., Lyons-Levy, G., Chino, K., Heuer, A. H., Liu, Z., Mrksich, M., Hodneland, C., Dillmore, W. S., Koob, T. J., Koob-Emunds, M. M., Kadler, K., and Holmes, D. (2000b). Towards a fibrous composite with dynamically controlled stiffness: Lessons from echinoderms. *Biochem. Soc. Trans.* **28**, 357–362.

Trotter, J. A., Kadler, K. E., and Holmes, D. F. (2000a). Echinoderm collagen fibrils grow by surface-nucleation-and-propagation from both centers and ends. *J. Mol. Biol.* **300**, 531–540.

Wang, W. M., Lee, S., Steiglitz, B. M., Scott, I. C., Lebares, C. C., Allen, M. L., Brenner, M. C., Takahara, K., and Greenspan, D. S. (2003). Transforming growth factor-beta induces secretion of activated ADAMTS-2. A procollagen III N-proteinase. *J. Biol. Chem.* **278,** 19549–19557.

Weinstock, M., and Leblond, C. P. (1974). Synthesis, migration, and release of precursor collagen by odontoblasts as visualized by radioautography after (3H) proline administration. *J. Cell Biol.* **60,** 92–127.

Wess, T. J. (2005). Collagen fibril form and function. *Adv. Protein Chem.* **70,** 341–374.

Wood, G. C., and Keech, M. K. (1960). The formation of fibrils from collagen solutions. I. The effect of experimental conditions: Kinetic and electron-microscope studies. *Biochem. J.* **75,** 588–598.

Zheng, S. Q., Keszthelyi, B., Branlund, E., Lyle, J. M., Braunfeld, M. B., Sedat, J. W., and Agard, D. A. (2007). UCSF tomography: An integrated software suite for real-time electron microscopic tomographic data collection, alignment, and reconstruction. *J. Struct. Biol.* **157,** 138–147.

Ziese, U., Janssen, A. H., Murk, J. L., Geerts, W. J., Van der Krift, T., Verkleij, A. J., and Koster, A. J. (2002). Automated high-throughput electron tomography by pre-calibration of image shifts. *J. Microsc.* **205,** 187–200.

CHAPTER 18

Visualization of Desmosomes in the Electron Microscope

Anthea Scothern and David Garrod

Faculty of Life Sciences
University of Manchester
Manchester, UK

Abstract
I. Introduction
II. Rationale
III. Methods
    A. Conventional Electron Microscopy
    B. Cryopreservation for Immunogold Labeling
    C. Immunogold Labeling
    D. Quantification of the Distribution of Gold Particles
IV. Materials
    A. Conventional Electron Microscopy
    B. Cryopreservation and Immunogold Labeling
    C. Immunogold Labeling
V. Discussion
VI. Summary
    References

Abstract

Desmosomes are intercellular junctions found in epithelia and some other tissues. Their primary function is strong cell–cell adhesion. They also link the intermediate filament (IF) cytoskeletons between cells and have roles in cell signaling, tissue morphogenesis, and wound repair. Because of their size (0.2–0.5 µm), details of their ultrastructure can only be resolved at the electron microscope (EM) level.

Desmosomes have been visualized using a variety of ultrastructural techniques including lanthanum infiltration, freeze-fracture, electron tomography, cryo-electron microscopy and immunogold labeling.

This chapter describes protocols for conventional transmission electron microscopy and for immunogold labeling of ultrathin cryosections. We also discuss the statistical analysis of immunogold particle distribution for low resolution molecular mapping.

I. Introduction

Desmosomes are intercellular junctions that are present in epithelia and some nonepithelial tissues such as cardiac muscle and the meninges. They are most abundant in tissues that withstand high degrees of mechanical stress such as the epidermis and esophageal lining. Their primary function is to provide strong intercellular adhesion and to link the IF cytoskeleton into a tissue-wide scaffolding thus lending great tensile strength to tissues. Although the tissue and cellular distribution of desmosomes can be determined by light microscopy and immunofluorescence, individual junctions can only be clearly identified and studied by electron microscopy because their size, usually 0.2–0.5 μm places them at the limit of light microscopic resolution.

There are five major protein components of desmosomes: the desmocollins (Dsc) and desmogleins (Dsg), two families of transmembrane glycoproteins belonging to the cadherin superfamily of cell–cell adhesion molecules; the constitutive desmosomal component desmoplakin (DP), and the armadillo gene family members plakophilin (PP) and plakoglobin (PG, also known as γ catenin) (Dusek et al., 2007; Garrod and Chidgey, 2007; Garrod et al., 2002a,b; Getsios et al., 2004; See Hatzfeld, 1999, 2007; Yin and Green, 2004, for reviews). Desmosomal components are involved in a number of diseases including the autoimmune skin blistering disease pemphigus, arrhythmogenic right-ventricular cardiomyopathy (ARVC) and cancer (see Chidgey and Dawson (2007) and Kottke et al. (2006) for reviews).

The ultrastructure of desmosomes is characteristic and highly organized (Fig. 1) (see Burdett, 1998 for review). They consist of two electron-dense plaques symmetrically arranged on the intracellular side of adjacent cell membranes. These plaques are joined to each other across the extracellular space by transmembrane adhesion molecules and within the cell to the IF network. The extracellular core domain (ECD) or desmoglea is approximately 30 nm wide between the two plasma membranes (PM) and has a central dense midline (DM). Each plaque consists of an outer dense plaque (ODP), 15–20 nm thick, adjacent to the PM, an electron lucent zone of about 8 nm and an inner dense plaque (IDP), less dense than the ODP and of more variable thickness but in the region of 15–20 nm. Thus the total thickness of the desmosomal plaque is 40–50 nm and the total thickness of a desmosome of the order of 130 nm. Stereoimaging form electron micrographs taken at a 10° tilt angle showed that the IFs insert into the IDP and then loop back out into the cytoplasm at a distance of 20–40 nm from the PM (Kelly, 1966).

18. Visualization of Desmosomes in the EM

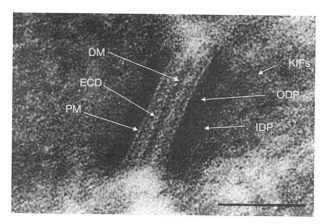

Fig. 1 Desmosomes as seen in the EM following conventional fixation and embedding. KIFs: Keratin Intermediate Filaments, IDP: Inner Dense Plaque, ODP: Outer Dense Plaque, ECD: Extracellular Core Domain, DM: Dense Midline, PM: Plasma Membrane. Scale bar = 100 nm.

Several studies using different ultrastructural techniques have attempted to analyze the structure of the desmoglea in more detail. Freeze-fracture studies have revealed a particulate composition within the PM and desmoglea, but have generally been rather uninformative (Breathnach and Wyllie, 1967; Kelly and Kuda, 1981; Kelly and Shienvold, 1976; Orwin et al., 1973; Staehelin, 1974). Fujimoto (1995) has shown that freeze-fracture replicas can be gold-labeled with desmosomal cadherin antibody. Lanthanum infiltration studies showed the ECD containing a staggered array of side-arms linking the midline to the PM (Rayns et al., 1969). In en face view, the side-arms appeared to be arranged in regular rows described as a quadratic array. Electron tomography carried out on freeze-substituted, plastic-embedded mouse epidermis showed the structure of the ECD as a tangle of molecules arranged in knots (He et al., 2003). However, cryo-electron microscopy of vitrified, unfixed epidermis have revealed straight, ordered, filamentous structures perpendicular to the PMs (Al-Amoudi et al., 2004). Garrod et al. (2005) argue that it is this high level of organization that confers their strongly adhesive property, referred to as "hyperadhesion," on desmosomes. Furthermore, they showed that the organized structure is dynamic and that the strength of adhesion varies in response to cellular events such as epidermal wounding, when desmosomes become more weakly adhesive. Immunogold labeling of epidermis has shown that different isoforms of the desmosomal cadherins, desmocollin and desmoglein, occur together in the same desmosomes and have exponentially graded distributions, and also that the N-terminus of desmoglein lies at the midline (North et al., 1996; Shimizu et al., 2005).

Electron microscopy also reveals a highly organized structure of the ODP. Conventional preparation techniques such as glutaraldehyde/formaldehyde fixation, plastic embedding, and uranyl/acetate/lead citrate staining tend to show merely a very dense structure. However, examination of ultrathin cryosections prepared by the Tokuyasu method revealed a regular transverse periodicity estimated at 2.9 Å (Miller *et al.*, 1987). Negative staining of polyvinyl alcohol-embedded material revealed lines of differential staining both parallel and perpendicular to the membrane that at high magnification appeared as two parallel arrays (North *et al.*, 1999). By quantitative analysis of immunogold particle distributions after immunolabeling of ultrathin cryosections with domain-specific antibodies to plaque proteins, it was possible to construct a molecular map of the entire desmosomal plaque showing the locations of the various components with respect to the plasma membrane (North *et al.*, 1999). Immunogold labeling has also revealed that protein kinase C α is a component of the ODP of weakly adhesive desmosomes in wound edge epidermis (Garrod *et al.*, 2005).

This chapter provides protocols for visualizing desmosomes using conventional transmission electron microscopy and for localization of desmosomal components using immunogold labeling of cryosections by a modified version of the Tokuyasu method (Griffiths *et al.*, 1984; Liou *et al.*, 1996; Peters *et al.*, 1991, 2006; Tokuyasu, 1973, 1980, 1986, 1989; Tokuyasu and Singer, 1976). The PVA-embedding technique is described by Small *et al.* (1986), North *et al.* (1993, 1994a,b), and North and Small (2000). Tissue prepared by cryopreservation or PVA embedding can be used to test the efficacy of an antibody prior to immunogold labeling by cutting semithin (200–250 nm) sections and processing them for immunofluorescence (Fig. 2). We recommend testing an unfamiliar antibody on semithin sections before attempting immunogold labeling of ultrathin sections.

In order to quantify the distribution of gold particles following immunogold labeling, we use a statistical software package called Simfit (Bardsley *et al.*, 1995a,b), which creates best fit curves of the gold distribution on labeled desmosomes. This software can be downloaded from http://www.simfit.man.ac.uk.

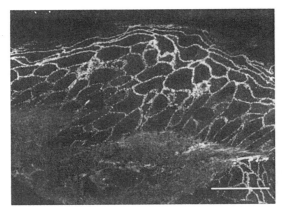

Fig. 2 Semithin (~200 nm thick) transverse section of human skin stained with Mab 33-3D (Vilela *et al.*, 1995) against the cytoplasmic domain of Dsg2. Note the characteristic punctate staining at the periphery of the cells in the epidermis. Scale bar = 35 μm.

We have used ImageJ to measure the distribution of gold particles. This software can be downloaded from http://rsb.info.nih.gov/ij/.

II. Rationale

At the light microscope level, desmosomes can be seen by immunofluorescence as dots or puncta, which may represent single desmosomes or small groups of desmosomes. Double immunolabeling combined with confocal microscopy can provide preliminary information on the colocalization of proteins to desmosomes. Electron microscopy is required to resolve individual junctions and to reveal details of their ultrastructure. Immunogold labeling of ultrathin sections is necessary to confirm desmosomal localization of proteins and can be quantified to enable low resolution molecular mapping of desmosome architecture.

III. Methods

A. Conventional Electron Microscopy

1. Fixation and Dehydration

- Place harvested tissue immediately into freshly made fixative: 2% PFA plus 2% glutaraldehyde in cacodylate buffer (see recipe in Materials). Mouse skin should be shaved prior to dissection. Although a little hair is helpful for orientating the tissue, too much prevents the tissue from sinking into the fixative and embedding solutions.
- Cut into pieces less than 1 mm^3 in a puddle of fixative under the dissection microscope using clean single edged razorblades or a scalpel. This step should be carried out in a fume hood.
- Transfer tissue to fresh fixative in glass vials and incubate for 2 h at room temperature, changing the fixative after 1 h. Leave in fix overnight at 4 °C.
- Wash in three changes of cacodylate buffer.
- Postfix in 1% osmium tetroxide in cacodylate buffer for 90 min at room temperature. The osmium turns the tissue black enabling easy visualization during embedding.
- Wash in three changes of distilled water.
- Stain with 2% uranyl acetate (UA) solution (see recipe in Materials), which has been centrifuged immediately prior to use, overnight at 4 °C.

Note: It is worth pointing out that some investigators have achieved better visualization of the DM by omitting this *en block* staining step and instead staining the sections with UA as described in the staining section of this protocol.

- Wash in three changes of distilled water
- Dehydrate through an acetone series:
  - 25%, 50%, 70%—20 min each
  - 80%, 90%, 95%, 3×100%—30 min each

Note: An ethanol series also gives excellent dehydration.

- Change into 100% propylene oxide for 10 min at room temperature

2. Embedding

Embedding is carried out at room temperature on a fixed tilt specimen rotator (Agar Scientific).

- Transfer tissue into 25% medium Spurr resin in propylene oxide overnight followed by:
- 50% resin in propylene oxide for 3–4 h
- 70% resin in propylene oxide for 3–4 h
- 100% resin overnight
- Three changes of 100% resin: morning, lunchtime, and evening. Leave in 100% resin overnight
- Embed in fresh resin at the trapezoid-shaped end of a flat embedding mould (Agar Scientific Catalogue number G530), taking care to orient the tissue so that it will be sectioned in the correct plane.
- Polymerize in an oven at 60 °C for 48 h

3. Sectioning

- Trim resin blocks first with a razorblade. Then finely trim on an ultramicrotome to remove excess resin and expose the tissue.
- Cut semithin sections (250–500 nm) and collect on glass slides for staining with toluidine blue. Examine under the light microscope to ensure that desmosome-bearing tissue (e.g., epidermis or wound edges) is included in the section.
- Cut ultrathin sections (50–70 nm) using a glass or diamond knife fitted with a boat filled with distilled water. The sections will float away from the knife edge on the water reservoir.
- Collect sections from the water on Formvar coated 100 mesh grids and leave to air dry.

4. Staining

- Float grids on 100 µl drops of distilled water laid out on a sheet of parafilm for 1 min.
- Float grids on 100 µl drops of 2% UA solution for 30 min at room temperature.

- Wash each grid three times on 100 μl drops of distilled water.
- Incubate on 100 μl drops of 0.3% lead citrate solution (see recipe in Materials) for 5 min.
- Wash on five distilled water drops for one minute per drop and then allow to air dry overnight before examination in the TEM.

B. Cryopreservation for Immunogold Labeling

1. Fixation and Embedding

- Prepare a fresh fixative solution of 2% PFA in 0.1 M PHEM buffer (see recipe in Materials).

Note: Glutaraldehyde can be added to a final concentration of 0.1%. This improves the appearance of the ultrastructure, but may interfere with the antigenicity of the tissue—we recommend running a trial with the antibodies to be used to determine whether glutaraldehyde should be used or not. Antibodies may be tested by carrying out immunofluorescent staining on cells or frozen sections that have been treated with the fixative (Fig. 2). If no staining is obtained, it is unlikely that the antibody will perform in immunogold labeling.

- Immerse the tissue in the fixative immediately after removal. As with conventional EM, shave mice prior to dissection.
- Use clean single edged razorblades and fine forceps to dissect the tissue into pieces of no more than 1 mm^3 in a puddle of fixative under a dissecting microscope in a fume hood
- Transfer to fresh fixative and incubate for 1 h at room temperature
- Remove the fixative and wash the tissue three times in phosphate buffer (PB, see recipe in Materials) containing 0.15 M glycine
- Remove the PB/glycine and replace with 2% gelatine in PB. Incubate at 42 °C for 30 min. Repeat with 6% and then 12% gelatine solutions, followed by a second 12% gelatine incubation again for 30 min.
- Using a plastic Pasteur pipette, remove the tissue pieces and drop them on to a piece of parafilm. Immediately press a second piece of parafilm on top to flatten out the gelatine and minimize the amount of gelatine above and below the tissue. Place this parafilm sandwich on ice and allow the gelatine to set completely.
- Peel back the upper layer of parafilm and then under a dissecting microscope, carefully cut out a cube of gelatine containing the tissue using a razorblade. Try to minimize the amount of gelatine surrounding the tissue block. Using fine forceps, transfer the gelatine-embedded tissue to a vial containing 1.15 M sucrose/PB/toluidine blue (see recipe in Materials). Leave on a rotator overnight at room temperature to allow the sucrose to infiltrate the blocks. The toluidine blue stains the epidermis and facilitates accurate orientation of the tissue when mounting on cryopins.

- Replace the 1.15 M sucrose with 2.3 M sucrose and leave on the rotator at room temperature overnight. This two stage infiltration improves the penetration of the sucrose

- Using fine forceps, remove one block of tissue and gelatine from the sucrose and transfer to the top of an aluminum cryopin under a dissection microscope. Orient the tissue so that the epidermis will be cut as a transverse section (Fig. 3). Ensure that the tissue is supported on all sides with sucrose, but avoid excessive quantities of sucrose, especially above the sample—the goal is a low pyramid shape. Samples that are too high are more fragile during sectioning and are more liable to "ping" off the cryopin when the block strikes the knife. Remove excess sucrose with the torn edge of a piece of filter paper and then plunge freeze the pin and tissue in liquid nitrogen.

- Store cryopins in cryovials as follows. In order to allow liquid nitrogen to flow into the cryovial, cut a small slot near the top using a razorblade and another near the bottom on the opposite side. Label the cryovial and then freeze in the liquid nitrogen. Using precooled forceps, transfer the cryopin with tissue sample into the cryovial and transfer to liquid nitrogen storage. The tissue will keep for years as long as it remains frozen.

2. Cryosectioning

This is a brief protocol for cryosectioning. Before proceeding, it is essential to be trained in the proper use of a cryoultramicrotome by a qualified person. The techniques require practice and judgment to be successful, particularly the manipulation of sections on the knife face, picking up sections and correct use of the antistatic device. It should be noted that the cutting edge of a diamond knife is extremely delicate and careless use can cause damage that will be very expensive to

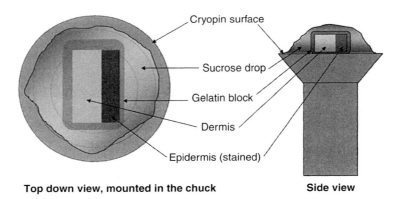

Fig. 3 Correct orientation of the tissue on the cryopin prior to plunge freezing. In order to minimize compression, the tissue should be sectioned vertically so that the epidermis is perpendicular to the edge of the knife.

repair. For this reason we use disposable glass knives for trimming and cutting semithin sections. Alternatively, diamond trimming knives are available from Diatome and other companies.

- Precool the cryoultramicrotome and a glass knife for trimming to $-80\,°C$ before removing the sample from liquid nitrogen storage.

- Remove the sample cryovial from storage and transfer immediately to a vacuum jar of liquid nitrogen. It is essential that the sample remains frozen at all times.

- Once the chamber of the cryoultramicrotome has reached the correct temperature, transfer the cryovial into the chamber, open it and fit the cryopin into the chuck using precooled forceps.

- Use an eyelash probe (see Materials) to brush any frost off the sample and check its orientation. Skin is best sectioned vertically, so the stained epidermis should be visible as a vertical blue line (Fig. 3).

- Fit and switch on the antistatic device.

- To trim the block, set the section thickness to 250 nm and the initial cutting speed to 5 mm/s. Advance the block towards the knife edge with the utmost caution! It is difficult to see exactly how close the knife edge is to the rounded surface of the block.

- Trim away the excess sucrose from the front of the block until the tissue is reached. As the area of the sections increases, they will start to crumple and compress.

- Trim the sides of the block face in order to reduce the area of the sections by moving the corner of the knife to one side of the block and trimming back by about 100 sections. Then withdraw the knife, rotate the chuck through 90° and trim again, repeating until the block face is square or rectangular in shape. The trimming speed may be increased up to 50 mm/s during this process, although a slower speed is recommended if the surface area of the trimmed sections is large. It is important to reduce the size of the block face as much as possible in order to avoid crumpled sections. Reducing the width rather than the height of the block face is more important for obtaining flat sections.

- Cut semithin sections of 200–250 nm thickness. The sections should come off the knife edge as a ribbon. Use an *eyelash probe* to guide the ribbon down the face of the knife and when the ribbon is 6–8 sections long, pick them up with a drop of 2.3 M sucrose in a 5 mm diameter wire loop. Transfer the sections to a glass microscope slide by touching the drop onto the surface. Stain with toluidine blue for 2 min, rinse in distilled, allow to dry and then examine under the light microscope to check that the epidermis is present and in the correct orientation. Rinse the loop in the beaker of distilled water and shake to dry.

- To cut ultrathin sections, fit a diamond knife and reduce the temperature to $-120\,°C$. Cut ultrathin sections (60–70 nm thick) at a speed of 1 mm/s, guiding the ribbon down the knife with an eyelash probe. The interference color should be gold/silver. When the ribbon reaches the bottom of the blade, pick up with a drop of 2.3 M sucrose/methyl cellulose (see recipe in Materials) in a 3 mm diameter loop. The drop will freeze when it touches the knife.

- Allow the drop to thaw and then examine under a dissecting microscope. The ribbon should only just be visible. Touch the drop onto an EM grid precoated with Formvar and carbon. The methyl cellulose will set over the top of the sections and protect them until they are labeled. Store in a covered Petri dish at 4 °C. Rinse and dry the loop before picking up sections again.

3. PVA Embedding for Immunogold Labeling

This method has been described previously. Please see the references mentioned above for a detailed description of the protocols involved.

C. Immunogold Labeling

The majority of this protocol involves transferring the grids between drops of reagents laid out on parafilm. With the exception of the antibodies, all drops should be about 100 µl in volume. It is very important to include controls which are performed in parallel with the test labeling. We use a known antibody as a positive control (generally the Mab 11–5F (Parrish et al., 1986) against the C terminus of DP) (Fig. 4) and as a negative control we omit the primary antibody, substituting blocking buffer, followed by the secondary gold-conjugated antibody.

- Invert grids containing sections in sucrose/methyl cellulose onto a Petri dish filled with PB and incubate for 30 min at 37 °C to dissolve the gelatine, sucrose and methyl cellulose.
- Using fine forceps, transfer the grids sequentially between five drops of PB/glycine, incubating for 2 min per drop. The glycine is to quench free aldehydes from the fixative.
- Block grids for 15 min on a drop of blocking buffer to prevent nonspecific labeling.
- Incubate the grids on 5 µl drops of primary antibody diluted in blocking buffer. The dilution and incubation time should be determined for each antibody. Generally speaking, the antibody dilution for immunogold labeling needs to be much lower than for immunofluorescence. If long incubations (i.e., more than 2 h) are required, we recommend carrying out the incubation on parafilm in a humidified Petri dish sealed with parafilm. It may also be appropriate to increase the volume of antibody to prevent its complete evaporation.
- Wash the grids on five drops of PB/glycine, 3 min per drop, followed by a final wash on PB
- Incubate grids on 5 µl drops of gold-conjugated secondary antibody diluted in blocking buffer for 20–30 min

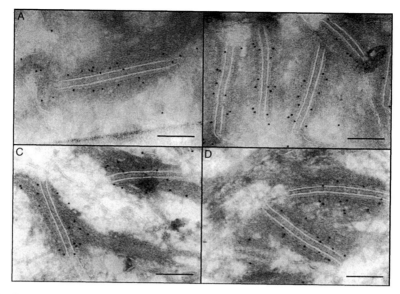

Fig. 4 Cryopreserved desmosomes of human skin. (A), (B): Labeled with Mab 33-3D against the cytoplasmic domain of Dsg2; (C), (D): Labeled with Mab 11-F against the C terminus of DP. Note in C the keratin intermediate filaments can be clearly seen. Scale bars = 230 nm.

- Wash on nine drops of PB—the first three washes are rapid, the next six for 3 min each.
- *For double immunogold labeling:* at this point wash the grids on PB/glycine, reblock and reincubate with the second primary antibody. Wash as described above and incubate with a secondary antibody conjugated to gold particles of a different size.
- Postfix the grids on drops of 1% glutaraldehyde in PB in the fumehood for five minutes
- Wash the grids on ten consecutive drops of water, 2 min each
- Incubate the grids on droplets of uranyl oxalate solution (see recipe in Materials) for 5 min, followed by a 1 min wash on water
- Wash the grids for 1 min on drops of UA/methyl cellulose (see recipe in Materials) and then transfer to fresh drops of UA/methyl cellulose on ice for 5 min.
- Using a steel loop mounted in the pointed end of a P1000 pipette tip, lift each grid out of the UA/methyl cellulose from underneath. Holding the side of the loop at a 90° angle touch it on a piece of filter paper and draw it along to drain the excess liquid away leaving the grid suspended on a film in the middle of the loop, but stopping before all of the liquid is drawn into the paper. Allow to air dry. The interference color of the film should be gold.

D. Quantification of the Distribution of Gold Particles

All micrographs should be taken at the same magnification. It is important to select desmosomes which have been sectioned transversely for quantification. "Good" desmosomes have clearly defined PM (Fig. 5). "Fuzzy" desmosomes, which are seen as two blurred lighter lines, often with abundant gold labeling (Fig. 6) are not appropriate for quantification. Electron micrographs will contain desmosomes that have been sectioned in various planes, so the experimenter should exercise judgment in the choice of the desmosomes that are to be measured. It is also important that the bias of the experimenter is minimized. Although selection is taking place, it must be on the basis of the orientation of the desmosome and not on the distribution of the gold!

We have utilized several methods for measuring the distribution of gold particles on desmosomes, including measuring with a ruler on printed micrographs and using image analysis software such as ImageJ. We have compared the data gathered from both methods and found the results to be identical.

If measuring on printed micrographs it is important to minimize bias by performing the measurements blind—i.e. without knowing which antibody was used and without knowing the magnification of the print.

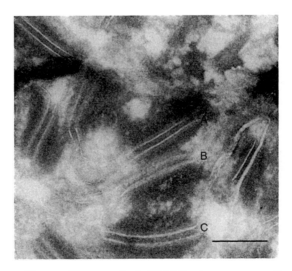

Fig. 5 Desmosomes of human skin as seen in the EM following cryopreservation. Note that there is less discernable ultrastructure than would be seen with conventional preservation, but that the PM are clearly defined. (A) An ideal desmosome for quantification following immunogold labeling as the membranes are sharp. (B) This desmosome has been sectioned at an oblique angle and is not suitable for quantification. (C) This desmosome can be quantified on one side only, where the membrane can be seen clearly. Scale bar = 300 nm.

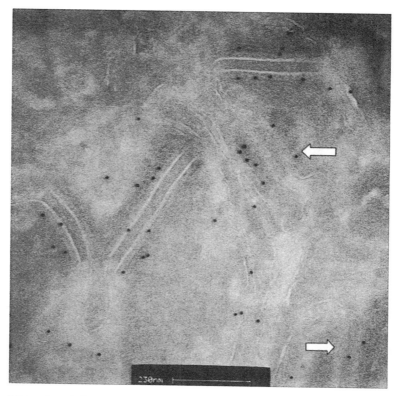

Fig. 6 Wider view of desmosomes labeled with 33-3D. Note the two desmosomes that have been sectioned at an oblique angle to the knife (*White arrows*) that initially appear to be non-specific labeling. Scale bar = 230 nm.

The advantage of using a program such as ImageJ is that data is collected by measuring the distance between two mouse clicks and the numbers are stored in a spreadsheet file for later analysis.

In both cases it is essential that the way distances are measured is consistent. Measurements may be taken from the intracellular or extracellular side of the PM and from the closer edge, centre or farther edge of the gold particle, but once the method of measurement is decided, it should be strictly adhered to.

Once a set of measurements has been obtained, their distribution may be statistically analyzed. We use the statistical software package Simfit to analyze gold particle distribution. The number of gold particles is plotted against the distance from the PM in nm and then a best fit curve is applied. If the analysis of particles shows a normal distribution, the peak of the distribution curve may be taken as the approximate location of the antibody epitope. If the distribution is not statistically normal, the analysis becomes more complicated.

When comparing the distribution of labeling with different antibodies, it is important to use an internal control to assess reproducibility between datasets. We use the Mab 11-5F either in double labeling with the antibody of interest or as a single labeled control in parallel with other antibodies. We have found that there is a high degree of variability in the results between different tissue samples, probably due to differences in tissue shrinkage during fixation, so if possible it is best to perform analyses on sections which have come from the same tissue block.

The number of desmosomes which are measured is in part dependent upon how dense the labeling is. With good labeling we consider a dataset to be 30 desmosomes per antibody.

IV. Materials

A. Conventional Electron Microscopy

SOLUTIONS:

Cacodylate buffer:

0.1 M sodium cacodylate
250 mM sucrose
2 mM calcium chloride
pH 7.5

Paraformaldehyde (PFA) solution:

Add 8 g of PFA powder to 100 ml distilled water in a glass flask. Heat with stirring to 60–70 °C then add a few drops of 4 M sodium hydroxide solution to clear the solution. Allow to cool and then dilute to 4% w/v with distilled water.

PFA in cacodylate buffer:

Prepare double strength cacodylate buffer above (i.e. 0.2 M sodium cacodylate) and mix 50:50 with fresh 4% aqueous PFA solution. Readjust the pH to 7.5.

Uranyl Acetate Stain:

Prepare a 4% (w/v) solution of UA dissolved in 77% ethanol
Centrifuge 1.5-ml aliquots at 3000 rpm for 20 min prior to use

Lead Citrate stain:

Prepare a 0.3% w/v solution of lead citrate in 0.1 M sodium hydroxide
Centrifuge 1.5 ml aliquots at 3000 rpm for 20 min prior to use
Spurr's resin (medium hardness)
Acetone concentrations 25%, 50%, 70%, 80%, 90%, 95%, 100%

Propylene oxide
0.1% Toluidine blue stain

EQUIPMENT

Fume hood
Benchtop microcentrifuge
Oven at 60 °C
Embedding mould (Agar Scientific)
Ultramicrotome (we use a Reichert-Jung Ultracut)
Probe (made by attaching an eyelash or eyebrow hair to a wooden stick approximately 15 cm long)
Glass knives
Knife making machine (We use a Leica)
Diamond knife (e.g., from Diatome) or glass knives with boats attached
Copper grids coated with Formvar

B. Cryopreservation and Immunogold Labeling

EMBEDDING

SOLUTIONS

PHEM buffer, 0.4 M Stock

240 mM PIPES
100 mM HEPES
8 mM $MgCl_2$
40 mM EGTA

Methyl cellulose/UA

Methyl cellulose stock: Prepare a 2% solution of 25 cP methyl cellulose (Sigma) in water at 90 °C. Cool on ice while stirring, until temperature has dropped to 10 °C. Stir overnight at low speed at 4 °C. Let the solution sit for 3 days at 4 °C. Centrifuge 95 min at 97,000 × g, 4 °C, then divide the supernatant into 10 ml aliquots and store for up to 2 months at 4 °C.

To prepare methyl cellulose/UA: Mix one part 4% UA solution with nine parts methyl cellulose and store at 4 °C.

Phosphate buffer (PB) 0.2 M Stock

Stock A: 35.6 g $Na_2HPO_4 \cdot 2H_2O$ in 1 l distilled water
Stock B: 31.2 g $NaH_2PO_4 \cdot 2H_2O$ in 1 l distilled water
Mix stock A and stock B in varying proportions for the correct pH:

pH 7.2 A:B 36 ml: 14 ml
pH 7 A:B 30.5 ml: 19.5 ml
pH 6.2 A:B 9.25 ml: 40.75 ml
Dilute the 0.2 M stock to 0.1 M for use.

Gelatine

Make fresh 2%, 6% and 12% solutions of gelatine in 0.1 M phosphate buffer pH 7.2 and store at 42 °C to keep liquid

16% paraformaldehyde solution

Glutaraldehyde solution

0.15 M glycine in 0.1 M phosphate buffer

1.15 M sucrose in 0.1 M phosphate buffer pH 7.2 + one drop of 0.1% toluidine blue stain per 50 ml

2.3 M sucrose in 0.1 M phosphate buffer

0.1% toluidine blue stain

Liquid nitrogen

EQUIPMENT

Dissecting microscope

Clean single edged razorblades

Fine forceps

Fixed angle specimen rotator (Agar Scientific)

Aluminum cryopins (Agar Scientific)

Cryovials and racks or canes

Liquid nitrogen storage facilities

SECTIONING

Solutions

2.3 M sucrose/PB

2.3 M sucrose/PB mixed 50:50 with methyl cellulose solution—keep on ice

Toluidine blue solution

Small beaker of distilled water

Equipment

Cryoultramicrotome (we use a Leica Ultracut S/FCS) or ultramicrotome with a cryo attachment

Antistatic device (we use a Diatome Static Line II)

Diamond knife (e.g. Diatome dry cryo 45°)

Knife making machine (we use a Leica)

Glass knives

Wire loops of 5 and 3 mm diameter

Eyelash probes (made by attaching an eyelash or eyebrow hair to a wooden stick approximately 15 cm long)

EM grids of copper or tungsten (we use 100 mesh thin bar, Agar Scientific) coated with Formvar and carbon

Light microscope

Glass microscope slides

C. Immunogold Labeling

SOLUTIONS

Uranyl oxalate solution

Mix 4% UA (see recipe) 1:1 with 0.15 M oxalic acid. Carefully adjust pH to 7 with 25% ammonium hydroxide, stirring well after adding each drop to ensure that any precipitates dissolve. Store in the dark at 4 °C.

M Phosphate Buffer

Blocking buffer (0.5% cold water fish gelatine, 1% Bovine Serum Albumen, 0.15 M glycine, 0.1 M PB, NaN_3)

PB containing 0.15 M glycine

Primary antibody diluted in blocking buffer

10 nm colloidal gold particles conjugated to an appropriate secondary antibody.

We use Goat anti-mouse IgG + IgM (from British Biocell International) diluted in blocking buffer (we dilute the BBI secondaries 1:20).

Note: For double labeling it is necessary to use different sized gold particles for each antibody, e.g. 10 and 20 nm particles.

PB containing 1% (v/v) glutaraldehyde

Uranyl oxalate solution

Methyl cellulose/UA solution (see recipe above)

EQUIPMENT

Whatman 50 filter paper

37 °C incubator

Tweezers with straight extra fine tips (we use Dumont No. 5 stainless steel, Agar Scientific)

Stainless steel loop slightly larger than grids attached to 1000 μl pipette tips

Parafilm

Fume hood
Benchtop microcentrifuge
Petri dishes

V. Discussion

The protocols described above have been used in our laboratory to localize desmosomal components, to identify calcium dependent and hyper-adhesive desmosomes and also to observe cellular events such as PKCα localization to desmosomes following wounding. This latter application using immunogold labeling holds a great deal of potential. The ability to label enzymes and other mobile molecules as well as structures within the cell will enable researchers to see changes in cellular activity far more precisely than with light microscopy alone.

There are still many unanswered questions about the structure of desmosomes and how that relates to desmosomal function e.g. how are the component proteins folded within the desmosomal plaque? What are the binding partners for each desmosomal component? There are also more EM-based techniques that can be used. He *et al.* (2003) used electron tomography to create a three dimensional map of the structure of a rapidly frozen, freeze substituted, plastic-embedded desmosome. The advent of cryo-electron microscopes that can accommodate unfixed frozen sections up to 200 nm in thickness should allow further insights into desmosome structure.

The molecular map of the desmosome created by North *et al.* (1999) looked at the distribution of the five major desmosomal proteins in the z-axis, but little or nothing is known about their arrangement in other dimensions—how are these molecules distributed across the face of the desmosomal plaque? EM methods will provide the means for investigators to observe more of the mechanism of desmosomal adhesion in cells and tissues.

VI. Summary

Protocols for conventional transmission electron microscopy and immunogold labeling of ultrathin cryosections of desmosomes are described. We also discuss methods for the quantitative analysis of the distribution of immunogold particles for molecular mapping and the appropriate controls that are required for the data to be valid.

Acknowledgments

The authors thank the members of the Garrod laboratory for helpful suggestions and in particular we acknowledge Dr. Mohammed Berika who contributed a picture from his thesis. We also thank all the staff of the Electron Microscopy Unit at the University of Manchester and especially Dr. Aleksandr Mironov Jr. for training and assistance. Work in this laboratory is funded by the Medical Research Council and the Wellcome Trust.

References

Al-Amoudi, A., Norlen, L. P., and Dubochet, J. (2004). Cryo-electron microscopy of vitreous sections of native biological cells and tissues. *J. Struct. Biol.* **148**, 131–135.

Bardsley, W. G., Ackerman, R. A., Bukhari, N. A., Deeming, D. C., and Ferguson, M. W. (1995a). Mathematical models for growth in alligator (*Alligator mississippiensis*) embryos developing at different incubation temperatures. *J. Anat.* **187**(Pt 1), 181–190.

Bardsley, W. G., Bukhari, N. A. J., Ferguson, M. W. J., Cachaza, J. A., and Burguillo, F. J. (1995b). Evaluation of model discrimination, parameter-estimation and goodness-of-fit in nonlinear-regression problems by test statistics distributions. *Comput. Chem.* **19**, 75–84.

Breathnach, A. S., and Wyllie, M. A. (1967). The problem of the Langerhans cells. *In* "Advances in Biology of Skin: The Pigmentary System" (W. Montagna, and F. Hu, eds.). Pergamon Press, Oxford.

Burdett, I. D. (1998). Aspects of the structure and assembly of desmosomes. *Micron* **29**, 309–328.

Chidgey, M., and Dawson, C. (2007). Desmosomes: A role in cancer? *Br. J. Cancer* **96**, 1783–1787.

Dusek, R. L., Godsel, L. M., and Green, K. J. (2007). Discriminating roles of desmosomal cadherins: Beyond desmosomal adhesion. *J. Dermatol. Sci.* **45**, 7–21.

Fujimoto, K. (1995). Freeze-fracture replica electron microscopy combined with SDS digestion for cytochemical labeling of integral membrane proteins. Application to the immunogold labeling of intercellular junctional complexes. *J. Cell Sci.* **108**(Pt 11), 3443–3449.

Garrod, D. R., Berika, M. Y., Bardsley, W. F., Holmes, D., and Tabernero, L. (2005). Hyper-adhesion in desmosomes: Its regulation in wound healing and possible relationship to cadherin crystal structure. *J. Cell Sci.* **118**, 5743–5754.

Garrod, D. R., and Chidgey, M. (2008). Desmosome structure, composition and function. *Biochim. Biophys. Acta* **1778**, 572–587.

Garrod, D. R., Merritt, A. J., and Nie, Z. (2002a). Desmosomal cadherins. *Curr. Opin. Cell Biol.* **14**, 537–545.

Garrod, D. R., Merritt, A. J., and Nie, Z. (2002b). Desmosomal adhesion: Structural basis, molecular mechanism and regulation (Review). *Mol. Membr. Biol.* **19**, 81–94.

Getsios, S., Huen, A. C., and Green, K. J. (2004). Working out the strength and flexibility of desmosomes. *Nat. Rev. Mol. Cell Biol.* **5**, 271–281.

Griffiths, G., McDowall, A., Back, R., and Dubochet, J. (1984). On the preparation of cryosections for immunocytochemistry. *J. Ultrastruct. Res.* **89**, 65–78.

Hatzfeld, M. (1999). The armadillo family of structural proteins. *Int. Rev. Cytol.* **186**, 179–224.

Hatzfeld, M. (2007). Plakophilins: Multifunctional proteins or just regulators of desmosomal adhesion? *Biochim. Biophys. Acta* **1773**, 69–77.

He, W., Cowin, P., and Stokes, D. L. (2003). Untangling desmosomal knots with electron tomography. *Science* **302**, 109–113.

Kelly, D. E. (1966). Fine structure of desmosomes, hemidesmosomes, and an adepidermal globular layer in developing new epidermis. *J. Cell Biol.* **28**, 51–72.

Kelly, D. E., and Kuda, A. M. (1981). Traversing filaments in desmosomal and hemidesmosomal attachments: Freeze-fracture approaches toward their characterization. *Anat. Rec.* **199**, 1–14.

Kelly, D. E., and Shienvold, F. L. (1976). The desmosome: Fine structural studies with freeze-fracture replication and tannic acid staining of sectioned epidermis. *Cell Tissue Res.* **172**, 309–323.

Kottke, M. D., Delva, E., and Kowalczyk, A. P. (2006). The desmosome: Cell science lessons from human diseases. *J. Cell Sci.* **119**, 797–806.

Liou, W., Geuze, H. J., and Slot, J. W. (1996). Improving structural integrity of cryosections for immunogold labeling. *Histochem. Cell Biol.* **106**, 41–58.

Miller, K., Mattey, D., Measures, H., Hopkins, C., and Garrod, D. (1987). Localisation of the protein and glycoprotein components of bovine nasal epithelial desmosomes by immunoelectron microscopy. *EMBO J.* **6**, 885–889.

North, A. J., Bardsley, W. G., Hyam, J., Bornslaeger, E. A., Cordingley, H. C., Trinnaman, B., Hatzfeld, M., Green, K. J., Magee, A. I., and Garrod, D. R. (1999). Molecular map of the desmosomal plaque. *J. Cell Sci.* **112**(Pt 23), 4325–4336.

North, A. J., Chidgey, M. A., Clarke, J. P., Bardsley, W. G., and Garrod, D. R. (1996). Distinct desmocollin isoforms occur in the same desmosomes and show reciprocally graded distributions in bovine nasal epidermis. *Proc. Natl. Acad. Sci. USA* **93**, 7701–7705.

North, A. J., Galazkiewicz, B., Byers, T. J., Glenney, J. R., Jr., and Small, J. V. (1993). Complementary distributions of vinculin and dystrophin define two distinct sarcolemma domains in smooth muscle. *J. Cell Biol.* **120**, 1159–1167.

North, A. J., Gimona, M., Cross, R. A., and Small, J. V. (1994b). Calponin is localised in both the contractile apparatus and the cytoskeleton of smooth muscle cells. *J. Cell Sci.* **107**(Pt 3), 437–444.

North, A. J., Gimona, M., Lando, Z., and Small, J. V. (1994a). Actin isoform compartments in chicken gizzard smooth muscle cells. *J. Cell Sci.* **107**(Pt 3), 445–455.

North, A. J., and Small, J. V. (2000). Immunofluorescent labelling of sections. *In* "Protein Localization by Fluorescence Microscopy" (V. J. E. Allen, ed.). Oxford University Press.

Orwin, D. F., Thomson, R. W., and Flower, N. E. (1973). Plasma membrane differentiations of keratinizing cells of the wool follicle. II. Desmosomes. *J. Ultrastruct. Res.* **45**, 15–29.

Parrish, E. P., Garrod, D. R., Mattey, D. L., Hand, L., Steart, P. V., and Weller, R. O. (1986). Mouse antisera specific for desmosomal adhesion molecules of suprabasal skin cells, meninges, and meningioma. *Proc. Natl. Acad. Sci. USA* **83**, 2657–2661.

Peters, P. J., Borst, J., Oorschot, V., Fukuda, M., Krahenbuhl, O., Tschopp, J., Slot, J. W., and Geuze, H. J. (1991). Cytotoxic T lymphocyte granules are secretory lysosomes, containing both perforin and granzymes. *J. Exp. Med.* **173**, 1099–1109.

Peters, P. J., Bos, E., and Griekspoor, A. (2006). Cryo-immunogold electron microscopy (Unit 4.7). "Current Protocols in Cell Biology". John Wiley & Sons, New York.

Rayns, D. G., Simpson, F. O., and Ledingham, J. M. (1969). Ultrastructure of desmosomes in mammalian intercalated disc; appearances after lanthanum treatment. *J. Cell Biol.* **42**, 322–326.

Shimizu, A., Ishiko, A., Ota, T., Saito, H., Oka, H., Tsunoda, K., Amagai, M., and Nishikawa, T. (2005). In vivo ultrastructural localization of the desmoglein 3 adhesive interface to the desmosome mid-line. *J. Invest. Dermatol.* **124**, 984–989.

Small, J. V., Furst, D. O., and De, M. J. (1986). Localization of filamin in smooth muscle. *J. Cell Biol.* **102**, 210–220.

Staehelin, L. A. (1974). Structure and function of intercellular junctions. *Int. Rev. Cytol.* **39**, 191–283.

Tokuyasu, K. T. (1973). A technique for ultracryotomy of cell suspensions and tissues. *J. Cell Biol.* **57**, 551–565.

Tokuyasu, K. T. (1980). Immunochemistry on ultrathin frozen sections. *Histochem. J.* **12**, 381–403.

Tokuyasu, K. T. (1986). Application of cryoultramicrotomy to immunocytochemistry. *J. Microsc.* **143**, 139–149.

Tokuyasu, K. T. (1989). Use of poly(vinylpyrrolidone) and poly(vinyl alcohol) for cryoultramicrotomy. *Histochem. J.* **21**, 163–171.

Tokuyasu, K. T., and Singer, S. J. (1976). Improved procedures for immunoferritin labeling of ultrathin frozen sections. *J. Cell Biol.* **71**, 894–906.

Vilela, M. J., Parrish, E. P., Wright, D. H., and Garrod, D. R. (1987). Monoclonal antibody to desmosomal glycoprotein 1-a new epithelial marker for diagnostic pathology. *J. Pathol.* **153**, 365–375.

Watt, F. M., Mattey, D. L., and Garrod, D. R. (1984). Calcium-induced reorganization of desmosomal components in cultured human keratinocytes. *J. Cell Biol.* **99**, 2211–2215.

Yin, T., and Green, K. J. (2004). Regulation of desmosome assembly and adhesion. *Semin. Cell Dev. Biol.* **15**, 665–677.

SECTION 4

The Nucleus

CHAPTER 19

A Protocol for Isolation and Visualization of Yeast Nuclei by Scanning Electron Microscopy

Stephen Murray* and Elena Kiseleva[†]

*TEM Service Facility
Paterson Institute for Cancer Research
University of Manchester
Manchester M20 4BX
United Kingdom

[†]Institute of Cytology and Genetics
Russian Academy of Science
Novosibirsk 630090
Russia

Abstract
I. Introduction
II. Materials and Instrumentation
    A. Chemicals
    B. Buffer Solutions
    C. Fixatives
    D. Equipment

III. Procedures
    A. Equipment Setup
    B. Preparation of Silicon Chips
    C. Yeast Growth
    D. Nuclear Isolation
    E. Preparation of Samples for FESEM
    F. Immunolabelling
    G. Ribosome Removal
IV. Comments and Problems
    A. Comments
    References

Abstract

This article describes a protocol that details methods for the isolation of yeast nuclei from budding yeast (*Saccharomyces cerevisiae*) and fission yeast (*Schizosaccharomyces pombe*), immunogold labelling of proteins, and visualization by Field Emission Scanning Electron Microscopy (FESEM). This involves the removal of the yeast cell wall and isolation of the nucleus from within, followed by subsequent processing for high resolution microscopy. The nuclear isolation step is performed by enzymatic treatment of yeast cells to rupture the cell wall and generate spheroplasts (cells that have partially lost their cell wall and their characteristic shape), followed by isolation of nuclei by centrifugation. This protocol has been optimized for the visualization of the yeast nuclear envelope (NE), nuclear pore complexes (NPCs), and associated cytoskeletal structures. Samples, once processed for FESEM, can be stored under vacuum for weeks, allowing considerable time for image acquisition.

I. Introduction

Using yeast as a model system has several advantages in the study of cellular structures and processes. The yeast genome has been fully sequenced and is small (approximately 1.25×10^7 base pairs in *Saccharomyces cerevisiae* and 1.38×10^7 in *Schizosaccharomyces pombe*) compared with other eukaryotic organisms (e.g., rice = 3×10^8 base pairs, human = 3.3×10^9 base pairs). Additionally, their short reproductive and growth cycles allows for rapid genetic manipulations and subsequent data acquisition. Yeast studies have yielded seminal work on cell-cycle events and molecular interactions (Fraser and Nurse, 1978; Nurse, 1975, 1983; Nurse and Bissett, 1981), and investigations of yeast nuclei and nuclear contents have facilitated the elucidation of cellular, nuclear, and chromatin structure. Early studies on yeast RNA (Sillevis Smitt *et al.*, 1970) have been followed by advances in our understanding of chromatin organization (Mason and Mellor, 1997) and more recently nucleosome mapping (Zhang and Reese, 2006). In addition, yeast has been highly important in the characterization of the structure of the nuclear envelope (NE) and nuclear pore complexes (NPCs). These analyses have been carried out using high resolution techniques such as negative stain electron microscopy and subsequent image reconstitution (Rout and Blobel, 1993; Winey *et al.*, 1997; Yang *et al.*, 1998), conventional transmission electron microscopy (TEM) (Allen and Douglas, 1989; Heymann *et al.*, 2006), new dual beam electron microscopy techniques (Kiseleva *et al.*, 2004), and field emission scanning electron microscopy (FESEM) (Rozjin and Tonino, 1964). Collection of structural data by negative stain or conventional TEM generates only images of sections through the cell of interest and additionally retention of the cell wall limits the penetration of many standard cellular stains, thereby decreasing the image quality and the information obtained. Although three-dimensional images can be obtained by serial sectioning of samples and image reconstitution, this will often miss rare structures and requires a degree of assumption about the continuity of structures that may be misleading.

Nuclei were first isolated from budding yeast in 1964 (Ho *et al.*, 2000). Previously used protocols for the isolation of yeast nuclei have been successful in generating a bulk material for biochemical analysis. However, these procedures involve successive centrifugation steps through Percoll and sucrose, which alters the nuclear morphology in two ways. Firstly, large numbers of ribosomes are maintained in tight association with the yeast NE and therefore mask direct visualization of the nuclear surface and secondly, the successive washing procedures can remove or dislodge filaments associated with the cytoplasmic aspect of the nuclear surface. Our aim has been to adapt these techniques to optimize preservation and visualization of the NE and associated NPCs. The protocol provided here is routinely used in our laboratory and has been used to reliably obtain high quality data for several key publications (Ho *et al.*, 2000; Rozjin and Tonino, 1964).

In comparison with the giant nuclei of the oocytes of the African clawed toad *Xenopus laevis* (Allen *et al.*, 2007) yeast nuclei cannot be manually dissected owing to their diminutive size—approximately 2 μm diameter in *S. cerevisiae* and approximately 3 μm diameter in *S. pombe*.

The methods described here have allowed the first direct visual confirmation that yeast NPCs share both the cytoplasmic and nucleoplasmic filamentous and annular structures (Kiseleva *et al.*, 2004) previously visualized within the vertebrate NPC. Previous investigations aimed at determining the structure of the yeast NPC involved stringent washing and high-speed centrifugation of yeast nuclear pellets (Yang *et al.*, 1998) and we suggest that this resulted in the initial assumption that yeast NPCs lacked the filamentous structures thought to be characteristic of NPCs in higher eukaryotes. This uniformity of NPC architecture between species highlights the high level of evolutionary conservation of the molecules and mechanisms involved in facilitated nucleoplasmic transport. In addition to this structural information, the identification of proteins or specific protein domains with immunogold, using either polyclonal antibodies or domain-specific monoclonal antibodies, has been crucial in determining the localization of several nuclear pore-specific proteins (nucleoporins; nups), specifically nup116 and nup159 (Ho *et al.*, 2000).

Here we describe a method for the isolation of budding and fission yeast nuclei, via spheroplast treatment and subsequent processing of these nuclei for scanning microscopy with immunogold labelling of component proteins.

II. Materials and Instrumentation

The following items are required for this protocol

A. Chemicals

- Acetone (VWR International)
- Argon cylinder, high purity >99.999% (Air Liquide)
- Bacto peptone (cat. no. 211830, Becton Dickenson)
- CO_2 cylinder, liquid with dip tube (<5 ppm water) (Air Liquide)
- Difco Bacto yeast extract (cat. no. 212750, Becton Dickenson)
- Dithiothreitol (cat. no. D0632, Sigma-Aldrich)
- Ethanol, 30%, 50%, 70%, 95%, solutions in dH_2O and 100% (cat. no. 101077Y, VWR International)
- Formaldehyde (37%) (cat. no. F8775, Sigma-Aldrich)
- Glucose (cat. no. 101176K, VWR International)
- Glutaraldehyde EM grade (cat. no. 49631, Fluka)
- Liquid nitrogen
- Lyticase 10,000 units (cat. no. L2524–10KU, Sigma-Aldrich)
- Magnesium chloride (cat. no. M1028, Sigma-Aldrich)
- NovoZym 234 (Novo Industri A/S, Bagsvaerd, Denmark)
- Osmium tetroxide, 4% in dH_2O (cat. no. R1017, Agar Scientific). Can be stored for several weeks at room temperature $20 \pm 2\ ^\circ C$
- Paraformaldehyde (Agar Scientific)
- Poly-L-lysine, 1 mg/ml in dH_2O (Sigma-Aldrich). Prepare fresh on day of use
- Polyvinylpyrrolidone PVP40 (Sigma-Aldrich). Polyvinylpyrrolidone 8% in 20 mM potassium phosphate pH 6.5, 0.5 mM $MgCl_2$, and 0.02% Triton X-100
- Potassium chloride, 1 M in dH_2O (cat. no. P9541, Sigma-Aldrich)
- Potassium Phosphate (Sigma-Aldrich) Potassium Phosphate Add 255 ml of 20 mM KH_2PO_4 to 245 ml of 20 mM K_2HPO_4 for pH7.4
- Sodium chloride, 1 M in dH_2O (cat. no. S3014, Sigma-Aldrich)
- Sorbitol (Sigma-Aldrich)
- Sucrose (Sigma-Aldrich)
- Tannic Acid (cat. no. T046, TAAB laboratories equipment)
- Tris–HCl (Sigma-Aldrich)
- Triton X-100 (Sigma-Aldrich). Prepare fresh on day of use
- Uranyl acetate, 1% in dH_2O (cat. no. R1260A, Agar Scientific). Can be stored at room temperature for several weeks. Protect from light

- Bovine serum albumin (BSA), 1% in 20 mM Tris–HCl (cat. no. A7030, Sigma-Aldrich). Prepare fresh on day of use
- Fish skin gelatin (FSG), 1% in 20 mM Tris–HCl (cat. no. G7765, Sigma-Aldrich). Prepare fresh on day of use
- Glycine, 0.1 M in PBS (cat. no. G7403, Sigma-Aldrich). Prepare fresh on day of use
- Potassium Hydroxide, 10 M in dH_2O (cat. no. 105021, VWR International)
- Primary antibody, as appropriate to research (Commercially available or custom made)
- Secondary antibody 10 nm gold conjugate EM grade (GE Healthcare)
- YPD medium 1% w/v Difco Bacto yeast extract, 2% w/v bacto peptone, 2% w/v glucose in dH_2O, sterilize by autoclaving and can be stored at 4 °C for several months

B. Buffer Solutions

- Buffer 1: 0.1 M Tris–HCl, 10 mM dithiothreitol in dH_2O pH 9.4. Prepare fresh on day of use
- Buffer 2: 1.2 M sorbitol, 20 mM potassium phosphate, 0.5 mM $MgCl_2$, in dH_2O pH 7.4. Prepare fresh on day of use
- Buffer 3 (for budding yeast): Defrost lyticase at room temperature for 15 min before use. Dissolve 10,000 units of lyticase in 800 μl of buffer 2 then further dilute 30 μl of this solution in 1 ml of buffer 2. Prepare fresh on day of use
- Buffer 4 (for fission yeast): Dissolve 2.5 mg/ml NovoZym 234 in buffer 2. Prepare fresh on day of use
- Buffer 5 (for immunolabelling): 20 mM Tris–HCl, 0.5 mM $MgCl_2$, 0.2 M Sucrose, pH 6.5
- Buffer 6 (for immunolabelling): 3.7% formaldehyde, 20 mM Tris–HCl, pH 6.5, 0.5 mM $MgCl_2$
- Buffer 7: 1.2M Sorbitol 2 M sucrose, 20 mM potassium phosphate, and 0.5 mM $MgCl_2$ pH 6.5

C. Fixatives

- Fixative 1 4% paraformaldehyde, 20 mM potassium phosphate pH 6.5, 0.5 mM $MgCl_2$, 0.2 M sucrose, freshly prepared and filtered using 0.4 μm syringe filter
- Fixative 2 2% glutaraldehyde, 0.2% tannic acid in 20 mM potassium phosphate, 0.5 mM $MgCl_2$, pH 7.4 freshly prepared and filtered using 0.4 μm syringe filter
- Fixative 3 3.7% formaldehyde, 20 mM Tris–HCl, pH 7.5, 0.5 mM $MgCl_2$
- Fixative 4 2% glutaraldehyde, 0.2% tannic acid in 20 mM Tris–HCl, 0.5 mM $MgCl_2$, pH 6.5

D. Equipment

- 100 ml conical glass flask
- 2 ml glass tubes (Biotrace International)
- 30 ml plastic tubes, (Biotrace International)
- 14 ml round bottomed centrifuge tubes, (Biotrace International)
- 50 ml plastic tubes, (Biotrace International)
- 30 mm plastic petri dishes (VWR International)
- 90 mm plastic petri dishes (VWR International)
- Back scattered Electron detector fitted to SEM
- Centrifuge with swing out rotor
- Coating unit with Chromium target capable of depositing 2–3 nm layers of chromium with a grain size of less than 0.4 nm. (e.g., Edwards BOC)
- Critical Point dryer (e.g., BAL-TEC, Liechtenstein)
- Dissecting microscope, variable zoom with a black base plate and two flexible light sources for incident illumination. (e.g., Leica Microsystems)
- Dumont tweezers 3 Dumostar (Agar Scientific)
- Filter paper Whatman grade no 50 (VWR International)
- High Resolution Scanning Electron Microscope
- Glass Homogenizer for 5 ml with glass plunger
- Light microscope with phase-contrast $100\times$ lens
- Microtube chambers see Fig. 1 and equipment setup
- Microtube 1.5 ml (VWR International)
- Mortar and Pestle
- Parafilm (VWR International)
- Nitrile gloves
- Razor blade
- Scalpel small
- Silicon chips 5 mm $\times$ 5 mm (Agar Scientific)
- Writing diamond (Agar Scientific)

III. Procedures

A. Equipment Setup

1. Coating unit set up to sputter chromium, using high purity Argon
2. Critical Point dry from liquid CO_2
3. Centrifuge with swing out rotor. Place a cotton wool support in the centrifuge tubes to keep the microtube chambers in vertical position during centrifugation.

4. Silicon chips: Number each chip using the writing diamond. Clean by dipping in a beaker of acetone and drying with tissue. Store in a petri dish to protect from contamination.

5. Microtube chambers: Remove the lid from a 1.5 ml microtube using a razor blade or scissors. Trim away any remaining lid attachment from the top of the tube and lid. Cut off the bottom of the tube to a diameter that will fit snugly inside the lid. When required, a 5 × 5 mm silicon chip should be placed inside the inverted lid and the tube pushed on to fit tightly without damaging the chip. When filled with liquid the chamber should not leak (Fig. 1).

B. Preparation of Silicon Chips

All samples for SEM analysis should be loaded onto silicon chips by centrifugation in pre-prepared chambers containing poly-L-lysine coated chips and appropriate buffer or fixative. Prepared chips should be used not longer that 30 min after preparation is completed.

1. Pipette 40 µl drops of 1 mg/ml poly-L-lysine in dH$_2$O onto the surface of 5 mm × 5 mm silicon chips (10–15 per experiment) in a small petri dish and incubate them at room temperature for 30 min.
2. Rinse poly-L-lysine coated chips in dH$_2$O and transfer to 10 mM potassium phosphate, 0.5 mM MgCl$_2$, pH 6.5 buffer if immunolabelling is to be performed. Otherwise rinse in fixative 1.
3. Place a single wet chip into each inverted microtube lid and carefully push the cut end of a microtube into the lid (see EQUIPMENT SETUP and Fig. 1).
4. Overlay chips with 20–40 µl buffer or fixative 1 as used in step 2. Store at room temperature until required but use within 30 mins.

C. Yeast Growth

Grow yeast cells in liquid YPD medium in 100 ml conical flask at 30 °C with moderate shaking (130 rpm) for 2 days. 50 mg of wet yeast cells in 50 ml YPD will be sufficient for 20 chips.

1. Pour 50 ml of fresh YPD medium into a fresh 100 ml flask and add 5 ml of the cell suspension from step 1. Grow for 3–4 h with strong shaking (170 rpm) at 30 °C to mid-log phase when most cells are elongated and have buds of different sizes. Check the stage of cell division between 3 and 4 h by phase-contrast in a light microscope at 100× magnification. Stop cell growth when approximately 80% of cells have buds. N.B. Yeast complete one round of the cell cycle in approximately 90 min at 30 °C. For examples of cell morphology please refer to Fig. 2.

It is important to generate mid-log phase cells to obtain a large number of high quality spheroplasts.

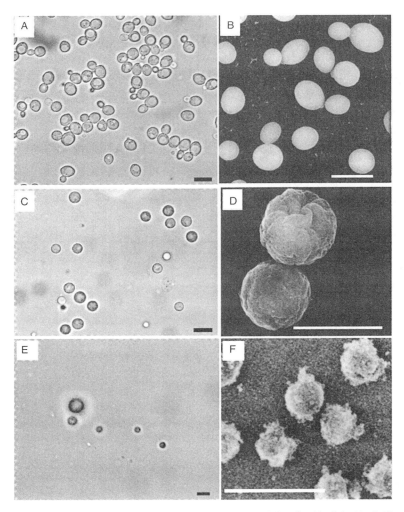

Fig. 1 Yeast cells (*S. cerevisiae*), spheroplasts and isolated nuclei visualized by light (A, C, E) and high resolution scanning (B, D, F) electron microscopy. (A, B) Budding yeast cells with small buds at logarithmic phase of growth; (C, D) spheroplasts obtained from yeast cells via removal of cell wall by zymolyse treatment; (E, F) free nuclei isolated from cells by spheroplasts centrifugation to chip (see Fig. 2). Bar equals 5 µm.

2. Transfer cells to 30–60 ml plastic tubes. Harvest ~3 g wet yeast cells by centrifugation (3000g) in a swing out rotor for 3 min at room temperature. Discard supernatant.

3. Resuspend (by pipetting) the pellet in 20 ml dH$_2$O for 1–2 min.

4. Pellet cells by centrifugation (3000g) in a swing out rotor for 3 min at room temperature. Discard supernatant.

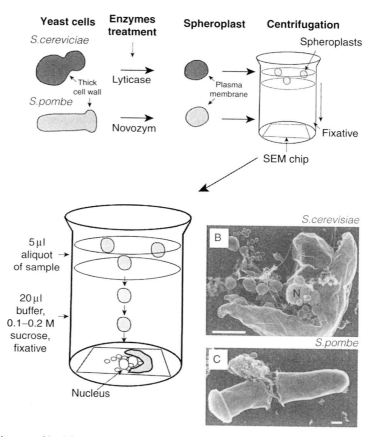

Fig. 2 Diagram of budding and fission yeast nuclear isolation procedure by spheroplast centrifugation onto an SEM silicon chip. (A) Yeast cells are treated with enzymes to generate spheroplasts and then isolated spheroplasts are spun down onto silicon chips through a sucrose and fixative cushion. (B, C) Examples of nuclear preparations revealing a nucleus emerging from an opened spheroplast of *S. cerevisiae* (upper panel) and *S. pombe* (lower panel). In both cases the rupturing of the spheroplast also reveals intranuclear contents. Bar equals 500 nm.

D. Nuclear Isolation

Isolate nuclei from the pellet using spheroplast centrifugation as follows.

1. Resuspend the pellet (from step 5) in 30 ml buffer 1 in a 50 ml plastic centrifuge tube for 25 min at 30 °C with moderate shaking.

2. Pellet cells by centrifugation (3000g) in a swing out rotor for 3 min at room temperature. Discard supernatant.

3. Resuspend pellet in 50 ml buffer 2.

4. Pellet cells by centrifugation (3000g) in a swing out rotor for 3 min at room temperature. Discard supernatant.

5. Resuspend the cell pellet in 20 ml buffer 3 (containing lyticase for budding yeast) or buffer 4 (containing Novo Zym for fission yeast).

6. Incubate cells in buffer 3 or in buffer 4 for 20–40 min at 30 °C with moderate shaking and swirling.

7. Check every 10 min for spheroplasts. Transfer 1 μl of sample to a glass slide, place a coverslip on top and observe under a phase microscope at 100×. Stop the incubation when about 70% of the cells are converted to spheroplasts. Spheroplasts have round forms and are less bright than yeast cells.

Note: Spheroplasts are very fragile therefore manipulate them carefully to avoid lysis.

8. Pellet the spheroplasts for 3 min at 3000g in a swing out rotor at room temperature and discard the supernatant. Resuspend the pellet in 10 ml buffer 2, pellet again, and repeat this step twice more to remove the enzyme.

9. Resuspend the spheroplast pellet in 10 ml buffer 2 and store at 4 °C for up to 2 h during which time silicon chips should be prepared as outlined in step **C** (also see Equipment Setup and Fig. 1).

10. Microtube centrifugation chambers containing silicon chips and fixative (step **C**) should be used within 30 min of preparation. Spheroplasts can be stored at 4 °C for several hours.

11. All subsequent steps should be completed at 4 °C.

12. Prepare three small glass bottles (1 ml screw cap vials) and label them 1, 2, and 3.

13. Transfer 200 μl of spheroplasts to each bottle and dilute carefully by adding 200, 400, or 600 μl of 0.5 mM $MgCl_2$ pH 6.5 to bottles 1, 2, and 3, respectively. Thus the spheroplasts:water ratios will be 1:1 in bottle 1, 1:2 in bottle 2, and 1:3 in bottle 3.

14. Flick the bottles several times manually to mix and incubate for approximately 5 min. Dilution of the spheroplast suspension in sorbitol solution (buffer 2) causes gentle osmotic shock to the spheroplasts.

15. Check 2 μl of each sample under a phase-contrast microscope (100× lens) and compare with untreated spheroplasts. Nuclei will be released from shocked spheroplasts and can be observed as small round phase dark objects. There should be a variation in size as those released from recently budded yeast will be smaller. Vacuoles are also released and are of a similar size but will be phase bright. Typically the observations might be as follows: 1:1 dilution—up to 50% of nuclei released; 1:2 dilution—many released but some nuclei lysed with chromatin leaking out; 1:3 dilution—generally fragmented, nuclei completely lysed or may be swollen. A lower dilution of 1:0.5 can be used if necessary.

16. Choose a sample dilution which gives a reasonable number of released nuclei without causing excessive lysis (usually 1:1 dilution) and prepare a freshly shocked sample containing 200 μl spheroplasts. If the nuclei are to be processed for

immunolabelling, continue as outlined in step G. Immunolabelling. If ribosome removal is required, continue as outlined in H. Ribosome removal.

To prepare samples for FESEM, continue immediately with step F.

E. Preparation of Samples for FESEM

1. To six previously prepared 1.5 ml microtube chambers (see Section III C. Preparation of Silicon Chips and Fig. 1) transfer 4 μ or 8 μl of the sample (from step 6) to the top of the fixative solution, as indicated in the Table I. Place the microtube chambers inside 14 mm centrifuge tubes (to hold them upright and contain any leakage) and spin in a swing out rotor for 3 min at 4 °C as detailed in the Table I below. During this centrifugation, the spheroplasts and nuclei make contact with the chip surface, break open, and release their contents, which become attached to the chip with immediate fixation (Fig. 3).

2. To visualize the nucleoplasmic side of the NPC, the samples should be pelleted for 5 min at 4000–5000g to break down the NE.

3. Isolated nuclei are very fragile and rupture very easily under any mechanical pressure; therefore all manipulations should be done carefully.

4. Take the microtube chambers out of the centrifuge tubes and carefully separate the chamber from the lid. Remove the chips with samples from the inverted lid using forceps and wash them for 10 min in petri dish with fixative 1 (without sucrose) in order to remove sucrose contamination.

5. Transfer the chips to a petri dish with Fixative 2 and incubate for a minimum of 10 min at room temperature.

It is possible to store samples in fixative 2 at 4 °C for up to 48 h. Cover the dishes to prevent them drying out. For the following steps use no. 3 tweezers to transfer chips manually into a petri dish containing the appropriate reagent. Keep the NE upwards. Do not allow to dry out at any point.

6. Wash chips briefly in dH_2O.

Table I
Preparation of Samples for FESEM

Amount of sample (μl)	Centrifugation[a] speed (g)
4	1000
4	2000
4	3000
8	1000
8	2000
8	3000

[a] Centrifugation of the samples was carried out for 3 min at 4 °C.

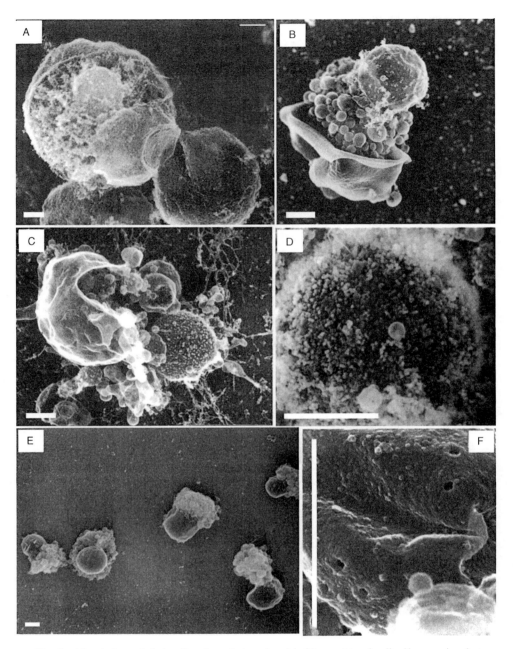

Fig. 3 Morphology of whole cells, spheroplasts and nuclei of *S. cerevisiae* visualized by scanning electron microscopy. (A) yeast cells at logarithmic phase of growth with partly removed plasma membrane. The centrally localized nucleus and cytoskeletal filaments are seen; (B, C) two examples of spheroplasts opened by centrifugation onto a chip. The nuclei and intracellular components partially released from an opened spheroplast; (D) nucleus densely covered by ribosomes at high magnification; (E) aggregates of intracellular components with nuclei (arrows) isolated from different spheroplasts; (F) several nuclear pores visualized over the NE after the incubation of isolated nuclei with 1 M NaCl. Bar equals 600 nm.

7. Post fix in aqueous 1% Osmium Tetroxide for 10 min.
8. Wash briefly in dH$_2$O.
9. Stain chips for 10 min in 1% aqueous uranyl acetate. This stains nucleic acids and proteins.
10. Dehydrate samples through an ethanol series: 2 min in each 30%, 50%, 70%, 95% (twice), 100% (three times).
11. Transfer chips to Critical Point drying apparatus containing ethanol as intermediate reagent. Critical point dry from high purity (<5 ppm water) liquid CO$_2$. Remove chips from the critical point drier and transfer them to a clean petri dish.
12. It is possible to store samples under vacuum for up to 1 week at room temperature.
13. Coat samples with 3 nm chromium.
14. It is possible to store samples under vacuum at room temperature for several weeks.
15. Visualize by High Resolution FESEM (for examples see Figs. 2 and 4–7).
16. Store samples under vacuum at room temperature for optimal preservation. They can be kept for several months and can be recoated with chromium (step 13), if required.

F. Immunolabelling

For the following steps use no. 3 forceps to transfer chips manually into a petri dish containing the appropriate reagent at room temperature. Keep the samples upwards. Do not allow to dry out at any point. All stages of immunolabelling are performed at room temperature unless otherwise stated.

1. Centrifuge yeast nuclei isolated from spheroplasts from step 6 (before or after removal of the ribosomes, see G) onto silicon chips for 3 min at 3000g through buffer 5.
2. Transfer samples into buffer 6 and incubate for 15 min.
3. Wash in buffer 6 (without formaldehyde) twice for 5 min.
4. Transfer to 0.1 M glycine in 20 mM Tris–HCl, pH 6.5, 0.5 mM MgCl$_2$ for 10 min (quenches unreacted aldehyde).
5. Block with 1% BSA in 20 mM Tris–HCl, pH 6.5, 0.5 mM MgCl$_2$ for 10–20 min. Alternatively 1% FSG can be used to minimize any nonspecific binding of the antibodies that may persist when BSA is used.
6. Prepare a wet chamber using a 9 cm Petri dish containing wet filter paper with Parafilm on top.
7. Dilute the primary antibody at an appropriate concentration in 20 mM Tris–HCl, pH 6.5, 0.5 mM MgCl$_2$.
8. Place chip on dry filter paper to dry the back.

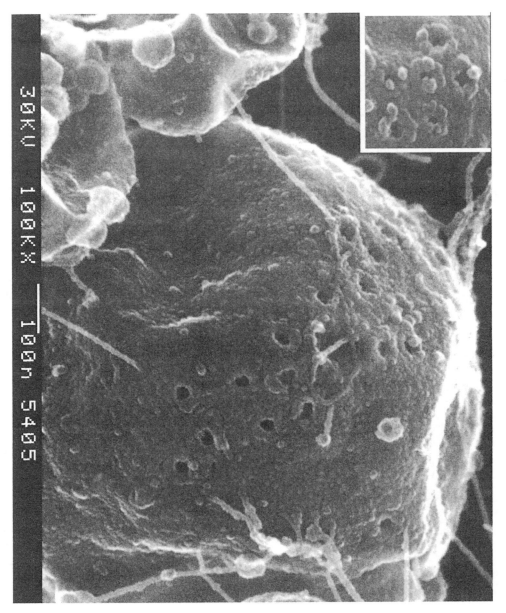

Fig. 4 Nuclear pore complexes (NPCs) in the nuclear envelope (NE) of *S. cerevisiae* visualized by scanning electron microscopy. Sample has been prepared by spheroplast production-centrifugation. Image was pseudocolored to highlight and differentiate the NE (green) and NPCs (blue). Insert demonstrates NPCs from the other nucleus. Bar equals 100 nm.

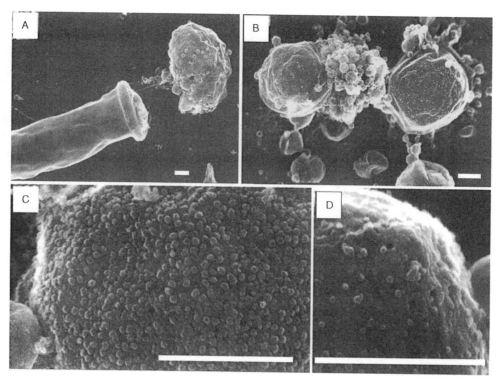

Fig. 5 Morphology of the spheroplasts and nuclei of *S. pombe* visualised by scanning electron microscopy. (A, B) Two examples of intracellular content isolation from spheroplasts. (C) A fragment of nuclear envelope covered with dense layer of ribosomes. (D) A fragment of nucleus after its treatment with 0.3 M NaCl. NPCs (arrows) are clearly seen at the NE. Bar equals 500 nm.

9. Place chip on the Parafilm in the wet chamber and quickly pipette on 10 μl antibody and incubate for 1 h at room temperature.
10. Wash in 20 mM Tris–HCl, pH 6.5, 0.5 mM $MgCl_2$ twice for 10 min.
11. Dilute the secondary antibody gold conjugate at a concentration of 1 in 10 (or 1:20) in 20 mM Tris–HCl, pH 6.5, 0.5 mM $MgCl_2$.
12. Place chip on dry filter paper to dry the underside of the chip.
13. Place chip on Parafilm in wet chamber and quickly pipette on 10 μl secondary antibody and incubate for 1 h at room temperature.
14. Wash 3 times for 10 min in 20 mM Tris–HCl, pH 6.5, 0.5 mM $MgCl_2$.
15. Fix in 2% glutaraldehyde, 0.2% Tannic acid, 20 mM Tris–HCl pH 6.5 (fixative 4).
16. Continue immediately with step 10 of F. Preparation of samples for FESEM.

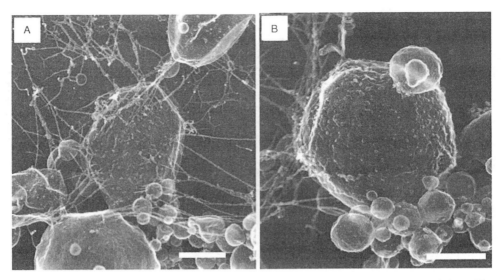

Fig. 6 Intracellular contents from two budding yeast spheroplasts demonstrate cytoskeleton filaments. (A) Filaments attach to the nuclear envelope all over the nucleus; B, filaments attach to the NE just at one side of the nucleus. Bar equals 500 nm.

G. Ribosome Removal

To analyze the structural organization of individual cytoplasmic components of NPCs it is necessary to remove the ribosomes from the nuclear surface. Sodium or potassium salts are usually used. These salt solutions remove ribosomes with similar efficiencies.

1. To remove ribosomes from the NE surface, incubate the samples for 10 min with 0.3 M–0.5 M NaCl or 0.3 M–0.5 M KCl in 20 mM potassium phosphate pH 6.5.
2. Proceed to step 7 without washing. Samples can subsequently be used for immunolabelling experiments in Box 2 step 2 if required.

Variation in ribosomal removal gives different observations by FESEM, for example:

1. 0.3 M salt—about 10% of ribosomes are removed.
2. 0.5 M salt—50% of ribosomes are removed.
3. 1 M salt—all ribosomes are removed.

In cases (1) or (2) it is possible to observe native NPCs with cytoplasmic components. In case (3) the cytoplasmic NPC components are damaged and just the central NPC components are preserved.

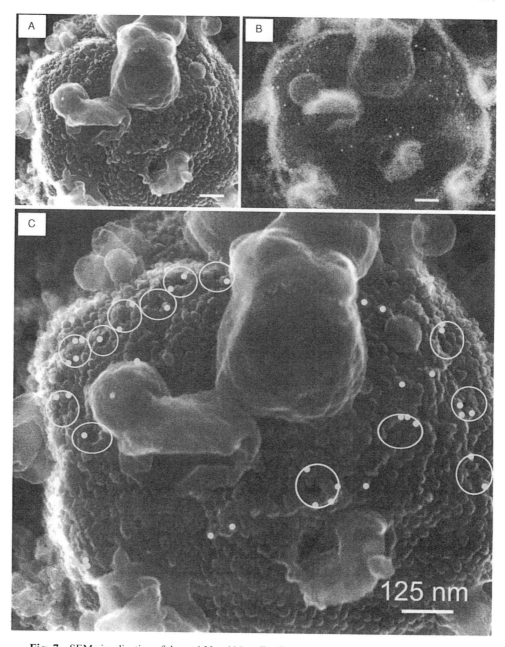

Fig. 7 SEM visualization of the anti-Nup 116 antibodies on the cytoplasmic surface of an *S. cerevisiae* nucleus. (A) Visualization of the yeast nucleus isolated from a spheroplast by centrifugation; (B) secondary image of the corresponding backscatter electron image of the same nucleus shows the position of the anti-Nup-116 immunogold labels; (C) images A and B are overlaid and the positions of the gold particles are pseudocolored yellow. Nuclear pores are marked by white circles. Bar equals 5 μm.

Table II
Troubleshooting

Step	Problem	Possible reason	Solution
Step D 7	Cells do not convert to spheroplasts	Enzyme does not work	Use new bottle of enzyme and fresh cells
Step D 7	Cells do not convert to spheroplasts	Cell wall of this yeast strain is thicker than that of wild type	Increase enzyme concentration in solution or the time of cell incubation with the enzyme
Step D 7	Cells convert to spheroplasts very quickly and most spheroplasts burst	Cell wall of this yeast strain is thinner than that of wild type	Take new cells and incubate them in solution with lower enzyme concentration
Step D 7	Cells convert to spheroplasts but lyse when transferred to slide for phase microscopy	Sorbitol concentration is too low	Take new cells and incubate them in solution with correct concentration of Sorbitol
Step D 15	No nuclei seen by phase microscopy	Cell wall was not removed completely because of incomplete/ineffective enzyme treatment	Prepare new spheroplasts
Step D 15	No nuclei seen by phase microscopy	Incomplete/weak osmotic shock of spheroplasts	Reduce concentration of spheroblasts being treated (add more water)
Step D 15	Spheroplast and nuclei lyse when transferred to microscope slide	Osmotic shock too strong	Increase spheroblast concentration (add less water)
Step E 15 observed Adjust at step 1	No material seen by FESEM on silicon chip	Sample volume was too small	Increase volume of samples (e.g., 10 µl)
Step E 15 observed Adjust at step 1	No material seen by FESEM on silicon chip	Speed of spheroplast centrifugation is insufficient to sediment material	Increase the speed of centrifugation
Step E 15 observed Adjust at C preparation of silicon chips	No material seen by FESEM on silicon chip	Poly-L-lysine does not adhere to sample therefore sample was washed away from chip during successive washing steps	Use fresh solution of poly-L-lysine for chip incubation
Step E 15 observed Adjust at step 1	Chip surface is covered by thick layer of material	Sample volume centrifuged onto chip is too great	Decrease volume of loaded samples
Step E 15 observed Adjust at step 1	Nuclei damaged	Centrifugation speed is too high	Decrease speed of centrifugation
Step E 15 observed Adjust at step 1	Structures do not look clean and are covered by layer of "melted" material	Chip was not washed sufficiently and is likely to be contaminated with sucrose	Increase the time of chip washing in buffer 2 (without sucrose)
Step E 15 observed	No gold label	Fixation may be inappropriate for primary antibody	Check fixation tolerance of an-tibody. Omit glutaraldehyde

(*continues*)

Table II (*continued*)

Step	Problem	Possible reason	Solution
Adjust at C Preparation of silicon chips Step E 15 observed	No gold label	Insufficient concentration of primary antibody	Increase concentration of primary antibody. EM protocols typically require higher concentrations
Adjust at F7 Immunolabelling Step E 15 observed	Nonspecific labelling	Insufficient blocking or washing	Check that blocking steps are suitable for primary antibody. Adapt protocol if required. Increase washing steps
Adjust at F7 Immunolabelling Step E15 observed	Nonspecific labelling	Nonspecific binding of the gold secondary	Always do a diluent control with no primary antibody to exclude the possibility of nonspecific binding of the gold secondary. Change supplier if this is a problem
Adjust at F (11) Immunolabelling Step E15 observed	Limited gold labelling	Steric hindrance. The gold particles are preventing access to the epitope	Use smaller colloidal gold. 5 nm gold typically gives more labelling than do 10 nm but can be more difficult to detect. 10 nm can be searched for at 20–40,000× magnification but 5 nm usually requires 60,000×
Adjust at F (11) Immunolabelling			

IV. Comments and Problems

A list of possible problems and causes together with solutions can be found in Table II.

A. Comments

It is critical to ensure that the starting material for this protocol is of the highest standard as this will have the single largest influence on the quality of images that can be captured by SEM. Therefore it is essential to ensure that cultured cells are kept appropriately. Figures 5–7 show several representative images of budding yeast; *S. cerevisiae* (Figs. 5 and 6) and fission yeast *S. pombe* (Fig. 7) when treated with enzymes to produce spheroplasts. Subsequent homogenization of spheroplasts liberates nuclei and these can be processed for high resolution visualization (Fig. 8). SEM images of the "ribosome-free" nuclear surfaces of yeast nuclei generated using the protocols herein reveal the density of NPCs within the NE and the macromolecular structural components of both the nucleoplasmic and cytoplasmic faces of the NPC. When immunogold labelling specific proteins and protein domains, the secondary electron image is acquired simultaneously and in exact register with the backscatter image. Using standard imaging software the backscatter image is superimposed upon the secondary image, thereby revealing the location of the gold particles (Fig. 7).

Acknowledgments

S.M. acknowledges the support of CR-UK. E.K. acknowledges the Wellcome Trust and Russian Federation for Basic Research. Victor Alvarez and Iain Hagan of the Cell Division group are acknowledged for the provision of *S. pombe* and *S. cerevisiae* cultures.

References

Allen, J. L., and Douglas, M. G. (1989). Organization of the nuclear pore complexes in *Saccharomyces cerevisiae*. *J. Ultrastruct. Mol. Struct. Res.* **102,** 95–108.

Allen, T., Rutherford, S. A., Murray, S., Sanderson, H. S., Gardiner, F., Kiseleva, E., Goldberg, M. W., and Drummond, S. P. (2007). A protocol for isolating *Xenopus* oocyte nuclear envelope for visualization and characterization by scanning electron microscopy (SEM) or transmission electron microscopy (TEM). *Nat. Protoc.* **2,** 1166–1172.

Fraser, R. S., and Nurse, P. (1978). Novel cell cycle control of RNA synthesis in yeast. *Nature* **23,** 726–730.

Heymann, J. A., Hayles, M., Gestmann, I., Gianuzzi, L. A., Lich, B., and Subramaniam, S. (2006). Site-specific 3D imaging of cells and tissues with a dual beam microscope. *J. Struct. Biol.* **155,** 63–73.

Ho, A. K., Shen, T. X., Ryan, K. J., Kiseleva, E., Levy, M. A., Allen, T. D., and Wente, S. R. (2000). Assembly and preferential localization of Nup116p on the cytoplasmic face of the nuclear pore complex by interaction with Nup82p. *Mol. Cell Biol.* **20,** 5736–5748.

Kiseleva, E., Allen, T. D., Rutherford, S., Bucci, M., Wente, S. R., and Goldberg, M. W. (2004). Yeast nuclear pore complexes have a cytoplasmic ring and internal filaments. *J. Struct. Biol.* **145,** 272–288.

Mason, J. A., and Mellor, J. (1997). Isolation of nuclei for chromatin analysis in fission yeast. *Nucleic Acids Res.* **25,** 4700–4701.

Nurse, P. (1975). Genetic control of cell size at cell division in yeast. *Nature* **256,** 547–551.

Nurse, P. (1983). Coordinating histone transcription and DNA replication. *Nature* **6,** 78.

Nurse, P., and Bissett, Y. (1981). Gene required in G1 for commitment to cell cycle and in G2 for control of mitosis in fission yeast. *Nature* **292,** 558–560.

Rout, M. P., and Blobel, G. (1993). Isolation of the yeast nuclear pore complex. *J. Cell Biol.* **123,** 771–783.

Rozjin, T. H., and Tonino, G. J. (1964). Studies on the yeast nucleus. I. The isolation of nuclei. *Bioch. Biophys. Acta.* **91,** 105–112.

Sillevis Smitt, W. W., Nanni, G., Rozijn, T. H., and Tonino, G. J. (1970). Sedimentation characteristics of RNA from isolated yeast nuclei. *Exp. Cell Res.* **59,** 440–446.

Winey, M., Yarar, D., Giddings, T. H., Jr., and Mastronarde, D. N. (1997). Nuclear pore complex number and distribution through the *Saccharomyces cerevisiae* cell cycle by three-dimensional reconstruction from electron micrographs of nuclear envelope. *Mol. Biol. Cell.* **8,** 2119–2132.

Yang, Q., Rout, M. P., and Akey, C. W. (1998). Three-dimensional architecture of the isolated yeast nuclear pore complex: Functional and evolutionary implications. *Mol. Cell.* **1,** 223–234.

Zhang, Z., and Reese, J. C. (2006). Isolation of yeast nuclei and micrococcal nuclease mapping of nucleosome positioning. *Methods Mol. Biol.* **313,** 245–255.

CHAPTER 20

Scanning Electron Microscopy of Nuclear Structure

Terence D. Allen,* Sandra A. Rutherford,* Stephen Murray,* Sheona P. Drummond,† Martin W. Goldberg,‡ and Elena Kiseleva§

*Department of Structural Cell Biology
Paterson Institute for Cancer Research
University of Manchester
Manchester M20 4BX
United Kingdom

†Post-transcriptional Research Group
Manchester Interdisciplinary Biocentre
University of Manchester
Manchester M1 7ND
United Kingdom

‡Department of Biological and Biomedical Sciences
University of Durham
Durham DH1 3LE
United Kingdom

§Institute of Cytology and Genetics
Russian Academy of Science
Novosibirsk 630090
Russia

    Abstract
  I. Introduction
      A. Microscopy of the Nucleus
      B. The Nuclear Envelope
      C. Scanning Electron Microscopy of the Nucleus and NE
 II. Rationale
III. Methods
 IV. Colloidal Gold in the SEM
  V. Immunolabeling Protocol

VI. CPD for High Resolution SEM
   A. Coating for High Resolution SEM
VII. Discussion
   A. The Value of FESEM Imaging for nuclear and subcellular structure
References

Abstract

Accessing internal structure and retaining relative three dimensional (3D) organization within the nucleus has always proved difficult in the electron microscope. This is due to the overall size and largely fibrous nature of the contents, making large scale 3D reconstructions difficult from thin sections using transmission electron microscopy. This chapter brings together a number of methods developed for visualization of nuclear structure by scanning electron microscopy (SEM). These methods utilize the easily accessed high resolution available in field emission instruments. Surface imaging has proved particularly useful to date in studies of the nuclear envelope and pore complexes, and has also shown promise for internal nuclear organization, including the dynamic and radical reorganization of structure during cell division. Consequently, surface imaging in the SEM has the potential to make a significant contribution to our understanding of nuclear structure.

I. Introduction

A. Microscopy of the Nucleus

The last decade has marked a renaissance in the study of nuclear organization, largely driven by advances in molecular biology which have facilitated tagging of specific proteins in living cells with novel probes such as green (and other) fluorescent proteins. These probes are visualized by conventional and confocal light microscopy (LM), allowing live cell imaging combined with time lapse. This approach has produced seminal advances in our understanding of both static and dynamic nuclear organizations (Lamond and Spector, 2003; Trinkle-Mulcahy and Lamond, 2007). However, there are limitations with LM; firstly, a resolution limited to around 100 nms and secondly, although LM fluorescence may be sensitive enough to pick up single molecule signals, it is always against a dark background (hence the term "nuclear speckles") and consequently lacks detail of the macromolecular context of the tagged structure. Both of these limitations are overcome by electron microscopy (EM), and although the nature of EM precludes live cell imaging, this limitation can be addressed by viewing the same cell by LM directly before EM preservation, allowing access to specific stages of biological processes (see Chapters 5 and 6, this volume). Transmission EM has produced fundamental nuclear structural information, particularly when combined with specific staining (Cremer et al., 2006; Hozák and Fakan, 2006; Chapter 22, this volume), but the sheer bulk and largely fibrous internal organization distributed over hundreds of cubic microns has made it difficult to produce three dimensional (3D) reconstructions of nuclear structure over large areas. The purpose of this chapter is to illustrate how field emission scanning electron microscopy (FESEM) can provide direct and unique visualization of the macromolecular details of structure both at the surface and within the core structure of the entire nucleus.

B. The Nuclear Envelope

The nuclear envelope (NE) is the barrier between the repository of genetic information within the nucleus and the sites of protein synthesis in the cytoplasm, and controlled transport across this barrier is crucial to cellular well-being. The NE is formed by a double membrane, which is regularly penetrated by areas of fusion between the outer and inner membranes, creating sites in the NE for insertion of nuclear pore complexes (NPCs) (Allen et al., 2000; Stewart, 2007).

On the outer (cytoplasmic) face of the NE there are proteins which link the NE and the cytoskeleton. All three components of the cytoskeleton, i.e., microtubules, actin filaments, and intermediate filaments are involved in nuclear positioning via KASH proteins on the outer nuclear membrane which are connected to SUN proteins on the inner nuclear membrane (Starr, 2007). Other nuclear surface proteins such as Nesprins have also been shown to link with both actin and intermediate filaments, and thus generating a continuous network linking the nucleus and its contents to cell surface and extracellular matrix (Wilhelmsen et al., 2005).

On the inner (nucleoplasmic) aspect of the NE is an underlying mesh-like structure termed the nuclear lamina (Bridger et al., 2007). The lamina interfaces with heterochromatin clusters at the nuclear periphery, indicating that the proteins of the lamina may participate in regulation of gene expression. Recent data suggests that lamina proteins regulate transcription by recruiting chromatin modifiers and transcription factors to the nuclear periphery (Heessen and Fornerod, 2007; Shaklai et al., 2005). Alterations in the lamina proteins have been implicated in several disease states (Grunebaum et al., 2005), and more recently in stem cell differentiation and development (Pajerowski et al., 2007).

C. Scanning Electron Microscopy of the Nucleus and NE

Although the use of a surface imaging technology such as scanning electron microscopy (SEM) might appear an unlikely route to study the nucleus, buried as it is within the cytoplasm, there are many advantages to this type of approach. The arrival of FESEM as a generally accessible technology has improved resolution in the instrument to around the nanometer level. Surface imaging by FESEM has proved particularly useful for studying the cytoplasmic/nuclear interface, for interaction of the NE with both cytoplasmic and nucleoplasmic elements, and also the structure of the NPCs themselves (Allen et al., 2000). Recent studies using FESEM show promise for both internal nuclear fiber organization and overall sub nuclear organization (Kiseleva et al., 2004).

Access to the nuclear surface and interior clearly requires invasive methods, and while these may have the potential to generate artifacts, in many cases they merely reflect long accepted approaches for nuclear studies in biochemistry (Kirschner et al., 1977). It is unlikely that any experimental approach is totally free of artifact, and this author takes the view that information which advances our understanding, even if subsequently modified at a future date by more sophisticated approaches, is still valuable and certainly worth the effort. It is now generally accepted that rapid freezing which results in amorphous ice can stabilize cells and tissues in milliseconds, providing a starting point of "undisturbed" ultrastructure for subsequent nuclear access via frozen fractures, which allow direct visualization of native structure using FESEM for observation in the cryohydrated state (Walther, 2003). There are, however, two areas in which direct rapid freezing has disadvantages. Firstly, it is incompatible with specific labeling, although this can be addressed by the use of frozen sections. Secondly, the "behavior" of frozen fractures through the nucleus can also limit this method, as the fractures travel along the planes of least resistance through the frozen cell, namely membrane surfaces, and consequently rarely expose the nuclear interior. At the NE the fractures only expose the central or core structure of NPCs, excluding access to the more peripheral structure. Pore complexes are consequently best visualized by SEM following removal of the cytoplasm/nucleoplasm. If giant nuclei such as those from amphibian oocytes are used, a single NE can be isolated by hand and mounted for surface imaging. In comparison to somatic cells in eukaryotic tissue which have around 5 thousand NPCs, the oocyte (germinal vesicle) NE from amphibians has around 50 million add NPCs after 'million' (Allen et al., 2007a,c). Although physical removal of the nucleus and NE isolation might be expected to generate structural alterations, the retention of functional transport through the NPCs has been observed in isolated envelopes (Peters, 2006). Subsequent "in situ rapid freezing" studies and cryotomographic reconstruction in Dictyostelium (Beck et al., 2004, 2007) have provided further insights into nuclear pore structure, extending rather than disproving previous SEM findings.

Other approaches to nuclear morphology by SEM have also exploited the fact that the amphibian oocyte nucleus contains relatively little peripheral chromatin compared to somatic nuclei, thus allowing exposure of internal filamentous structures that would otherwise be obscured. We have shown that on the nucleoplasmic side of the NE, NPCs are connected by an actin-dependent filament network which also interacts with Cajal bodies and chromatin (Kiseleva et al., 2004).

The amphibian "in vitro nuclear formation system," which is generated from isolated egg cytoplasm seeded with demembranated sperm heads, has also proved ideal for SEM. This cell free system models mitotic events, providing visualization of intermediate NPC structure during both their formation and breakdown (Allen *et al.*, 2007a,c; Cotter *et al.*, 2007; Goldberg *et al.*, 1997). Tissue culture cell nuclei can also be exposed *in situ*, generating information not only about NPC structure and distribution, but also the dynamics of chromosome condensation and NE reformation at cell division (Allen *et al.*, 2007a,c). In each case, the use of FESEM allied to cell and molecular biological approaches has generated unique structural information. More recently, similar approaches to the changes in nuclear morphology at cell division in tissue culture cells by FESEM, have been reported (Drummond *et al.*, 2006). As will be apparent throughout this volume, biological structural information allied to function and the involvement of specific proteins is unlikely ever to have a single optimum methodology, and advances in our understanding in the molecular mechanisms of life will result from a synthesis of information gleaned from a variety of experimental approaches.

II. Rationale

Surface imaging by high resolution SEM offers direct, 3D imaging and immunolocalization at a resolution of a few nanometers. This chapter briefly reviews areas of progress made with this approach, and outlines the methodologies and choice of materials required to image the nuclear periphery, including NE and nuclear pores, the nuclear interior, and the dynamics of nuclear structural changes at cell division.

III. Methods

1. Imaging oocyte NE and NPC (Fig. 1).
2. Imaging internal structure in the oocyte nucleus (Fig. 2).
3. Imaging somatic nuclei and mitotic events (Figs. 3–5).
4. Imaging *in vitro* nuclear formation (see Allen *et al.*, 2007a,c; Cotter *et al.*, 2007; Goldberg *et al.*, 1996, 1997).

1. Imaging Oocyte NE and NPC

1. Remove ovary from *Xenopus* (Smith *et al.*, 1991). Isolated tissue can be stored in amphibian Ringer's at 4 °C for up to 1 week or 20 °C for up to 1–2 days.

2. Wash oocytes in fresh amphibian Ringer's, remove small clusters of approximately 20 oocytes, and transfer to 5:1/HEPES buffer. All incubations should be done at room temperature (approximately 20 °C) taking care at all times to prevent the oocytes from drying out.

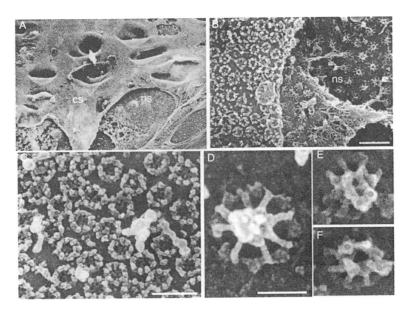

Fig. 1 SEM of isolated nuclear envelope (NE) from *Xenopus* oocyte. (A) Survey micrograph of approximately 1 mm² of isolated envelope on Si chip. The cytoplasmic face (cs) is characterized by several NE "pockets" that penetrate the nuclear interior. The cytoplasmic side overlays the majority of the nucleoplasmic side (ns). Field width = 1.5 mm. (B) Detail of area showing both cytoplasmic (cs) and nucleoplasmic (ns) faces, clearly demonstrating the differences in structure between the outer and inner faces of the nuclear pore complex (NPC). Scale bar = 0.3 μm. (C) Detail of cytoplasmic aspect of NPCs. Each has eightfold radial symmetry with cytoplasmic filaments surrounding a wagon wheel configuration of radial spokes and hub-like central transporter. Central granules may represent transport complexes "in transit" at the point of preservation. Scale bar = 0.25 μm. (D) Detail of a single NPC basket, showing eight filaments joined at their distal ends, projecting into the nucleoplasm. Scale bar = 0.05 μm (this was imaged at 300,000× in the SEM). (E, F) NPC Baskets after labeling with an antibody to basket filament protein, visualized with backscattered electron imaging, then superimposed (pseudocolored yellow) onto the secondary electron image. Field width = 120 nm. (See Plate no. 40 in the Color Plate Section.)

3. Transfer oocytes to a 35-mm petri dish containing fresh 5:1/HEPES buffer and observe under a dissecting microscope with fiber optic illumination. Carry out all subsequent manipulations under the binoculars. Select oocytes at the required stage of development, usually stage 5 or 6, which are 0.8–1.0 mm in diameter. Use No. 5 tweezers to manipulate the oocytes (see Movie 1, supplementary material; Allen *et al.*, 2007a,c).

4. Pierce the oocyte with a dissecting needle avoiding the centre of the dark hemisphere where the nucleus is located. Gently squeeze the oocyte to extrude the cytoplasm. Look for the nucleus, visible as a transparent sphere, amongst the cytoplasm as it comes out.

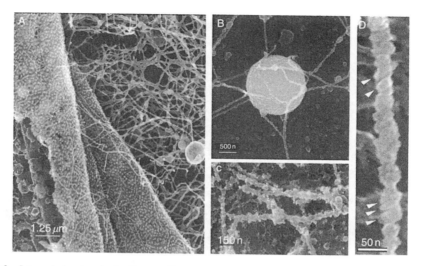

Fig. 2 Internal structure in the *Xenopus* Oocyte nucleus exposed by rapid (within 5–10 s of extraction) opening of freshly extracted nucleus into fresh fix with gentle spreading. (A) The folded back NE is visible at the left covered by close packed NPC baskets. The baskets are only visible as light dots at this magnification (given in detail in (C)). Intranuclear fibers of varying diameters up to 300 nm and formed of aggregates of thinner filaments are associated both with the inner aspect of the NE and with spherical bodies. (B) Detail of spherical body interacting with intranuclear filaments, most likely a nucleolus, but alternatively a snurposome or Cajal body—as determined by suitable antibody labeling (see Kiseleva et al., 2004). (C) The filamentous network is specifically decorated by antibodies against actin, tagged with 10 nm gold conjugated secondary antibodies visualized with backscattered electron imaging and pseudocolored yellow in the micrograph. Aggregation of smaller filaments is apparent within the larger fibers. Although the actin is abundantly labeled, it is discontinuous, possibly suggesting the presence of other structural proteins. (D) High magnification of a filament, showing substructure with periodicity (arrows indicate repeat elements). Scale bars in this fig are on the micrographs. (See Plate no. 41 in the Color Plate Section.)

5. Clean the nucleus of excess cytoplasm by pippetting carefully up and down in a fine glass pipette which has an internal diameter slightly larger than the nucleus, taking care not to allow any air bubbles to touch the nucleus otherwise it will lyse and disappear.

6. Carefully transfer the nucleus to a silicon chip (shiny side) in fresh 5:1/HEPES buffer in an adjacent petri dish.

7. Allow the nucleus to settle. It will attach to the surface of the chip within a few seconds. Very gently pipette the buffer against the nucleus to remove any remaining cytoplasm. Carefully break open the nucleus using fine glass needles and continue to pipette the buffer against the NE to remove the nuclear contents. The envelope will attach to the chip. Use the fine glass needles to further spread and affix the envelope, ensuring that there is sufficient buffer above the NE and cover the dish with its lid to prevent it from drying out. Repeat steps 4–7 until the required number of NEs have been obtained. At least two envelopes per subsequent treatment up to a maximum of 16 chips is a reasonable number

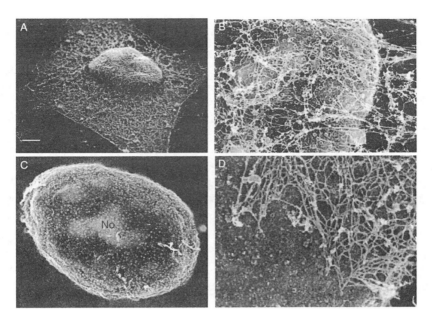

Fig. 3 Exposure of the nuclear surface after mild detergent extraction and dry fracturing *in situ*. (A) Extracted HeLa cell, before dry fracture, showing Triton X-100 removal of outer cell membrane and soluble cytoplasmic elements, leaving a cytoskeletal meshwork which overlies the nucleus. Scale bar = 2.0 μm. (B) Detail of surface of nucleus underneath the cytoskeletal fibers, characterized by numerous NPCs over its surface, which retains a largely smooth membranous appearance. Scale bar = 0.4 μm. (C) In this fracture, the entire nucleus has been plucked out of the cell and is viewed held in the adhesive side of the fracture, thus revealing the nuclear underside, in this case free of all cytoskeletal material. The lighter central area (No) shows the underlying signal from the Nucleolus, as a result of beam penetration generating and increased backscattered signal from this most dense part of the nucleus. Scale bar = 1.5 μm. (D) Detail of the nuclear surface from an area in which the dry fracture has only produced a partial removal of the cytoskeleton, leaving an edge which could indicate sites of attachment (arrowed) between the nuclear surface and cytoskeleton, which could be confirmed by antibody labeling for proteins known to be involved in these interactions (see Wilhelmsen *et al.*, 2005). Scale bar = 0.3 μm.

to process at one time. At this point a choice of fixation must be made depending on whether or not immunolabeling is required.

8. If immunolabeling is required, follow the protocol outlined below before proceeding with step 9. Otherwise, for standard SEM preparation, continue immediately with step 9.

9. Transfer the chip with NE (quickly without air-drying) to a 0.5-ml tube containing the fixative (2% glutaraldehyde, 0.2% tannic acid in 100 mM HEPES, pH 7.4). It is helpful to mark the position of the envelope by circling the area with a writing diamond. Fix for a minimum of 10 min at room temperature. This fixative has been optimized for visualization of NPC ultrastructure by FESEM and also works well for TEM. If necessary preps can be stored in fixative at 4 °C for up to 1 week at this point.

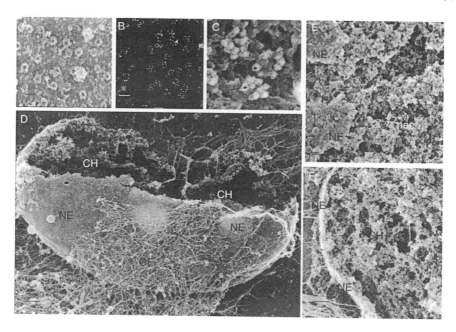

Fig. 4 Nuclear surface and internal structure exposed after dry fracturing. (A) Secondary electron imaging of nuclear surface of HeLa cell after infection with HIV, showing potential aggregations of viral capsid protein associated with NPCs (arrowed); see Arhel et al. (2007). These are rare events, maybe one or less on average per nucleus and consequently extremely unlikely to be found in thin section TEM. Scale bar = 0.2 μm. (B) Backscattered electron image of similar area to (A) with gold label to mAb 414, an antibody which is specific for several NPC proteins. This BSE image shows the distribution of the label as specific circular patches for each NPC. Scale bar = 0.2 μm. (C) High magnification of gold label at the NPC. Each gold particle can be resolved to show the central metallic core and thin surface shell of antibody protein (arrows). Scale bar = 0.1 μm. (D) Low power of typical nuclear exposure produced by the dry fracture technique. On the lower half of this nucleus there still remains an area of undisturbed cytoskeleton, with "clean" NE on either side. In the upper half of the micrograph, the fracture has opened up the entire depth of the nucleus, exposing areas of internal chromatin organization (CH) at various levels down to the base of the nucleus near the substratum (top edge). Scale bar = 1.0 μm. (E) In this area of fracture, the removal of material has been at or near the nuclear surface, as indicated by the remaining patches of NE, characterized by NPCs (arrows). Adjacent areas of peripheral chromatin have been exposed, in some cases showing potential interaction with the NPC baskets (double arrow). Scale bar = 0.6 μm. (F) Detail of an area similar to that exposed at the mid left in (D). There is an area of the NE seen in profile (NE), exposing the relationship to the adjacent chromatin, which shows clear variation in the density of condensation close to the NE in comparison to the more open packaging of the internal chromatin. This fracture represents a 3D "sectioned" profile towards the centre of the nucleus in comparison to the surface profile seen in Fig D. Scale bar = 0.5 μm.

It is important, for the following steps, to use No. 3 tweezers to transfer chips manually into a petri dish containing the appropriate reagent. Keep the NE upwards. Do not allow to dry out at any point.

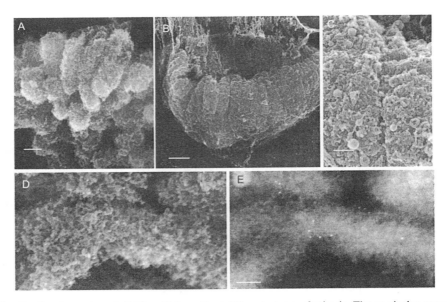

Fig. 5 Dry fracturing of dividing HeLa cells at different stages of mitosis. The particular stage is readily apparent from the extent of chromosome condensation in prophase, positioning of chromosomes around the metaphase plate, then separation of chromatid cohorts in anaphase and telophase. As the chromosomes themselves rarely fracture separately, the overall appearance of the mitotic figure is clearly diagnostic of the progression through division. Using unsynchronized HeLa cells the 4% spontaneous incidence of mitosis is sufficient to find dividing cells with a reasonable frequency, by searching subconfluent monolayers in the SEM at low magnifications similar to LM. (A) Group of chromosomes arranged on the metaphase plate. Individual chromosome morphology is clearly resolved together with chromatin fiber organization. Both kinetochore and telomere regions are readily accessed. Scale bar = 0.5 μm. (B) Low power image of a "bowl" of chromatids viewed as half a separating pair at anaphase/telophase. Although now quite closely packed, the individual chromatid profiles are quite apparent, and 20 can be seen from this viewpoint, as expected for chromosome numbers in this cell line. Scale bar = 1 μm. (C) Detail of chromatid surface from (B). Despite the absence of reforming NE membranes at this stage, NPC profiles are already apparent (highlighted in green) at the surface of the chromatin. Scale bar = 0.2 μm. (D, E) Paired secondary (D) and backscattered (E) electron images from a centromere region of a metaphase chromosome labeled for a kinetochore antigen to Cenp-f. The chromatin fiber organization in the secondary images is readily resolved, but the strong secondary signal from the chromatin obliterates the gold, which is still clearly visible by backscattered imaging in (E). In practice, these synchrously acquired images would be superimposed in exact register to show the position of the gold particles *in situ* on the chromatin (see Drummond *et al.*, 2006). Scale bars = 0.19 μm. (See Plate no. 42 in the Color Plate Section.)

10. Wash briefly in dH_2O.

11. Post fix in aqueous 1% Osmium Tetroxide for 10 mins. Osmium preserves lipid and phospholipid content and also enhances contrast in the specimen.

12. Wash briefly in dH_2O.

13. Stain for 10 min in 1% aqueous uranyl acetate. This stains nucleic acids and proteins.
14. Dehydrate through the following ethanol concentrations for 2 min each: 30%, 50%, 70%, 95% (×2), and 100% (×3).
15. Transfer to the critical point drying (CPD) apparatus containing ethanol as intermediate reagent. Use a suitable holding device. A Leica basket stem assembly with small baskets is suitable for the Baltec CPD. (It may be necessary to shorten the stem slightly to fit in the chamber.) Critical point dry from high purity (<5 ppm water) liquid CO_2.
16. Coat samples with 3 nm chromium and observe by high resolution SEM (Fig. 1). Store samples under vacuum for optimal preservation. If possible coat just before examination in the SEM. Coated specimens should last for several months and can be recoated with chromium if required (Fig. 1). See section on coating.

Immunolabeling (from step 8 above)

For the following steps use No. 3 forceps to transfer chips manually into a petri dish containing the appropriate reagent. Keep the NE upwards. Do not allow it to dry out at any point.

1. Fix with 2% paraformaldehyde, 0.25% glutaraldehyde (optional). Glutaraldehyde gives better ultrastructural preservation but should be avoided if it inhibits binding of the primary antibody. Fix in 5:1/HEPES buffer for 10 min at room temp or overnight at 4 °C.
2. Wash in 5:1/HEPES buffer 2 × 5 min.
3. Transfer to 0.1 M glycine in 5:1/HEPES for 10 min (quenches unreacted aldehyde).
4. Block with 1% Bovine serum albumin (BSA) in 5:1/HEPES buffer for 10–20 min.
5. Prepare a wet chamber using a 9-cm petri dish containing wet filter paper with Nescofilm on the top.
6. Dilute the primary antibody at an appropriate concentration in 5:1/HEPES.
7. Place chip on a dry filter paper to dry the back.
8. Place chip on parafilm in wet chamber and quickly pipette on 10 μl antibody, incubate for 1 h at room temperature.
9. Wash in 5:1/HEPES buffer 2 × 10 min.
10. Block with 1% fish skin gelatin in 5:1/HEPES for 10 min.
11. Dilute the secondary antibody gold conjugate (see below) at a concentration of 1 in 10 in 5:1/HEPES. Run a control prep in which the primary antibody has been left out, but still incubate with the gold/secondary complex to check for nonspecific binding.
12. Place chip on dry filter paper to dry the back. Take care that the upper surface remains wet.

13. Place chip on parafilm in wet chamber and quickly pipette on 10 µl antibody, incubate for 1 h at room temperature.
14. Wash three times for 10 min in 5:1/HEPES.
15. Fix in 2% glutaraldehyde, 0.2% Tannic acid, 0.1 M HEPES as described in step 9 in the main protocol and continue from there (Fig. 1).

IV. Colloidal Gold in the SEM

For visualization of the colloidal gold marking the antibodies, there are various options available dependent upon the FESEM used, and the operating conditions chosen. Later, instruments will allow direct visualization of 5-nm gold particles (or smaller) by using low accelerating voltages (below 3 kV), at which the gold particles will stand out by virtue of their significantly increased contrast in comparison to that generated by the biological material. Coating in this situation may be carbon, but 3 nm of Chromium will enhance structural detail without obscuring contrast differences (see note on coating for FESEM). Earlier instruments, however, may generate optimum resolution at higher voltages (10–30kV), in which case the use of a backscattered electron detector will allow the same contrast separation to be imaged. In our own case, using an instrument no longer commercially available, we chose to image simultaneously, on separate viewing screens, the secondary and backscatter images, optimizing each, secondary for structure and backscatter for gold location, and then we superimposed the images in Photoshop, usually with the gold particles pseudocolored to provide extra contrast (see Figs. 1E, 2C, 4B and C, 5D and E). It should also be pointed out that two different sizes of colloidal gold, e.g., 5 nm and 10 nm in diameter, suitably bound to specific secondary antibodies, will allow direct comparison of the distribution of two antibodies on the same preparation.

2. Imaging Internal structure in the amphibian Oocyte nucleus

1. Surgically remove ovaries and place in amphibian Ringer's solution (111 mM NaCl, 1.9 mM KCl, 1.1 mM $CaCl_2$, 2.4 mM $NaHCO_3$).
2. Place small pieces of ovary in petri dishes in buffer A (83 mM KCl, 17 mM NaCl, 10 mM Hepes, 250 mM sucrose, 3 mM (or 0.3 mM) $MgCl_2$, pH 7.4). Pre-coat Silicon chips with poly-L-lysine and place near the oocytes in the same dish. All solutions should be kept ice-cold.
3. Select *Xenopus* oocytes at developmental stages 3–4 that are of 480–500 µm diameters and are light brown to brown in color. Although PLFs are present in nuclei from oocyte stages 3–6, stages 3–4 are highly active for transcription and nuclear isolation is easier and quicker than at stage 6, largely because stages 3–4 oocytes have fewer yolk granules.

4. Manually isolate nuclei (gently) from each oocyte using glass needles, and quickly separate from surrounding cytoplasm.

5. Transfer two to three isolated nuclei to a silicon chip, and carefully open the NEs manually with a glass needle. For nuclear content "spreads," displace the nucleoplasm slightly using the glass needle and carefully spread the nucleoplasm over the surface of the chip.

6. To examine intact filaments under gentle conditions, first release each nucleus from the oocyte and transfer to a Si chip. Carefully strip the NE away from the underlying gelated nucleoplasm using watchmaker's forceps. Fix the intact demembranated ball-like nuclear gel within 2–3 s after isolation by adding fix solution (see below) directly to the chip in the petri dish, and then transfer the chip to another petri dish with fresh fix.

7. For both "spread" and "intact" samples, fix the chips and treat as follows, using ice-cold solutions. Transfer the chips from dish to dish without drying, by keeping a drop of the previous solution on the chip. Fix samples for 10 min in 2% glutaraldehyde, 0.2% tannic acid, 10 mM Tris–HCl (pH 7.4) and 3 mM $MgCl_2$. It is most important that the elapsed time from nuclear isolation to placing in fix is always 5–10 s or less.

8. Post fix samples for 30 min in 1% OsO_4, wash in double distilled H_2O, and stain for 20 min in 1% uranyl acetate.

9. Dehydrate in ethanol and critical point dry from CO_2 using 3M Novec HFE 7100, (3M UK, Bracknell) as the transitional solvent.

10. Sputter coat specimens with a 4-nm layer of chromium, and examine in the FESEM (Fig. 2).

Immunolabeling Protocol

1. For immunogold localization, nuclei were manually isolated from stages 3–4 *Xenopus* oocytes using glass needles, transferred to a silicon chip and NEs opened manually as described above.
2. Fix nuclei for 15 min at 22–24 °C in 3.7% formaldehyde, 3 mM $MgCl_2$, 0.2% tannic acid, and 10 mM Tris–HCl pH 7.0.
3. Wash fixed samples three times (1 min each) in PBS, incubate in PBS containing 1% BSA for 30 min, and wash in PBS and incubate for 1–3 h with primary antibody.
4. Remove unbound primary antibodies by washing three times in PBS, and incubate for 1 h with 10 nm gold-conjugated secondary goat anti-rabbit antibody (Amersham Corporation). As negative controls, we used gold-conjugated secondary antibody diluted 1:20 in PBS.
5. Post fix for 30 min in 1% OsO_4, then follow from step 8 above.
6. Colloidal gold particles are visualized using a solid state backscatter electron detector, or low kV accelerating voltage (see Section IV; Fig. 2).

3. Imaging somatic nuclei and mitotic events

We have mainly used *Xenopus* (XTC) and HeLa cells (Figs. 3–5), but the methodology should be suitable for almost any tissue culture cell.

1. Grow cells on silicon chips in cell culture petri dishes. Cells should be plated on day 1, medium changed on day 2, and the cells used on day 3 when they should be approximately 50% confluent. They should be grown on the shiny side of the chip (prior to this, chips are rinsed in 70% alcohol and dried in a 200 °C oven). Check density by observing cells growing on the surrounding petri dish surfaces, which grow at the same rate.
2. Rinse cells on chips briefly with culture fluid without serum at room temperature.
3. Fix cells with 2% paraformaldehyde, 0.01% glutaraldehyde in PBS for 10 s. Hold each chip individually with forceps during immersion in fixative.
4. Transfer to a petri dish containing 0.1 M glycine in PBS for 10 min to quench unreacted aldehyde.

For the following steps use No. 3 tweezers to transfer chips manually into a petri dish containing the appropriate reagent. Keep the cells upwards. Do not allow to dry out at any point.

5. Rinse in PBS.
6. Transfer to 0.5% triton in PBS for 30 min.
7. Rinse in PBS.
8. If immunolabeling is required, follow the protocol provided below before proceeding to step 9. Otherwise, continue immediately with step 9.
9. Fix with 3% glutaraldehyde in PBS for 1 h at room temperature.

If necessary, specimens can be stored at this point at 4 °C (up to 1 week).

10. Wash briefly in Sorensen's buffer.
11. Post fix in 1% Osmium tetroxide in Sorensen's buffer for 10 min. Osmium preserves lipid and phospholipid content and also enhances contrast in the specimen.
12. Wash briefly in dH_2O.
13. Dehydrate through ethanol, 5 min each wash: 30%, 50%, 70%, 95% (×2), and 100% (×3), and ×2 in 100% Arklone (trichlorotrifluoroethane; ICI) or other suitable transitional solvents.
14. Transfer to CPD apparatus containing 100% ethanol as intermediate reagent. Use a suitable holding device. A Leica basket stem assembly with small baskets is suitable for the Baltec CPD.
15. Critical point dry from high purity (<5 ppm water) liquid CO_2.
16. Dry fracture cells. Attach a 40-mm length of 13-mm wide adhesive tape to the bench so that it is secured by its ends but the adhesive is uppermost in the middle. Invert chips onto the adhesive and lightly tap with the points

of the forceps. Remove chip by carefully pulling from the tape avoiding lateral movements (see Movie 1, supplementary material; Allen *et al.*, 2007a,c). If necessary, samples stored under vacuum should be stable for up to 1 month at this point.

17. Coat samples with 3 nm chromium and observe with High Resolution SEM. Store samples under vacuum for optimal preservation. They should last for several months and can be recoated with chromium if required (Figs. 3–5).

V. Immunolabeling Protocol

1. Block with 1% BSA in PBS buffer, 10–20 min.
2. Prepare a wet chamber using a 90-mm petri dish containing wet filter paper with Nescofilm on the top.
3. Dilute the primary antibody at an appropriate concentration in PBS.
4. Place chip on dry filter paper to dry the back.
5. Place chip on parafilm in wet chamber and quickly pipette on 10 μl antibody and incubate for 1 h at room temperature.
6. Wash in PBS buffer 2 × 10 min.
7. Block with 1% fish skin gelatin in PBS for 10 min.
8. Dilute the secondary antibody gold conjugate at a concentration of 1 in 10 in PBS.
9. Place chip on dry filter paper to dry the back.
10. Place chip on parafilm in wet chamber and quickly pipette on 10 μl antibody and incubate for 1 h at room temperature.
11. Wash three times for 10 min in PBS.
12. Continue from step 9 of the main protocol (Figs. 4,5).

4. Imaging *in vitro* nuclear formation

   In Vitro Nuclear Assembly and Isolation

   It is outside the remit of this chapter to provide a full methodology for the preparation of the cytoplasmic extract and sperm heads required for *in vitro* nuclear formation. For these details the reader is referred to Allen *et al.* (2007b). Here we pick up the method from the point of addition of demembranated sperm heads to the cytoplasmic extract.

   1. To initiate nuclear formation, assemble reactions containing either LSS or membranes resuspended in cytosol in eppendorf tubes.
   2. Add E-mix to a final concentration of 1X and mix gently, but thoroughly.
   3. Add *Xenopus* demembranated sperm to a final concentration of 1000 sperms/l and incubate at 22 °C for 90 min.

It is worth noting that every extract will vary in the efficiency of nuclear assembly and therefore it is necessary to monitor nuclear assembly to determine optimal incubation times for each extract. Assembly can be assayed using visual and biochemical markers such as observation by conventional phase microscopy and incorporation of suitably labeled dUTPs indicating DNA replication competency. Assembled nuclei must be processed immediately for SEM.

Nuclear processing for visualization by SEM

1. Dilute a 4-l aliquot of each assembly reaction and dilute in 500 l of membrane wash buffer (MWB).
2. Transfer diluted sample into clean flat-based tube holding silicon chip at the bottom and sediment nuclei onto acetone-washed silicon chips by centrifugation at 3000g for 10 min at 4 °C.
3. Transfer chips and incubate for 10 min in Fix Buffer ensuring a minimum of MWB dilutes the fixative. Should immunolabeling be required switch to protocol 2 immunolabeling variation, and then rejoin this protocol at step 4 (below).
4. Remove Fix buffer whilst ensure that chips do not air dry and wash chips in 0.2 M sodium cacodylate (pH 7.2 in water) to remove residual Fix Buffer.
5. Post fix samples in 1% osmium tetroxide (in 0.2 M sodium cacodylate pH 7.2) for 10–20 min.
6. Remove osmium tetroxide and wash in water for 10 s.
7. Remove water and stain for 10 min in 1% uranyl acetate (in water). Caution: Uranyl acetate must be kept and used within a fume cabinet.
8. Dehydrate through a graded ethanol series (30%, 50%, 70%, 95%, twice at 100%) and twice in 100% Arklone (trichlorotrifluoroethane; ICI) or other suitable transitional solvents.
9. Critical point dry the samples from high purity (<5 ppm water) CO_2.
10. Sputter coat the samples with 2–3 nm of chromium.
11. Examine in FESEM.

VI. CPD for High Resolution SEM

CPD for SEM was introduced to overcome distortion of surface structure associated with drying material from aqueous environments (Cohen, 1976). The distortion is due to surface tension effects during drying as the phase barrier between liquid and gas passes through the specimen. "Air drying" from water produces severe flattening of surface features at the macromolecular level when observed by FESEM (Allen, unpublished data). These surface tension effects are avoided by the use of CPD, but because high pressure is involved CPD has been considered controversial. However, in our own hands, macromolecular structure of the NPC at a few nanometers resolution has been well preserved and also retained subtle differences in structure deliberately induced by experimental alteration (Goldberg and Allen, 1996).

CPD relies on the physical properties of CO_2 around the "critical point," which is 31.1 °C at 1073 psi. At the critical point, liquid and vapor states coexist; hence if the temperature is raised slightly above the critical point, the specimen is transferred from a liquid to gaseous environment without the phase barrier (and distorting surface tension) having passed through its surface. Clearly, the specimen needs to be placed in a suitable pressure chamber and infiltrated with CO_2 at 800 psi, but at this stage it is in a liquid state, and liquids are incompressible, so that fine structure is not compromised. Before CPD the specimen is dehydrated in exactly the same way as it would be for routine resin sectioning for TEM, but transferred into a nonaqueous intermediate solvent (3M Novec HFE 7100, 3M UK, Bracknell). The intermediate solvent is miscible with liquid CO_2, so that when several changes of liquid CO_2 are made, the specimen becomes infiltrated with pure liquid CO_2. The temperature is raised gently above the critical point in the apparatus, and the gaseous CO_2 is gently released over 15–20 min, so that at no stage is the specimen subject to any rapid alteration in pressure. In theory, it should be possible to critical point dry from water, avoiding both dehydration and intermediate solvents, but the critical point for H_2O (373 °C and 3200 psi) is too high a temperature and pressure to be regularly achieved on the laboratory bench! However, for those recoiling in horror at this suggestion, rapid freezing also involves instant temperature changes of approximately 300 °C.

In our own experience, CPD of a variety of what could be considered the most delicate biological structures—such as isolated membranes, organelles such as mitochondria, and fine protein structures such as NPCs—have all been through the CPD process without serious deformation when compared to other types of preparation including rapid freezing and examination cryohydrated in both FESEM (Walther, 2003) and TEM (Beck et al., 2004, 2007) and aqueous state with atomic force microscopy (Maco et al., 2006). The most distorting aspect of specimen preparation for EM in general is dehydration, crucial to all TEM sectioning from resin embedded material, and only avoided in cryohydrated protocols which bring another category of potential artifacts not completely explored at this point in time. Cryohydrated material in the SEM also needs to be fractured and/or etched to expose structure, and consequently brings limitations in terms of accessing specific surfaces for high resolution SEM (see Section I). Clearly all preparation methods have limitations, but CPD in general is still the best starting point for new material.

A. Coating for High Resolution SEM

One of the consequences of the enhanced resolution in Field Emission instruments is that conventional coatings such as sputtered gold have a characteristic morphology of their own when viewed at magnifications greater than 100,000×.

This "decoration artifact" is clearly incompatible with subnanometer resolution, and alternative coating methods are required. Ideally, observation in the SEM without coating should be the optimal procedure, but this is prevented by the following considerations. Biological material is formed of C, H, N, and O, which are light elements and virtually nonconductive when bombarded with a primary electron beam, leading to specimen charging, and they produce relatively little secondary electron emission which is required for image formation. These limitations can be reduced by osmium fixation which partially "metalizes" biological specimens by the incorporation of osmium dioxide throughout the specimen. Use of osmium mordants such as tannic acid have generated "conductive staining" protocols which may be useful in some circumstances (Naguro et al., 1990). Frozen biological material should be conductive when viewed cryohydrated, but in practice specimen charging and lack of contrast make high resolution imaging difficult to impossible. Although specimen charging reduces with decreasing accelerating voltage, this does not appear to be the case in cryo SEM, and the best imaging results from cryohydrated imaging come from Chromium sputter coated material (Apkarian, 1994; Apkarian et al., 1999) or Platinum shadowing combined with carbon coating, as originally used for TEM imaging of surface replicas (Walther, 2003).

Our own experience has been that Chromium coating provides the lowest grain size consistent with the thinnest films and sufficient contrast for detailed imaging at magnifications in excess of 200,000×. We have used planar magnetron sputtering, but deposition by ion beam or electron beam is also suitable. The best coating requires a vacuum environment with absolutely minimal concentrations of oxygen and water vapor. These parameters can be fulfilled in a variety of vacuum environments, but our own solution was to use a standard 12-inch coating unit fitted with a cryopumped vacuum system, which is highly efficient for the removal of water vapor, with O_2 reduced by pumping to a vacuum of 10^{-7} millibar prior to introducing ultra pure Argon (99.999%) at lower vacuum (10^{-3} millibar) to generate a sputter deposition rate of approximately 0.1 nm/s, as measured by a film thickness monitor. For sputter coating, electron bombardment of the specimen should be avoided by the use of magnetic deflectors in the sputter source. The first 30 s of sputtering should take place with the specimen covered with a removable shield, to allow removal of the film of chromium oxide which forms on the target surface each time it is exposed to air. It may be possible to see the color of the plasma change from pinkish blue to blue as this coating is removed. Target oxidation can be avoided if the vacuum system incorporates an air lock for specimen exchange. Although ion or electron beam deposition both produce similar coatings, a limitation of these methods is that coating is directional, which can lead to shadowing artifacts unless the specimen is rotated in a planetary motion. Sputter coating of Chromium deposits evenly over the surface, so that a continuous film forms by 1 nm of deposition. Even a 3-nm Cr layer allows backscattered imaging of two sizes of colloidal gold particles to be easily discriminated for antibody labeling, while still allowing good secondary electron imaging.

It should be apparent from these considerations that coating equipment for high resolution SEM in biology is almost as important as the microscope itself, and should be a major element in any budgeting considerations.

VII. Discussion

A. The Value of FESEM Imaging for nuclear and subcellular structure

For the first 30 years of biological EM, the NPC was considered to be structurally symmetrical about the plane of NE, i.e., the same at both the nucleoplasmic and cytoplasmic faces. However, the first FESEM imaging of isolated NE by Ris (1991), revealed a novel and characteristic basket structure on the nucleoplasmic face, which was apparent as a physical projection on the isolated envelope and appeared when viewed by surface imaging. Previous results from TEM (thin sections from whole nuclei and isolated envelopes) had failed to identify this structure because the basket structure had insufficient contrast to be identified. Once the structure had been identified, however, similar images to those from FESEM were produced in the TEM using rotary shadowing techniques (Maco *et al.*, 2006). This example indicates why FESEM is extremely useful in frontline studies for new structure and for structural relationships over a scale that could be massively labor intensive for transmission EM either by serial section reconstruction of computer aided tomography. For example, to cover the area illustrated in Fig. 1C would require reconstructions involving 60 serial sections—not a practical realization. It is the sheer accessibility over large areas offered at all magnifications that makes SEM such an effective tool for cell biology. Most modern instruments offer a magnification range of × 10 to × 1 million allowing wide ranging and easy searching over large areas of material, and then molecular imaging of areas of interest. Identification of a single HIV particle at the surface of the nucleus has proved feasible with this technology (Arhel *et al.*, 2007; see Fig. 4A).

Another example of the value of surface imaging has been in studies of the mechanisms and structural assembly pathways in NE and NPC. Here we used both biochemical manipulation of *in vitro* nuclear formation together with molecular resolution for NE and NPC assemblies (Goldberg *et al.*, 1997), where FESEM imaging identified intermediate structures present on the NE during NPC formation. Imaging by FESEM allowed extensive areas of membrane surface to be searched, finding rare (and transient) structures that would be difficult, if not impossible, to detect using conventional thin-section TEM. The earliest intermediate structures in NPC assembly are termed "dimples" which are the sites where the inner and outer nuclear membranes approach each other, prior to fusion to form a pore. These structures have an initial diameter of only 5 nm, which is around one-tenth of the thickness of a thin section for TEM, and consequently unlikely to be recognized. Thus detailed structure stages in the formation of the NPC have been identified by FESEM, which currently could not have been found by any other microscopies. This is spite of the perceived "limitations" of fixation, CPD, metal coating, and high accelerating voltages for imaging required for FESEM.

The response to these limitations largely rests on the future role of rapid freezing as a starting point for FESEM, which has been discussed above, and will develop further, but as in all things, the "more than one way to skin a cat" principle will always apply. Other surface techniques such as scanning probe microscopies show great potential for unfixed and nondehydrated samples, but require further development and understanding of specimen/probe interactions at this point in time. There is also a tendency for results in these technologies to be generated from the averaging of data sets, and with the danger that rare structures are likely to be missed. Consequently, the choice of FESEM as a first approach to visualize molecular interactions occurring at surfaces within the cell would seem to be a sensible option.

References

Allen, T. D., Cronshaw, J. M., Bagley, S., Kiseleva, E., and Goldberg, M. W. (2000). The nuclear pore complex: Mediator of translocation between nucleus and cytoplasm. *J. Cell. Sci.* **113,** 1651–1659.

Allen, T. D., Rutherford, S. A., Murray, S., Sanderson, H. S., Gardiner, F., Kiseleva, E., Goldberg, M. W., and Drummond, S. P. (2007a). A protocol for isolating *Xenopus* oocyte nuclear envelope for visualization and characterization by scanning electron microscopy (SEM) or transmission electron microscopy (TEM). *Nat. Protoc.* **2,** 1166–1172.

Allen, T. D., Rutherford, S. A., Murray, S., Sanderson, H. S., Gardiner, F., Kiseleva, E., Goldberg, M. W., and Drummond, S. P. (2007b). Generation of cell-free extracts of *Xenopus* eggs and demembranated sperm chromatin for the assembly and isolation of *in vitro*-formed nuclei for Western blotting and scanning electron microscopy (SEM). *Nat. Protoc.* **2,** 1173–1179.

Allen, T. D., Rutherford, S. A., Murray, S., Gardiner, F., Kiseleva, E., Goldberg, M. W., and Drummond, S. P. (2007c). Visualization of the nucleus and nuclear envelope *in situ* by SEM in tissue culture cells. *Nat. Protoc.* **2,** 1180–1184.

Apkarian, R. P. (1994). Analysis of high quality monatomic chromium films used in biological high resolution scanning electron microscopy. *Scanning Microsc.* **8,** 289–299.

Apkarian, R. P., Caran, K. L., and Robinson, K. A. (1999). Topographic imaging of chromium-coated frozen-hydrated cell and macromolecular complexes by in-lens field emission scanning electron microscopy. *Microsc. Microanal.* **5,** 197–207.

Arhel, N. J., Souquere-Besse, S., Munier, S., Souque, P., Guadagnini, S., Rutherford, S., Prévost, M. C., Allen, T. D., and Charneau, P. (2007). HIV-1 DNA flap formation promotes uncoating of the pre-integration complex at the nuclear pore. *EMBO J.* **26**(12), 3025–3037.

Beck, M., Förster, F., Ecke, M., Plitzko, J. M., Melchior, F., Gerisch, G., Baumeister, W., and Medalia, O. (2004). Nuclear pore complex structure and dynamics revealed by cryoelectron tomography. *Science* **306,** 1387–1390.

Beck, M., Lucić, V., Förster, F., Baumeister, W., and Medalia, O. (2007). Snapshots of nuclear pore complexes in action captured by cryo-electron tomography. *Nature* **449,** 611–615.

Bridger, J. M., Foeger, N., Kill, I. R., and Herrmann, H. (2007). The nuclear lamina. Both a structural framework and a platform for genome organization. *FEBS J.* **274**(6), 1354–1361.

Cohen, W. D. (1976). Simple magnetic holders for critical point drying of microspecimen suspensions. *J. Microsc.* **108,** 221–226.

Cotter, L., Allen, T. D., Kiseleva, E., and Goldberg, M. W. (2007). Nuclear membrane disassembly and rupture. *J. Mol. Biol.* **369**(3), 683–695.

Cremer, T., Cremer, M., Dietzel, S., Müller, S., Solovei, I., and Fakan, S. (2006). Chromosome territories—a functional nuclear landscape. *Curr. Opin. Cell. Biol.* **18,** 307–316.

Drummond, S. P., Rutherford, S. A., Sanderson, H. S., and Allen, T. D. (2006). High resolution analysis of mammalian nuclear structure throughout the cell cycle: Implications for nuclear pore complex assembly during interphase and mitosis. *Can. J. Physiol. Pharmacol.* **84,** 423–430.

Goldberg, M. W., and Allen, T. D. (1996). The nuclear pore complex and lamina: three-dimensional structures and interactions determined by field emission in-lens scanning electron microscopy. *J. Mol. Biol.* **257,** 848–865.

Goldberg, M. W., Wiese, C., Allen, T. D., and Wilson, K. L. (1997). Dimples, pores, star-rings, and thin rings on growing nuclear envelopes: Evidence for structural intermediates in nuclear pore complex assembly. *J. Cell. Sci.* **110,** 409–420.

Gruenbaum, Y., Margalit, A., Goldman, R. D., Shumaker, D. K., and Wilson, K. L. (2005). The nuclear lamina comes of age. *Nat. Rev. Mol. Cell. Biol.* **6**(1), 21–31.

Heessen, S., and Fornerod, M. (2007). The inner nuclear envelope as a transcription factor resting place. *EMBO Rep.* **8**(10), 914–919.

Hozák, P., and Fakan, S. (2006). Functional structure of the cell nucleus. *Histochem. Cell. Biol.* **125** (1–2), 1–2.

Kirschner, R. H., Rusli, M., and Martin, T. E. (1977). Characterization of the nuclear envelope, pore complexes, and dense lamina of mouse liver nuclei by high resolution scanning electron microscopy. *J. Cell. Biol.* **72,** 118–132.

Kiseleva, E., Drummond, S. P., Goldberg, M. W., Rutherford, S. A., Allen, T. D., and Wilson, K. L. (2004). Actin- and protein-4.1-containing filaments link nuclear pore complexes to subnuclear organelles in *Xenopus* oocyte nuclei. *J. Cell. Sci.* **117,** 2481–2490.

Lamond, A. I., and Spector, D. L. (2003). Nuclear speckles: a model for nuclear organelles. *Nat. Rev. Mol. Cell. Biol.* 605–612.

Naguro, T., Inaga, S., and Iino, A. (1990). The tannin–osmium conductive staining after dehydration: an attempt to observe the chromosome structure by SEM without metal coating. *J. Electron. Microsc. (Tokyo).* **39,** 511–513.

Maco, B., Fahrenkrog, B., Huang, N. P., and Aebi, U. (2006). Nuclear pore complex structure and plasticity revealed by electron and atomic force microscopy. *Methods Mol. Biol.* **322,** 273–288.

Pajerowski, J. D., Dahl, K. N., Zhong, F. L., Sammak, P. J., and Discher, D. E. (2007). Physical plasticity of the nucleus in stem cell differentiation. *PNAS* **104,** 15619–15624.

Peters, R. (2006). Use of *Xenopus laevis* oocyte nuclei and nuclear envelopes in nucleocytoplasmictransport studies. *Methods Mol. Biol.* **322,** 259–272.

Shaklai, S., Amariglio, N., Rechavi, G., and Simon, A. J. (2005). Gene silencing at the nuclear periphery. *FEBS J.* **274**(6), 1383–1392.

Smith, L. D., Xu, W., and Varnold, R. L. (1991). Oogenesis and oocyte isolation. *Methods Cell Biol.* **36,** 45–60.

Starr, D. A. (2007). Communication between the cytoskeleton and the nuclear envelope to position the nucleus. *Mol. Biosyst.* **3**(9), 583–589.

Stewart, M. (2007). Molecular mechanism of the nuclear protein import cycle. *Nat. Rev. Mol. Cell. Biol.* 195–208.

Ris, H. (1991). The three dimensional structure of the nuclear pore complex as seen by high voltage electron microscopy and high resolution low voltage scanning electron microscopy. *EMSA Bull.* **21,** 54–56.

Trinkle-Mulcahy, L., and Lamond, A. I. (2007). Toward a high-resolution view of nuclear dynamics. *Science* **318**(5855), 1402–1407.

Walther, P. (2003). Recent progress in freeze-fracturing of high-pressure frozen samples. *J. Microsc.* **212,** 34–43.

Wilhelmsen, K., Litjens, S. H., Kuikman, I., Tshimbalanga, N., Janssen, H., van den Bout, I., Raymond, K., and Sonnenberg, A. (2005). Nesprin-3, a novel outer nuclear membrane protein, associates with the cytoskeletal linker protein plectin. *J. Cell. Biol.* **171**(5), 799–810.

CHAPTER 21

Electron Microscopy of Lamin and the Nuclear Lamina in *Caenorhabditis elegans*

Merav Cohen,[*,‡] Rachel Santarella,[†] Naama Wiesel,[‡] Iain Mattaj,[†] and Yosef Gruenbaum[‡]

[*]Division of Cell Biology
MRC-Laboratory of Molecular Biology
Cambridge CB2 0QH
United Kingdom

[†]European Molecular Biology Laboratory
Heidelberg
Germany

[‡]Department of Genetics
The Institute of Life Sciences
The Hebrew University of Jerusalem
Jerusalem 91904
Israel

Abstract
I. General Introduction
   A. The Nuclear Lamina
   B. The Nuclear Lamina in *C. elegans*
II. *In Vitro* Assembly of Ce-Lamin Filaments
   A. Protein Expression and Purification
   B. Comments
   C. Assembly of Ce-Lamin Filaments
   D. Comments
   E. Assembly of Ce-Lamin Paracrystals
III. Preparation of Embryos and Adults for Transmission Electron Microscopy Using Microwave Fixation
   A. Comments
IV. Preparation of *C. elegans* Embryos and Adults for Conventional Transmission Electron Microscopy by High Pressure Freezing Combined with Freeze Substitution

   A. High-Pressure Freezing of *C. elegans*
   B. Freeze Substitution for Morphological Studies
   C. Infiltration and Embedding
   D. Sectioning and Staining
  V. Preembedding Immunogold EM Staining of Lamina Proteins in *C. elegans* Embryos
  VI. Postembedding Immunogold EM Staining of Lamina Proteins in *C. elegans* Embryos
   A. High Pressure Freezing of *C. elegans*
   B. Freeze Substitution for Immunolabeling
   C. Sectioning
   D. Labeling and Staining
  VII. Summary
   References

Abstract

The nuclear lamina is found between the inner nuclear membrane and the peripheral chromatin. Lamins are the main components of the nuclear lamina, where they form protein complexes with integral proteins of the inner nuclear membrane, transcriptional regulators, histones and chromatin modifiers. Lamins are required for mechanical stability, chromatin organization, Pol II transcription, DNA replication, nuclear assembly, and nuclear positioning. Mutations in human lamins cause at least 13 distinct human diseases, collectively termed laminopathies, affecting muscle, adipose, bone, nerve and skin cells, and range from muscular dystrophies to accelerated aging. *Caenorhabditis elegans* has unique advantages in studying lamins and nuclear lamina genes including low complexity of lamina genes and the unique ability of bacterially expressed *C. elegans* lamin protein to form stable 10 nm fibers. In addition, transgenic techniques, simple application of RNA interference, sophisticated genetic analyses, and the production of a large collection of mutant lines, all make *C. elegans* especially attractive for studying the functions of its nuclear lamina genes. In this chapter we will include a short review of our current knowledge of nuclear lamina in *C. elegans* and will describe electron microscopy techniques used for their analyses.

I. General Introduction

A. The Nuclear Lamina

In eukaryotic cells, the nuclear and cytoplasmic compartments are separated by the nuclear envelope (NE). The NE consists of inner and outer nuclear membranes (IM and OM, respectively). The two membranes are separated by a lumenal space that is continuous with the ER lumen. Communication between the nucleoplasm and cytoplasm occurs through pores in the NE, where the IM and OM join. In the metazoan cell, underlying the IM is a meshwork of nuclear-specific intermediate filaments, termed the nuclear lamina. The lamina is composed of lamin proteins and a growing number of lamin-associated proteins (Gruenbaum *et al.*, 2005). Lamins are evolutionarily conserved nuclear intermediate filaments (Herrmann and Aebi, 2004; Stuurman *et al.*, 1998). They have a short globular N-terminal "head" domain, an α-helical coiled-coil "rod" domain, and a globular "tail" domain. Lamins are grouped as A- or B-types based on their biochemical and structural properties and their behavior during mitosis. B-type lamins are expressed in all cell types throughout development and remain attached to membranes in mitosis (Stuurman *et al.*, 1998). A-type lamins are expressed in differentiated cells, usually at the onset of gastrulation, in a tissue-specific manner and are soluble during mitosis (Stuurman *et al.*, 1998). It is unclear whether A- and B-type lamins can copolymerize, or how they are organized within the nucleus. Polymerizing lamins *in vitro* is not a trivial task but in a recent breakthrough, Ce-lamin, the only lamin in *Caenorhabditis elegans* (B-type), has been polymerized *in vitro* into 10 nm intermediate filaments (Foeger *et al.*, 2006; Karabinos *et al.*, 2003).

Genetic analyses of lamins in both vertebrates and invertebrates has suggested that they support a wide range of functions through their interactions with a growing number of IM proteins (Mattout *et al.*, 2006). These functions include: maintaining the nuclear shape, spacing of nuclear pores, replication of DNA, regulation of gene expression, transcription elongation by Pol II, nuclear positioning, segregation of chromosomes, meiosis and apoptosis (Broers *et al.*, 2006; Goldman *et al.*, 2002; Gruenbaum *et al.*, 2005; Worman and Courvalin, 2005).

B. The Nuclear Lamina in *C. elegans*

The free-living soil nematode, *C. elegans*, is a useful model organism for investigating the biological functions of the nuclear lamina. It offers the great advantages of a tractable genetic system, easily grown in the lab. The animals are translucent throughout their life cycle, have a 4 days reproductive cycle, during which they lay nearly 300 eggs, and an established cell lineage, making microscopy and time-lapse microscopy highly feasible. In addition, transgenic studies are relatively simple and rigorous when done in *C. elegans*. A main advantage of *C. elegans* in studying the nuclear lamina is the low complexity, yet highly conserved nuclear lamina (Cohen *et al.*, 2001). For example, the *C. elegans* genome contains only a single lamin gene, annotated *lmn-1*, encoding a single lamin isoform, termed Ce-lamin, as opposed to 2 lamin genes in *Drosophila* and 3 lamin genes encoding 7 isoforms in mammals (Melcer *et al.*, 2007). Ce-lamin is ubiquitously expressed during development and is localized at both the nuclear periphery and the nucleoplasm (Liu *et al.*, 2000).

Lamins interact with integral and peripheral proteins of the IM, they also interact with specific proteins at the nucleoplasm and with DNA (Mattout-Drubezki and Gruenbaum, 2003). These interactions (direct and indirect) are responsible for many of the nuclear functions mediated by the nuclear lamina. Proper localization and retention of many proteins at the IM requires interaction with lamins, chromatin, or both. The number of these proteins and their chromatin partners has grown significantly over the past few years (Broers *et al.*, 2006; Goldman *et al.*, 2002; Gruenbaum *et al.*, 2005; Worman and Courvalin, 2005). Known protein partners of Ce-lamin include Ce-emerin, LEM-2, UNC-84, matefin/SUN-1, and BAF-1 (Gorjánácz *et al.*, 2006; Gruenbaum *et al.*, 2002; Lee *et al.*, 2000, 2002; Liu *et al.*, 2003; Margalit *et al.*, 2005). The list of Ce-lamin binding partners is expected to increase in light of the recent discovery of many novel mammalian proteins of the IM that have *C. elegans* orthologs (Schirmer *et al.*, 2003).

II. *In Vitro* Assembly of Ce-Lamin Filaments

A. Protein Expression and Purification

To make filaments, Ce-lamin should be bacterially expressed and purified to near homogeneity. Any bacterial expression system can be used. The bacterially expressed Ce-lamin is always found in the inclusion body preparations and is solubilized in 8.0 M urea.

We routinely use the pET24d vector (Novagen, Merck, Germany) modified to have an N-terminal His tag followed by a TEV protease recognition site, and the *Escherichia coli* strain BL21(DE3)-(codon plus-RIL) (Stratagene, La Jolla, CA). An overnight bacterial culture is diluted 1:100 in 2xYT medium and grown to an OD_{600} of 0.6–0.9. IPTG is added to a final concentration of 0.1 mM and cells are incubated at 37 °C for 1.5–2 h and harvested by centrifugation. The bacterial pellet is resuspended in resuspension buffer containing 20 mM Tris–HCl pH 7.5, 200 mM NaCl, 1 mM EDTA, 200 μg/ml lysozyme, and protease inhibitors (0.1 mM AEBSF, 3.4 μg/ml Aprotinin, and 5 mM Benzamidine). The bacterial suspension is then sonicated three times, 30 s each at 40% pulse and centrifuged at 13,000 rpm for 20 min at 4 °C. The inclusion bodies pellet is washed twice in the resuspension buffer containing 1% Triton X-100 without lysozyme and then resuspended in urea buffer (8.0 M urea, 50 mM Tris–HCl pH 8, 300 mM NaCl and protease mix) and centrifuged to separate soluble Ce-lamin from non-solubilized material. The supernatant containing the expressed Ce-lamin is purified in Ni-NTA batch. The Ni-NTA beads are washed with washing buffer (8.0 M urea, 50 mM Tris–HCl pH 8, 500 mM NaCl and 20 mM imidazole) and the protein is eluted with elution buffer (8.0 M urea, 20 mM Tris–HCl pH 7.5, 100 mM NaCl and 200 mM imidazole). For removal of the His-tag, protein fractions are pooled and dialyzed against a buffer containing 50 mM Tris–HCl pH 8.0, 150 mM NaCl, 1 mM DTT. Ten milligrams of the His-tagged Ce-lamin are then incubated at 4 °C overnight with 1 mg His-tagged TEV protease, dialyzed again into the urea buffer described above and separated from TEV protease and uncleaved material via Ni-NTA chromatography (Foeger *et al.*, 2006).

B. Comments

1. In general, bacterially expressed lamins are stable and not toxic to bacteria.
2. Freshly transfected bacteria give higher levels of expression.

C. Assembly of Ce-Lamin Filaments

Bacterially expressed and purified *C. elegans* lamin (0.1–0.2 mg/ml) in urea buffer is dialyzed at room temperature for 4 h against a buffer containing 2 mM Tris–HCl pH 9, 1 mM DTT, and then for up to 16 h against a buffer containing 15 mM Tris–HCl pH 7.4, 1 mM DTT. Assembly is terminated and the filaments are fixed by adding an equal volume of double distilled water (DDW) containing 0.2% (w/v) EM grade glutaraldehyde (Agar Scientific, Essex, UK). After 3–5 min, 5–7 µl aliquots are applied on glow-discharged formvar and carbon-coated EM copper grids and left to absorb for 1 min. The grids are washed with two drops of DDW and negatively stained with 2% uranyl acetate (Fluka, #73943). Under these conditions, Ce-lamin makes stable 10-nm thick intermediate filaments. These filaments exhibit rather irregular and complex branched patterns (Fig. 1A) (Foeger *et al.*, 2006; Karabinos *et al.*, 2003). The *in vitro* association reactions are extremely fast and filaments assemble within seconds (Foeger *et al.*, 2006).

D. Comments

1. Use only highly purified water for all EM procedures. We use Milli Q water (Millipore, Billerica, MA).
2. Glutaraldehyde is toxic and should be used only in the fume hood.
3. Uranyl acetate is radioactive and the solution is prepared in a fume hood with gloves and a lab coat.

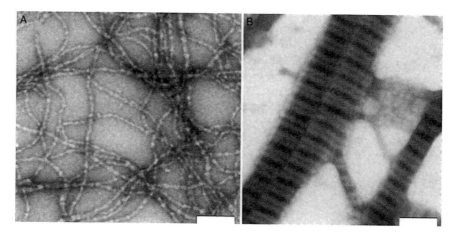

Fig. 1 *In vitro* assembly of invertebrate lamins. Depending on assembly conditions, *C. elegans* Ce-lamin assembles either as 10 nm wide filaments (A) or paracrystals (B). See text for details. Bars indicate 100 nm.

E. Assembly of Ce-Lamin Paracrystals

Bacterially expressed and purified *C. elegans* lamin (0.1–0.2 mg/ml) in urea buffer is dialyzed at room temperature for 2.0 h against a buffer containing 15 mM Tris–HCl pH 7.4, 1 mM DTT, followed by 3 h dialysis against a buffer containing 25 mM Tris—Cl pH 9, 20 mM $CaCl_2$, 1 mM DTT. The salt is then washed, dialyzed for 15–30 min against a buffer containing 25 mM Tris–HCl pH 9, 1 mM DTT. Assembly is stopped by adding an equal volume of DDW containing 0.2% (w/v) glutaraldehyde. Sample preparation and visualization are similar to those of filaments, the only difference being that grids are not washed with DDW. Under these conditions, Ce-lamin forms long paracrystalline fibers exhibiting a pronounced transverse banding pattern with a 24–25 nm axial repeat (Fig. 1B). *In vitro* association reactions are extremely fast, and paracrystalline arrays are assembled within seconds (Foeger *et al.*, 2006).

III. Preparation of Embryos and Adults for Transmission Electron Microscopy Using Microwave Fixation

C. elegans embryos are encapsulated in a hard chitin shell, which is difficult to penetrate. To view these embryos by transmission electron microscopy (TEM), an efficient fixation protocol is required in which the chitin shell is penetrated and the integrity of organelles and membranes is maintained.

TEM analysis of the nuclear envelope can reveal changes in the morphology of the nucleus and nuclear envelope. For example, downregulation of Ce-lamin causes chromatin condensation, clustering of nuclear pore complexes, cell death associated nuclear membrane blebbing resulting from decreasing/diminishing lamin expression in *C. elegans* embryos. An example of phenotypes observed in lamin downregulated embryos is shown in Fig. 2 (Cohen *et al.*, 2002).

Preparation of *C. elegans* embryos for TEM is essentially as described (Cohen *et al.*, 2002; Margalit *et al.*, 2005). Embryos are collected by the standard sodium hypochlorite/M9 technique (Chen *et al.*, 2000). A drop of 100–200 µl of 0.1% polylysine (Sigma P8920) is placed on a ~22 × 22 mm^2 marked at the center of a Falcon 3001 petri dish (35 × 10 mm). The plate is incubated for 5 min at room temperature and excess polylysine is removed. The petri dish is then safely attached, with double-stick tape, to the lid of a 50 ml conical tube (petri dish facing up). Washed and pelleted embryos are placed on the polylysine-treated area of the petri dish in a drop of M9 buffer (42 mM Na_2HPO_4, 22 mM KH_2PO_4, 85 mM NaCl, 100 µM $MgSO_4$). The tubes are centrifuged at 1000*g* for 1 min in a swinging bucket. This step ensures that the embryos remain on the polylysine for the rest of the procedure. The petri dish is then removed from the tube, immediately placed on ice, and the embryos are fixed with 1 ml of 0.1 M HEPES pH 7.0 containing 1% freshly made formaldehyde and 2.5% EM grade glutaraldehyde (Agar Scientific, Essex, UK). The plate is sealed with parafilm and immediately microwaved (Shef exclusive

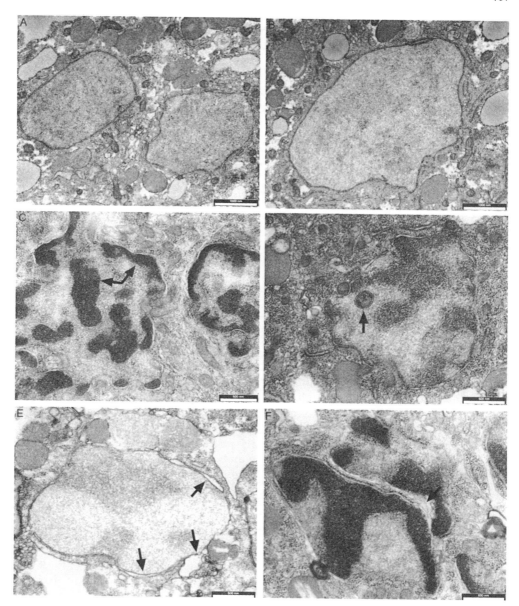

Fig. 2 Visualization of wildtype and *lmn-1* (RNAi) embryos by thin section transmission electron microscopy. (A, B) Nuclei from control animals fed with an empty L4440 vector have normal nuclear shape and chromatin distribution, and typical NE with two membranes and a uniform lumen. (C–F) Nuclei of *lmn-1* (RNAi) embryos, which have a drastic condensation of chromatin (arrows in C, see also F), membrane "vesicles" inside the nucleus (arrow in D), blebbing of membranes (arrows in E) and nuclear membrane shared by two nuclei (arrow in F). Bars represent 1.0 μm in panels A–B and 0.6 μm in panels C–F. (Reproduced with permission; Cohen *et al.* (2002)).

microwave oven SF-1875E at 70% microwave power) on ice for 20 s ON, 10 s OFF, and then again 20 s ON. The sample is then cooled on ice for 5 min in a fume hood and the microwave cycle is repeated once more. A Pelco-brand microwave, which provides a continuous power of 550 W or an equivalent microwave, can also be used. Fixation is completed by further incubation of the sample for 1 h at room temperature. Excess fixative solution is then removed and the embryos are washed three times, 10 min each, with 1 ml of 0.2 M HEPES at pH 5.6. One milliliter of solution containing 1% OsO_4, 0.5% reduced $K_4Fe(CN)_6$, 0.1 M HEPES pH 5.6 is then added, and the sample is incubated in the dark for 1 h at room temperature. The sample is then washed twice, 10 min each, with DDW and then incubated for 10 min in 50% ethanol, 10 min in 70% ethanol and 10 min in 90% ethanol, followed by three washes, 30 min each, in dry 100% ethanol. The sample is then incubated in a series of graded Epon:ethanol solutions, 2 h each, at room temperature, with Epon:ethanol (1:2), Epon/ethanol (1:1), Epon:ethanol (2:1), and left overnight in fresh 100% Epon at 4 °C in a desiccating chamber. The samples are exchanged three more times, 2 h each, at room temperature with 100% Epon. The sample is embedded in Epon and polymerized for 2 days at 60 °C in a dry oven. The Epon block is then separated from the petri dish by soaking it in liquid nitrogen. Use a sharp razor blade (e.g., Wilkinson razors) cleaned with acetone, to make a pyramid shape around the worm. For sectioning, any type of microtome should work. Sections can be cut with a glass knife or a diamond knife. The Epon block is sectioned to give thin sections, 80–90 nm each, and the sections are picked up on 200 mesh thin bar copper grids and stained with uranyl acetate and lead citrate. Samples are then viewed with an electron microscope.

The above protocol is modified to prepare *C. elegans* adults for TEM (Haithcock *et al.*, 2005). Adult *C. elegans* are collected in PBS and washed twice in PBS. The animals are cut with sharp forceps behind either the pharynx or the tail to allow rapid penetration of solutions and immediately subjected to fixation overnight at 4 °C in PBS (pH 7.5) containing 2.5% paraformaldehyde and 2.5% glutaraldehyde. The fixed animals are washed three times with PBS and mounted in 3.4% low melting point agarose (Sigma) to control the orientation of the worms in the final Epon block. The agarose blocks are incubated at 22 °C in 0.1 M sodium cacodylate (pH 7.5) for 10 min, and for 1 h in 0.1 M sodium cacodylate containing 1% OsO_4 and 1.5% reduced $K_4Fe(CN)_6$. The blocks are then washed four times, 10 min each, with 0.1 M sodium cacodylate, washed twice, 10 min each, with DDW and incubated in 30%, 50%, 70%, 80%, 90%, and 95% ethanol, 10 min each, followed by three washes, 20 min each, in 100% ethanol. Samples are then washed twice, 10 min each, in propylene oxide and incubated in a series of graded Epon in propylene oxide solutions each overnight at 4 °C (1:4, 1:1, 3:1), followed by an overnight incubation at 4 °C in fresh 100% Epon. The Epon blocks are polymerized for 2 days at 60 °C. The Epon blocks are sectioned horizontally, using a Diatome diamond knife to give 80–90-nm-thick sections containing a longitudinal view of each worm from head to tail. The sections are picked up on 200 mesh thin bar copper grids and stained with uranyl acetate and lead citrate. Samples are then ready for viewing with an electron microscope.

A. Comments

1. Osmium is very toxic and should be treated in a fume hood with gloves and a lab coat. All waste should be disposed of properly under the fume hood.
2. Fixation can be performed without the microwave treatment. In that case embryos are fixed of overnight at 4 °C in 2.5% paraformaldehyde and 2.5% glutaraldehyde in PBS followed by freezing the vial in liquid nitrogen, thawing the vial under tap water and fixing for an additional 1 h at room temperature. The quality of the data obtained by this procedure is still reasonable.

IV. Preparation of *C. elegans* Embryos and Adults for Conventional Transmission Electron Microscopy by High Pressure Freezing Combined with Freeze Substitution

Another powerful technique to penetrate the robust eggshell while maintaining an outstanding structural preservation is high-pressure freezing (HPF) in combination with freeze substitution (FS).

A. High-Pressure Freezing of *C. elegans*

We are using the Leica (Vienna, Austria) EMPACT2 HPF machine. However, any HPF machine (BALTEC, EMPACT HPF 01 or Wohlwend) should be compatible with this protocol.

A half microliter of 0.5% phosphatidylcholine in chloroform is placed onto a HPF carrier (Fig. 3). The carrier is then left to dry for 5–10 min. The phosphatidylcholine creates a lipid layer on the surface of the carrier allowing the worms to readily slip out when processed. A drop of M9 buffer containing 20% BSA (a cryo-protectant) is pipetted onto the carrier and 10–15 worms are placed in the drop, using a worm pick (Muller-Reichert *et al.*, 2003). A piece of Whattman filter paper is used to soak up excess buffer, so that the level of buffer on the carrier surface is flat. This is critical, since a meniscus causes loss of worms during freezing, due to squeezing of the holder onto the actual carrier (Fig. 3). This step should be done relatively quickly before the worms dry out. Next, the holder is placed in the HPF machine and is subjected to freezing according to the manufacturer's instructions. The carrier is then released from the holder under liquid nitrogen. These carriers can also be stored in a liquid nitrogen tank for long-term storage.

B. Freeze Substitution for Morphological Studies

Prepare the FS machine according to the manufacturer's instructions. Program the machine to maintain the carriers at −90 °C for 48 h and then slowly raise the temperature to −30 °C in increments of 5 °C/h. Once the temperature reaches

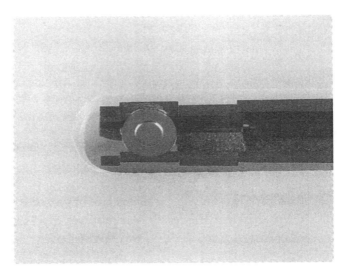

Fig. 3 Image of a Leica membrane carrier used for HPF.

−30 °C, keep it for 3 h and then raise the temperature again to 0 °C in increments of 5 °C/h. This process takes about 69 h.

The FS cocktails are prepared as follows: Dissolve a 100 mg crystal grade OsO_4 (Serva, #31251) in 10 ml of dried acetone (EMS, #10015). Pipette 950 μl of the 1% OsO_4 solution into a cryo tube (Sarstedt, #72.609). Add 50 μl of a 2% uranyl acetate stock in water to the OsO_4 solution. This will produce a cocktail containing 1% OsO_4, 0.1% uranyl acetate, and approximately 5% water. Close the cryo tube and place it under liquid nitrogen to freeze. The excess water is added to the solution to enhance membrane quality (Walther and Ziegler, 2002) and indeed this cocktail gives striking membrane staining in which the double bilayer of the NE is visible (Fig. 4).

Keeping the carriers under liquid nitrogen, transfer them from the HPF machine to the box of liquid nitrogen with the FS cocktail cryo tubes. Under liquid nitrogen, unscrew the top of the cryo tube and place 1–2 carriers inside. Screw the cap back on and transfer the tubes directly into the Freeze Substitution machine. Once all the cryo tubes are placed in the machine, start the machine and let it run through the entire program.

C. Infiltration and Embedding

Once the machine has reached zero degrees, take the cryo tubes out and place them in a rack in the fume hood. Pipette out the OsO_4 solution and wash the carriers twice with dried acetone. Rinse the carriers twice more with acetone letting them sit for 10 min in between each rinse. To begin infiltration, prepare a solution

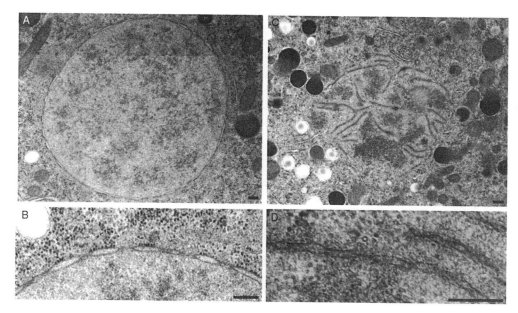

Fig. 4 Visualization of wildtype and *baf-1* mutant embryos by thin section transmission electron microscopy. (A) A wild-type nucleus. (B) A blown up region of the periphery of a wild-type nucleus. (C) A nucleus from a mutant *baf-1* embryo (D) A blown up region of the *baf-1* affected nucleus. Eliminating *baf-1* activity caused abnormal assembly of the NE (compare to A, B) Bars represent 200 μm.

of acetone:Epon (3:1) in a 15 ml Falcon tube. Pipette off the acetone and add 1 ml of the acetone:Epon to each cryo tube. Place the cryo tubes for 1 h on a turning wheel in the fume hood. The wheel is used for all additional infiltration steps. As the cryo tubes are turning on the wheel, worms are highly likely to fall out of the carrier. Hence, be cautious when pipetting the next set of solutions. The worm pellet should be free from the carrier at this point, therefore pipette out the pellet and place it into a new Eppendorf tube. If the pellet is not loose, carefully tap the side of the tube until it becomes free. Carefully exchange the 3:1 solution with the 1:1 acetone:Epon solution and incubate for 2 h. Next, exchange with a 1:3 solution of acetone:Epon and leave overnight. In the next morning, exchange with 100% Epon. Repeat this step three more times every 30 min and leave the cryo tubes on the wheel for an additional 3–4 h.

For embedding the worms, clean two glass microscope slides and spray one side of each slide six times with Teflon Spray (EMS). Using a Kimwipe, polish the Teflon until the glass is clear. Make sure you mark the side you spray for easier detection later. Place one glass slide under the binocular microscope, Teflon side up. Using a 200 μl pipette, cut off the end of the tip and pull up a small amount of Epon from your Eppendorf tube to coat the inside of the pipette tip. Next, pipette out the worm pellet and put it onto the glass slide. Using a very fine needle and forceps, try and separate the worms from each other as best you can without breaking them into small pieces. Pipette the excess Epon off of the slide; the total amount of Epon on the slide should not exceed 100 μl. Cut two Parafilm pieces of approximately 2 × 2 cm, fold them in half and place one on each end of the slide. Place the other slide on top with the Teflon side facing down onto the Epon to make a "sandwich" (Fig. 5). Put the slide overnight in a 60 °C dry oven for polymerization. Save a small portion of the left over Epon at −20 °C to be used for mounting the worms.

After taking the slides out of the oven, place a razor blade between the two slides and carefully try to separate them. Using the binocular microscope, choose a worm and cut it out of the Epon using a razor blade. For this next step you will need dummy blocks, which are plain Epon blocks you can make ahead of time with any type of mold. To mount the worms, thaw the left over Epon from earlier and position the dummy block so it is standing upwards in the cap of a 0.5 ml

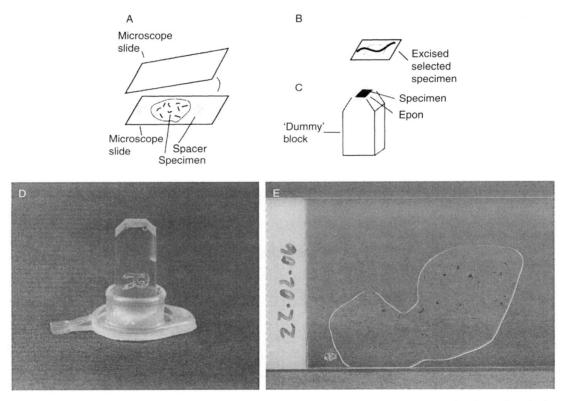

Fig. 5 Preparing Epon blocks following high pressure freezing combined with Freeze Substitution. (A–C) Illustrations showing the steps involved in flat embedding *C. elegans* worms between two slides, excising them from Epon and mounting them onto dummy blocks. (D) An example of an Epon block placed in the cap of an Eppendorf tube. (E) The worms embedded between the two glass slides.

Eppendorf tube (Fig. 5D). Using a toothpick put a small drop of Epon on top of the dummy block. Lay the worm that you have cut out on top of the Epon block and cover it with a small amount of excess Epon (Fig. 5). Place the blocks in a glass Petri dish and incubate them overnight in the 60 °C oven.

D. Sectioning and Staining

Use a sharp razor blade (e.g., Wilkinson razors) cleaned with acetone to trim around the worm making a pyramid shape. Sections can be cut with a glass knife or a diamond knife. We use a Diatome Ultra 35° knife and cut 55-nm-thick sections and collect the sections on a Copper/Palladium slot grid (Plano) coated with 1% Formvar (Serva) in chloroform.

Post staining is performed on a piece of parafilm. Incubate the grids face down on a drop of 50% methanol for 5 min, then on a drop containing 2% uranyl acetate in 70% methanol for 10 min. Rinse the grids quickly four times in 50% methanol and then twice in DDW for 5 min. Use a new piece of parafilm to make a row of KOH pellets and incubate the grids on a drop of Reynolds lead citrate next to the row of pellets. Lead citrate combines readily with CO_2 therefore, place a cover over the KOH pellets and the grids to reduce excess CO_2. Finally, wash the grids twice, 5 min each, in DDW. Pull off the excess water on the grids by holding a piece of Whattman filter paper to the edge. These grids can then be stored for years before further processing.

V. Preembedding Immunogold EM Staining of Lamina Proteins in *C. elegans* Embryos

Preparation of embryos for preembedding immunogold staining is essential as described (Cohen *et al.*, 2002). *C. elegans* embryos are collected using the sodium hypochlorite/M9 technique (Chen *et al.*, 2000). The isolated embryos are washed three times with DDW, resuspended in 400 µl DDW and fixed by adding 1 ml of a solution containing 80 mM KCl, 20 mM NaCl, 1.3 mM EGTA, 3.2 mM spermine, 7.5 mM sodium HEPES pH 6.5, 25% methanol, 2% formaldehyde, and 1.5% glutaraldehyde. The embryos are immediately frozen in liquid nitrogen, quickly thawed under tap water and incubated at room temperature for additional 45 min in the same fixative. The embryos are then washed once in 1 ml of 100 mM Tris–HCl pH 7.5 and three times, 15 min each, in PBS containing 0.1% BSA and 1 mM EDTA (PBSB) and pelleted by centrifugation at 2000*g*. Primary antibodies are diluted to the desired concentration in PBS containing 1 mM EDTA and 1% BSA (PBSEB) and added to the embryo pellet. Embryos are resuspended and incubated overnight at 4 °C. Excess antibody is then removed by four washes, 1 h each, in PBSB. The embryo pellet is then resuspended in secondary gold-conjugated antibodies diluted to the desired concentration and incubated overnight at 4 °C.

We usually use 6 nm gold-conjugated antibodies and dilute the antibodies according to manufacturer instructions. Embryos are washed as above, pelleted and fixed again for 1 h at room temperature in PBS containing 2.5% glutaraldehyde. After washing in PBS and centrifugation at 2000g, the embryo pellet is mounted in 15 μl of 3.4% low-melting agarose. The agarose block is washed four times, 30 min each, in PBS. Embryos are postfixed in 1% buffered OsO4 containing 1.5% reduced $K_4Fe(CN)_6$ for 1 h at room temperature, dehydrated in graded series of ethanol (50, 70, and 100%), and subjected to two changes of propylene oxide, Epon infiltration, embedding, and polymerization at 60 °C for 2 days, as described above for TEM sample preparation. Sectioning, staining with uranyl acetate and lead citrate, and viewing are also performed as described above, for TEM sample preparation. An example of such an experiment is shown in Fig. 6A.

For experiments in which the nuclear membranes were removed, washed embryos are spun down and resuspended in 1.25 ml of fixation solution containing 0.4% freshly made formaldehyde, 18% methanol, and 10 mM EGTA. After 20 min incubation on ice, embryos are subjected to two rounds of freezing in liquid

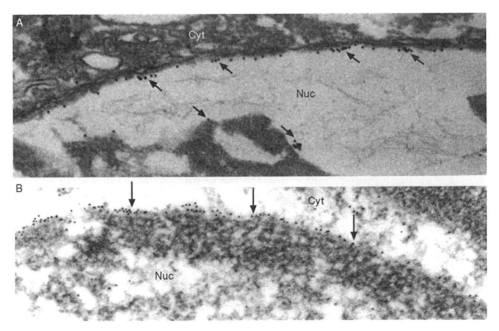

Fig. 6 Preembedding immuno-gold TEM labeling of Ce-lamin in *C. elegans*. (A) Nucleus with intact nuclear membranes. (B) Nucleus in which nuclear membranes were removed prior to labeling. Rabbit antibodies against the rod and tail regions of Ce-lamin were used as primary antibodies and 12-nm gold-conjugated goat–anti-rabbit IgG were used as secondary antibodies to localize Ce-lamin in embryos. The positions of the gold particles near the inner nuclear membrane and in the nuclear interior are indicated by arrows. Nucleus, Nuc; Cytoplasm, Cyt. (Panel A was reproduced with permission; Cohen *et al.* (2002)).

nitrogen, thawing under tap water and then shaken for 20 min at room temperature. Embryos are washed once with 100 mM Tris–HCl pH 8.0 containing 1 mM EDTA and 0.5% Triton X-100 and once with PBS containing 1 mM EDTA and 0.5% Triton X-100 (PBST) and incubated for 15 min at room temperature with PBST. Embryos are then incubated overnight at room temperature with primary antibodies diluted in PBS containing 0.5% Triton X-100 and 0.1% BSA (PBSTB). After four washes, 30 min each, with PBSTB, embryos are incubated with gold-conjugated secondary antibodies as above. Embryos are washed four times, 30 min each, with PBST and fixed for 1 h at room temperature in 2.5% glutaraldehyde. Immunogold labeling, post-fixation with OsO4, Epon embedding, sectioning, staining, and viewing are performed as described above. An example of such an experiment is shown in Fig. 6B.

VI. Postembedding Immunogold EM Staining of Lamina Proteins in *C. elegans* Embryos

A very powerful technique for postembedding immuno-gold staining is the HPF combined with FS. This technique is used to gain better insight on the function and subcellular localization of NE proteins. As an example we used the monoclonal antibody 414 (mAb414), which recognizes the conserved FXFG repeats found in many NPC proteins (Davis and Blobel, 1986). This antibody specifically localizes to *C. elegans* NPC's (Fig. 7A and B).

A. High Pressure Freezing of *C. elegans*

High pressure freezing is performed as described in Section IV.

B. Freeze Substitution for Immunolabeling

It is important to choose a resin that preserves ultrastructure, while retaining antigenicity. We are using Lowicryl HM20 (Polysciences Inc., #15924), which is a nonpolar resin that polymerizes at very low temperatures and produces high contrast with minimal damage of ultrastructure. Since Lowicryl HM20 is highly toxic and must polymerize at low temperatures ($-45\ °C$), the FS setup should be designed to avoid direct contact with the chemicals. Our Leica AFS2 is a fully automated system allowing easy handling of the specimen.

First, cool the machine with liquid nitrogen. Prepare a 2% uranyl acetate stock solution in dried methanol and use it to prepare a 0.5 ml FS solution containing 0.5% uranyl acetate in acetone. Put the FS solution in a glass bottle and place it in the machine to cool down. This process takes about 10–15 min. Once cooled, the solution is used to fill up the cryo tubes. It is important to make sure the solution is cooled down to the correct temperature, otherwise ice crystals form in the specimen. Next, put an Eppendorf tube in liquid nitrogen, the carrier in the Eppendorf

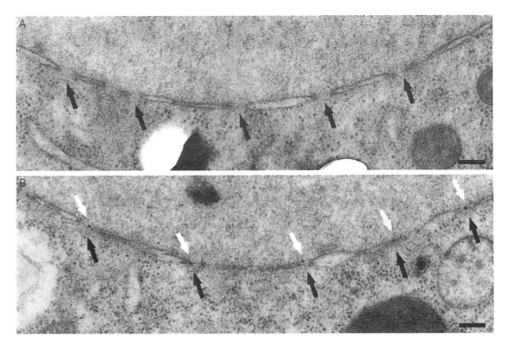

Fig. 7 Post-embedding Immuno-gold TEM labeling of mAb414 in *C. elegans*. (A) A micrograph displaying a portion of a wild-type nucleus. (B) Wild-type nuclei labeled with mAb414 as a primary antibody and a 15 nm Protein-A gold conjugate as secondary antibody. Black arrows point to nuclear pores and white arrows point to mAb414 gold markers. Bars represent 200 μm.

tube and transfer the tube to the FS machine keeping the carrier under liquid nitrogen at all times.

Program the machine according to the manufacturer's instructions. The program we use is as follows: leave the carriers for 50 h at −90 °C. Next, raise the temperature 5 °C per hour until it reaches −45 °C and leave the carriers for 5 h at this temperature, followed by three acetone washes, 30 min each. Next, infiltrate with 25% Lowicryl for 2 h, 50% Lowicryl for 2 h, 75% Lowicryl for 2 h and 100% Lowicryl for 12 h. Change the Lowicryl again and leave it for an additional 2 h. Polymerize at −45 °C for 48 h with the UV lamp on and then raise the temperature 5 °C per hour until the machine reaches 20 °C for the final polymerization steps. The carriers should stay at 20 °C for another 48 h until polymerization is complete. This whole process takes about 8 days. When finished, place the blocks under the fume hood for the rest of the day to allow further evaporation. The blocks are ready for sectioning when they become transparent.

C. Sectioning

Sectioning is performed as described in Section IV.

D. Labeling and Staining

Labeling is performed directly on the grids, while the grids are placed face down on top of the drops of solution. On a piece of parafilm, place the grids for 20 min on a drop of PBG buffer (0.8% BSA, 0.1% Fish Skin Gelatin, 0.01% Tween 20 in PBS buffer). Then place the grids on a 5 μl drop of primary antibody diluted in PBG for 1 h. We suggest testing a 10-fold range of different dilutions. Next, rinse the grids on drops of PBS, three times, 5 min each, followed by 5 min incubation on a drop of PBG. Place the grid on a 7 μl drop of secondary antibody for 1 h. Dilute the secondary antibody in PBG according to the manufacturer's instructions, while verifying that the secondary antibodies match the primary antibodies. In case of rabbit, guinea pig, dog, cat, mouse IgG2a or human IgG1, 2 or 4, Protein-A coupled to gold can be used as well. We recommend using 15-nm gold-conjugated antibodies or protein A when labeling *C. elegans* embryos, as it is easier to visualize. Next, rinse the grids three times, 5 min each, in PBS followed by a 5 min fixation on a drop of 1% glutaraldehyde in PBS. Rinse the grids five times, 1 min each, in DDW. Post staining is performed as described in Section IV.

VII. Summary

The electron microscopy techniques described above, combined with the powerful genetic tools in *C. elegans*, have helped establish lamins as essential components of nuclear architecture, adding significant knowledge and helping understanding the functions of lamins and their associated proteins. These techniques, however, are also useful for analyzing other cellular proteins in *C. elegans*.

The structure of the nuclear lamina in somatic cells is mostly unknown. Applying powerful EM techniques, including cryo-EM tomography to *C. elegans* vitreous sections or isolated nuclei, should provide better understanding of how lamins and nuclear lamina proteins are organized *in vivo*.

Acknowledgments

We thank Thomas Mueller-Reichert for the diagram in Fig. 5. We also thank Shai Melcer, Mátyás Gorjánácz, Gareth Griffiths, Heinz Schwarz, Claude Antony and Uta Haselmann for helpful discussions. The Gruenbaum lab is funded by grants from the USA-Israel Binational Science Foundation (BSF), Israel Science Foundation (ISF), Israel Ministry of Health and the European Union's FP6 Life Science, Genomics and Biotechnology for Health. The Mattaj lab is funded by grants from the European Molecular Biology Laboratory (EMBL), Federation of European Biochemical Societies (FEBS) and the Human Frontier Science Program Organization (HFSPO).

References

Broers, J. L., Ramaekers, F. C., Bonne, G., Yaou, R. B., and Hutchison, C. J. (2006). Nuclear lamins: Laminopathies and their role in premature ageing. *Physiol. Rev.* **86**, 967–1008.

Chen, F., Hersh, B. M., Conradt, B., Zhou, Z., Riemer, D., Gruenbaum, Y., and Horvitz, H. R. (2000). Translocation of *C. elegans* CED-4 to nuclear membranes during programmed cell death. *Science* **287,** 1485–1489.

Cohen, M., Lee, K. K., Wilson, K. L., and Gruenbaum, Y. (2001). Transcriptional repression, apoptosis, human disease and the functional evolution of the nuclear lamina. *Trends Biol. Sci.* **26,** 41–47.

Cohen, M., Tzur, Y. B., Neufeld, E., Feinstein, N., Delannoy, M. R., Wilson, K. L., and Gruenbaum, Y. (2002). Transmission electron microscope studies of the nuclear envelope in *Caenorhabditis elegans* embryos. *J. Struct. Biol.* **140,** 232–240.

Davis, L. I., and Blobel, G. (1986). Identification and characterization of a nuclear pore complex protein. *Cell* **45,** 699–709.

Foeger, N., Wiesel, N., Lotsch, D., Mucke, N., Kreplak, L., Aebi, U., Gruenbaum, Y., and Herrmann, H. (2006). Solubility properties and specific assembly pathways of the B-type lamin from *Caenorhabditis elegans*. *J. Struct. Biol.* **155,** 340–350.

Goldman, R. D., Gruenbaum, Y., Moir, R. D., Shumaker, D. K., and Spann, T. P. (2002). Nuclear lamins: Building blocks of nuclear architecture. *Genes Dev.* **16,** 533–547.

Gorjánácz, M., Klerkx, E. P. F., Galy, V., Santarella, R., López-Iglesias, C., Askjaer, P., and Mattaj, I. W. (2006). *C. elegans* BAF-1 and its kinase VRK-1 participate directly in postmitotic nuclear envelope assembly. *EMBO J.* **26,** 132–143.

Gruenbaum, Y., Lee, K. K., Liu, J., Cohen, M., and Wilson, K. L. (2002). The expression, lamin-dependent localization and RNAi depletion phenotype for emerin in *C. elegans*. *J. Cell Sci.* **115,** 923–929.

Gruenbaum, Y., Margalit, A., Shumaker, D. K., and Wilson, K. L. (2005). The nuclear lamina comes of age. *Nat. Rev. Mol. Cell Biol.* **6,** 21–31.

Haithcock, E., Dayani, Y., Neufeld, E., Zahand, A. J., Feinstein, N., Mattout, N., Gruenbaum, Y., and Liu, J. (2005). Age-related changes of nuclear architecture in *Caenorhabditis elegans*. *Proc. Natl. Acad. Sci. USA* **102,** 16690–16695.

Herrmann, H., and Aebi, U. (2004). Intermediate filaments: Molecular structure, assembly mechanism, and integration into functionally distinct intracellular scaffolds. *Annu. Rev. Biochem.* **74,** 749–789.

Karabinos, A., Schunemann, J., Meyer, M., Aebi, U., and Weber, K. (2003). The single nuclear lamin of *Caenorhabditis elegans* forms *in vitro* stable intermediate filaments and paracrystals with a reduced axial periodicity. *J. Mol. Biol.* **325,** 241–247.

Lee, K. K., Gruenbaum, Y., Spann, P., Liu, J., and Wilson, K. L. (2000). *C. elegans* nuclear envelope proteins emerin, MAN1, lamin, and nucleoporins reveal unique timing of nuclear envelope breakdown during mitosis. *Mol. Biol. Cell* **11,** 3089–3099.

Lee, K. K., Starr, D., Liu, J., Cohen, M., Han, M., Wilson, K., and Gruenbaum, Y. (2002). Lamin-dependent localization of UNC-84, a protein required for nuclear migration in *C. elegans*. *Mol. Biol. Cell* **13,** 892–901.

Liu, J., Lee, K. K., Segura-Totten, M., Neufeld, E., Wilson, K. L., and Gruenbaum, Y. (2003). MAN1 and emerin have overlapping function(s) essential for chromosome segregation and cell division in *C. elegans*. *Proc. Natl. Acad. Sci. USA* **100,** 4598–4603.

Liu, J., Rolef-Ben Shahar, T., Riemer, D., Spann, P., Treinin, M., Weber, K., Fire, A., and Gruenbaum, Y. (2000). Essential roles for *Caenorhabditis elegans* lamin gene in nuclear organization, cell cycle progression, and spatial organization of nuclear pore complexes. *Mol. Biol. Cell* **11,** 3937–3947.

Margalit, A., Segura-Totten, M., Gruenbaum, Y., and Wilson, K. L. (2005). Barrier-to-autointegration factor is required to segregate and enclose chromosomes within the nuclear envelope and assemble the nuclear lamina. *Proc. Natl. Acad. Sci. USA* **102,** 3290–3295.

Mattout, A., Dechat, T., Adam, S. A., Goldman, R. D., and Gruenbaum, Y. (2006). Nuclear lamins, diseases and aging. *Curr. Opin. Cell Biol.* **18,** 1–7.

Mattout-Drubezki, A., and Gruenbaum, Y. (2003). Dynamic interactions of nuclear lamina proteins with chromatin and transcriptional machinery. *Cell Mol. Life Sci.* **60,** 2053–2063.

Melcer, S., Gruenbaum, Y. and Krohne, G. (2007). Invertebrate lamins. *Exp. Cell. Res.* **10,** 2157–2166.

Muller-Reichert, T., Hohenberg, H., O'Toole, E. T., and McDonald,, K. (2003). Cryoimmobilization and three-dimensional visualization of *C. elegans* ultrastructure. *J. Microsc.* **212**, 71–80.

Schirmer, E. C., Florens, L., Guan, T., Yates, J. R., and Gerace, L. (2003). Nuclear membrane proteins with potential disease links found by subtractive proteomics. *Science* **531**, 1380–1382.

Stuurman, N., Heins, S., and Aebi, U. (1998). Nuclear lamins: Their structure, assembly, and interactions. *J. Struct. Biol.* **122**, 42–66.

Walther, P., and Ziegler, A. (2002). Freeze substitution of high-pressure frozen samples: The visibility of biological membranes is improved when the substitution medium contains water. *J. Microsc.* **208**, 3–10.

Worman, H. J., and Courvalin, J. C. (2005). Nuclear envelope, nuclear lamina, and inherited disease. *Int. Rev. Cytol.* **246**, 231–279.

CHAPTER 22

Visualization of Nuclear Organization by Ultrastructural Cytochemistry

Marco Biggiogera[*,†] and Stanislav Fakan[*]

[*]Centre of Electron Microscopy
University of Lausanne
Bugnon 27
CH-1005 Lausanne
Switzerland

[†]Laboratorio di Biologia Cellulare e Neurobiologia
Dipartimento di Biologia Animale
Università di Pavia
Piazza Botta 10
I-27100 Pavia
Italy

I. Introduction
II. Cytochemical Contrasting Approaches
    A. DNA Staining with Osmium Ammine
    B. The EDTA Staining Method for Nuclear Nucleoproteins
    C. Terbium Staining for RNA
    D. Bismuth Staining for Proteins
    E. Ag-NOR-Staining for Nucleoli
III. High Resolution Autoradiography
IV. Immunocytochemistry
V. Molecular in situ Hybridization
VI. Identification of Nucleic Acids by Means of Enzymatic Reactions
    A. TdT Reaction
    B. PnT Reaction
VII. Targeting of Intranuclear Substrates using Enzyme-Colloidal Gold Complexes
VIII. Concluding Remarks
    References

I. Introduction

Ultrastructural cytochemistry has been, for more than four decades now, the major tool in investigating the functional organization of the cell nucleus. While different methods of light microscopic analysis may offer a faster approach and a possibility of three-dimensional visualization, they often give rise to alteration of cellular structure, as shown by subsequent examination by electron microscopy (Solovei *et al.*, 2002; Visser *et al.*, 2000). Moreover, keeping in mind the relatively low resolution of the confocal laser scanning microscopy, which under best conditions can reach 220 nm in x/y directions (Stelzer, 1995), the light microscopic data can often be approximate and will not answer fundamental questions regarding nuclear topology when dealing with "closely" located nuclear domains. Modern methods of ultrastructural cytochemistry usually provide a good structural preservation thanks to the stabilization of the ultrathin section by the presence of a resin, while offering a resolution about two orders of magnitude superior to light microscopy. Furthermore, two additional points should be stressed when findings regarding different nuclear domains are interpreted based solely on light microscopic observations. When an accumulation of a factor is stated, possibly generating an idea of a new nuclear domain, such observation does not represent a sufficient evidence about a real occurrence of a structural support and it can only reflect an accumulation site of the factor. On the other hand, a diffuse, apparently "structureless" signal can be supported by structural constituents which, however, are far below the resolving power of light microscopy.

High resolution features of the functional nuclear architecture have been extensively described in recent review articles (Biggiogera and Pellicciari, 2000; Fakan, 2004; Raska *et al.*, 2006; Spector, 2003; Woodcock, 2006). We will, therefore, mention details regarding different nuclear structural domains only in relation with the various cytochemical approaches we will deal with.

In this chapter, we will mainly pay attention to the application of different ultrastructural cytochemical methods that contributed in an important way to our present understanding of the functional nuclear architecture. We are providing the working protocol for some of them and specify pertinent references for those where a detailed procedure has already been described elsewhere.

II. Cytochemical Contrasting Approaches

A major goal for EM cytochemistry has always been the possibility to detect a single molecule or component within a biological structure in a cell *in situ*. Specificity is one of the basic concepts for cytochemistry: an ideal reaction should be specific for a single reactive chemical group or for a single component. In practice, it is especially important to know the specificity limits of a reaction. These limits may be defined in terms of pretreatments (required for revelation or blockade of specific groups) within the range of temperature, incubation time or pH in which the reaction takes place under optimal conditions. The final goal is to obtain interpretable, reproducible and useful results.

In this section, a choice of methods for nucleic acid, nucleoprotein and protein contrasting on thin sections is presented. The main requirements for EM cytochemistry include reaction specificity and electron density of the end product as well as its fineness in order to reach the highest possible resolution. However, differences between molecules can be minor, at least from a cytochemical standpoint. Thus for DNA and RNA, both nucleic acids may take a similar structural conformation (ssDNA and dsRNA), the sugar differing by an OH group, and thymine and uracil only by a methyl group.

A. DNA Staining with Osmium Ammine

A specific DNA staining for electron microscopy can be accomplished by a Feulgen-type reaction. The incubation of thin sections with 5 N HCl engenders pseudo-aldehyde groups on purines of DNA in the sample, which are subsequently revealed by a Schiff-type, aldehyde-specific reagent. RNA, on the other hand, does not give rise to aldehyde groups, but is depolymerized by HCl and may be extracted from the sections (Biggiogera et al., 2001).

The reaction has been widely used for light microscopy since 1924, but the quest for an electrondense Schiff-type reagent applicable in transmission electron microscopy has lasted for decades. A Schiff (or Schiff-type) reagent must fulfill several requirements before being used to specifically detect aldehyde groups. These requirements include the presence of at least one amino group, "activation" by SO_2 (Kasten, 1960) and electrondensity (Gautier, 1976). The reagent should not give rise to staining if the preliminary, aldehyde-engendering treatment (acid hydrolysis or periodic oxidation) is omitted. Moreover, no staining should be obtained when the aldehyde groups are blocked prior to exposure to the reagent.

Many reagents have been proposed (see Gautier, 1976) before the synthesis of OAC (Osmium Ammines Complex) by Cogliati and Gautier (1973). OAC is a polyammine complex whose structure is still unclear. Its synthesis has always been capricious until the modification of its synthesis procedure (Olins et al., 1989) which has allowed for an easier way of obtaining the reagent, now commercially available.

Activation of the reagent has routinely been carried out by bubbling with SO_2; however, this can be accomplished also by adding SO_2-generating chemicals like for light microscopy (Vazquez-Nin et al., 1995). This preparation method has the advantage of avoiding the use of gaseous SO_2 for the activation of the compound, thus rendering the technique readily available to laboratories for which could be difficult to store a gas bottle containing SO_2.

The main advantages of this technique are its specificity after formaldehyde fixation, and simplicity, since it involves just two steps. Activated osmium ammine behaves like a Schiff-type reagent and the staining can be selectively and completely abolished by aldehyde blocking compounds.

Moreover, this reaction can be applied to tissue sections embedded in any kind of resin and provides a good contrast (in particular on acrylate sections) and an excellent resolution. The osmium ammine reaction can also be used after immunolabeling (Fig. 1), since acid hydrolysis does not remove or displace gold grains (Biggiogera *et al.*, 2001).

The applications of this technique have been numerous. In terms of high resolution, the distinct visualization of nucleosomal fibers in which DNA can be seen to encircle an unstained core (Derenzini *et al.*, 1983) and the detection of viral particles (Puvion-Dutilleul, 1991) can be quoted. More recently, the localization of condensed and loose chromatin in relationship to the interchromatin space has been analyzed by combining OAC with immunocytochemical labeling of DNA and of a histone-GFP fusion protein (Albiez *et al.*, 2006).

1. Preparation of the Reagent (Vazquez-Nin *et al.*, 1995)

For acrylic resin sections, dilute 10 mg of osmium ammine-B in 4.8 ml of bidistilled water. After the complete dissolution of the reagent, add 0.2 ml of 5 N HCl (to reach a final concentration of 0.2 N HCl), stir and finally add 190 mg of sodium metabisulfite ($Na_2S_2O_5$) to give a final concentration of 0.2 M $Na_2S_2O_5$.

For epoxy resin sections, dilute 10 mg of osmium ammine-B in 2.5 ml of bidistilled water and, after the dissolution of the reagent, add 2.5 ml of acetic acid (to

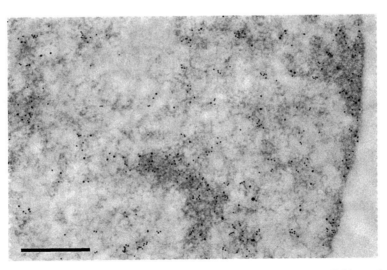

Fig. 1 HeLa cell stably expressing histone H2B-GFP fusion protein. LR White resin section immunolabeled for GFP (small gold particles) and DNA (large gold particles) and stained with osmium ammine. Both labels are predominantly localized on osmium ammine-stained areas reflecting chromatin domains, while interchromatin space is virtually devoid of staining and immunocytochemical signal. Courtesy of L. Vecchio. For more details, see Albiez *et al.*, 2006. Bar represents 0.5 μm.

give a final concentration of 8 N acetic acid), stir and add 38 mg of sodium metabisulfite (40 mM final concentration).

The solutions are ready for use 30 min after dissolving the metabisulfite and can be stored for several weeks in a tightly stoppered vial at room temperature.

2. Feulgen Reaction

Thin sections are collected on naked gold grids or on homemade plastic Marinozzi rings for floating sections. Avoid copper or nickel grids which are attacked by acids. The grids are floated onto freshly prepared 5 N HCl for 30 min at room temperature, rinsed with distilled water and floated onto one of the above solutions for 60 min at room temperature. They are then thoroughly rinsed with water and dried.

3. Blockade Reaction (Control)

After 30 min HCl hydrolysis, epon and acrylic sections are incubated on saturated dimedone in 1% acetic acid for 60 min at 60 °C to block specifically the aldehyde groups, rinsed with water and then stained with osmium ammine as above.

B. The EDTA Staining Method for Nuclear Nucleoproteins

Few cytochemical techniques for electron microscopy can boast a yield of results comparable with the EDTA regressive staining. Since its introduction (Bernhard, 1969; Monneron and Bernhard, 1969) this simple method has been used in a large number of studies dealing with the cell nucleus.

Although its mechanism has never been completely clarified, this staining is based on the same principle of the differentiation as used in light microscopic cytochemistry. The biological sample is fixed with aldehydes (without osmium postfixation) and embedded in epoxy or acrylic resin. The thin sections are then stained with uranyl acetate, treated with a solution of neutral EDTA and finally stained with lead citrate. The first step allows uranyl ions to bind both to deoxyribonucleoproteins and RNPs. EDTA, being a chelating agent, is capable of more easily removing uranyl from the sites on deoxyribonucleoproteins.

Lead then binds to the uranyl ions still present on RNPs and increases the contrast.

The final result is a cell nucleus in which condensed chromatin is bleached and RNP-containing structures are well contrasted. Among these, perichromatin fibrils were first observed thanks to this technique, and perichromatin and interchromatin granules could be studied in more detail. Other nuclear structures, such as coiled (Cajal) bodies, interchromatin granule associated zones (IGAZ) or PML bodies can also be observed.

The method is simple and can be applied to ultrathin resin sections or cryosections (Fig. 2). Another staining method derived from the EDTA technique can also be applied to ultrathin frozen sections (Puvion and Bernhard, 1975).

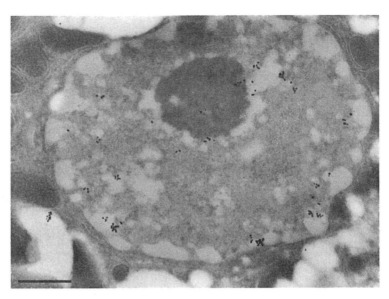

Fig. 2 Ultrathin frozen section of an isolated rat hepatocyte labelled in culture for 2 min with ^{3}H-uridine. The radioactive label was detected by high resolution autoradiography and the section stained by the EDTA method. Silver grains indicating newly synthesized RNA are mainly localized on the periphery of the condensed chromatin regions (light gray) and in the nucleolus. For more details, see Fakan *et al.*, 1976. Bar represents 1 µm.

It is important to keep in mind that the technique is preferential, and not specific, for nuclear RNPs. In the original paper it is stated that this method could be considered as a reproducible tool only for nuclear RNP-containing structural constituents. However, some RNP-containing structures, obviously of nuclear origin, are EDTA positive also in the cytoplasm (Biggiogera *et al.*, 1997).

There are only few exceptions to the efficacy of the technique: isolated nuclei or necrotic cells often fail to reveal bleached chromatin, and mature sperm heads are never bleached. In these cases, however, chromatin is not in "normal" condition, both for ions and associated proteins.

The EDTA technique must be adapted to the different biological samples. The delicate step is represented by the EDTA treatment; the results are influenced by the incubation time and temperature. After a short incubation, chromatin is not bleached, while with a too long incubation, all nuclear structures can loose contrast. Moreover, epoxy resin requires undiluted EDTA, while acrylic sections can be treated with a 1:10 diluted solution. The differentiation time with EDTA therefore needs a monitoring depending on different factors, such as the cell type, fixation, sort of resin, etc.

One of the most important results reported with this technique is the original observation of perichromatin fibrils, together with the first description of the ultrastructure of the coiled body, later renamed Cajal body (Monneron and Bernhard, 1969). The studies on the localization of transcription sites have benefited from this technique (Cmarko *et al.*, 1999; Fakan and Bernhard, 1971; Fakan *et al.*, 1976). More recently, it has allowed the distinction of RNP structures from chromatin in apoptotic cells (Biggiogera *et al.*, 1997).

C. Terbium Staining for RNA

The contrasting methods for RNA are less numerous than for DNA; in fact, the structural difference between the two nucleic acids can be exploited in the case of a Feulgen type reaction but not in the opposite sense. A useful technique represents the use of terbium citrate (Biggiogera and Fakan, 1998).

Terbium (III) ions can bind guanosine monophosphate in RNA (Ringer *et al.*, 1980); in the citrate form (which must be synthesized in the laboratory) it can selectively stain RNA in ultrathin sections and be subsequently observed at the electron microscopic level. The end product, although weak in contrast, is very fine and allows for high resolution visualization (Fig. 3A).

The staining can be abolished by a previous treatment with RNase or nuclease S1, but not by DNase or pronase. Since staining is prevented by single stranded RNA digestion, and Tb citrate does not react with ssDNA in adenovirus 5-infected HeLa cells (Biggiogera and Fakan, 1998), terbium stains only ssRNA.

The technique can be applied to ultrathin sections of aldehyde-fixed and epoxy or acrylate resin embedded samples, and is not influenced by the different fixatives. Most interestingly, terbium staining can also be performed after immunolabeling (Fig. 3B).

It must be noted that, for acrylic embedded tissues, the sections should be cut at a white-silver interference color thickness, since acrylic resins confer a higher inherent contrast to cell structures; reducing the section thickness is therefore a simple way of increasing the signal-to-noise ratio.

1. Preparation of the Reagent

435 mg of terbium (III) nitrate pentahydrate are dissolved in 5 ml of double distilled or ultrapure water to obtain a 0.2 M solution; then 5 ml of a 0.3 M solution of tri-sodium citrate 2-hydrate are prepared. Under continuous stirring, the terbium solution is added dropwise to the citrate and a white precipitate is formed. Then 1 N NaOH is added dropwise until the solution becomes transparent again and the precipitate is completely dissolved (this happens at pH 6.7–7.0). pH is adjusted to 8.2–8.5 with 1 N NaOH. At pH lower than 8.0, a precipitate is formed within 2–3 days. For this reason, the pH must be checked again after 24 h and readjusted if necessary. The final solution of terbium citrate is stable for several months at room temperature and can be kept in a plastic syringe fitted with a 0.22 μm Millipore filter.

2. Staining Procedure for Epon Sections

- Incubate the grids on drops of a 0.5 M solution of sodium citrate brought to pH 12 with 1 N NaOH (this solution is also very stable) for 1 h at room temperature.

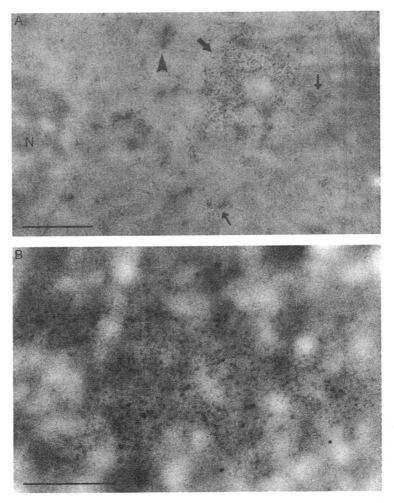

Fig. 3 (A) Mouse liver, glutaraldehyde fixation, LR White embedding and staining with terbium citrate. The RNA present within interchromatin granules (large arrow), perichromatin fibrils (small arrows), perichromatin granules (arrowhead) and the nucleolus (N) is contrasted, while chromatin remains unstained. Bar represents 1 µm. (B) Section from HeLa cell nucleus, paraformaldehyde-fixed and LR White embedded. The micrograph shows a cluster of interchromatin granules. Terbium staining is combined with immunolabeling for the interchromatin granule marker protein PANA (large grains) and the cleavage factor CFIm (small particles). Bar represents 0.5 µm.

- Blot rapidly the grid with filter paper to remove the excess of the solution.
- Stain on terbium citrate for 1 h at room temperature.
- Rinse with distilled water for 15–30 s, blot and dry.

3. Staining Procedure for Acrylic Sections

- Stain on terbium citrate diluted 1:5 with distilled water for 1 h at 37 °C.
- Blot rapidly the grid with filter paper to remove the excess of the solution.
- Put the grid on a large drop of water for 10–15 s.
- Blot and dry.

4. Controls

The following control reactions can be carried out on paraformaldehyde-fixed, acrylate-embedded tissues:

- 0.1% RNase A in triethanolamine buffer (pH 7.0) for 5 h at 37 °C;
- 0.1% DNase I in phosphate buffer containing 2 mM $MgCl_2$, for 5 h at 37 °C.
- Nuclease S1 from *Aspergillus oryzae*, specific for single stranded nucleic acids, at the concentration of 100 units/ml in acetate buffer (pH 4.5), containing 50 mM NaCl and 3 mM $ZnCl_2$, for 5 h at 37 °C.
- 0.1% Pronase in water for 15–30 min at room temperature.

The grids are then stained with terbium citrate as above.

D. Bismuth Staining for Proteins

The method, originally developed by Locke and Huie (1977) and modified by Wassef (1979), represents a tunable technique for staining proteins at the EM level. Bi^{3+} can react with amine, amidine and phosphate groups, but the results are dependent on the type of fixative utilized.

After formaldehyde fixation, nonesterified phosphate groups of proteins (but not phosphate groups linked to nucleic acids) and free amino groups are available to bismuth. Bismuth staining thus reveals nucleoli and reacts with histones and protamines.

When a glutaraldehyde fixation is carried out, $-NH_2$ groups are blocked by the fixative and only phosphates can react. In this way, a selective stain of several nuclear structures can be obtained. Perichromatin and interchromatin granules, coiled (Cajal) bodies and the dense fibrillar component of the nucleolus are stained. On the other hand, condensed chromatin, ribosomes and preribosomal particles in the nucleolar granular component are not stained.

Originally used as an *en bloc* method, alkaline bismuth can be successfully used on acrylate thin sections. The detailed technique is described by Wassef (1979) and Moyne (1980).

The contrast obtained is medium, but the staining provides a high resolution. The staining can be prevented by proteinase K treatment and may also be used after immunolabeling (Puvion-Dutilleul, 1991).

E. Ag-NOR-Staining for Nucleoli

Silver staining of proteins can be accomplished by a variety of pre- and post-embedding methods. It has essentially been used for visualizing argentophilic proteins in the nucleolus. In this chapter, we will refer to the modification proposed by Moreno *et al.*, 1985), representing a simple and reproducible procedure applied directly to ultrathins sections. Silver salts can be reduced to metallic silver by different chemical groups such as PO_4, $-SH$ and $-CHO$. When the reaction is carried out in the right, controlled conditions (in order to exclude the influence of light, temperature, pH variations) the silver grains can be easily visualized mainly at the nucleolar level. In particular, silver deposits are found on the fibrillar centers and on the dense fibrillar component, while the granular component remains practically devoid of silver precipitates (Fig. 4).

The silver staining of nucleoli has been considered strictly related to the cell proliferation activity also in terms of marker of malignancy (Derenzini *et al.*, 1990a), and by the end of the 1980s several papers have correlated the staining of isolated proteins with their EM localization and function (Biggiogera *et al.*, 1990; Lischwe *et al.*, 1979). Nucleolin and B23 protein seem to be the major responsible for the staining; however, the distribution of silver grains did not overlap completely with the EM immunolabeling (Biggiogera *et al.*, 1989); it has since been proposed that another protein, p135 (Vandelaer *et al.*, 1999) might be the real responsible of the staining.

Interestingly, the silver staining obtained on thin sections from Lowicryl or LR White embedded material can be "erased" by floating the grid on a solution of

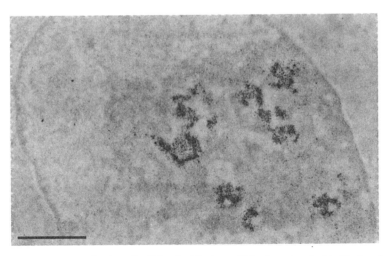

Fig. 4 A section of a P815 cultured cell fixed with glutaraldehyde and embedded in Lowicryl K4M resin. The section was stained with the AgNOR method (Moreno *et al.*, 1985). The silver deposit mostly occurs in the fibrillar components of the nucleolus i.e. the dense fibrillar component and the fibrillar centers. Bar represents 1 μm.

Exargent (Tetenal) diluted 1:10 in water. The time of incubation is dependent on the amount of silver.

Finally, silver stained sections can also be further submitted to DNA staining with osmium ammine (Derenzini *et al.*, 1990b).

III. High Resolution Autoradiography

High resolution autoradiography (EM-ARG) has been used for many years to investigate the intracellular distribution of nucleic acids, proteins, sugars and other compounds. The introduction of halogenated precursors of nucleic acids combined with immunogold detection of the halogen-containing epitopes enabled one to replace radioactive, mostly tritiated precursors of DNA and RNA in many types of experiments. It offered, at the same time, a better resolution compared to EM-ARG, where this is essentially determined by the diameter of silver halide crystals in autoradiographic emulsion and by the thickness of the specimen. Nevertheless, it still remains a valuable tool, for instance in ultrastructural studies of RNA kinetics, as a frequent occurrence of Br-uridine along a newly synthesized RNA molecule can apparently interfere with normal RNA-processing (Wansink *et al.*, 1994). EM-ARG application protocols have been extensively described in a number of previous publications (e.g. Fakan and Fakan, 1987; Granboulan, 1965, Salpeter and Bachmann, 1965). It can be combined with different cytochemical techniques used for investigating the nuclear architecture, such as the EDTA regressive staining (Fakan and Bernhard, 1971), the terminal deoxynucleotidenyl transferase DNA detection (Fakan and Modak, 1973) or in situ hybridization (Morel, 1993). It has been applied on both resin sections or ultrathin cryosections (e.g. Fakan *et al.*, 1976; Fig. 2). EM-ARG offers a possibility to quantitatively evaluate the density of labeling and the statistical significance of the data, taking into account basic parameters inherent to this methodological approach (e.g. Moyne, 1977).

IV. Immunocytochemistry

Since the end of the 1970s, immunocytochemistry has very rapidly taken over the techniques of protein cytochemistry. Nowadays it is possible to find an antibody for any given protein or, if not, it is not difficult to produce a probe *ex novo*. The success obtained by this technique is directly related to its easiness of use, provided that a few requirements are fulfilled.

Immunocytochemical reactions can be carried out following either a preembedding or a postembedding technique. The former method is preferred when the antigen can be altered or damaged by procedures other than fixation, such as dehydration or resin embedding and curing. In this method, the cells are blandly fixed, permeabilized by adding a detergent and incubated with the primary antibody. The secondary antibody is then used together with e.g. the peroxidase–anti peroxidase DAB-osmium system. The result is an electron dense and easily detectable deposit which localizes where the antigen should be. One of the first examples of this technique applied to the cell nucleus is the work of Spector *et al.* (1983) describing the localization of snRNPs. More rarely, the secondary probe is labeled with colloidal gold. Among the potential disadvantages of preembedding methods there is the possibility that either the antigen or the complex might be displaced due to the treatment with detergents. Moreover, the biological sample submitted to this technique is committed to be used for the localization of a single antigen solely.

In contrast, postembedding techniques allow one to localize more that one antigen on the same ultrathin sections or, alternatively, to use different sections from the same block with a battery of antibodies and are presently the most frequently used. Although a good result can be obtained also on epoxy sections, especially after etching of the surface (Bendayan and Puvion, 1984), acrylic resins (in particular Lowicryl K4M and LR White) (Fig. 5) or cryosections offer the best yield of labeling (e.g. Fakan *et al.*, 1984).This is at least partially due to irregularities of the surface which exposes more epitopes than in epoxy sections. The latter resins, in fact, bind covalently to the tissue, and the sectioning results in a real cutting making the surface smooth.

A number of papers and books have been devoted to EM immunomethods, and the protocols are easily available. These describe the use of secondary antibody or

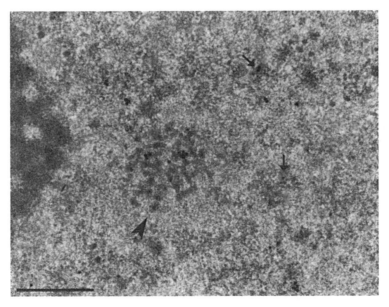

Fig. 5 Section of a mouse liver cell, embedded in Lowicryl K4M. Sm antigen of snRNPs was detected by an autoimmune antibody visualized with protein A-colloidal gold complex. The signal is mainly associated with a Cajal (coiled) body (arrowhead) and perichromatin fibrils (some indicated by arrows) in the nucleoplasm. For more details, see Fakan *et al.* (1984). Bar represents 0.5 μm.

protein A or protein G complexes with colloidal gold (e.g. Bendayan, 1995). Nevertheless, we would like to underline some caveat when applying immunotechniques in the studies of the cell nucleus. Background (or unspecific) labeling should always be kept to a minimum and, in this sense, condensed chromatin represents a good internal control. The high local concentration of DNA and proteins can attract immunoglobulins. When labeling a nuclear component (with the obvious exception of DNA and histones) condensed chromatin represents a suitable negative internal control. The second possible source of background is the nucleolus, where proteins and nucleic acids are highly packed. Background can be reduced by modifying pH (toward alkalinity), incubation temperature or adding surface charge saturating agents (skimmed milk, glycin).

V. Molecular in situ Hybridization

When a specific nucleic acid sequence has to be localized in a cell in situ, one must use methods of molecular in situ hybridization (ISH). First developed at the level of light microscopy (Gall and Pardue, 1969), it has later been adapted for the application to electron microscopic specimens. It is based on the formation of hybrids by virtue of base pairing between a target nucleic acid sequence occurring in the specimen and complementary bases of an oligo- or polynucleotide probe. This can be applied either in a form of DNA or RNA, essentially following the nature of the target. The probe is labeled either by a radioactive isotope detectable by using high resolution autoradiography or with a marker molecule (e.g. biotin or digoxigenin), which can typically be detected by specific antibodies and immunoelectron microscopy. The latter has become widely used in most protocols, thanks especially to a better detection resolution and the possibility to avoid work with radioactive material. Two procedures have been proposed. In a preembedding mode, the whole hybridization and revealing reaction take place in the specimen following a usually mild fixation and finally the specimen is dehydrated and embedded into a resin. Since this method requires permeabilization of the cells and, consequently, leads to a degradation of cellular structure, different techniques have been developed allowing one to hybridize the probe directly to the nucleic acid sequences exposed on the surface of ultrathin resin sections or cryosections. The use of resin sections, although giving rise to a lower hybridization yield limited by the number of available target sites exposed on the surface of the section, offers a better protection for the specimen, especially when rather harsh conditions must be used before and during the hybridization reaction. When antisense-RNA probes are applied for identifying RNA in the section (Fig. 6), this can be done under most favorable conditions usually offering a sufficiently good signal (e.g. Cmarko et al., 2002, Fig. 6). When detecting RNA using DNA probes, the specimen is also not submitted to a prior denaturation (Vazquez-Nin et al., 2003). Different variants of nonradioactive ISH protocols have been described and extensively reviewed (e.g. Puvion-Dutilleul and Puvion, 1996).

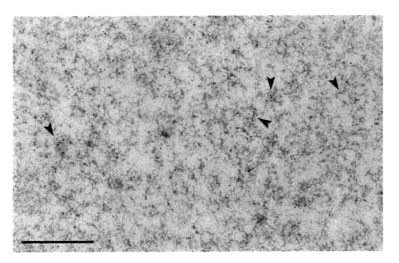

Fig. 6 A section of an HIV-1 transfected COS cell embedded into LR Lowicryl K4M. Viral RNA detected by in situ hybridization with a digoxigenin-labelled antisense RNA probe revealed with colloidal gold particles is associated with ribonucleoprotein-containing fibrils (some indicated by arrowheads) occurring in the nucleoplasm. Micrograph courtesy of D. Cmarko. For more details see Cmarko et al., 2002. Bar represents 0.5 μm.

VI. Identification of Nucleic Acids by Means of Enzymatic Reactions

Besides methods involving contrasting agents, EM cytochemistry can also be carried out by means of enzyme-related techniques. In this section, two methods are presented, involving the use of specific enzymes and a subsequent localization with gold-conjugated secondary probes.

A. TdT Reaction

The terminal deoxynucletidyl transferase (TdT) is based on the original technique by Fakan and Modak (1973) which utilized this particular DNA polymerase to add deoxyribonucleotide triphosphates to the 3′-end of either ds- or ssDNA on thin sections.

The rationale is based on the fact that, during the sectioning process, DNA ends are present at the surface of the sections, and can be recognized by exogenous TdT which will subsequently add tritiated dATP. The sections will then be submitted to autoradiography.

This methods has been modified by Thiry (1992) with the use of nonisotopic labeled nucleotide analogues such as BudR or biotinylated dUTP or dCTP. The incorporated analogue is then revealed by an immunogold procedure.

The advantages of this version of the technique include a better resolution and a shorter procedure in comparison with EM-ARG.

B. PnT Reaction

In 1993, Thiry proposed a new technique for RNA detection based on the use of polyadenylate nucleotidyl transferase (PnT). PnT from *Escherichia coli* can catalyze the addition of 5′-adenosine monophosphate to the 3′-end of ssRNA. The reaction is commonly used for in vitro labeling of RNA.

The technique consists in incubating the grids with the thin sections on a solution containing PnT and biotinyl-17-ATP for 5 min at 37 °C; the step is followed by the immunocytochemical reaction with an anti-biotin antibody (amplification of the signal) and a final secondary antibody conjugated with gold (Fig. 7) (Thiry, 1993).

The method is compatible with different fixatives and embedding media, although glutaraldehyde fixation and epon embedding give the best results.

In comparison with other labeling methods for RNA, PnT offers a higher resolution than autoradiography, in addition to being easier to be carried out. Moreover, when considering anti-RNA antibodies, the problem of masking or conformational alteration of the antigenic determinants can be avoided.

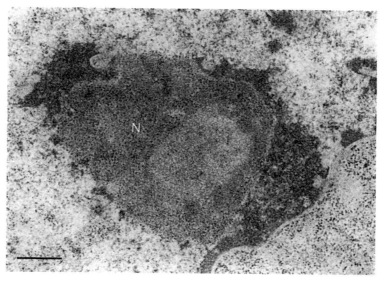

Fig. 7 Section of a HeLa cell labelled by the PnT method visualizing a general distribution pattern of RNA. N, nucleolus. Courtesy of M. Thiry. For more details, see Thiry, 1993. Bar represents 0.5 μm.

VII. Targeting of Intranuclear Substrates using Enzyme–Colloidal Gold Complexes

The fact that hydrolytic enzymes are able to recognize their substrate molecules in fixed and resin embedded cells offer the possibility to use them as probes for visualizing different nuclear structural constituents. Nucleases were especially used for this purpose. Bendayan (1982) carried out pioneering work in preparing DNase and RNase-gold complexes and evaluating their efficiency under conditions of different fixation and embedding procedures. He was able to show that, while RNase offers reliable and reproducible detection of RNA after using different preparation protocols, DNase-gold complex is much more sensitive to the specimen preparative conditions and gives rise to reproducible results, especially when glycol methacrylate-embedded samples are used.

In addition, other enzymes proved to be useful in this kind of approach; thus Fraschini *et al.* (1999) were able to analyze phospholipase A2 and its association with different nuclear domains.

VIII. Concluding Remarks

Ultrastructural cytochemistry represents a powerful tool in the analysis of the functional architecture of the cell nucleus. The possibility of combining different approaches offers the opportunity to follow dynamic processes and to determine, at the same time, the chemical nature of the structural constituents. Although reaching only two-dimensional imaging, the high resolution of electron microscopic methods makes them invaluable complement to any type of light microscopic visualization. This is especially important when one carries out parallel analysis by in vivo light microscopy where the same cell, first recorded by fluorescence microscopy using vital fluorochromes, can then be analyzed by methods of electron microscopic cytochemistry. This new approach avoids harsh treatments and artefacts occurring during different light microscopic labeling protocols and opens certainly a new avenue in cell biology. The diversity of electron microscopic preparative procedures, extending from good chemical fixation to cryofixation and cryopreparative techniques, offers to the microscopist a plethora of methodological approaches, without which light microscopic observations often remain only indicative.

Acknowledgments

We thank Drs. Dusan Cmarko, Marc Thiry and Lorella Vecchio for kindly providing their micrographs. We are indebted to Mrs. Liliane Hautle for helping with the preparation of the manuscript and Mr. Willy Blanchard for photographic work.

The work in the authors' laboratories has been supported by the Swiss National Science Foundation and by the Fondo di Ateneo per la Ricerca.

References

Albiez, H., Cremer, M., Tiberi, C., Vecchio, L., Schermelleh, L., Dittrich, S., Kupper, K., Joffe, B., Thormeyer, T., Von Hase, J., et al. (2006). Chromatin domains and the interchromatin compartment form structurally defined and functionally interacting nuclear networks. *Chromosome Res.* **14**, 707–733.

Bendayan, M. (1982). Ultrastructural localization of nucleic acids by the use of enzyme-gold complexes: Influence of fixation and embedding. *Biol. Cell* **43**, 153–156.

Bendayan, M. (1995). Colloidal gold post-embedding immunocytochemistry. *Progr. Histochem. Cytochem.* **29**, 1–159.

Bendayan, M., and Puvion, E. (1984). Ultrastructural localization of nucleic acids through several cytochemical techniques on osmium-fixed tissues: Comparative evaluation of the different labelings. *J. Histochem. Cytochem.* **32**, 1185–1191.

Bernhard, W. (1969). A new staining procedure for electron microscopical cytology. *J. Ultrastruct. Res.* **27**, 250–265.

Biggiogera, M., Burki, K., Kaufmann, S. H., Shaper, J. H., Gas, N., Amalric, F., and Fakan, S. (1990). Nucleolar distribution of proteins B23 and nucleolin in mouse preimplantation embryos as visualized by immunoelectron microscopy. *Development* **110**, 1263–1270.

Biggiogera, M., Bottone, M. G., Martin, T. E., Uchiumi, T., and Pellicciari, C. (1997). Still immunodetectable nuclear RNPs are extruded from the cytoplasm of spontaneously apoptotic thymocytes. *Exp. Cell Res.* **234**, 512–520.

Biggiogera, M., and Fakan, S. (1998). Fine structural specific visualization of RNA on ultrathin sections. *J. Histochem. Cytochem.* **46**, 389–395.

Biggiogera, M., Fakan, S., Kaufmann, S. H., Black, A., Shaper, J. H., and Busch, H. (1989). Simultaneous immunoelectron microscopic visualization of protein B23 and C23 distribution in the HeLa cell nucleolus. *J. Histochem. Cytochem.* **37**, 1371–1374.

Biggiogera, M., Malatesta, M., Abolhassani-Dadras, S., Amalric, F., Rothblum, L. I., and Fakan, S. (2001). Revealing the unseen: the organizer region of the nucleolus. *J. Cell Sci.* **114**, 3199–3205.

Biggiogera, M., and Pellicciari, C. (2000). Heterogeneous ectopic RNP-derived structures (HERDS) are markers of transcriptional arrest. *FASEB J.* **14**, 828–834.

Cmarko, D., Boe, S. O., Scassellati, C., Szilvay, A. M., Davanger, S., Fu, X. D., Haukenes, G., Kalland, K. H., and Fakan, S. (2002). Rev inhibition strongly affects intracellular distribution of human immunodeficiency virus type 1 RNAs. *J. Virol.* **76**, 10473–10484.

Cmarko, D., Verschure, P. J., Martin, T. E., Dahmus, M. E., Krause, S., Fu, X. D., van Driel, R., and Fakan, S. (1999). Ultrastructural analysis of transcription and splicing in the cell nucleus after bromo-UTP microinjection. *Mol. Biol. Cell.* **10**, 211–223.

Cogliati, R., and Gautier, A. (1973). Mise en évidence de l'ADN et des polysaccharides à l'aide d'un nouveau réactif 'de type Schiff'. *C. R. Acad. Sci. Paris* **276**, 3041–3044.

Derenzini, M., Hernandez-Verdun, D., and Bouteille, M. (1983). Visualization of a repeating subunit organization in rat hepatocyte chromatin fixed in situ. *J. Cell Sci.* **61**, 137–149.

Derenzini, M., Pession, A., and Trere, D. (1990a). Quantity of nucleolar silver-stained proteins is related to proliferating activity in cancer cells. *Lab. Invest.* **63**, 137–140.

Derenzini, M., Thiry, M., and Goessens, G. (1990b). Ultrastructural cytochemistry of the mammalian cell nucleolus. *J. Histochem. Cytochem.* **38**, 1237–1256.

Fakan, S. (2004). The functional architecture of the nucleus as analysed by ultrastructural cytochemistry. *Histochem. Cell Biol.* **122**, 83–93.

Fakan, S., and Bernhard, W. (1971). Localization of rapidly and slowly labelled nuclear RNA as visualized by high resolution autoradiography. *Exp. Cell Res.* **67**, 129–141.

Fakan, S., and Fakan, J. (1987). Autoradiography of spread molecular complexes. In "Electron Microscopy in Molecular Biology: A Practical Approach" (J. Sommerville, and U. Scheer, eds.), pp. 201–214. IRL Press, Oxford.

Fakan, S., Leser, G., and Martin, T. E. (1984). Ultrastructural distribution of nuclear ribonucleoproteins as visualized by immunocytochemistry on thin sections. *J. Cell Biol.* **98,** 358–363.

Fakan, S., and Modak, S. P. (1973). Localization of DNA in ultrathin tissue sections incubated with terminal deoxynucleotidyl transferase, as visualized by electron microscope autoradiography. *Exp. Cell Res.* **77,** 95–104.

Fakan, S., Puvion, E., and Spohr, G. (1976). Localization and characterization of newly synthesized nuclear RNA in isolated rat hepatocytes. *Exp. Cell Res.* **99,** 155–164.

Fraschini, A., Biggiogera, M., Bottone, M. G., and Martin, T. E. (1999). Nuclear phospholipids in human lymphocytes activated by phytohemagglutinin. *Eur. J. Cell Biol.* **78,** 416–423.

Gall, J. G., and Pardue, M. L. (1969). Formation and detection of RNA-DNA hybrid molecules in cytological preparations. *Proc. Nat. Acad. Sci. USA* **63,** 378–383.

Gautier, A. (1976). Ultrastructural localization of DNA in ultrathin tissue sections. *Int. Rev. Cytol.* **44,** 113–191.

Granboulan, P. (1965). Comparison of emulsions and techniques in electron microscope radioautography. *In* "Use of Radioautography in Investigating Protein Synthesis" (C. P. Leblond, and K. B. Warren, eds.), pp. 43–63. Academic Press, New York.

Kasten, F. H. (1960). The chemistry of Schiff's reagent. *Int. Rev. Cytol.* **10,** 1–93.

Lischwe, M. A., Smetana, K., Olson, M. O., and Busch, H. (1979). Proteins B23 and C23 are the major nucleolar silver staining proteins. *Life Sci.* **25,** 701–708.

Locke, M., and Huie, P. (1977). Bismuth staining for light and electron microscopy. *Tissue Cell* **9,** 347–371.

Monneron, A., and Bernhard, W. (1969). Fine structural organization of the interphase nucleus in some mammalian cells. *J. Ultrastruct. Res.* **27,** 266–288.

Morel, G. (1993). "Hybridization Techniques for Electron Microscopy." CRC Press, Boca Raton.

Moreno, F., Hernandez-Verdun, D., Masson, C., and Bouteille, M. (1985). Silver staining of the nucleolar organizer regions (NORs) on Lowicryl and cryo-ultrathin sections. *J. Histochem. Cytochem.* **33,** 389–399.

Moyne, G. (1977). An application of the "hypothetical grain method": Autoradiographic localization of transcriptional events in isolated rat liver cells. *Eur. J. Cell Biol.* **15,** 126–134.

Moyne, G. (1980). Methods in ultrastructural cytochemistry of the cell nucleus. *Prog. Histochem. Cytochem.* **13,** 1–72.

Olins, A. L., Moyer, B. A., Kim, S. H., and Allison, D. P. (1989). Synthesis of a more stable osmium ammine electron-dense DNA stain. *J. Histochem. Cytochem.* **37,** 395–398.

Puvion, E., and Bernhard, W. (1975). Ribonucleoprotein components in liver cell nuclei as visualized by cryoultramicrotomy. *J. Cell Biol.* **67,** 200–214.

Puvion-Dutilleul, F. (1991). Simultaneous detection of highly phosphorylated proteins and viral major DNA binding protein distribution in nuclei of adenovirus type 5-infected HeLa cells. *J. Histochem. Cytochem.* **39,** 669–680.

Puvion-Dutilleul, F., and Puvion, E. (1996). Non-isotopic electron microscope in situ hybridization for studying the functional sub-compartmentalization of the cell nucleus. *Histochem. Cell Biol.* **106,** 59–78.

Raska, I., Shaw, P. J., and Cmarko, D. (2006). Structure and function of the nucleolus in the spotlight. *Curr. Opin. Cell Biol.* **18,** 324–334.

Ringer, D. P., Howell, B. A., and Kizer, D. E. (1980). Use of terbium fluorescence enhancement as a new probe for assessing the single-strand content of DNA. *Anal. Biochem.* **103,** 337–342.

Salpeter, M. M., and Bachmann, L. (1965). Assessment of technical steps in electron microscope radioautography. *In* "Use of Radioautography in Investigating Protein Synthesis" (C. P. Leblond, and K. B. Warren, eds.), pp. 23–41. Academic Press, New York.

Solovei, I., Cavallo, A., Schermelleh, L., Jaunin, F., Scasselati, C., Cmarko, D., Cremer, C., Fakan, S., and Cremer, T. (2002). Spatial preservation of nuclear chromatin architecture during three-dimensional fluorescence in situ hybridization (3D-FISH). *Exp. Cell Res.* **276,** 10–23.

Spector, D. L. (2003). The dynamics of chromosome organization and gene regulation. *Annu. Rev. Biochem.* **72,** 573–608.

Spector, D. L., Schrier, W. H., and Busch, H. (1983). Immunoelectron microscopic localization of snRNPs. *Biol. Cell* **49,** 1–10.

Stelzer, E. H. K. (1995). The intermediate optical system of laserscanning confocal microscopes. *In* "Handbook of Biological Confocal Microscopy" (J. B. Pawley, ed.), pp. 139–154. Plenum, New York.

Thiry, M. (1992). Ultrastructural detection of DNA within the nucleolus by sensitive molecular immunocytochemistry. *Exp. Cell Res.* **200,** 135–144.

Thiry, M. (1993). Immunodetection of RNA on ultra-thin sections incubated with polyadenylate nucleotidyl transferase. *J. Histochem. Cytochem.* **41,** 657–665.

Vandelaer, M., Thiry, M., and Goessens, G. (1999). AgNOR proteins from morphologically intact isolated nucleoli. *Life Sci.* **64,** 2039–2047.

Vazquez-Nin, G. H., Biggiogera, M., and Echeverria, O. M. (1995). Activation of osmium ammine by SO2-generating chemicals for EM Feulgen-type staining of DNA. *Eur. J. Histochem.* **39,** 101–106.

Vazquez-Nin, G. H., Echeverria, O. M., Ortiz, R., Scassellati, C., Martin, T. E., Ubaldo, E., and Fakan, S. (2003). Fine structural cytochemical analysis of homologous chromosome recognition, alignment, and pairing in Guinea pig spermatogonia and spermatocytes. *Biol. Reprod.* **69,** 1362–1370.

Visser, A. E., Jaunin, F., Fakan, S., and Aten, J. A. (2000). High resolution analysis of interphase chromosome domains. *J. Cell Sci.* **113,** 2585–2593.

Wansink, D. G., Nelissen, R. L. H., and de Jong, L. (1994). *In vitro* splicing of pre-mRNA containing bromouridine. *Mol. Biol. Rep.* **19,** 109–113.

Wassef, M. (1979). A cytochemical study of interchromatin granules. *J. Ultrastruct. Res.* **69,** 121–133.

Woodcock, C. L. (2006). Chromatin architecture. *Curr. Opin. Struct. Biol.* **16,** 213–220.

CHAPTER 23

Scanning Electron Microscopy of Chromosomes

Gerhard Wanner and Elizabeth Schroeder-Reiter

Department of Biology I
Ludwig-Maximilians-Universität München
Menzinger Straße 67
80638 Munich
Germany

Abstract
I. Introduction
    A. History
    B. Where do We Find Mitotic Metaphase Chromosomes in Plants and Animals?
II. Materials and Methods
    A. Fixation of Plant Material
    B. Drop/Cryo Technique
    C. Formaldehyde Fixation
    D. DNA and Protein Staining
    E. DNaseI Treatment
    F. Controlled Decondensation with Proteinase K
    G. Immunogold Labeling for SEM
    H. *In Situ* Hybridization
    I. Dehydration, Critical Point Drying and Mounting for SEM
III. Chromosome Preparation
    A. Isolation of Chromosomes
    B. Drop/Cryo Technique
    C. Formaldehyde-Fixed Chromosomes
IV. Chromosome Structure in SEM
    A. Ultrastructure of Mitotic Plant Chromosomes
    B. Where do Spindle Fibers Attach?
    C. Chromosome Condensation and Decondensation During Mitosis: Dynamic Matrix Model
    D. Meiosis

V. Chromosome Analysis in SEM
    A. Metal Impregnation of Chromosomes
    B. Specific Staining of DNA in Chromosomes with Platinum Blue
    C. Staining of Protein in Chromosomes
    D. Controlled Decondensation of Chromosomes
    E. Immunogold Labeling of Chromosomal Proteins and ISH
VI. Outlook
    References

Abstract

Scanning electron microscopic analysis is an indispensable tool for high-resolution visualization of chromosomes and their ultrastructural details. It allows a three-dimensional structural approach for elucidating higher-order chromatin structure and chromosome architecture. Artificial decondensation under a variety of conditions shows that structural elements of chromosomes are composed of matrix fibers and chromomeres. Currently, chromosome labeling methods include DNA contrasting with platinum blue, silver contrasting of proteins, and immunolabeling with Nanogold®. With these techniques, DNA and protein distribution can be determined, and functionally relevant elements (e.g., epigenetic modifications, specific proteins, DNA sequences) can be located to structural elements of chromosomes with, at present, local resolution of ~30 nm.

I. Introduction

A. History

Mitotic chromosomes in plant cells were described in impressive structural detail as early as 1888 by Strasburger (1888), while the basic function of chromosomes was postulated (Boveri, 1902; Sutton, 1902). Much later, the elucidation of the structure of DNA (Watson and Crick, 1953) initiated a molecular focus to chromatin structural studies, as a result of which the the 10 nm elementary fiber (Thoma et al., 1979), the 30 nm solenoid (Finch and Klug, 1976), and further resolution of the structure of the nucleosome (Arents et al., 1991) has been described. To date, although several models exist, beyond the solenoid there is no ruling consensus on the higher-order structure of chromatin, leaving marked compaction factors that must be achieved during condensation in mitosis and meiosis essentially unexplained. The establishment of high-resolution scanning electron microscopy (SEM) in the field of Biology in the 1980s was finally the breakthrough for auspicious three-dimensional investigation of chromosomes (e.g., Allen et al., 1986; Harrison et al., 1982; Sumner and Ross, 1989).

B. Where do We Find Mitotic Metaphase Chromosomes in Plants and Animals?

Mitotic metaphase chromosomes are preferentially isolated from dividing cells. Animal chromosomes are typically isolated from cultured lymphocytes. Mitotic plant chromosomes can be found in all meristems, particularly in the apex or root tips (Fig. 1A). Root tips are the most suitable for chromosome isolation, as they regenerate rapidly and can be harvested easily from plants or germinating seeds. To increase the mitotic index, both plant and animal cells can be synchronized (accumulating nucleii at the terminal phase of replication) and subsequently arrested by hindering spindle apparatus formation. During mitosis, the nuclear membrane dissociates, so that cytoplasm and nuclear contents comingle, leaving the plasma membrane and the cell wall (in plants) as barriers to chromosome isolation.

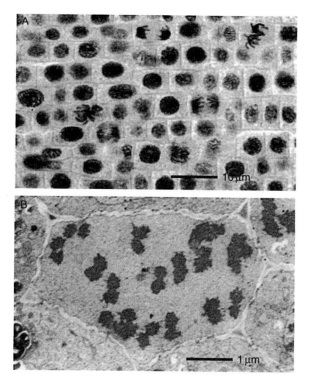

Fig. 1 Light micrograph (A) of a semi-thin section of onion root tip meristem (stained with toluidin blue) and a transmission electron micrograph (B) of an ultrathin section of a mitotic cell in barley (stained with osmium/uranyl acetate/lead citrate). Although chromosomes can be seen in all mitotic stages, light microscopic imaging is limited by optical resolution and section thickness (500–700 nm) (A). In spite of high resolution in TEM, three-dimensional information of chromosomes is lost, and it is difficult to interpret whether their contrast derives from protein and/or DNA (B).

Approaches to visualizing chromosomes have changed with the development of microscopic techniques. Electron microscopy allows investigation of chromosome subunits with >100-fold higher resolution compared to light microscopy. In conventional transmission electron microscopy (TEM) of specimens (fixed with glutaraldehyde/osmium; post-stained with lead citrate and uranyl acetate), it is difficult to distinguish between DNA and protein (Fig. 1B). This is especially true when structural elements like solenoids (30 nm) are superimposed due to section thickness (50–70 nm). Stereoimaging with high-voltage TEM (1000 kV) was successfully applied to chromosomes, but is limited by the thickness of sections (max. 1 μm, Ris and Witt, 1981). SEM allows visualization of chromosomes in their structural entirety, and is especially appropriate for investigation of isolated chromosomes. Three-dimensional visualization of different modes of chromatin condensation allows investigation of higher-order structure of chromatin and processes required for mitosis and meiosis. With labeling techniques, various kinds of signals can be correlated to chromosome substructures. Two signals are routinely used in SEM: detection of secondary electrons (SE) shows topography and ultrastructural details at high resolution, and detection of backscattered electrons (BSE) allows in-depth visualization of markers.

II. Materials and Methods

A. Fixation of Plant Material

Barley or rye seeds are germinated on moist filter paper. Root tip cells are synchronized by 18 h incubation in hydroxy urea (1.25 mM), incubated 4 h in water, and arrested in amiprophosmethyl (0.2 μM; 4 h). After washing, root tips are dissected, incubated overnight in water at 0 °C, fixed in 3:1 (v/v) ethanol:acetic acid, and stored at −20 °C. For chromosome isolation, fixed root tips are washed in distilled water for 30 min prior to dissection of the meristematic tips and maceration in 2.5% pectolyase and 2.5% cellulase (w/v in 75 mM KCl, 7.5 mM EDTA) 1 h at 30 °C. The mixture is filtered through 100 μm gauze, hypotonically treated for 5 min in 75 mM KCl, washed by low-speed centrifugation in 3:1 ethanol/acetic acid fixative and stored at −20 °C.

B. Drop/Cryo Technique

Cell suspensions are dropped onto cold laser-marked slides (Laser Marking, Fischen, Germany). Just prior to fixative evaporation, 20 μl 45% acetic acid is applied, the specimens are covered with a cover slip, and frozen upside-down on dry ice for 15 min. The cover slip is pried off, and specimens immediately transfered into fixative (2.5% glutaraldehyde, 75 mM cacodylate buffer, 2 mM $MgCl_2$, pH 7.0). For details see Martin *et al.* (1994).

C. Formaldehyde Fixation

Root tips are fixed (2% formaldehyde in 10 mM Tris, 10 mM Na_2EDTA, 100 mM NaCl), mechanically dispersed (Polytron® 5 mm mixer, Kinematica, Switzerland) in isolation buffer (15 mM Tris, 2 mM Na_2EDTA, 0.5 mM Spermin, 80 mM KCl, 20 mM NaCl, 15 mM Mercaptoethanol, 1% Triton-X-100) and spun ("swing out"-centrifugation) onto laser marked slides. For details see Schubert et al. (1993).

D. DNA and Protein Staining

For DNA staining, chromosomes are stained for 30 min at room temperature with platinum blue ($[CH_3CN]_2Pt$ oligomer, 10 mM, pH 7.2), washed with distilled water, and then with 100% acetone prior to critical point drying. For silver staining, chromosomes are stained with 20% aqueous silver nitrate solution or with an aqueous solution of colloidal silver containing 0.1 M elementary silver at pH 8. For details see Wanner and Formanek (1995, 2000).

E. DNaseI Treatment

Glutaraldehyde-fixed chromosomes are washed in distilled water, incubated in aqueous solutions of DNase-I (10–100 µg/ml) at 37 °C for 30 min. For any enzymatic treatment it is beneficial to block free aldehydes on fixed chromosomes with a glycine solution (1%) to ensure optimal enzymatic activity.

F. Controlled Decondensation with Proteinase K

Glutaraldehyde-fixed chromosomes are washed in distilled water, blocked with 1% glycine solution, incubated with proteinase K (1 mg/ml, ICN Biochemicals) for 2 h at 37 °C, and washed in distilled water. For details see Wanner and Formanek (2000).

G. Immunogold Labeling for SEM

Fixed chromosomes are incubated for 1 h with the rabbit anti-serine 10 phosphorylated histone H3 (diluted in blocking solution 1:250; Upstate Biotechnologies, USA) and for 1 h with anti-rabbit Nanogold® Fab'-fragments (diluted in blocking solution 1:20; Nanoprobes, USA). Specimens are washed (PBS + 0.1% Tween) and post-fixed with 2.5% glutaraldehyde in PBS and subsequently silver enhanced (HQ Silver, Nanoprobes, USA) according to the manufacturer's instructions. For details see Schroeder-Reiter et al. (2003).

H. *In Situ* Hybridization

45SrDNA was labeled by nick translation with biotin-16-dUTP, and for *in situ* hybridization (ISH), 20-ng probe is applied per slide. Conventional ISH procedure has been shortened: enzymatic (pepsin, RNase) treatment and intermediate air-drying and dehydration steps are omittted. For immunodetection, AlexaFluor® 488-FluoroNanogold™-Streptavidin (diluted 1:100, Nanoprobes, NY, USA) is applied for 1 h, after which specimens are washed and silver enhanced (see earlier). For details see Schroeder-Reiter *et al.* (2006).

I. Dehydration, Critical Point Drying and Mounting for SEM

Specimens are washed in distilled water, dehydrated in 100% acetone and critical point dried from CO_2. Slides are mounted onto stubs and either sputter-coated (3–5 nm with Pt) or carbon-coated by evaporation. Specimens are examined at various accelerating voltages with an Hitachi S-4100 field emission scanning electron microscope equipped with a YAG-type BSE detector (Autrata). For BSE detection acceleration voltages from 15 to 30 kV are used. Secondary electron (SE) and BSE images are recorded simultaneously with Digiscan™ hardware (Gatan, USA).

III. Chromosome Preparation

A. Isolation of Chromosomes

The establishment of the "drop technique" by Ambros *et al.* (1986) was a major advancement for chromosome isolation. Critical for chromosome spreading is the volatility of the fixative (3:1 methanol/acetic acid) and adhesive forces to the glass slide during fixative evaporation and air drying. However, the air drying step results in a thin artificial surface layer of only a few nanometers in thickness which completely obscures all ultrastructural details of chromosomes (Allen *et al.*, 1988; Sumner, 1991). We speculate that this layer results from esterification of ethanol and acetic acid. In addition, air drying frequently leads to a flattening of chromosomes due to capillary forces that prevents three-dimensional structural investigation.

B. Drop/Cryo Technique

A breakthrough for SEM investigation of chromosomes was the modification of the drop technique by addition of a freezing step and omission of all air-drying. This "drop/cryo" technique was established for plants in 1994 by Martin *et al.* Fixation with ethanol/acetic acid remains critical for this technique. Ideally, individual chromosomes are distinguishable from each other but lay recognizably in their orignial chromosome complements. Under favorable conditions, all stages of condensation and decondensation of mitosis can be observed with SEM (Fig. 2). Although the drop/cryo technique is applicable to all plants tested so far, the quality of chromosome spreading and the amount of nucleoplasmic residue depends on the plant species and the degree of chromatin condensation. A significant factor for chromosome spreading seems to be chromosome size. Chromosomes within the range of 5–10 µm spread well; very small or very large chromosomes (e.g., *Arabidopsis* with length of 1–3 µm, or *Trillium* with a

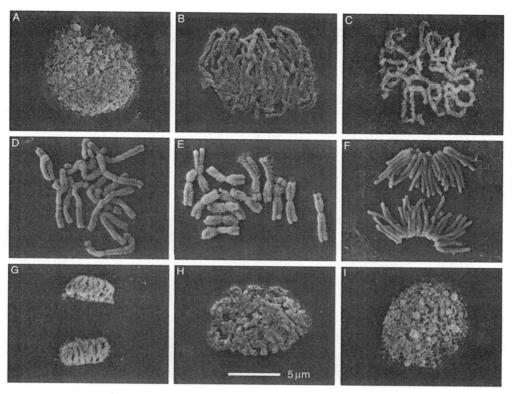

Fig. 2 Mitotic cycle of rye (*Secale cereale*) in SEM. Chromatin in interphase (A) shows a compact fibrillar network. In prophase (B) individual chromosomes become visible. In late prophase (C) chromosomes condense typically into a S-like conformation. In prometaphase (D) primary and secondary constrictions become visible. In late metaphase (E) chromosomes are highly condensed, and separation of chromatids becomes apparent; primary and secondary constrictions are clearly visible. In anaphase (F) chromosomes are bundled at the centromeres. In late telophase (G) chromosome arms start to decondense. During transition to interphase (H) individual chromosomes are no longer discernible. In interphase (I) several compact regions, presumably heterochromatin, can be recognized. (Zoller *et al.*, 2004a; reproduced by permission of S. Karger AG, Basel).

length of 20 μm) tend to cluster or overlap. All parameters for this technique must be optimized for different species. The drop/cryo technique can also be applied to animal cells, but, with only few exceptions, animal chromosomes exhibit significantly more residual nucleoplasm than plants. We assume that this is due to a different protein compostion and/or concentration of animal nucleoplasm.

C. Formaldehyde-Fixed Chromosomes

As an alternative to 3:1 alcohol/acetic acid fixation, chromosomes can be isolated from formaldehyde-fixed root tips by mechanical dispersion of tissues and cells and low-speed centrifugation onto glass slides. A disadvantage is, however, that chromosomes can only be investigated individually rather than within their chromosome complements and that chromosome morphology can be slightly distorted, obviously due to centrifugal force. This isolation method provides a method for investigation of ethanol/acid-sensitive structures and/or antigens.

IV. Chromosome Structure in SEM

A. Ultrastructure of Mitotic Plant Chromosomes

Investigations with different techniques and of a variety of organisms have shown that basic chromosome substructures are universal. Throughout all stages of the cell cyle chromosomes are composed of fibrous structures that vary in diameter and orientation. The predominant fibers observed have a diameter of approximately 30 nm (Martin et al., 1996). Some of these fibers are "bunched," forming chromomeres with diameters ranging from 200 to 400 nm (Fig. 3A). The surface topography of chromosomes observed in SEM could be confirmed with scanning force microscopy (SFM) of hydrated chromosomes (Schaper et al., 2000).

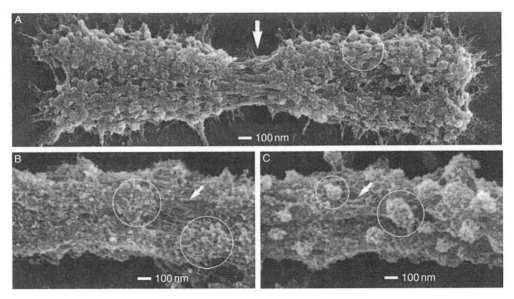

Fig. 3 Scanning electron micrograph of plant chromosomes in metaphase (A; barley), early prometaphase (B; barley) and prometaphase (C; rye) isolated with the drop/cryo technique. In metaphase the chromatids exhibit numerous chromomeres (A; circle) and begin to separate. The centromere (A, arrow) is characterized by exposed parallel fibers. Parallel fibers (B/C; arrows) and chromomeres of different sizes (B/C; circles) are seen in all stages of condensation. (C from Zoller et al., 2004a; reproduced by permission of S. Karger AG, Basel)

During condensation from prophase to prometaphase, chromomeres are interspersed with longitudinally oriented fibers, which are defined as parallel matrix fibers (Fig. 3B and C; Wanner and Formanek, 2000). At metaphase chromosomes show two characteristic features: coherent, more or less cylindrical sister chromatids which appear "knobby" at low magnifications due to tightly packed chromomeres, and a centromeric region seen as a constriction characterized mainly by parallel fibers (Fig. 3A). These fibers are particularly exposed in chromosomes isolated from arrested cells; in nonarrested metaphase chromosomes the parallel fiber region in the centromere is less exposed. Arresting by disruption of spindle formation seems to influence chromosome morphology: in general it causes (artificial) shortening and, consequently, compaction of the chromosomes in addition to extremely pronounced primary and secondary constrictions. Secondary constrictions, generally the structural manifestation of nucleolus organizing regions (NORs), are also characterized by parallel matrix fibers. In the case of the genus *Luzula* (rush, family Juncaceae) which has holocentric chromosomes, SEM investigations confirmed their lack of centromeric constriction, and revealed parallel fibers interspersed with chromomeres along the entire chromosome.

B. Where do Spindle Fibers Attach?

Visualization of a characteristic spindle attachment site, or kinetochore, is in principle problematic since cold treatment and arrestation of root tips prevents formation of microtubules. In nonarrested metaphase chromosomes, however, a minor fraction of chromosomes show an additional structure, which may be extrachromosomal elements of the kinetochore. Nonarrested chromosomes fixed with formaldehyde tend to exhibit bundles of filamentous structures attached predominantly in the pericentric region of both chromatids (Fig. 4).

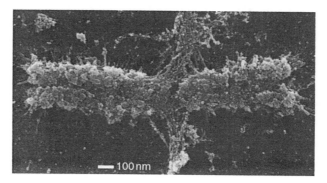

Fig. 4 Scanning electron micrograph of a nonarrested barley metaphase chromosome fixed with formaldehyde and spread by centrifugation. At the centromere fibrillar bundles are visible which are interpreted as parts of the kinetochore and the spindle apparatus.

C. Chromosome Condensation and Decondensation During Mitosis: Dynamic Matrix Model

According to structural data from SEM investigations, the "Dynamic Matrix Model" has been postulated, explaining chromosome condensation and the higher order of chromatin (Wanner and Formanek, 2000). Condensation of barley and rye chromosomes in mitosis starts in the centromeric region and then continues toward the telomeres (Martin *et al.*, 1996; Zoller *et al.*, 2004a). From prophase to prometaphase, solenoids (30 nm) form chromomeres which are interspersed with parallel matrix fibers (Fig. 5). The number of chromomeres increases with further condensation. At maximum condensation—and typically for arrested chromosomes—chromomeres are tightly compacted and parallel fibers are exposed at the centromere. The basis for chromosome condensation is an antiparallel movement of the matrix fibers, an energy-dependent process which is limited sterically in binding sites, and which is counteracted by elastic tension (potential energy) in the condensed chromomeres. Under normal conditions these two forces reach an equilibrium, which can be shifted depending on preparative treatment (such as interruption of spindle apparatus formation by "arresting"). In telophase constrictions disappear and decondensation starts at the telomeres by successive separating and loosening of chromomeres. Matrix fibers can still be observed, but are no longer in parallel arrangement. In physical terms, decondensation of chromosomes could be described as a release of potential energy in the chromomeres and matrix fibers.

D. Meiosis

SEM has been implemented for high resolution imaging of chromatin organization in meiosis I and II of rye (*Secale cereale*) (Fig. 6). Although the basic structural elements (30 nm fibers, chromomeres) are observed, in general the chromosome surface structure during meiosis is smoother than during mitosis (Zoller *et al.*, 2004a, b). There are several indications in SEM of how recombination manifests itself structurally. Chromatin interconnecting threads are typically observed in prophase I between homologous and nonhomologous chromosomes. In zygotene, chromatid strands between homologous counterparts are observed (Fig. 7A and B). In pachytene segments of synapsed and nonsynapsed homologs alternate (Zoller *et al.*, 2004b). At synapsed regions pairing is so intimate that homologous chromosomes form a filament-like structural entity. Chiasmata are characterized by chromatid strands that cross from one homolog to its counterpart. Bivalents are characteristically fused at their telomeric regions. In metaphase I and II, primary and secondary constrictions are not observed (Zoller *et al.*, 2004a).

V. Chromosome Analysis in SEM

A. Metal Impregnation of Chromosomes

Most biological specimens are electrical insulators. To increase specimen conductivity in order to avoid charging effects, specimens with an elaborate surface topography must be coated by metals to a thickness that does not obscure fine

Fig. 5 Illustration of the dynamic matrix model for chromatin condensation. DNA assembles with histone octamers to form nucleosomes and elementary fibrils which wind up to solenoids. Solenoids attach to matrix fibers by matrix fiber binding proteins. Dynamic matrix fibers move in an antiparallel fashion (arrows), possibly assisted by linker proteins. As condensation progresses, chromomeres are formed by loop stabilizing proteins, and the chromosomes become shorter and thicker as more chromomeres are formed. This causes, for sterical reasons, a tension vertical to the axial direction which forces the chromatids apart. (Wanner et al., 2005; reproduced by permission of S. Karger AG, Basel)

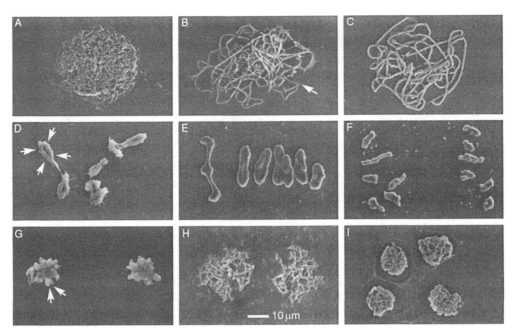

Fig. 6 SEM micrographs of the meiotic cycle of rye (*S. cereale*). In leptotene (A) chromatin forms a fibrillar network. In zygotene (B) "pairing forks" at sites of paired homologs can be discriminated (B, arrow). In pachytene (C) homologs are completely paired. In diakinesis (D chromosomes are highly condensed and paired at their chiasmata at the telomeric regions (arrows). In metaphase I (E) chromosome surface is smooth; neither primary nor secondary constrictions are detectable; the telomeric regions are characteristically fused. In anaphase I (F) homologs separate individually. In telophase I (G) forked telomeres become evident (each arrow represents the telomere of one chromatid). In prophase II (H) chromosomes appear "kinky". In telophase II of early tetrad stage (I) chromosomes decondense again. (Zoller *et al.*, 2004a; reproduced by permission of S. Karger AG, Basel)

structural details. Metal impregnation also enhances signal contrast, and in some cases allows substance analysis due to specific binding properties to metal compounds, for which different detectors (e.g., BSE, X-ray analysis) can be implemented for signal localization. Osmium-(TCH-Os)$_n$ impregnation, first described by Ip and Fishman (1979) for cytoskeletal elements, was commonly used in early SEM studies for chromosomes (Allen *et al.*, 1986; Harrison *et al.*, 1982; Sumner, 1991). A few reports of osmium-impregnated preparations of human chromosomes with adequate contrast exist (Rizzoli *et al.*, 1994; Sumner, 1991), but as human chromosome preparation improved, impregnation by osmium has been implemented less frequently. It has never been successfully applied to plant chromosomes. We interpret the earlier observations of impregnation of human chromosomes to derive from the presence of osmiophilic nucleoplasmic residues on the chromosome surface.

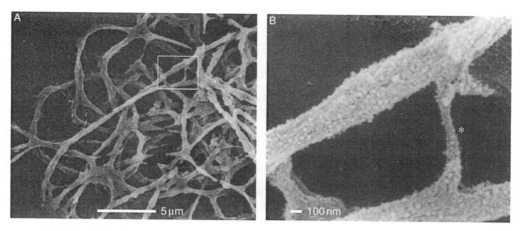

Fig. 7 SEM micrographs of meiotic chromosomes of rye (*S. cereale*) in the zygotene stage (A; detail of Fig. 6B). Fibrillar interactions characterize synapsis of homologs (B, detail of framed area in A). Part of a chromatid of one homolog switches to its homologous counterpart (B; asterisk). (Zoller *et al.*, 2004b; reproduced by permission of S. Karger AG, Basel)

B. Specific Staining of DNA in Chromosomes with Platinum Blue

Platinum blue (Pt blue; $[CH_3CN]_2Pt$ oligomer; Wanner and Formanek, 1995) and related platinum-organic compounds selectively react with nucleic acids, with especially high affinity to DNA (Aggarwal *et al.*, 1975, 1977; Frommer *et al.*, 1990; Köpf-Maier and Köpf, 1986; Lippert and Beck, 1983). When viewing Pt blue-stained chromosome preparations with BSE microscopy at low magnification, the nuclei and chromosomes are clearly visible at all stages of condensation. Platinum signals from interphase chromosomes exhibit a fibrous network and highly contrasted foci, indicating areas of high chromatin density presumably corresponding to heterochromatin. In mitotic chromosomes, the BSE signal from contrasted DNA clearly reflects characteristic chromosome structures, including primary and secondary constrictions and chromatids (Fig. 8A). At higher magnification, regions of different signal density are observed throughout metaphase chromosomes. Regions corresponding to chromomeres (recognizable from SE signal) have "bright" BSE signals, whereas fibers can be correlated with either strong or weak BSE signals, indicating that these structures are either DNA-rich or DNA-poor. A striking difference between SE and BSE images occurs in the primary and secondary constrictions: the BSE signals of these regions are narrower and weaker than that of the chromatids, whereas the SE signal reveals unmistakable structural elements (Fig. 8B). This indicates that the parallel structures in the centromere and satellite regions contain less DNA than the chromosome arms. Pt blue contrasting of chromosomes treated with increasing concentrations of DNaseI shows a progressive decrease in BSE signal intensity due to DNA reduction, but the basic

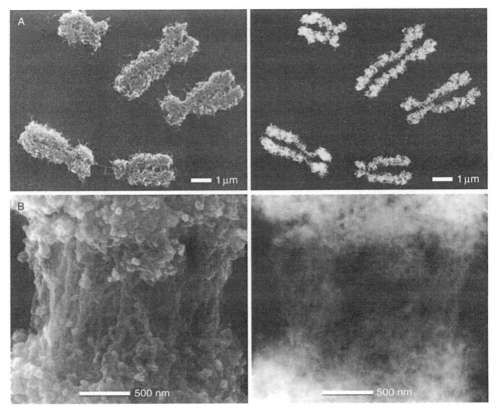

Fig. 8 Scanning electron micrographs of human (A) and barley (B) metaphase chromosomes stained for DNA with platinum blue. The SE image (A, left) shows overall structural substance resulting from both protein and DNA, whereas the BSE image shows that the DNA content is lower between the sister chromatids (A, right). At the centromere, chromomeres border exposed parallel fibers (B, left) which contain less DNA than surrounding areas (B, right).

chromosome morphology (SE signal) was maintained (Fig. 9A). At the highest DNaseI concentration the chromosome structure collapses to a flat network of fibers (Fig. 9B). The DNase-insensitive nature of chromosome structure supports the existence of a three-dimensional protein matrix.

C. Staining of Protein in Chromosomes

Proteins as a substance class can be stained by certain heavy metals (e.g., mercury, silver, and lead) using their specific affinities to cysteine-SH-group and/or complex formation with different amino acids such as glutamine and asparagine. Silver nitrate reacts with chromatin, but preferentially with centromeres and

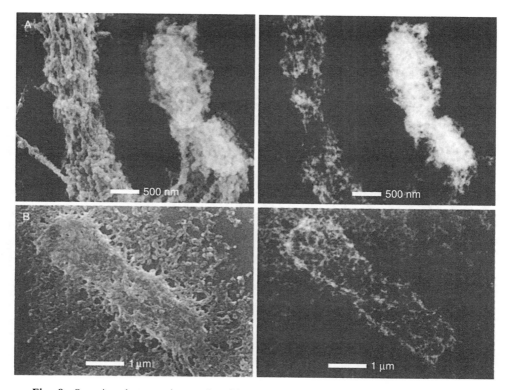

Fig. 9 Scanning electron micrographs of human metaphase chromosomes treated with moderate concentrations (10 μg/ml) of DNaseI and stained with platinum blue. A significant decrease in DNA judged from the BSE signal (A, right) does not result in corresponding structural degradation (A, left). At high DNase concentrations (100 μg/ml) chromosomes collapse and spread to a flat network (B, left) with much less DNA (B, right).

the telomeres (Fig. 10A). Colloidal silver with a tannic acid "coating" reacts intensely with chromosomal proteins. However, structural preservation of chromosomes is restricted because they cannot be fixed with glutaraldehyde before colloidal silver staining. In contrast to DNA-staining, the BSE image of silver-stained chromosomes shows a strong signal in the centromeric region, and no separation is seen between the sister chromatids (compare Fig. 10B with Fig. 8A and B). Dependent on the stringency of the conditions for staining with colloidal silver, it can be shown that the centromere is protein-enriched (Fig. 10B). Efforts are ongoing to increase the efficiency of silver staining, as the resulting BSE signal is still generally weaker than that of DNA staining with Pt blue. This discrepancy may be due to sterical hindrance of the rather large diameter of the colloid particles (approx. 5 nm) for reaction with chromosomal proteins.

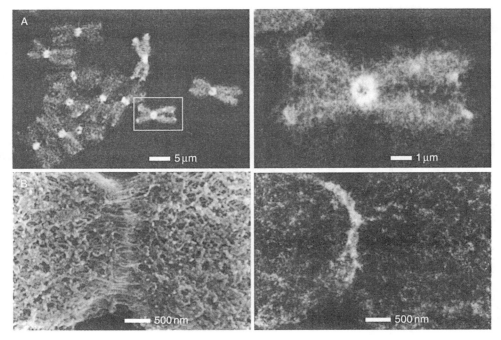

Fig. 10 SEM micrographs of barley metaphase chromosomes stained with silver compounds showing general protein distribution. BSE images show that silver nitrate reacts preferentially with chromatin in the centromeres and telomeres (A, left and detail, right). Colloidal silver/tannic acid staining results in a strong BSE signal contrast (B, right) corresponding to the parallel fibers in the centromere (B, left) indicating that the centromere is protein-enriched.

D. Controlled Decondensation of Chromosomes

Studies on the influence of pH and different buffers, initially investigated to monitor artificial structural changes, led to the development of methods useful in simulating decondensation. In combination with SEM analytic techniques, this provides insight into internal chromosome structure and composition. Chromosome structure appears stable over a broad range of pH, from acidic to neutral conditions (pH 3 to pH 7), but loosens with increasing alkalinity (Fig. 11A and B). Chromosomes incubated in citrate buffer typically are somewhat stretched and "fan out" at all four telomeres (Fig. 11C). The centromeric region remains relatively unaffected and stains intensely for DNA, but the spread telomeric regions reveal underlying non-DNA structural elements (Fig. 11C).

In general, the stronger the fixation, the less influence of subsequent treatment with buffers or detergents. Glutaraldehyde fixes proteins so well that the 3D structure is unaffected by most reagents. If, however, glutaraldehyde-fixed chromosomes are treated with high concentrations of proteinase K (0.1–1.0 mg/ml), there is a marked (time-dependent) effect on the chromosomes: they stretch in

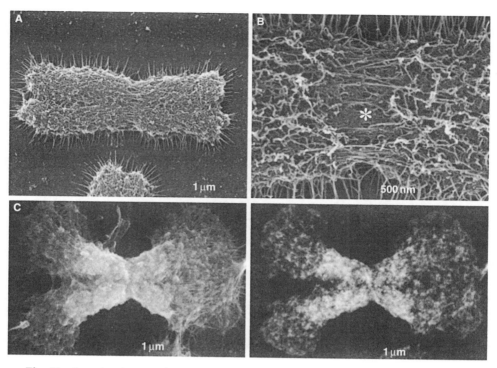

Fig. 11 Scanning electron micrographs of barley metaphase chromosomes after artificial decondensation under different pH and buffer conditions. Under basic (pH 10) conditions, typical chromomeres are no longer visible (A). Chromomeres loosen, revealing predominant 30 nm fibers (B); the parallel fibers in the centromere remain exposed (B; asterisk). Chromosomes treated with citrate buffer fan out at the telomeres (C, left); staining for DNA with platinum blue indicates that there are matrix structures throughout the chromatids (C, right).

length and/or fan out in breadth. In some cases, chromosomes reach an extraordinary length of up to 200 μm. In such extreme stages of stretching, chromatin begins to elongate to elementary chromatin substructures: "bunched" solenoids (chromomeres, on average 200 nm), solenoids (30 nm) and elementary fibers (10 nm) (Figs. 12–14). This artificial loosening of the compact metaphase chromosomes is referred to as "controlled decondensation" (Wanner and Formanek, 2000). The loosening effect seems to be a combined result of partial digestion of chromatin proteins, physical and mechanical forces (e.g., surface tension) during preparation and dehydration.

DNA staining of proteinase K-treated chromosomes and three-dimensional visualization reveals a loosened chromomere configuration, with interspersed and underlying non-DNA matrix fibers (Fig. 15). Consistent with all cases of stretching or loosening of chromosomes, parallel matrix fibers can be observed over the total length of the chromosome. Controlled decondensation can be

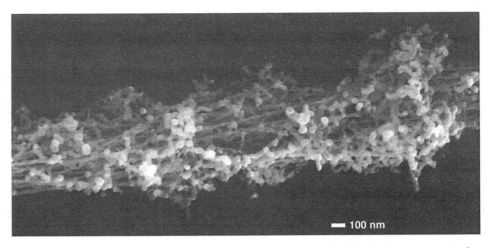

Fig. 12 SEM micrograph of a barley metaphase chromosome (region distal to the centromere) after artificial decondensation (fixation with glutaraldehyde and treatment with proteinase K). During stretching, chromomeres loosen and can be recognized as bundles of solenoid fibers. Three-dimensional arrangement of parallel matrix fibers is visible throughout. (See Plate no. 43 in Color Plate Section for 3D anaglyph for red/blue glasses.) (Wanner and Formanek, 2000; reproduced by permission of Elsevier).

interpreted according to the dynamic matrix model as follows: Since chromatin and protein matrix fibers are associated tightly by loop binding and matrix binding proteins, even partial digestion of these proteins releases tension created by chromatin compaction, causing the observed stretching and spreading.

E. Immunogold Labeling of Chromosomal Proteins and ISH

Immunodetection in SEM is a promising method for identifying functional elements on chromosomal substructures with high resolution. It is critical to use smallest possible markers (e.g., Nanogold® labels, Fab' fragments, haptens when applicable), as conventional labeling with colloidal gold has been unsuccessful to date for SEM investigation of chromosomes. Different kinds of immunogold labels have been applied for SEM investigations ISH and immunodetection of specific proteins: Nanogold® (NG), a 1.4 nm gold cluster, and Fluoronanogold™ (FNG), a combined fluorescein and 1.4 nm gold marker (Hainfeld and Powell, 2000). FNG allows direct correlation of fluorescent LM and SEM signals on the same specimen (Powell *et al.*, 1998; Schroeder-Reiter *et al.*, 2006). Since the size of the 1.4 nm gold clusters for both FNG and NG is at the resolution limit of the SEM, both FNG and NG must be either gold or silver enhanced, a time-dependent process of controlled metallonucleation, which gradually increases the size of the bound gold cluster (Hainfeld *et al.*, 1999). Using ultrasmall markers, it is currently possible to localize epitopes down to the solenoid level of chromatin. High resolution immunogold labeling with Nanogold® allowed quantification and three-dimensional

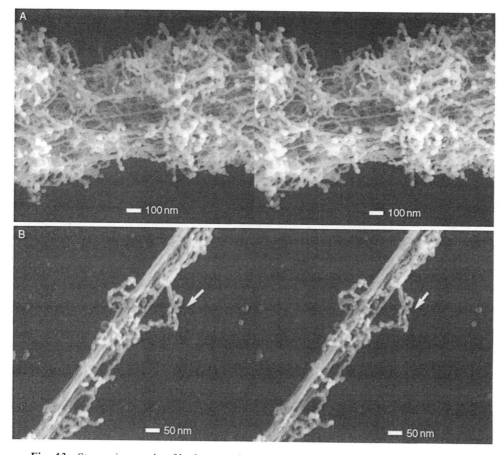

Fig. 13 Stereomicrographs of barley metaphase chromosomes at different stages of artificial decondensation (fixation with glutaraldehyde and treatment with proteinase K). During stretching, chromomeres loosen and can be recognized as bundles of solenoid fibers. The architecture of the centromere also loosens, showing three-dimensional arrangement of parallel matrix fibers (A). After ultimate stretching, the number of matrix fibers is decreased and the chromomeres are stretched and decondensed to the level of the individual solenoid fibers (B, arrow). (Wanner and Formanek, 2000; reproduced by permission of Elsevier)

localization of histones, histone modifications, and centromere proteins (Fig. 16) (Houben et al., 2007; Schroeder-Reiter et al., 2003). ISH was recently improved for SEM studies in terms of signal localization, labeling efficiency and structural preservation, allowing for the first time 3D SEM analysis an unusual NOR structure and its rDNA distribution (Fig. 17) (Schroeder-Reiter et al., 2006). Dependent on signal size (according to thickness of enhancement layer) and accelerating voltage, signals can be detected from within chromosome structures, allowing visualization of three-dimensional signal distribution (Fig. 18).

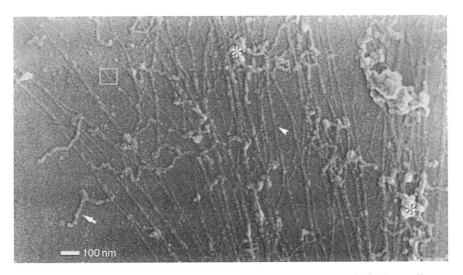

Fig. 14 SEM micrograph of chromatin of a barley metaphase chromosome in final spreading stages after fixation with glutaraldehyde and treatment with proteinase K. After extreme spreading, the basic levels of chromatin condensation can be observed: coiled solenoids (asterices), solenoids (white arrow), elementary fibers (white arrowhead) and DNA (frame). (Wanner and Formanek, 2000; reproduced by permission of Elsevier)

VI. Outlook

Preservation of 3D structure of chromosomes in combination with high-resolution SEM and analytical methods is an ongoing challenge, as fixation (Lat. *fixare*=to fasten) and analysis (Gr. *analysis*=loosening up) are essentially contradictory concepts. Dependent on the scientific goals—whether priority is on high-resolution ultrastructure or rather on sensitivity of analysis—a compromise must be accepted.

With modern equipment, structural investigations of chromatin should already be possible with an instrumental resolution approaching 1 nm at low voltage (1 kV), thus requiring thinner metal coating (down to 1 nm) and enabling a significant improvement of specimen resolution down to 2–3 nm. This would allow for a more precise visualization of solenoids, elementary fibers and even ideally of DNA. Modern BSE detectors (e.g., in-lens, semiconductor) are highly sensitive even at very low accelerating voltages, so that analytical SEM can be performed under conditions that are optimal for visualization of structural details. This would allow significant improvement of visualization of specific staining, in particular of DNA staining with Pt blue, for structures at or below 30 nm and would enable more precise differentiation between DNA-containing and proteinaceous structural elements. Optimizing metallonucleo-enhancement and development of direct labeling systems, preferentially with Fab' fragments and 5–8 nm gold markers which could be detected without enhancement would permit

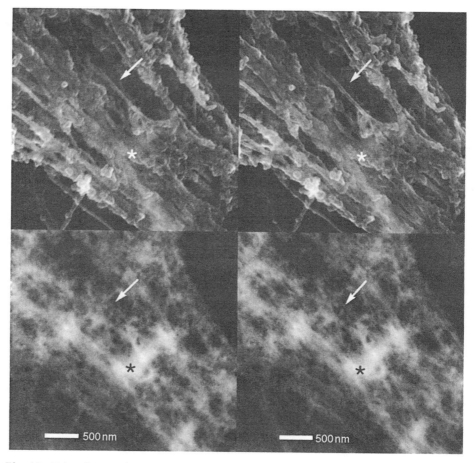

Fig. 15 Pairs of stereomicrographs of barley metaphase chromosomes after artificial decondensation (fixation with glutaraldehyde and treatment with proteinase K) and staining with platinum blue. The overall structure illustrates the fibrous architecture of the stretched chromatin (upper SE image pair). DNA is distributed mainly in a three-dimensional network with highest concentration in chromomeres (asterices, lower BSE image pair). Some fibers (arrows) are not stained (lower BSE image pair), indicating that these are of proteinaceous nature. (Wanner and Formanek, 2000; reproduced by permission of Elsevier)

localization of markers even to the elementary fibril (10 nm). As low-kV microscopy severely limits depth information, e.g., for immunolabeling and DNA staining, "SEM tomography" with focused ion beam milling will allow direct visualization of both internal architecture and markers within chromosomes. Despite increasing instrumental and analytical resolution, the limiting factor remains the achievement of adequate chromosome preparations—an ongoing challenge.

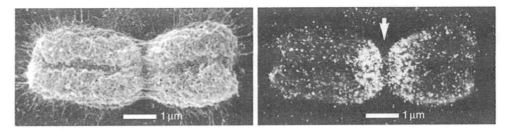

Fig. 16 SEM micrographs of a barley metaphase chromosome showing immunolabeling of phosphorylated histone H3 (serine 10) detected with silver-enhanced Nanogold®. The exposed parallel matrix fibers at the centromere (left) are coincident with a BSE signal gap (arrow, right).

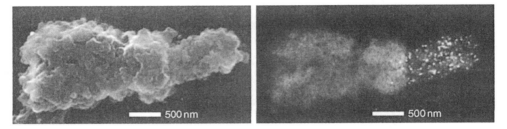

Fig. 17 SEM micrographs of an *Oziroë biflora* (plant family Hyacinthaceae) metaphase chromosome labeled for 45SrDNA with *in situ* hybridization and silver-enhanced Nanogold®. The chromosome exhibits an elongated peg-like terminal constriction (left) that harbors 45SrDNA sequences (right), identifying it as a nucleolus organizing region.

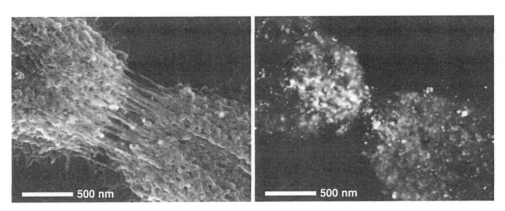

Fig. 18 SEM SE (left) and BSE (right) micrographs showing the centromeric region of a barley metaphase chromosome labeled for phosphorylated histone H3 (serine 10) with Nanogold® and enhanced with silver. (See Plate no. 44 in Color Plate Section for 3D anaglyph for red/blue glasses of superimposed SE and BSE images.) Signals (BSE image; yellow in color image) are detected from different depths from within the chromatin, especially on the chromomeres bordering the 30 nm parallel fibers at the centromere. The signal "gap" corresponds to the parallel fibers.

Acknowledgments

The authors acknowledge contributions from Dr. Helmut Formanek, Dr. Jutta Zoller, and Ursula Wengenroth in data presented in this manuscript, and thank Sabine Steiner for excellent technical assistance. Gerhard Wanner was supported in part by the Deutsche Forschungsgesellschaft (DFG).

References

Aggarwal, S., Ofosu, G., and Waku, Y. (1977). Cytotoxic effects of platinum-pyrimidine complexes on the mammalian cell *in vitro*: A fine structural study. *J. Clin. Hematol. Oncol.* **75,** 47–561.

Aggarwal, S., Wagner, R., McAllister, P., and Rosenberg, B. (1975). Cell-surface-associated nucleic acid in tomorigenic cells made visible with platinum-pyrimidine complexes by electron microscopy. *Proc. Natl. Acad. Sci. USA.* **72,** 928–932.

Allen, T. D., Jack, E. M., Harrison, C. J., and Claugher, D. (1986). Scanning electron microscopy of human metaphase chromosomes. *Scan. Electron Microsc.* **1,** 301–308.

Allen, T. D., Jack, E. M., and Harrison, C. H. J. (1988). The three dimensional structure of human metaphase chromosomes determined by scanning electron microscopy. *In* "Chromosomes and Chromatides" (K. W. Adolph, ed.), pp. 51–72. CRC Press, FL.

Ambros, P. F., Matzke, M. A., and Matzke, A. J. M. (1986). Detection of a 17 kb unique sequence (T-DNA) in plant chromosomes by *in situ* hybridization. *Chromosoma* **89,** 243–253.

Arents, G., Burlingame, R. W., Wang, B. C., Love, W. E., and Moudrianakis, E. N. (1991). The nucleosomal core histone octamer at 3.1 Å resolution: A tripartite protein assembly and a left-handed superhelix. *Biochemistry* **88,** 10148–10152.

Boveri, T. (1902). Über mehrpolige Mitosen als Mittel zur Analyse des Zellkerns. *Verhandlungen der physikalisch-medizinischen Gesellschaft zu Würzburg* **35,** 67–90.

Finch, J. T., and Klug, A. (1976). Solenoidal model for superstructure in chromatin. *Proc. Natl. Acad. Sci. USA.* **73,** 1897–1901.

Frommer, G., Schöllhorn, H., Thewalt, U., and Lippert, B. (1990). Platinum(II) binding to N7 and N1 of guanine and a model for a purine-N^1,Pyrimidine-N^3 cross-link of cis-platin in the interior of a DNA duplex. *Inorg. Chem.* **29,** 1417–1422.

Hainfeld, J. F., and Powell, R. D. (2000). New frontiers in gold labeling. *J. Histochem. Cytochem.* **48,** 471–480.

Hainfeld, J. F., Powell, R. D., Stein, J. K., Hacker, G. W., Hauser-Kronberger, C., Cheung, A. L. M., and Schöfer, C. (1999). Gold-based autometallography. *In* "Proceedings of the 57th Annual Meeting, Microscopy Society of America", pp. 486–487. Springer, New York.

Harrison, C. J., Allen, T. D., Britch, M., and Harris, R. (1982). High-resolution scanning electron microscopy of human metaphase chromosomes. *J. Cell Sci.* **56,** 409–422.

Houben, A., Schroeder-Reiter, E., Nagaki, K., Nasuda, S., Wanner, G., Murata, M., and Endo, T. R. (2007). CENH3 interacts with the cetnromeric retrotransposon cereba and GC-rich satellites and locates to centromeric substrtuctures in barley. *Chromosoma* **116,** 275–283.

Ip, W., and Fishman, D. (1979). High resolution scanning electron microscopy of isolated *in situ* cytoskeletal elements. *J. Cell Biol.* **83,** 249–253.

Köpf-Maier, P., and Köpf, H. (1986). Cytostatische Platin-Komplexe: Eine unerwartete Entdeckung mit weitreichenden Konsequenzen. *Naturwissenschaften* **73,** 239–247.

Lippert, B., and Beck, W. (1983). Platinkomplexe in der Krebstherapie. *Chemie u Z* **17,** 190–197.

Martin, R., Busch, W., Herrmann, R. G., and Wanner, G. (1994). Efficient preparation of plant chromosomes for high-resolution scanning electron microscopy. *Chromosome Res.* **2,** 411–415.

Martin, R., Busch, W., Herrmann, R. G., and Wanner, G. (1996). Changes in chromosomal ultrastructure during the cell cycle. *Chromosome Res.* **4,** 288–294.

Powell, R. D., Halsey, C. M., and Hainfeld, J. F. (1998). Combined fluorescent and gold immunoprobes: Reagents and methods for correlative light and electron microscopy. *Microsc. Res. Tech.* **42**(1), 2–12.

Ris, H., and Witt, P. L. (1981). Structure of the Mammalian Kinetochore. *Chromosoma* **82,** 153–170.

Rizzoli, R., Rizzi, E., Falconi, M., Galanzi, A., Baratta, B., Lattanzi, G., Vitale, M., Manzoli, L., and Mazzotti, G. (1994). High resolution detection of uncoated metaphase chromosomes by means of field emission scanning electron microscopy. *Chromosoma* **103**(6), 393–400.

Schaper, A., Rößle, M., Formanek, H., Jovin, T. M., and Wanner, G. (2000). Complementary visualization of mitotic barley chromatin by field-emission scanning electron microscopy and scanning force microscopy. *J. Struct. Biol.* **129**, 17–29.

Schroeder-Reiter, E., Houben, A., Grau, J., and Wanner, G. (2006). Characterization of a peg-like terminal NOR with light microscopy and high-resolution scanning electron microscopy. *Chromosoma* **115**, 50–59.

Schroeder-Reiter, E., Houben, A., and Wanner, G. (2003). Immunogold labeling of chromosomes for scanning electron microscopy: A closer look at phosphorylated histone H3 in mitotic metaphase chromosomes of *Hordeum vulgare*. *Chromosome Res.* **11**, 585–596.

Schubert, I., Dolezel, J., Houben, A., Scherthan, H., and Wanner, G. (1993). Refined examination of plant metaphase chromosome structure at different levels made feasible by new isolation methods. *Chromosoma* **102**, 96–101.

Strasburger, E. (1888). "Über Kern- und Zelltheilung im Pflanzenbereich, nebst einem anhang über Befruchtung." Gustav Fischer, Jena.

Sumner, A. T. (1991). Scanning electron microscopy of mammalian chromosomes from prophase to telophase. *Chromosoma* **100**, 410–418.

Sumner, A. T., and Ross, A. (1989). Factors affecting preparation of chromosomes for scanning electron microscopy using osmium impregnation. *Scanning Microsc.* **3**, 87–99.

Sutton, W. S. (1902). The chromosomes in heredity. *Biol. Bull.* **4**, 231–251.

Thoma, F., Koller, T., and Klug, A. (1979). Involvement of histone H1 in the organization of the nucleosome and the salt-dependent superstructures of chromatin. *J. Cell Biol.* **83**, 402–427.

Wanner, G., and Formanek, H. (1995). Imaging of DNA in human and plant chromosomes by high-resolution scanning electron microscopy. *Chromosome Res.* **3**, 368–374.

Wanner, G., and Formanek, H. (2000). A new chromosome model. *J. Struct. Biol.* **132**, 147–161.

Wanner, G., Schroeder-Reiter, E., and Formanek, H. (2005). 3D Analysis of chromosome architecture: Advantages and limitations with SEM. *Cytogen. Gen. Res.* **109**, 70–78.

Watson, J. D., and Crick, F. H. C. (1953). A structure for deoxyribose nucleic acid. *Nature* **171**, 737–738.

Zoller, J. F., Herrmann, R. G., and Wanner, G. (2004a). Chromosome condensation in mitosis and meiosis of rye (*Secale cereale* L.). *Cytogen. Gen. Res.* **105**, 134–144.

Zoller, J. F., Hohman, U., Herrmann, R. G., and Wanner, G. (2004b). Ultrastructural analysis of chromatin in meiosis I+II of rye (*Secale cereale* L.). *Cytogen. Gen. Res.* **105**, 145–156.

PART III

Cells and Infectious Agents

CHAPTER 24

Infection at the Cellular Level

Christian Goosmann, Ulrike Abu Abed, and Volker Brinkmann

Max-Planck-Institut für Infektionsbiologie
Charitéplatz 1
10117 Berlin
Germany

Abstract
I. Introduction
II. Methods and Materials
   A. Negative Staining of Isolated Bacteria
   B. Resin Embedding of Bacteria-Infected Cell Cultures
   C. Resin Embedding of Infected Tissue
   D. Rapid Processing for TEM
   E. Preembedding Immunodetection Methods
   F. Immunodetection Using Ultrathin Cryosections
   G. Scanning Electron Microscopy
III. Overview and Conclusion
   References

Abstract

Fine structural analysis of the infection process is indispensable for understanding the relation between microorganisms and host cells. This chapter focuses on standard techniques for transmission as well as scanning electron microscopy that will be of benefit even to researchers new to the field.

I. Introduction

After the "descriptive" era (1960–1970) of electron microscopy, the use of EM in the study of microorganisms declined. During the last 15 years however, tools were developed to modify tissues and cultured cells as well as pathogens on a molecular basis either by expressing additional genes or by silencing genes by knock-out techniques (somatic gene deletion, interfering RNA). Thus, precisely defined alterations were introduced, and the interest in fine structural analysis of these genetically modified systems gave rise to a new era of electron microscopy. The expression of proteins with fluorescent tags in living cells allows the correlation between live cell imaging and subsequent fine structure analysis of the same cell, this time identifying the transfected proteins on an EM level either by immunogold techniques, or more directly, by photoconversion inducing an electron-dense precipitate.

In this chapter, we present preparation methods that will enable workers from other research areas to process their samples and have them analyzed in the electron microscope. Most of the methods we will present can be successfully used without great manual or technical skills provided that access to the usual EM periphery (ultramicrotome, critical point dryer, sputter coater, etc.) is given.

Although a more detailed and probably more live-like appearance of microorganisms can be obtained by specialized techniques like cryo-EM of vitreous sections (CEMOVIS; Al-Amoudi et al., 2004), these methods are far beyond reach of a conventional laboratory and are not topics of this chapter.

II. Methods and Materials

Choosing which method is most suitable for the fine structure analysis of a microorganism depends on the available source. The source could be

- isolated microorganisms from cultures or clinical samples
- infected primary cells or cell cultures, or
- tissue from infected animals or humans

For isolated microbes, suitable methods range from negative staining techniques, ultrathin sections to scanning electron microscopy (SEM). Cell cultures are most often analyzed with transmission electron microscopy (TEM) or SEM depending on the underlying question, and most tissue samples are embedded and viewed using a TEM. All techniques can be combined with immunodetection methods. On the following pages we will give an overview on the techniques we use to study infection at the cellular level.

A. Negative Staining of Isolated Bacteria

Bacteria are harvested from plates or liquid cultures, washed twice in PBS, and fixed using 1–2.5% glutaraldehyde/PBS for best preservation of fine structure, or 2–4% PFA dissolved in PBS for immunodetection. If the fine structure is not sufficient, glutaraldehyde at concentrations between 0.05% and 0.1% can be added (Hayat, 1981). The following sample preparation can also be used for fixed suspensions of fractions of bacterial or cellular components or enriched suspensions of viruses, particles, etc.

1. Sample Adhesion to Carbon-Coated Formvar Films

Small droplets (20–50 µl) of the bacterial suspensions are placed on Parafilm® in a humid chamber (plastic boxes with a lid having wet filter paper along the edges). Grids (200 mesh) with carbon-coated formvar or pioloform films are placed with the coated side on the droplets. The grids have to be glow discharged (Dubochet *et al.*, 1971) to ensure even wetting of the surface and to assure binding of bacteria to the grids which normally takes a couple of minutes. Clinical samples often do not contain high amount of particles, and are normally contaminated with other materials, so several washing cycles or gradient purification may be required before negative staining.

If the desired density of particles on the EM-grid is not reached by adhesion to the floating grid, the suspension can be enriched by centrifugation prior to the adhesion step. Improved adhesion to the grid can also be achieved by carefully sticking the grids on Parafilm® in a humid chamber with the carbon-film side facing upwards and by placing droplets of 3–4 µl of the suspension on the film for up to 1 h. It is important for the following steps that only the filmed side of the grid is wetted. This will ensure that the grids will float on drops in all the following steps, which prevents contamination of the noncoated side of the grid as well as chemical reactions of the copper with other reagents.

2. Negative Stain

For the conventional negative stain the grids are removed from the specimen drops, washed five times on drops of distilled water to remove salts, and then placed on drops of the aqueous solution of heavy metal salts (2% uranyl acetate, 2% phosphotungstic acid, or 3% ammonium molybdate) for 10 s–2 min (Harris and Horne, 1994). The concentration of the contrasting solution is not critical, but it is advisable to centrifuge the contrasting solution for some minutes at high speed to sediment any precipitates before use. After that, the grids are removed from the droplets with a fine forceps, and the negative stain is partly removed by holding the coated surface of the grid against filter paper. The contrast achieved depends on the thickness of the remaining negative contrast solution and varies with the angle between grid and filter paper: a small angle will result in a thin layer and lower contrast. This is especially desirable if isolated bacterial compounds like flagella are to be analyzed. After air drying the grids can be examined in TEM.

If complete bacteria are to be analyzed, the negative stain density can be adequate for small structures like flagella, but too dense to show details of the bacterial cells (see Fig. 1A). In this case, a series of specimens contrasted under different conditions has to be prepared. Negative contrast is a good way to give information on purity of bacterial fractionations prepared for nonmicroscopic techniques like protein gel electrophoresis (see Fig. 1B).

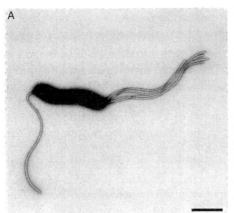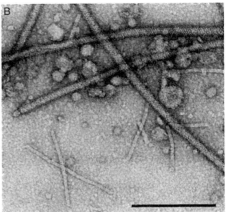

Fig. 1 (A) TEM image of *Helicobacter pylori*, uranyl acetate negative contrast. Although details of the flagella are nicely visible, the stain is too dense to reveal details of the bacterial cell. Scale bar: 1 µm. (B) TEM image of a *Salmonella* cell-fractionation preparation, uranyl acetate negative contrast. The image shows fine structure of flagella and pili. Scale bar: 200 nm.

3. Immunogold Labeling Combined with Negative Staining

For antigen localization studies negative staining can easily be combined with immunogold detection. This method is quick and gives nice results especially when epitopes on the surface of thin structures like flagellae, pili, etc. are to be examined. Following attachment to the coated side, the grids are washed three times on drops of PBS and transferred to a blocking solution (1% BSA, 0.02 M glycine, 10% cold water fish gelatine in PBS). After blocking for 30 min, grids are transferred to the primary antibody (1–10 µg/ml in blocking buffer) and incubated for 30–60 min. Following washes on PBS drops (6 × 2 min), the grids are incubated for 30 min with secondary antibodies coupled to gold colloids (in blocking buffer).

If correlative studies including samples with fluorescent immunostains are planned it is a good idea to use gold coupled secondary antibodies of the same source as the fluorescent coupled secondary antibody. The choice of colloid diameter depends on the planned magnification: while 6 nm colloids require a magnification of at least 20,000×, 18 nm colloids are already visible in low magnification overviews. As shown earlier, the particle density of the label is higher with smaller colloid diameter (Fig. 2B and C.) (Slot and Geuze, 1981). The specimens are incubated with antibodies in a moist chamber at 37 °C. After washing (6 × 2 min PBS, 5 × 2 min distilled water), the negative staining is performed as mentioned previously. A low contrast is desirable to clearly depict the gold colloids.

4. Labeling with Quantum Dots

Semiconductor nanocrystals, commercially available as Quantum Dots or Qdots, play an increasing role as fluorescent probes in biomedical research (Michalet *et al.*, 2005). Owing to their metal core they appear rather electron

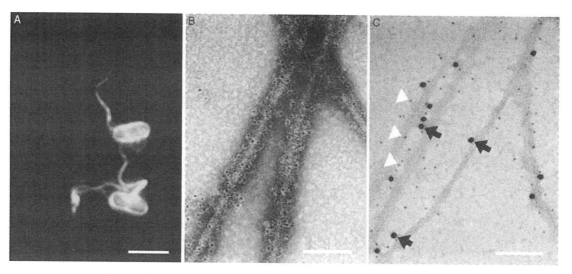

Fig. 2 Correlating fluorescence and negative staining of bacterial flagellae with polyclonal antibodies. (A) Light microscopic fluorescence image of *Salmonella typhimurium* labeled using secondary antibodies coupled to Quantum Dots. Scale bar: 2.5 µm; (B, C) negative contrast TEM images: (B) labeled with the same primary antibody, detected with secondary antibody coupled to 6 nm gold particles. Scale bar: 100 nm; (C) labeled simultaneously as before with Quantum Dots (λ: 565 nm, arrowheads) and with 12 nm gold particles (arrows). The labeling density achieved with Quantum Dots is significantly greater. Scale bar: 100 nm.

dense, and due to their uniform size they can also be used as immunoprobes for TEM structure analysis. They give good labeling densities and are easily visualized in TEM images if a low contrast of the specimen is achieved (see Fig. 2). After incubation of specimens with the primary antibodies and washing, the samples are incubated with a suspension of Quantum Dots coated with secondary antibody. After washing, a low contrast negative stain is applied as mentioned earlier. Note that the electron microscope may reveal even more Quantum Dots than the light microscope in a comparable staining experiment because a subfraction of a given Quantum Dot preparation is sometimes in a permanent dark state and thus does not emit fluorescence light (Yao *et al.*, 2005).

The evaluation of staining with Quantum Dots at high magnifications on pioloform-carbon-coated grids may be hindered by the structure of the film, which may prevent detection of all Quantum Dots owing to their low contrast compared to gold particles. Qdots are better visible, if the pioloform film is partially or in total dissolved in a way that only the structureless carbon layer remains (Fig. 3).

Coat grids with a thin layer of pioloform. Coat the other side of the grid with 20–30 nm carbon. Remove the film by dipping the grid carefully into chloroform for some seconds (Pontefract and Bergeron, 1981). To preserve the beam stability of the carbon layer we recommend not to remove the film completely, but to induce holes. Figure 3 shows that Quantum Dots are better visible in areas where the pioloform film is missing, and obscured in areas with undissolved film layer.

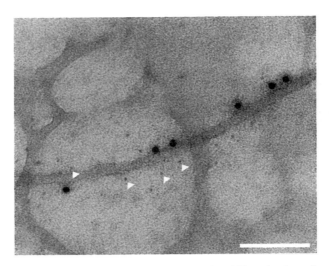

Fig. 3 *Salmonella* flagellae stained with polyclonal antibody rabbit-anti-*salmonella*, secondary antibodies goat-anti-rabbit Qdot 565 together with goat anti-rabbit, conjugated with 12 nm gold. Scale bar: 100 nm.

5. Correlative Light- and Transmission-Electron Microscopy of Negative Stained Samples

Before visualization of a structure by negative staining techniques in the TEM it may be interesting to study it at light-microscopy level using fluorescent dyes or phase contrast to screen a number of samples quickly or to find a suitable specimen detail before TEM analysis. Often, correlative microscopy combining light and electron microscopy gives a more complete picture than using TEM alone. It can be useful to identify certain fine-structural phenotypes with patterns occurring at light-microscopy level and vice versa. Host cell–pathogen interactions are one example. The appearance of mammalian cells in a negative contrasted TEM image reveals little more than the cytoskeleton due to the harsh extracting and air drying conditions during the staining procedure, but extracellular structures can be nicely preserved.

Light microscopic examination of adherent cells and/or bacteria can be carried out on the EM-grid after adhesion and before the negative staining step. For this, cells are cultivated on the carbon surface of a film-coated copper grid. Again it is important for later processing to keep the back surface of the grid dry. This can be achieved by applying very small volumes of media on a grid stuck to Parafilm®. Another method is to place drops of 10 µl of cell suspension in medium on the wells of a 12-well teflon-coated diagnostic slide and to invert the slide after applying the TEM-grids to the drops so that the cells settle on the grid, suspended under the hanging droplet. Note that the grids will always float to the rim of the droplet when the slide is inverted. The droplets should be small enough to have the grid suspended nearly parallel to the slide under the medium. The slide can be kept in a moist chamber supported on Plasticine® beads at 37 °C to allow for cell adhesion and interaction. It is good to start with around 10^3 cells per grid. Fixation can be carried out after the slide has been turned upward again by keeping the grids floating on the droplets and carefully adding an appropriate amount of fixative stock solution.

After fixation and, depending on the experiment, extraction, permeabilizing and fluorescent staining of the grids, the grids can be stuck to a large glass-coverslip with a small drop of glycerol with the carbon-film side facing the glycerol. Excess glycerol should be removed, so that the layer of glycerol is thin enough to view the cells through the other side of the glass with an immersion lens of a light microscope. After acquiring light microscopic images using phase contrast or fluorescence illumination the grids are brought back into floating condition by carefully adding distilled water to the glycerol layer between the glass and the grid. Alternatively, if light microscopic images of lower magnification are sufficient, the grids can be viewed and imaged floating on a culture medium droplet on the 12-well-slide with a long working distance lens. Note that Qdots may not fluoresce in this setting, possibly because of their sensitivity to quenching effects. The negative staining procedure is carried out as before starting with the five washing steps on drops of distilled water.

Figure 4 shows human neutrophil granulocytes infected with *Staphylococcus aureus*. After stimulation, the cells flatten down (Fig. 4A) and release neutrophil extracellular traps (NETs; Brinkmann *et al.*, 2004). The fine structure can be studied by negative contrast in TEM (Fig. 4C and D).

B. Resin Embedding of Bacteria-Infected Cell Cultures

For resin embedding, the microbe to be analyzed directs the choice of the resin. Bacteria with a very hydrophobic cell wall, like mycobacteria, cannot successfully be embedded with standard resins like the Epon derivative Polybed (Fig. 6A). We compared the ultra structure of different bacteria embedded into different resins as well as ultra-thin cryosections.

1. Fixing Specimens

For the analysis of infection at the cellular level, in most cases we fix by adding the desired amount of fixative stock solution into the warm tissue culture medium. This ensures that no alterations of the cell culture due to media changes are introduced. Bacteria on the surface of the cells are not washed away, but will be kept in place even if they are not firmly attached to the cell surface. We then keep cell culture plates at RT for two hours after which period even pathogenic bacteria can be processed in a regular S1 lab.

To maintain physical integrity of the cells, we do not scrape them off the plate after fixation. This induces membrane breaks, distortion of plasma fine structure and regularly leads to loss of the basal domains of the cells. Especially analysis of

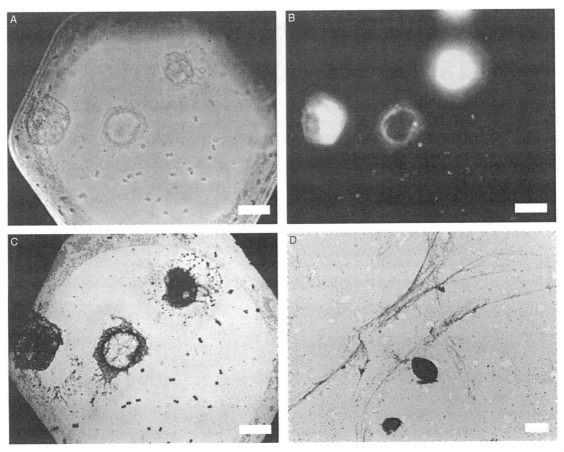

Fig. 4 Correlation of whole mounts of cells cultivated on filmed grids. Human neutrophils and *Staphylococcus aureus* on a film-coated EM-grid, (A, B) light microscopic images of fixed cells and bacteria, detection of DNA with SYTOX Green, (C) negative contrast TEM image of the same mesh, (D) negative stain image of *S. aureus* and neutrophil extracellular traps (Brinkmann *et al.*, 2004) from the same experiment. Scale bars: (A–C) 10 μm, (D) 1 μm.

the membrane integrity of i.e., phagosomes is severely aggravated if scraped cells are used. Instead, we leave the cells on the plate during postfixation, contrasting, and dehydration. We usually use styrene as intermediate step before embedding into resin. Styrene dissolves the tissue culture plates, so if timed correctly, the cells become detached without physical damage. We use a shaker to move the styrene over the cells. The time necessary to detach the cells varies with the cell type between 1 and 10 min; confluent epithelial layers remain on the plastic for a long time and then detach as multicellular aggregates. It can be useful to cut an X into the layer with a scalpel before styrene is added to allow a faster detachment. If the styrene is on the plates for too long, the bottom of the plate dissolves into a smear that contaminates the cell preparation. Should this happen, the resulting suspension should be transferred to a 15 ml bluecap that is filled with styrene. After vortexing and centrifugation, the cells separate from the dissolved plastic and can be pelleted. The plastic-contaminated supernatant is removed carefully, and the sediment is then resuspended in fresh styrene and embedded using styrene-resin mixtures. A 1:2 mixture is left in an open cylindrical glass vessel overnight to allow complete infiltration and increasing resin concentration due to partial evaporation of the styrene. The next morning, three changes using undiluted resin finish the embedding. With increasing resin concentration, the cells do not sediment spontaneously. Thus, to ensure that during replacement the supernatant no cellular material gets lost, the specimens are transferred to Eppendorf® tubes and centrifuged for about 10 min in a swing out rotor at about $350 \times g$. With the resin changes, the cells become resuspended. After the final change, the cells are sedimented, a paper tag with the specimen identification is inserted into the cap, and the specimens are polymerized for 1–2 days. For cultured cells, we prefer this method to flat embedding, since the concentration of cells remains high, while during flat embedding, cell clusters tend to float apart thus reducing the number of cells per section.

2. Staining of Semithin Sections for Light Microscopic Overviews

After polymerization, semithin (200–500 μm) sections are prepared with a histology diamond knife. Transfer them to a drop of distilled water on a microscope slide with an eyelash or a grid, and let the drop dry on a hot plate at about 80 °C. The sections are then stained with filtered solution of 1% toluidine blue in 1% borax on the hot plate. The drop of staining solution remains on the sections until it starts to dry out at the edges. Before crystals are formed, the slide is washed in tap water and transferred to distilled water until no further stain washes out and the water remains clear. The slide can then be mounted and analyzed with a light microscope. If necessary, the block can be retrimmed to the area of highest interest, before ultra-thin sections are prepared. Figure 5 shows the comparison of light- and electron microscopic images of macrophages infected with mycobacteria.

For fine structure analysis, we routinely use uranyl acetate block staining, so ultra-thin sections normally carry enough contrast to be photographed with a digital camera, which sets the boundaries of the image histogram in a way that areas of highest brightness will be white while pixels of lowest brightness will be black. Thus, a high-contrast image is created, which can be evaluated without further processing. Nonetheless, for print quality images further contrasting of the section (e.g., with lead citrate) is indispensable to avoid the introduction of background noise that is created by stretching the histogram boundaries on a weakly contrasted image. Furthermore it is uncomfortable to evaluate a weakly or noncontrasted section on the TEM screen.

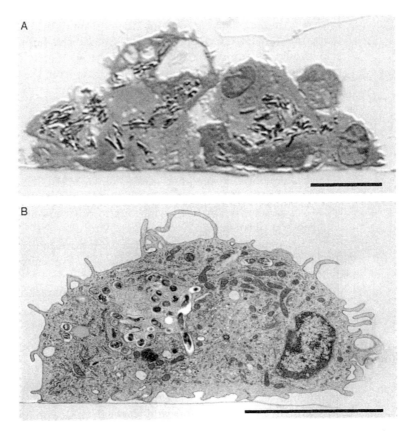

Fig. 5 Comparison of light microscopy of semi-thin section and low power TEM. The images show macrophages infected with mycobacteria. The semi-thin section (A) was stained with toluidine blue and shows a lot of features visible in detail in the TEM micrograph (B). Scale bars: 5 μm.

3. Comparison of Different Embedding Media

Polybed 812 has displaced classical Epon 812 and is now the probably most widely used resin. Completed by NMA (Nadic Methyl Anhydride) as epoxy resin, DDSA (Dodecenylsuccinic Anhydride) as hardener, and DMP-30 (2,4,6 Tris (dimethylaminomethyl)phenol as accelerator it gives hard blocks, easy to section and stable under the electron beam even under high emission current (~35 μA). Of course, the ratio of components can be varied to alter the hardness of the blocks. Although Polybed 812 has many advantages, it has limited value for preparations of microorganisms with waxy cell walls (mycobacteria, Fig. 6A) or for high density organisms like *Staphylococcus* (Fig. 6E) or for elementary bodies of *Chlamydia* (Fig. 6M). Polybed will not penetrate the particles sufficiently; the resulting sections are instable and may even have holes. For embedding of dense material, the use of media with reduced viscosity is advisable.

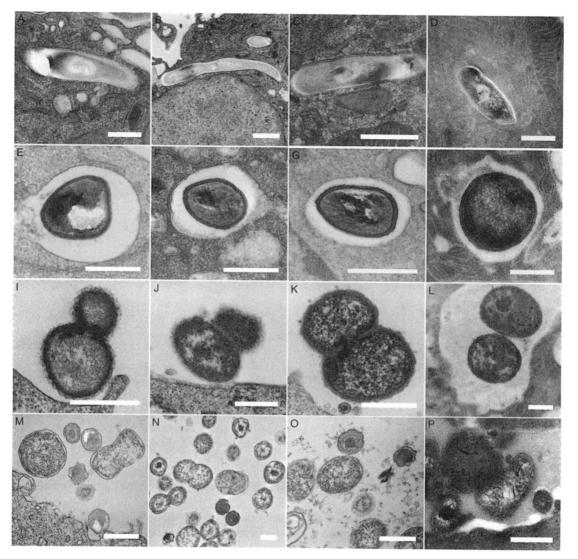

Fig. 6 Comparison of different embedding media. The widely used Polybed 812 will not penetrate efficiently into mycobacteria and staphylococci resulting in holey unstable sections (A, E). More reliable results are obtained with low-viscosity media like Spurr's (B, F) and Embed-It (C, G). Due to reduced lipid extraction, cryopreparations can reveal more details (D, H). (columns from left to right: Polybed, Spurr's, Embed-It, Cryopreparation). A–D: Mycobacteria (BCG), E–H: *Stapylococcus aureus*, I–L: *Neisseria meningitides*, M–P: *Chlamydia trachomatis*. Scale bars: 500 nm.

"Spurr's" low-viscosity embedding medium was developed in the late 1960s primarily to meet the needs of botanists for a low-viscosity resin that would more easily penetrate the cell walls. It consists of four components: Based on the cycloaliphatic diepoxide vinylcyclohexene oxide (VCD), it uses the epoxy resin D.E.R. 736 (diglycidyl ether of polypropylene glycol) as flexibilizer, in addition to NSA and accelerator dimethylaminoethanol (DMAE). The single components allow to specify the properties of the cured blocks depending on the respective purpose, like rapid polymerization or desired block hardness. "Spurr's" viscosity of 60 cps leads to a better and quicker penetration of hydrophobic material, such as cell walls of Gram positive bacteria (Fig. 6F) and very dense structures such as spores, elementary bodies of *Chlamydiae* (Fig. 6N), and yeast cell walls. Its liquidity makes it easy to work with. Specimens sediment quickly, so less material is lost during resin exchange. Spurr resin unfortunately has less than ideal characteristics for ultrathin sectioning. It tends to stick to the cutting edge of the diamond knife and produces folded aggregates of sectioned materials. To resolve this problem antistatic devices can be used during cutting to obtain a consistent and steadily produced ribbon of sections. Another disadvantage is the instability of Spurr resin in the electron beam. It rips very quickly on exposure to the beam and is even more sensitive after lead citrate staining. To overcome this problem, the sections can be relatively stabilised by starting observations at low magnification ($1000\times$) and low emission current (ca. 15 μA).

Embed-It™ (Polysciences, Inc.) was created as an easy-to-mix polymer of low-viscosity (65 cps) and also to avoid inconsistency between blocks. The "Spurr's" derivative consists of two nonhazardous components, its viscosity is similar to Spurr's, while the stability of the sections in the TEM is comparable to Polybed.

If the Polybed mixture does not penetrate samples sufficiently, it may help to prolong incubation times before changing to a different resin. A last incubation step overnight in 100% resin will produce better results, as will additional steps using different intermediate/resin ratios (3:1, 2:1, 1:1, 1:1.5, 1:2, 1:3). Changing the ratios too quickly may cause excessive shrinking and lead to holes in the specimen at, e.g., the space of a phagosome, between the phagosomal membrane and enclosed bacteria.

The comparison of the different resins shows that although all three are suitable for gram negative bacteria (e.g. *Neisseria*, Fig. 6I–K), Polybed is recommended because of easier section handling. Besides that, it is less toxic than VCD. Very dense structures such as elementary bodies of *Chlamydiae*, spores, gram positive bacteria (e.g., *Staphylococcus*) are better and quicker penetrated by Spurr's resin, although even Polybed works sufficiently well in most cases. Polybed embedding is however completely unsuitable for highly hydrophobic structures such as the waxy cell walls of mycobacteria (Fig. 6A). These structures need resins like Spurr's (Fig. 6B and F) or Embed-It (Fig. 6C and G). The latter is easier to handle than Spurr's and is stable under the electron beam; however it cannot totally replace Spurr's resin, since it needs longer incubation times due to its higher viscosity. Spurr's is definitely the resin of choice for quick embeddings of dense or hydrophobic structures. It is noticeable that Spurr's produces a slightly weaker contrast than Polybed or Embed-It sections. Contrast can be improved by longer incubation times of lead citrate but this may produce artefacts due to lead precipitation.

This is the standard embedding protocol for glutaraldehyde-fixed tissue culture specimens:

10′	three washings with PBS
60′	0.5% osmium tetroxide in distilled water
10′	4 washings with distilled water
60′	0.1% tannin in 20 mM HEPES buffer
10′	four washings with 1% Na_2SO_4 in 20 mM HEPES
10′	four washings with distilled water
60′	2% uranyl acetate in distilled water
5′	each in 30/50/70/80/90% ethanol
3 × 5′	100% ethanol
30′	styrene, over night 1:2 styrene/resin
60′	1:3 styrene/resin
3 × 60′	freshly prepared resin over night resin 1–2 days embedding at 60 °C (don't exceed 24 h for "Spurr's")

C. Resin Embedding of Infected Tissue

For embedding of infected tissue, it is essential to keep the size of the tissue specimen small (less than 2 mm × 2 mm × 2 mm). It is advisable to select the areas of an infected organ that will be embedded with a stereo microscope and only dissect parts than promise to be highly interesting, for the EM analysis does not allow screening of larger tissue sections. Because fixative and solvent penetration takes place over considerably larger distances in tissue, incubation times should be at least five times longer than for cell culture specimens.

It is possible to reanalyze tissue that has been embedded into paraffin for routine histology. Of course, fixation in formalin as well as heating to 65 °C are detrimental for tissue fine structure, but in some instances the reembedding into resin and subsequent EM analysis can be helpful. An advantage of this method is that particularly interesting tissue areas can be identified in sections using histological staining or light-microscopy immunodetection methods. Small cubes including these areas are then cut out of the paraffin block using a scalpel, rehydrated slowly, postfixed with glutaraldehyde and osmium tetroxide and reembedded using the standard embedding protocol. Care must be taken to remove the paraffin wax completely (Gonzalez-Angulo *et al.*, 1978), otherwise penetration with the resin will not be successful.

After embedding, tissue blocks are transferred to cavities of flat embedding forms, laser printed identification tags are inserted and the cavity is filled with fresh resin. The sample is aligned in parallel to the tip of the cavity to ensure quick trimming and the option to section the entire tissue block surface.

D. Rapid Processing for TEM

It can be desirable to shorten the preparation of TEM specimens, e.g., of diagnostic samples. Resin embedding using microwave-assisted tissue processing reduces embedding time down to 4–5 h (Schroeder *et al.*, 2006); alternatively cryomethods can be used which can provide better ultra structure (compare Fig. 6I–K to Fig. 6L) since less material gets extracted compared to treatment with organic solvents (Korn and Weisman, 1966).

E. Preembedding Immunodetection Methods

Conventional resin techniques are compatible with immunodetection if the antibodies are employed before embedding. If the antigens of interest are extracellular, antibodies can be incubated with living cells on ice (to limit internalization of antibody complexes) or with cells fixed with formaldehyde (2–4% in PBS) for 30–120 min. Cells are then washed (living cells with ice-cold PBS, and then fixed with PFA), before incubation with the secondary antibody which is coupled to gold colloids (30–60 min at RT). After washing, cells are postfixed using 2.5% glutaraldehyde in PBS, and embedded conventionally. This method results in good ultrastructure combined with excellent detection of surface antigens.

Intracellular antigens are only accessible after limited permeabilization. The choice and concentration of the detergent depends on the localization of the antigen, i.e., how many membrane systems have to be crossed by the antibody before it can bind to its antigen. Our first choice is saponin at concentrations around 0.1%. Owing to the permeabilization, loss of ultrastructure cannot be avoided. To assure good penetration, antibody fragments (Fab$_2$) should be employed, and the diameter of the gold should be less than 10 nm. Alternatively, ultra small gold (1 nm diameter or less) can be used which has to be silver-intensified before embedding (Danscher, 1981). In this case, osmium tetroxide should be omitted. Quantum Dots can be used in a similar way as ultra small gold probes. They offer the advantage that the staining can be analyzed on the light microscopic level, before the Quantum Dots are silver enhanced and processed for electron microscopy (Stoltenberg *et al.*, 2007).

F. Immunodetection Using Ultrathin Cryosections

For more than thirty years, ultra-thin sections of frozen material have been used for immunodetection (Tokuyasu, 1973, 1978). Although in the meantime antibody labeling methods with specialized resin techniques have been developed, cryosections are still widely used since the antigen preservation in weakly fixed frozen material is superior to resin techniques. Furthermore, extraction of cytoplasmic material is reduced. The structural preservation in ultra-thin cryosections is often of poorer quality than in resin sections mainly due to the softer fixation methods employed.

Structural preservation and antigen reactivity have to be balanced to obtain optimal results. The contrast in TEM images of cryosections is weaker and less differentiated than in osmicated samples of resin sections. Although fine-structure resolution in resin sections is often superior, EM-sample preparation by cryosectioning takes less time. Also, problems with resin polymerization in bacteria that show a particularly resistant cell wall will not come up under cryosectioning conditions.

Samples of infected cells for cryosectioning can be prepared from adherent cells grown in 6-well culture plates. One sample should consist of at least one confluent well, or preferably two wells ($>10^6$ cells). It is important to achieve a high infection rate in order to obtain enough representative cross sections of pathogens in the resulting ultra-thin sections. Cell density and infection can be monitored with an inverted light microscope using phase contrast illumination.

Fixation of the infected cells should take place at incubation temperature to prevent artefacts due to temperature shock. Fixation in PBS or in the growth medium works well in most cases. In some cases further improvement in structural preservation is observed when fixing the cells in cytoskeleton stabilizing buffer (1 mM EGTA, 4% polyethylene glycol 6000 (8000), 100 mM PIPES pH 6.9, (Lindroth et al., 1992)). A combination of 2–4% PFA with 0.05% glutaraldehyde, applied for two hours is a good starting point for many antigens and results in acceptable structure preservation combined with sufficient antigen reactivity. This is discussed in detail in the classic book by Griffith (1993).

Infected cells are harvested with a rubber scraper and suspended in 10% gelatine in PBS at 37 °C and immediately centrifuged to sediment as a pellet. This is excised from the tube after cooling and gelatinizing and cut into smaller pieces fit for infiltration in sucrose solution (2.3 M sucrose in sodium phosphate buffer with 1% PFA, pH 7.4) After infiltrating preferably in a cold room and on a slow inverting shaker overnight the gelatin bits can be mounted on aluminium stubs and frozen in liquid nitrogen for cryosectioning. Sectioning is carried out at −120 °C. Sections are transferred to carbon-coated pioloform films on copper or nickel EM-grids and kept section side down on 2% gelatine until immunolabeling and contrasting. For a detailed description of cryosectioning techniques see Chapter 8, this volume.

A basic immunogold labeling of ultra-thin cryosections is carried out as follows: After blocking the sections by floating the grids on drops of blocking buffer (1% BSA, 0.02 M glycine, 10% cold water fish gelatine in PBS) for at least 30 min. Primary antibodies are applied at 1–10 µg/ml diluted in blocking buffer. After incubating for 1 h at 37 °C or over night at 4 °C in a humid chamber, excess antibody is removed by washing 6 × 3 min on drops of PBS. Secondary antibody coated gold colloids are applied for 30 min–1 h at 37 °C diluted in blocking buffer. The labeled grids are washed 6× in PBS and 5× in distilled water prior to negative contrasting and embedding in methyl cellulose. For this, the sections are floated on three consecutive drops of 0.2% uranyl acetate and 2% methyl cellulose in distilled water for 2.5 min each before blotting off excess contrasting solution and air drying the grids.

Again, this technique can be correlated with light microscopy. Ribbons of semithin sections can be transferred to glass coverslips together with the pick-up solution. After washing the coverslip on three consecutive drops of PBS, the sections can be stained with a DNA dye and mounted for quick viewing with a fluorescence microscope or used for immunofluorescence staining. The comparison between immunogold labeling and immunofluorescence can help to secure staining specificity if the antigen is only present in small amounts. While intense immunogold staining that correlates with morphology (Fig. 7) is easy to interpret, specificity of a weak staining especially of cytoplasmic antigens can more reliably assessed if correlated with immunofluorescence.

G. Scanning Electron Microscopy

1. Scanning Electron Microscopy of Cultured Cells

For Scanning Electron Microscopy (SEM) analysis, we routinely grow cells on round coverslips (dia. 13 mm). Cells are fixed by adding the fixative stock solution to the medium at 37 °C. If no immunodetection is intended, we use glutaraldehyde at a final concentration of 2.5% for 120 min at room temperature. The fixative is directly added to the cell cultures to avoid loss of material or changes in the pathogen/host interplay. Thus, even fragile structures like NETs (see Fig. 8) can be sufficiently stabilized to survive the consecutive preparation steps.

After several washes with H_2O, the specimens are postfixed for 30 min with OsO_4 to stabilize membranes. Cells are washed repeatedly with H_2O, and then transferred for 30 min to a buffered solution of tannic acid (0.5% in 20 mM HEPES). Repeated treatment with osmium and tannic acid will deposit a

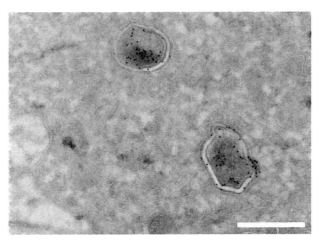

Fig. 7 Detection of mycobacterial antigens on ultrathin cryosections of infected macrophages. The indirect immunogold method using 12 nm gold colloids reveals that mycobacterial antigens are exclusively located inside the phagosome. Scale bar: 500 nm.

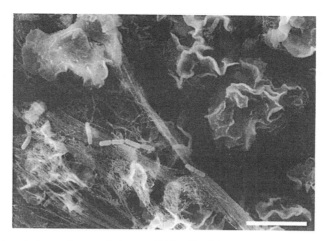

Fig. 8 Scanning electron microscopic image of *Shigella*—infected human neutrophil granulocytes with neutrophil extracellular traps (NETs). By adding fixative directly into the culture without exchanging media, even fragile structures like NETs as well as bacteria only lightly adhering to cells surfaces can be retained. Scale bar: 5 μm.

conductive layer on the cells which helps minimizing surface charging to a degree that no additional metal layer will be necessary if the specimens are to be analyzed at low acceleration voltage (<1 kV). After the last treatment with osmium, cells are dehydrated using a graded ethanol series. We use molecular sieves to keep the 100%-ethanol water free which requires the filtration of the dry ethanol to remove debris of the molecular sieves. After three dehydration steps using 100% ethanol, cells are transferred to a critical point drying apparatus equipped with a holder for the coverslips, and dried. Since we regularly examine specimens at 20 kV, we coat the surface of the dry cells with 3–5 nm of platinum/carbon. This results in a coating with a finer grain than sputtering them with gold or gold–palladium.

2. Immunodetection Using SEM

For detection of surface antigens, we fix the cells by adding stock solutions of PFA to the warm medium to a final concentration of 4%. After 30 min at room temperature, the specimens are washed with PBS and blocked using 1% BSA in PBS. Primary antibodies are diluted in the same buffer at a concentration between 1 and 10 μg. The coverslips with the cells facing the bottom is placed on 50 μl drops of the antibody solution on parafilm in a humid chamber and incubated at 37 °C for 30–60 min. After repeated washing, the secondary antibody coupled to gold colloids is applied correspondingly. Although the labeling density is lower, we prefer medium-sized gold particles (12–15 nm), because they are more readily identified in the SEM than smaller particles (6 nm or less).

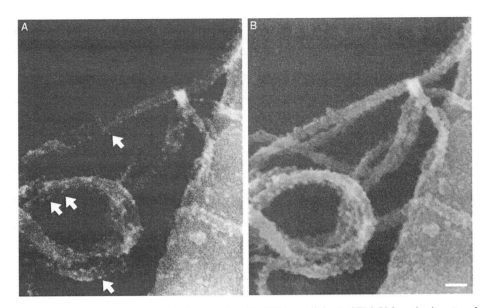

Fig. 9 Immunogold detection of antigens on flagella of *Salmonella* in the SEM. Using a backscattered electron detector, 12 nm gold colloids can be clearly identified by their element contrast (A). In the surface topographic image of the secondary electron detector (B), gold colloids cannot be differentiated from bacterial details of the same diameter. Scale bars: 100 nm.

We take material contrast views with the backscattered electron detector (BSE) (Fig. 9A) and topographic images with the secondary electron (SE) detector (Fig. 9B). Even when the specimens are coated with a layer of 2–3 nm platinum/carbon, the gold signal is clearly visible with the BSE detector. Using individual images from both type of detectors normally allows easier identification of gold signal than images created with mixed signal from both detectors at the same time. For presenting both images in an overlay, SE and BSE micrographs can be pseudocolorized using different look-up tables, (e.g., green and red) (Brinkmann *et al.*, 2004).

III. Overview and Conclusion

In this article we have presented some standard preparations for both TEM and SEM, which will be helpful to infection biologists who would like to get a more precise view on the process of infection. Yet certain caveats have to be kept in mind. Compared with that in light microscopy, in electron microscopy the number of individual cells in a given sample is normally much smaller. Great care has to be taken to assure that an adequate number of individual cells are analyzed to represent the entire sample; thus, it is of great importance to elaborate infection parameters that ensure homogenous infection patterns in cell culture or organ samples. It is a good idea to correlate light and electron microscopy to be able to monitor the infection status of the entire sample before concentrating on fine structure analysis of a relatively small number of cells.

If knock-down techniques are used to repress expression of a certain gene, it is necessary to carefully monitor the degree of gene silencing. While primary cells or organs from mice with properly induced KO germline mutations will surely not express the protein under investigation, silencing genes with RNA techniques never repress protein expression entirely. Often, 20–40% of residual expression remains, which may be enough to yield a readout using light microscopic techniques, but is not satisfactory for electron microscopy, which concentrates on a limited number of individual cells. Thus, for EM studies RNA-induced gene silencing is only adequate if a high repression rate (>90%) is accomplished, and the number of analyzed cells is big enough to be statistically significant.

For a researcher new to transmission electron microscopy it is often dissatisfactory that the number of infected cells in a given sample seems to be much smaller than in a corresponding sample processed for light microscopy. It has to be kept in mind that while light microscopy always displays the entire cell and even a single bacterium infecting this cell can easily be visualized, an ultra-thin section of 60 nm only represents about 0.5–1% of the cell, and thus many cell profiles erroneously appear uninfected. This again is an argument for correlative studies.

The multiple facets of electron microscopy offer indispensable tools for studying infection processes. We would like to encourage more biologists to use electron microscopy in their infection models to gain a more precise insight into the biology of infection.

References

Al-Amoudi, A., Chang, J. J., Leforestier, A., McDowall, A., Salamin, L. M., Norlen, L. P. O., Richter, K., Blanc, N. S., Studer, D., and Dubochet, J. (2004). Cryo-electron microscopy of vitreous sections. *EMBO J.* **23**, 3583–3588.

Brinkmann, V., Reichard, U., Goosmann, C., Fauler, B., Uhlemann, Y., Weiss, D. S., Weinrauch, Y., and Zychlinsky, A. (2004). Neutrophil extracellular traps kill bacteria. *Science* **303**, 1532–1535.

Danscher, G (1981). Histochemical demonstration of heavy-metals—A revised version of the sulfide silver method suitable for both light and electron-microscopy. *Histochemistry* **71**, 1–16.

Dubochet, J., Ducommun, M., Zollinger, M., and Kellenberger, E. (1971). A new preparation method for dark-field electron microscopy of biomacromolecules. *J. Ultrastruct. Res.* **35**, 147–167.

Gonzalez-Angulo, A., Ruiz de Chavez, I., and Castaneda, M. (2003). A reliable method for electron microscopic examination of specific areas from paraffin-embedded tissue mounted on glass slides. *Am. J. Clin. Pathol.* **70**, 697–699.

Griffith, G. (1993). *In* "Fine structure immuno-cytochemistry", pp. 26–45. Springer, Berlin.

Harris, J. R., and Horne, R. W. (1994). Negative staining—a brief assessment of current technical benefits, limitations and future possibilities. *Micron* **25**, 5–13.

Hayat, M. A. (1981). Fixation for Electron Microscopy Academic Press, London.

Korn, E. D., and Weisman, R. A. (1966). I. Loss of lipids during preparation of amoebae for electron microscopy. *Biochim. Biophys. Acta* **116**, 309.

Lindroth, M., Bell, P. B. J., Frederiksson, B.-A., and Liu, X.-D. (1992). Preservation and visualization of molecular structure in detergent-extracted whole mounts of cultured cells. *Microsc. Res. Tech.* **22,** 130–150.

Michalet, X., Pinaud, F. F., Bentolila, L. A., Tsay, J. M., Doose, S., Li, J. J., Sundaresan, G., Wu, A. M., Gambhir, S. S., and Weiss, S. (2005). Quantum dots for live cells, *in vivo* imaging, and diagnostics. *Science* **307,** 538–544.

Pontefract, R. D., and Bergeron, G. (1981). A simple technique for the removal of formvar from carbon-formvar coated grids. *Stain Technol.* **56,** 51–53.

Schroeder, J. A., Gelderblom, H. R., Hauroeder, B., Schmetz, C., Milios, J., and Hofstaedter, F. (2006). Microwave-assisted tissue processing for same-day EM-diagnosis of potential bioterrorism and clinical samples. *Micron* **37,** 577–590.

Slot, J. W., and Geuze, H. J. (1981). Sizing of protein A-colloidal gold probes for immunoelecton microcopy. *Journal of Cell Biology* **90,** 533–536.

Stoltenberg, M., Larsen, A., Doering, P., Sadauskas, E., Locht, L. J., and Danscher, G. (2007). Autometallographic tracing of quantum dots. *Histol. Histopathol.* **22,** 617–622.

Tokuyasu, K. T. (1973). Technique for ultracryotomy of cell suspensions and tissues. *J. Cell Biol.* **57,** 551–565.

Tokuyasu, K. T. (1978). Study of positive staining of ultrathin frozen sections. *J. Ultrastruct. Res.* **63,** 287–307.

Yao, J., Larson, D. R., Vishwasrao, H. D., Zipfel, W. R., and Webb, W. W. (2005). Blinking and nonradiant dark fraction of water-soluble quantum dots in aqueous solution. *Proc. Natl. Acad. Sci. USA* **102,** 14284–14289.

CHAPTER 25

Electron Microscopy of Viruses and Virus-Cell Interactions

Peter Wild

Electron Microscopy
Institutes of Veterinary Anatomy and of Virology
University of Zürich
CH-8057 Zürich
Switzerland

Abstract
I. Introduction
II. Methods
    A. Isolated Virus Particles
    B. Cell Cultures
III. Material
IV. Discussion
V. Summary
    References

Abstract

Electron microscopy is a powerful tool to visualize viruses in diagnostic as well as in research settings for investigating viral structure and virus-cell interactions. Here, a simple but efficient method is described for demonstrating viruses by negative staining, and its limit is discussed. A prerequisite to obtain reliable information on virus-cell interactions is excellent preservation of cellular and viral ultrastructure. The crux is that during fixation and embedding, by applying conventional protocols about 50% of the lipids are lost, which results in loss of integrity of cell membranes. To achieve good preservation of cellular architectures, good contrast, and both high spatial and temporal resolution, methods for freezing, freeze-substitution, and freeze-etching are described and their applicability discussed mostly taking complicated built herpes viruses as examples.

I. Introduction

Electron microscopy is an efficient tool to demonstrate viruses for diagnostic purposes, for structural analysis, and for studying virus-cell interaction. In diagnostic virology, electron microscopy employing negative staining is often the first step yielding results within minutes (Figs. 1 and 2). If electron microscopy reveals viruses, selective investigations can be initiated for characterization of the virus detected. Rapidity is important in the diagnosis of emerging infectious agents in diseased men and animals but even more in discrimination between harmful and

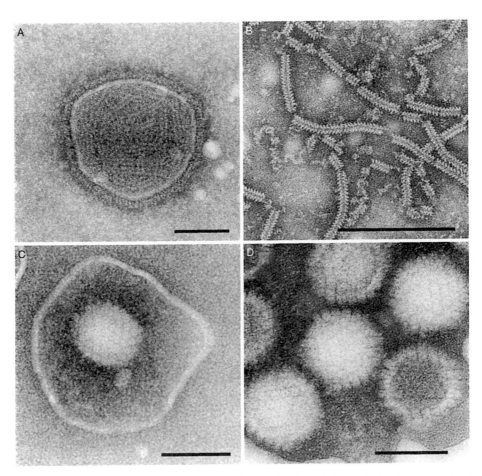

Fig. 1 (A): Intact Sendai virus (paramyxovirus). (B): Capsids of parainfluenza 3 virus (paramyxovirus). (C): Intact herpes simplex virus 1 exhibiting envelope, tegument and capsid with capsomers. (D): Capsids of herpes simplex virus 1 with and without (black) DNA. All are stained with PTA. Images by E.M. Schraner. Bars = 100 nm.

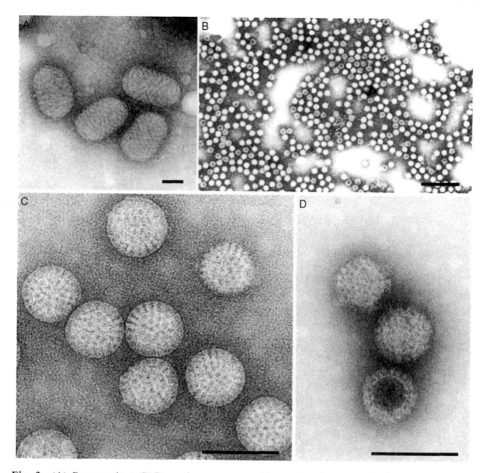

Fig. 2 (A): Parapoxvirus. (B) Parvovirus concentrated by coating the supporting film with antibodies prior to adsorption of virus. (C): Rotavirus (reovirus) stained with uranyl acetate. D: Rotavirus of the same suspension as in panel C stained with PTA. Images by E.M. Schraner. Bars = 100 nm.

harmless particles in criminality. Negative staining is also used for control of preparation steps in virus research. Negative staining may lead to severe artifacts that depend, to some extent, on the negative stain used. Another technique to visualize viruses is shadowing with heavy metals like platinum. By doing so, artifacts due to negative stains are omitted but not collapsing of structures unless viruses are frozen (and freeze-dried) prior to shadowing. Artifacts can be omitted by embedding viruses in vitreous ice that involves rapid freezing of thin samples and examination in the frozen hydrated state. This technology is highly suitable for structural analysis.

Viruses can also be detected within tissue, isolated cells, or cell cultures for diagnostic purposes or for research. For simple detection of viruses, preparation

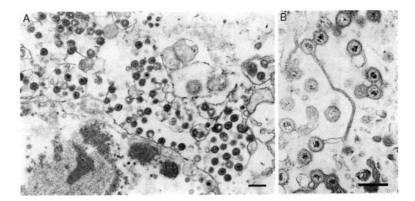

Fig. 3 Herpes virions within nucleus and cytoplasm of poorly preserved cells. Most of the membranes probably represent Golgi membranes. Images by F. Mettler. Bars = 500 nm.

protocols for routine electron microscopy can be employed because most viral structures can be easily recognized even in cells with pure preservation of the ultrastructure. Some virus structures can even be identified to some extent, e.g. herpes virus (Fig. 3), adenovirus, and papilloma virus. To study viral behavior, it must be kept in mind that cellular structures like cell membranes may be severely disturbed while applying routine techniques involving immersion into fixatives, dehydration, and embedding. The organelle to be disrupted first and most severely is the Golgi complex (Fig. 3), which is heavily involved in intracellular trafficking but also in maturation of viruses. Furthermore, processes may be fast requiring high temporal resolution. Therefore, the method of choice to study virus-cell interactions is rapid freezing followed by freeze substitution (Fig. 4) or freeze-etching, and final preparation for transmission electron microscopy and (cryo)-scanning electron microscopy, respectively.

To demonstrate isolated virus particles, a simple protocol is described. To achieve reliable structural preservation to study cellular responses to viral infection, protocols for freezing, freeze-substitution, and freeze-etching of cell monolayers and isolated cells are presented.

II. Methods

A. Isolated Virus Particles

1. Negative Staining

Negative staining is simple, fast, and effective. It needs a supporting film on copper grids and negative stains. Among those stains, uranyl acetate (UAc) and Na-phosphotungstic acid (PTA) are most widely used. Uranyl vanadate, uranyl formiate, uranyl oxalate, ammonium molybdate, and others may be used for

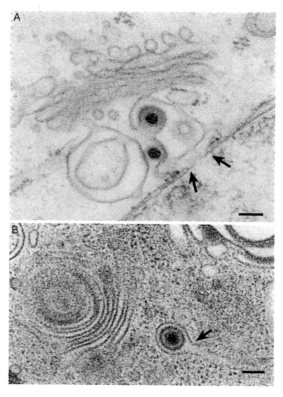

Fig. 4 Golgi complex after high-pressure freezing and freeze-substitution (A) and after pre-fixation with 0.25% glutaraldehyde followed by freezing and substitution (B). (A) Budding of bovine herpes virus 1 capsids at the *cis* face of an intact Golgi complex. Note the connection between outer nuclear membrane and Golgi membrane and the dilated (180 nm) nuclear pore with intact bordering by intact membranes (arrows). (B) Membranes have distinct contrast and are well preserved. An intact herpes simplex virus 1 is in a Golgi cisterna that show indications for fission (arrow). Bars = 200 nm.

special occasions. Viruses can be brought to the supporting film by spraying, by placing a drop onto it, or by floating the grid on a drop of virus suspension. A highly sophisticated method to prepare viruses for negative staining is by agar filtration (Kellenberger and Bitterli, 1976). Apart from agar filtration, the floating method is probably the most effective concerning the number of virions that adheres to the supporting film because virions accumulate at the surface of the drop owing to the surface tension (Johnson and Gregory, 1993), and thus will be described in detail.

a. Negative Staining of Viruses

1. Place a drop (20 to 50 μl) of suspension onto a Parafilm mounted on a smooth surface.
2. Place the grid covered with a supporting film upside-down onto the drop.

3. Allow adsorption of the virus to the supporting film for 3–15 min, depending on the concentration and purity of the suspension.
4. Optional: Wash quickly by placing the grid onto a drop (40 μl) of pure water.
5. Place the grid onto a drop (40 μl) of negative stain for 1–10 min.
6. Blot the grid from the side using a blotting paper (e.g. Whatmann No. 1) cut in triangles.
7. Examine the specimen immediately after blotting because the quality of negatively stained viruses reduces with time.

Important: DO NOT blot after adsorption or washing. You may lose viruses and/or staining quality will be poor. Grids with supporting film coated with carbon possibly need to be glow discharged immediately prior to use to make them hydrophilic.

b. Preparation of Supporting Films

The supporting film with good adsorption property is the basic prerequisite for successful negative staining. The supporting film is made of formvar, collodion, or parlodion. Formvar is more stable than collodion and parlodion. Collodion and parlodion films are easier to produce and are good for negative staining. All three films are sensitive to electron beam irradiation and thus need to be coated with carbon. Alternatively, pure carbon film can be used as for DNA spreading. For demonstration of viruses, we succeeded in using parlodion films prepared as follows:

1. Fill a Petri dish (16 cm in diameter, 4.5 cm in height) completely with filtered (0.22 μm) water.
2. Break away the tip of a Pasteur pipette prior to use it to place a drop of parlodion onto the water surface in the centre of the Petri dish. Avoid bubbles.
3. Parlodion will form a thin film – let it dry until Newton rings have disappeared.
4. Remove the first parlodion film in order to clean the water surface.
5. Produce a second film as described earlier.
6. Place carefully grids cleaned in acetone onto the parlodion film without touching it. Avoid vibrations.
7. Place a piece of parafilm onto the grids. Wait until it is completely attached to the film.
8. Remove the parlodion around the parafilm with clean forceps.
9. Remove the parafilm-grids-parlodion sandwich from the water surface.
10. Put this sandwich onto a blotting paper within a Petri dish for drying.
11. Coat films with 2–3 nm carbon prior to use.

Important: Dishes and tools must be clean and free of lipids. Use acetone or chloroform for cleaning. Do not touch anything with bare hands. This procedure can be repeated many times using the same water. Work on a quiet place under a small hood made of Plexiglas.

3% parlodion: 0.6 g of parlodion cleaned with chloroform are cut into small pieces and solved in 20 ml acetic acid-isoamylester. After parlodion is solved, which needs 2–3 days, molecular sieve is added.

Copper grids (300 mesh/inch) are cleaned twice in acetone for 15 min in an ultrasonic bath. If films do not stick to the grids, they need to be glow discharged immediately prior to use.

c. Na-Phosphotungstic Acid 2%

Solve 0.2 g Na-PTA in 8 ml·H_2O. Adjust pH with 1 N and 0.2 N NaOH to 6.6, 7.0, and 7.4, and add H_2O ad 10 ml. Filter the solution (0.22 μm) into Eppendorf tubes for storage at 4 °C.

d. Uranyl Acetate

Make a saturated solution using Milipore filtered (0.22 μm) water in an Eppendorf tube, centrifuge it, and keep it cool in a light-protected container, e.g., in a box for 35 mm films. For changing staining intensity, stain for different periods of time.

2. Shadowing

Adsorption of viruses to the supporting film is exactly as for negative staining. However, quick washing is often necessary unless the viral suspension is very pure. After blotting, the grids are transferred into an evaporation unit and shadowed preferentially with platinum-carbon at an angle between 45° and 85° depending on the virus to be visualized. Shadowing can be unidirectional or multidirectional by rotating the specimen. We often combine the two starting with rotary shadowing (80–90%) followed by unidirectional shadowing (10–20%). The total thickness of the platinum-carbon layer measured by the measuring quartz located beside the specimen but perpendicular to the platinum source may vary between 2.5 and 3 nm. Shadowing may be used to verify whether structures found in negatively stained preparations are artificially introduced by negative stains. We use shadowing more for demonstration of macromolecules of unknown structure or of pili of bacteria *in situ* rather than for demonstration of viruses, and in combination with freeze-drying.

To avoid collapsing of viruses during air-drying, viruses adsorbed onto the supporting film can be freeze-dried in the evaporation unit used for shadowing. Good freezing quality (no formation of ice crystals disturbing viral structures) is achieved by plunge-freezing as described in detail later for freezing of cell monolayers. The crucial step is blotting prior to freezing. Alternatively, freeze-drying of negatively stained viruses can be done within the electron microscope provided the microscope is equipped for it.

3. Embedding in Ice

Artificial changes can be almost completely avoided by plunge-freezing of thin films of viral suspensions (Adrian et al., 1984). Images can be used for three-dimensional reconstruction (Baker et al., 1999). Recently, structures of nonrotational virus particles were reconstructed on the basis of images obtained by tomography (Cyrklaff et al., 2005; Grunewald et al., 2003).

B. Cell Cultures

For transmission electron microscopy or scanning electron microscopy to examine the cell surface, cells such as MDBK cells, HeLa cells, or Vero cells, can be grown on sapphire disks with a diameter of 3 mm and a thickness of 50 µm (Bruegger, Minusio, Switzerland) to use it for fixation at ambient temperatures or for freezing by any methods discussed later except for high-pressure freezing using the freezing unit made by Leica, which allows only disks of 2 mm in diameter. Sapphire disks are covered with 10 nm carbon (Eppenberger-Eberhardt et al., 1997) permitting excellent cell growth and, very important, easy removal from the epon after embedding and polymerization. Furthermore, a "1" can be scratched into the carbon allowing recognition of that surface cells are attached to. It is important to clean the sapphires prior to carbon coating with a detergent for laboratory dish washing, e.g. neodisher LM3. Cleaning, best done in an ultrasonic bath, must be followed by rigorous rinsing in hot water and finally in acetone (twice in an ultrasonic bath). Clean and dry sapphire disks can be coated by carbon, e.g. by electron gun evaporation in a vacuum unit at a pressure of about 10^{-5} mbar. After carbon coating, sapphire disks should be stored in a dry place (30–40 °C) for at least 24 h to allow firm attachment of the carbon to the smooth sapphire surface. Omitting this step may result in detachment of the carbon film at the very moment disks are placed into the medium. The disks are UV irradiated prior to seeding. For seeding cells we use commonly 6 well plates because monolayers are more evenly distributed than in smaller wells, and so is the distribution of virus during attachment. Seeding must be done first. Disks are then inserted into the medium containing the cells to avoid floating of the disks.

For cryoscanning electron microscopy after freeze-etching, cells can be grown in petriperm dishes. The bottom of these dishes can be cut out and then mounted in a specimen holder of a high-pressure freezer. Alternatively, cells can be grown in any culture dish, and removed by trypsinization or scraping. Pellets obtained by slow centrifugation can be used for any method described later. We use them only for fixation at ambient temperature and for high-pressure freezing followed by freeze-etching.

An elegant method is to grow cells in cellulose capillary tubes of 200 µm in diameter (Hohenberg et al., 1994). Cells within these tubes can be fixed by conventional methods or cryo-fixed by high-pressure freezing followed by freeze-substitution or by cryo-sectioning. Alternatively, cells can be grown on Aclar

(Jimenez et al., 2006), a copolymer with similar properties as tissue culture plastic, or on golden grids covered by a formvar film and coated with carbon (Osborn et al., 1978). They can be used for conventional fixation or for high-pressure freezing (Murk et al., 2003). Because golden grids are soft they are difficult to handle for plunge-freezing. This problem can be overcome by using gold-coated copper grids.

1. Conventional Transmission Electron Microscopy

Cells grown on sapphire disks can be fixed simply by immersion into glutaraldehyde. We commonly use 2.5% glutaraldehyde in 0.1 M Na/K-phosphate because K^+ have a stabilizing effect on membranes (Kuhn and Wild, 1992). To reduce stress we let cells cool to room temperature and add 2.5% buffered glutaraldehyde to the medium in which cells were grown in giving a final concentration of 1.25%. Then the cells are kept at 4 °C for 30 min, briefly washed with 0.1 M Na/K-phosphate, postfixed with 1% OsO_4 in 0.1 M Na/K-phosphate at 4 °C for 1 h, dehydrated in graded series of ethanol, infiltrated in epon/acetone (1/1) and embedded in epon. We use rather acetone than propylene oxide because many structures, e.g. tegument and envelope of herpes viruses are better preserved when propylene oxide is omitted.

2. Low-Temperature Transmission Electron Microscopy

a. Freezing of Cells Grown as Monolayers

Cells grown as monolayers on sapphire disks are ideal objects for successful freezing because monolayers are thin, and easy to handle enabling freezing within seconds after removing from culture medium. Keeping the culture medium under incubation conditions, virtually no disturbance of cellular processes and of ultrastructure will be expected to occur in the short time needed to prepare cells for freezing. The thickness of monolayers of established cell lines probably does not exceed 20 µm. The thickness of Vero cells, HeLa cells, MDBK cells, or MDCK cells is less than 10 µm. Therefore, monolayers can be frozen by any freezing method.

b. Plunge Freezing

The simplest way to freeze cells is probably by plunging cells grown as monolayers on sapphire disk into a cryogen such as propane cooled by liquid nitrogen. The setup consisting of a guillotine (Adrian et al., 1984) and a container for the cryogen can be self made (Wild et al., 2001a). However, there are two problems to overcome. The first, severe problem is removing surplus medium from the monolayers after the sapphire disks are removed from the medium. This has to be done rapidly but carefully and effectively so that only minimal amounts but enough of medium covers the cell and that they do not suffer from osmotic shock or even dry out. Medium covering the monolayer surface thicker than a few micrometer hinders rapid freezing of the monolayer surface. Taking into account that the distance of excellent freezing is in the order of 10 µm, the ideal medium thickness would be 1–3 µm, an approach difficult to achieve and therefore needing a great deal of experience. FEI has developed the *Vitro-Bot* for freezing thin films of suspension for cryoelectron microscopy. This instrument can probably be used for freezing cell monolayers because the removal of medium immediately prior to freezing can be exactly controlled. This instrument also controls the temperature of the cryogen, the second, minor problem of plunge-freezing. The melting point of propane is at $-189.6\,°C$, that of ethane at $-183.5\,°C$. Heat conductivity of ethane is better than that of propane. Propane is easier to handle. We overcame the problem of keeping the cryogen in its liquid form by mixing ethane and propane (about 8:2) in a partially air insulated container (Fig. 5A). Constant stirring permits the mixture to cool down to about $-194\,°C$ without freezing. Propane (or ethane in the propane/ethane mixture) lost during plunging can be easily replaced via a hypodermic needle within a few seconds. For the first filling of the container by starting with ethane, a 200 µl pipette tip is used, whose tip is cut off obliquely.

The rate of successful freezing depends to a large extent on the experience of the operator, particularly on its ability to remove surplus medium, but also on the relative water content of the cells. Old (3 days) cells are richer in proteins than young cells. The rate of successful freezing is thus much higher than in young cells except for the nuclei, which are often not satisfactorily frozen. Although removal of surplus medium is crucial for successful freezing of monolayers, the region next

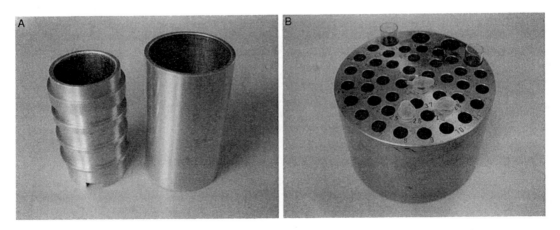

Fig. 5 (A) Partially insulated container to be filled with propane/ethane for plunge-freezing. For details see Wild *et al.* (2001a). (B) Block of stainless steel with holes with the size that 700 µl Eppendorf tubes fit exactly into it providing good heat conductivity. In the back there are four holes for glass containers for cooling the substitution medium prior to use. Note that there is a gradient of temperature of 20 °C at least from bottom (where the indicated temperature is measured to the top. Verify the temperature at the place where the specimen is located).

to the sapphire disk are mostly well frozen because the heat conductivity of the sapphire is high – and much better than that of the medium covering monolayers – permitting rapid freezing of the first few micrometers.

Plunge freezing can be employed for studying rapid membrane-bound processes as is expected during cell entry of viruses. For this purpose a humid chamber is build around the guillotine (Bailey et al., 1991) allowing incubation of infected cells for seconds prior to freezing (Wild et al., 1998). This methodology revealed for the first time exact data for viral cell entry (Figs. 8 and 9).

c. Metal Mirror Freezing

Freezing by slamming the specimen onto a highly polished copper block cooled by liquid nitrogen is not a good option to freeze cell monolayers because of difficulties to remove surplus medium although it has been successfully applied (Dalen et al., 1992). However, metal mirror freezing is an excellent tool to freeze bacteria (Wild et al., 1997). Bacterial colonies grown on Agar are cut out using a razor blade or a scalpel, placed on a cigarette paper that is mounted on a specimen holder for immediate slamming. Bacteria can be easily frozen because of their low water content. This is also true for viruses. Even when cells show severe segregation artifacts viral structure is commonly reasonably well preserved.

d. High-Pressure Freezing

With any method described earlier cell nuclei might not be well frozen or the rate of successful freezing is low. Under high pressure crystallization time is increased, allowing freezing of much thicker specimen than cell monolayers without formation of ice crystals perturbing cellular ultrastructure (Mueller, 1992) as described in chapter 11.

Commonly, cells or tissue to be frozen are placed in an aluminum platelet with a central cavity of 50–150 µm. Cells grown on sapphire disks can be placed on the flat side of an aluminum platelet and covered with an aluminum platelet with a 50 µl cavity keeping the amount of medium low. As mentioned earlier, heat conductivity of sapphire is high. 50 µm thick sapphire disks of 3 mm in diameter are mechanically stable as long as the impact forces are evenly distributed on their surface. Therefore, sapphire disks can be directly mounted into the specimen holder of a high-pressure freezer. In doing so, one disk is placed with its monolayer on the upper side, the second one is placed upside down. The two disks are separated by a thin copper spacer (one hole grid) giving a sandwich of about 100 µm with minimal but sufficient medium covering the cells. The sapphire disks remain intact during freezing under high pressure provided that all surfaces the sapphires are hold in place with are smooth. Cells are well frozen in more than 95% of the trials.

A crucial step is loading the specimen holder, insofar as care must be taken that cells do not dry. To avoid drying of cells, we quickly blot the sapphire disk after removal from the medium with a wet blotting paper prior to placing it into hexadecane (Studer et al., 1989). Hexadecane is not miscible with water. The little medium remaining on cells will be covered by hexadecane, and thus cannot evaporate that would result in osmotic shock or even in drying of cells. There are two other simple advantages using hexadecane. First, sapphires and spacers can be very easily handled for mounting onto the specimen holder when they are covered with hexadecane. Second, the more important, the two sapphires of a sandwich can be easily separated after cells have been frozen. Cells grown in cellulose tubes (Hohenberg et al., 1994) can rapidly mounted into a special holder with very short delay between removal from the culture medium and freezing.

e. Jet Freezing

A cheaper way of freezing cell monolayers than high-pressure freezing is by jet-freezing. The specimen is protected and kept in place by a thin specimen holder made of copper. Then liquid propane is propelled simultaneously from opposite sites onto the thin specimen holder. Mounting the sapphire sandwich, as described for high-pressure freezing, into the specimen holder allows rapid freezing of the monolayers because heat conductivity of both sapphire and copper is low.

f. Fixation prior to Freezing

Fixation with aldehydes results in distortion and shrinkage (Murk et al., 2003). Nevertheless, pre-fixation with low concentrations (0.1–0.5%) of glutaraldehyde may be a good option to study virus-cell interactions. This is especially true when dealing with pathogens allowing transportation of infected samples to the laboratory, and handling for freezing. The main distortions are not due to the fixative *per se*, but they are caused during dehydration. Hence, structural distortions due to fixatives are probably of minor importance in studying virus-cell interactions (Fig. 4B).

g. Freeze-Substitution

To visualize the interior of cells they must be cut into slices of 30–300 nm depending on the methods intended to use for investigating cellular architecture. Cutting a thin section requires embedding in an appropriate medium, which also allows visualization of cellular structure in great detail. Prior to embedding, cellular water must be removed and replaced by a solvent such as ethanol or acetone. Specimen fixed by conventional techniques, e.g. by immersion into glutaraldehyde, are transferred into solvents for dehydration through graded series of ethanol or acetone. Ice of a frozen specimen is substituted by solvents at temperatures just above the melting point. There are probably hundreds of protocols. The choice of the protocol depends on what structures are aimed to be visualized. To study cell membrane-bound processes, e.g., envelopment of herpes viruses, a protocol must be employed that leads to high resolution of membranes combined with appropriate contrast. Note that good contrast in biological specimen depends on the amount of heavy metal atoms bound to the structure to be visualized. Therefore, good contrast may be negatively correlated with good resolution.

To find a protocol yielding both good resolution and good contrast of membranes, we tested dozens of protocols (Wild *et al.*, 2001b) coming up with the following:

Substitution medium: acetone containing 0.25% glutaraldehyde and 0.5% osmium tetroxide.

1. Substitution: 6 h (or over night) at −90 °C that is followed by slow temperature rise (5°/h) to 0° or 2 °C. To improve contrast, cells can be kept at 2 °C for 1 or 2 h.
2. Washing: three short washings (5 min) in pure acetone at 4 °C.
3. Infiltration: epon/acetone (1/1) for 6 h at 4 °C, in open vials so that acetone can evaporate
4. Embedding: Place the sapphire disks onto a clean glass slide under a stereomicroscope to ensure that cells are on the upper side. For orientation use the '1.'

Fill completely a beam capsule with epon, and place it upside down onto the sapphire disk (Fig. 6). Epon does not leak, even at 60 °C.

5. Polymerization: at 60 °C for 2.5 days.
6. Removal of disks: Immersion in liquid nitrogen followed by mechanical forces.

Note: Sapphires break into small pieces; some of them may stick to the carbon or epon where the '1' was scratched.

Sapphire disks require a specimen holder so that the substitution medium can readily contact with the cells. We use 700 μl Eppendorf tubes. They fit exactly into holes drilled in a steel block (Fig. 5B) that is placed into the chamber of the substitution unit. Upon intimate contact of the Eppendorf tube, the wall and the bottom of the hole allows rapid heat exchange between the steel block and the

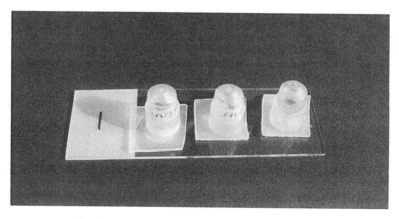

Fig. 6 Beam capsules filled with epon on a glass slide.

substitution medium. The temperature of the substitution medium at the bottom has to be measured to know the actual temperature at the very place of the specimen.

Important:

1. For good contrast combined with good resolution, the final temperature must be between 0 and +2 °C. NOTE: The temperature indicated in any freeze-substitution unit tested so far is not the actual temperature of the specimen because the temperature is measured below the location of the specimen. The actual temperature may differ from the indicated temperature by more than 12 °C!

2. Contrast may be enhanced by using higher concentrations of osmium tetroxide and/or increasing the temperature to 4 °C or even higher. Keep in mind that this kind of contrast enhancement results in decline of resolution.

3. Contrast enhancement without substantial loss of resolution is achieved by adding up to 5% water to the substitution medium (Walther and Ziegler, 2002).

4. Glutaraldehyde can be purchased as solution in acetone or in water. We use 25% glutaraldehyde in water and not in acetone as described (Wild *et al.*, 2001a) to make a 0.25% solution in acetone. Osmium tetroxide, however, is dissolved in pure acetone.

5. Contrast can be enhanced by substitution in acetone containing glutaraldehyde and tannic acid (Giddings, 2003).

6. Contrast can be enhanced by isobutanol during staining (see below).

7. Contrast can be enhanced by reducing substitution time down to 60 min (Hawes *et al.*, 2007).

h. Freeze-Substitution for Immunolabeling

Preparation methods for immunolabeling are discussed elsewhere. We thus just introduce a recently introduced protocol (Matsko and Mueller, 2005) that leads to good preservation and high resolution of cellular ultrastructure. The protocol is based on the fact that epoxy resins *per se* are fixatives (Sung *et al.*, 1996). Thus frozen cells can be substituted in acetone containing 20% epon as described earlier and embedded in epon. The result is a well preserved ultrastructure (Fig. 7) and reasonably well preserved antigenicity. The disadvantage of unspecific binding of antibodies may be prevented by incubation in acetylated bovine serum albumin prior to immunolabeling.

i. Staining

Sections of well frozen and freeze-substituted cells in the presence of low concentrations of glutaraldehyde and osmium tetroxide are poor in contrast but of good resolution. Good contrast is achieved by double staining with lead citrate

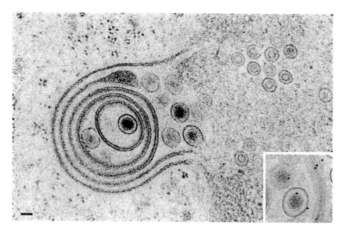

Fig. 7 High-pressure frozen and freeze-substituted cells in 20% pure epon dissolved in acetone. Note the distinct visibility of membranes associated with viral proteins and the specific although sparse labeling with antibodies against VP16, a herpes simplex virus 1 tegument protein. Compare structural integrity and labeling intensity with that published by Naldinho-Souto *et al.* (2006). Images by E.M. Schraner. Bars = 100 nm.

(Reynolds, 1963) and uranyl acetate. To increase contrast, 9% isobutanol (Roberts, 2002) is added to both lead citrate and uranyl acetate. Furthermore, Mg-uranyl acetate is used. The staining solutions are prepared as for use in pure water. Filter the solutions through Milipore filters (0.22 μm) and store it in Eppendorf tubes at 4 °C. The time period of staining is between 5 and 20 min at room temperatures. Prolonged staining leads to more precipitation rather than to more contrast.

j. Electron Tomography

High resolution in combination with good contrast is a prerequisite for in-depth investigation of membrane-bound processes. Freeze-substitution of cells as described earlier is highly suitable for studying membrane interactions in thin (40–60 nm) sections resolving the bilayer nature of membranes. Thick (150–300 nm) sections can be used for tomography. Iso-butanol in the presence of stains increases visibility, and stains penetrate deeper into epon than stains dissolved only in H_2O. Staining both sides of thick section is also useful. The use of electron microscopes equipped with high contrast lenses will result in images of sufficient high contrast to visualize the membrane bilayer (Fig. 8).

3. Field Emission Scanning Electron Microscopy

To investigate viruses and cell-virus interactions by scanning electron microscopy, instruments are needed that provide high resolution although we were able to demonstrate cell entry of herpes viruses using the secondary electron detector of a

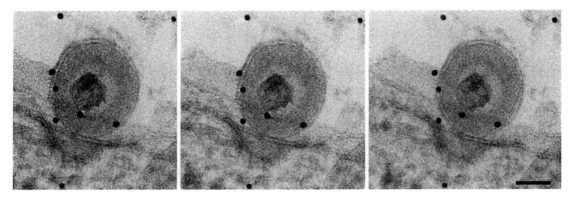

Fig. 8 Tomographic series of plunge-frozen and freeze-substituted herpes simplex virus 1 entering a cell. Note the clear visibility of the bilayer in this 200 nm thick section. Images by B.M. Humbel, University of Utrecht. Bars = 100 nm.

scanning transmission electron microscope Philips CM 12 (Wild et al., 1998). Details of viruses, however can only be visualized using a field emission electron source (FESEM) combined with adequate coating of the biological specimen (Hermann and Muller, 1991). This methodology can also be employed to study viral cell entry in detail. To do so, cells are grown on sapphire disks. Viruses are then led to attach to the cell surface at 4 °C for 1–2 h prior to incubating at 37 °C for seconds to a few minutes in a humid chamber mounded around a guillotine for plunge-freezing. After freezing, cells are freeze-substituted using the same protocol as for transmission electron microscopy, critical point dried, coated with 3 nm platinum-carbon at an angel of 45° by electron beam evaporation followed by 8 nm carbon perpendicular (Walther et al., 1995). Examination in a FESEM reveals structural details of virus interactions with the cell surface, e.g., of herpes simplex virus 1 (Fig. 9).

4. CryoField Emission Scanning Electron Microscopy

For *in Situ* examination by cryo-scanning electron microscopy, cells can be grown in petriperm dishes. Disks of 2 mm in diameter are cut out of the petriperm dishes with an ophthalmologic punch, transferred into hexadecane prior to placing them into an aluminum chamber with a central cavity (height 50 μm, diameter 2 mm) for high-pressure freezing. The frozen samples can be stored in liquid nitrogen until use. For freeze-fracturing, the sandwich with the cells on the petriperm disks between the two aluminum platelets are clamped in a special holder (Eppenberger-Eberhardt et al., 1997) and fractured in a freeze-etching machine by removing one aluminum platelet with the microtome (temperature −110 °C, vacuum about 2×10^{-7} mbar). Optional, samples can be "etched" (sublimation of some water at the fracture face) for about 2 min prior to double layer coating

Fig. 9 Herpes simplex virus 1 particles on the surface of a cell after plunge-freezing, freeze-substitution, critical point drying, and coating with platinum and carbon visualized in a FESEM Hitachi S 700 using the backscattered electron signal. Bar = 100 nm.

(Walther et al., 1995), by electron beam evaporation with 3 nm of platinum-carbon, from an angle of 45°, and about 8 nm of carbon, perpendicularly. The frozen sample is then mounted in a cryo-holder and transferred into the microscope. The whole center of the punched-out petriperm disks (about 1 mm^2) can be analyzed at a temperature of −130 °C and an acceleration voltage of 10 kV using the backscattered electron signal (Walther et al., 1995). This technique allows examination of cellular processes with little disturbance between sampling and examination (Fig. 10). Unfortunately, the rate of good specimen is low because fracturing of the 10 μm thick layer is tricky, and the petriperm often detaches from the platelet. To overcome this latter problem, cells can be grown an aluminum platelets coated with matrigel (Sawaguchi et al., 2003). Cells cannot easily be monitored by light microscopy necessary to estimate the progress of infection. Hence, growing cells on platelets is not a suitable method for studying cellular processes such as the morphogenesis of viruses.

An alternative way is to use isolated cells with the disadvantage that the delay between removal of cells from culture dishes and freezing is substantial and involves trypsinization or scraping. After trypsinization and an initial centrifugation (800–1000 rpm), cells must be resuspended in fresh medium and centrifuged again (800 rpm) to obtain a soft pellet of which aliquots can easily be removed. Aliquots are then placed between the flat surfaces of two aluminum platelets coated with 20 nm copper to render the surface hydrophilic for improved adherence of cells to the platelet surface. This may also achieved by scratching the platelet surface. A 15 μm thick 100 mesh/inch grid serves as a spacer. The sandwich is placed onto the specimen holder for high-pressure freezing. Frozen samples can be stored in liquid nitrogen until use. For freeze-fracturing the sandwich is transferred to a freeze-fracture unit. After adjusting the temperature to −115 °C at a

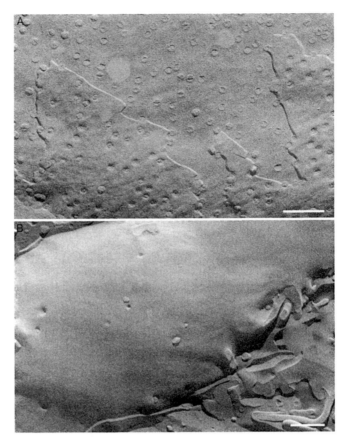

Fig. 10 Nuclear Surface of bovine herpes virus 1 infected cells grown in petriperm dishes, high-pressure frozen, freeze-etched, platinum-carbon coated and examined in a Hitachi S 5200 using the backscattered electron signal. Nuclear pores are irregularly distributed early in infection whereas they are almost absent late (15 h) of infection. Images by P. Walther, University of Ulm. Bar = 500 nm.

vacuum of 4×10^{-7} mbar, the top platelet is pushed off with the microtome, and the fracture surface coated with 3 nm platinum-carbon as described earlier for analysis at $-130\,°C$ and an acceleration voltage of 10 kV using the backscattered electron signal. Alternatively, the fracture surface may be coated with 2.5 nm platinum-carbon at an angle of 45° followed by 1 nm platinum-carbon rotating from 0 to 90° for analysis using the inlens secondary electron detector.

To reduce the delay between removal and freezing, cells may be fixed with 0.1–0.3% glutaraldehyde prior to scraping or after trypsinization. Cells fracture perfectly along membranes except nuclear membranes (Fig. 11). Fixation is especially recommended for safety reasons when working with pathogens.

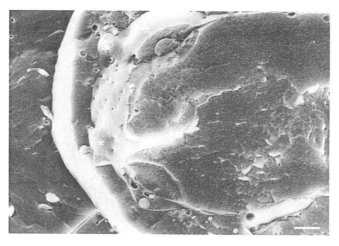

Fig. 11 Isolated cell after pre-fixation with 0.25% glutaraldehyde, high-pressure freezing, freeze-etching, and platinum coated and visualized in a Gemini 1530 FESEM. The fracture plane is through the nucleus, whose surface has protrusions (asterisks) devoid of nuclear pores. These protrusions might be the result of nuclear pore impairment as shown in Fig. 14. Image by A. Käch, Federal School of Technology, Switzerland. Bar = 500 nm.

III. Material

Cells: HeLa cells, Vero cells, MDBK cells: American Type Culture Collection
Virus: Isolates: Institute of Virology, University of Zürich, Switzerland
Dishes: 6 well test plates: Techno Plastic Products, Trasadingen, Switzeland
Petriperm dishes: Sartorius: Göttingen, Germany
Medium: Dulbecco's modified Eagle's medium, Gibco, Bethesda, MD, USA
Grids: Electron Microscopy Sciences, Ft. Washington, PA, USA
Parlodion: Electron Microscopy Sciences, Ft. Washington, PA, USA
Acetylated bovine serum albumin: Aurion, Wageningen, The Netherlands
Sapphire disks: Bruegger, Minusio, Switzerland
Fixatives: Glutaraldehyde, Osmium tetroxide: Electron Microscopy Sciences, Ft. Washington, PA, USA
Stains: Na-phosphotungstic acid (negative stain), Uranyl acetate (negative stain), Lead citrate (positive stain): Fluka, Buchs, Switzerland
Mg-uranyl acetate (positive stain): Polysciences, Warrington, PA, USA
Detergent: neodisher LM3, Dr.Weigert, Hamburg, Germany
Microscopes:
CM12, Philips, Eindhoven, The Netherlands, equipped with a CCD camera Ultrascan 1000, Gatan, Pleasanton, CA, USA
Tecnai 20, FEI Company, The Netherlands, equipped with a CCD camera, Temcam F214, TVIPS GmbH, Germany

S-5200 FESEM, Hitachi, Tokyo, Japan
S-700 FESEM, Hitachi, Tokyo, Japan
Gemini 1530, Zeiss, Oberkochen, Germany
High-pressure freezer: HPM010; BAL-TEC Inc., Balzers, Liechtenstein
EM HPM, Wohlwend Engineering, Sennwald, Switzerland
Freeze-substitution-units: CS auto, Reichert-Jung,
FS 7500, Boeckeler Instruments, Tucson, Arizona, USA
Freeze-etching units: BAF060; BAL-TEC, Balzers, Liechtenstein
BAF 300, BAL-TEC, Balzers, Liechtenstein
Critical point drying: CPD 030; BAL-TEC, Balzers, Liechtenstein

IV. Discussion

Negative staining is a powerful tool to demonstrate viruses for both diagnosis and research. It is rapid and gives an immediate answer on the presence of viruses or other particles. It is hence the primary tool for rapid diagnosis of infectious agents in emergent situations (Hazelton and Gelderblom, 2003). However, it must be kept in mind that the limit of detection of particles is about 10^6/ml. Viruses can be concentrated within a sufficient short period of time by centrifugation using an air-centrifuge or by adsorption of virus particles onto a supporting film that has been treated with antibodies to enhance specific adsorption.

Viruses may change structure during staining and subsequent drying as has been shown e.g. for vaccinia virus (Dubochet *et al.*, 1994). Thus for structural analysis viruses are rapidly frozen and examined in the frozen hydrated state (Adrian *et al.*, 1984). Imaging of vitrified virus particles can be used for 3-dimensional reconstruction (Baker *et al.*, 1999). The impact of UAc is probably less than that of PTA. This is certainly true for RNA viruses such as rotaviruses (Fig. 2C and D), which are severely affected by PTA. On the other hand, PTA is a versatile stain that can be used at various concentrations and pH for various periods of time. As a rule, the smaller the particle to be visualized the lower the concentration. Staining for 30 s with 1% PTA is sufficient to demonstrate pili (8 nm in diameter) of bacteria (Hahn *et al.*, 2002). The optimal pH, time and concentration have to be found out in a given sample starting with pH 7.4.

Generally, viruses collapse during air-drying. The degree of collapse is minimal in small viruses such as parvovirus (18 nm), and increases with size of the particles. Large complicated built viruses, like iridovirus and herpes virus, are most severely affected. The tegument surrounding the capsid in herpes viruses completely collapses given the feature of a fried egg (Fig. 1C), or the viral envelop disrupts so that both tegument and envelope are lost (Fig. 1D). Disruption of the viral envelope is not only a matter during negative staining but also during virus preparation (freezing-thawing, centrifugation). Differences in size and shape of multi-componential viruses may also originate during morphogenesis. Normally herpes viruses are sphere like structures (Grunewald *et al.*, 2003; Zhou *et al.*,

1999). However, the longer virus production proceeds in a given cell, the more heterogeneous herpes virus particles (Fig. 12) are produced considering shape and size (Schraner *et al.*, 2004). The capsid of herpes viruses is built of hexons and pentons (Baker *et al.*, 1990) assembled in the nucleus and then filled with DNA. Filling of capsids with DNA fails probably late in infection resulting in accumulation of empty capsids within the nucleus (Fig. 14 inset) and/or virions with empty capsids within the cytoplasm. Empty capsids may collapse during negative staining and stains may penetrate resulting in dark appearance of capsids whereas collapse of DNA filled capsids is minimal allowing demonstration of structural details. Collapsing of viral structures can be avoided to a large extent by freezing followed by freeze-drying, or by shadowing of unstained samples. For structural analysis negative staining can be combined with preparation of thin aqueous films (Adrian *et al.*, 1998) for freezing and examining in the frozen hydrated state (Harris, 2007).

Though many viruses can be demonstrated within cells fixed and processed in the conventional way studies on virus membrane interactions has led to misinterpretation. The two main reasons are that membrane-bound processes are fast and hence difficult to capture and that membranes loose integrity during conventional processing. Loss of membranes can be reduced e.g. by microwave enhanced fixation (Wild *et al.*, 1989) and certainly by cryofixation followed by freeze-substitution that results in markedly improved retention of lipids (Weibull and Christiansson, 1986; Weibull *et al.*, 1983; Weibull *et al.*, 1984). Loss of Golgi

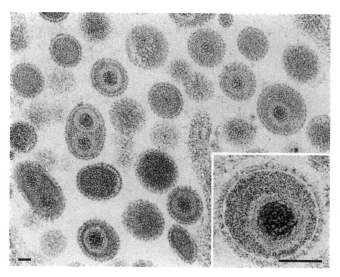

Fig. 12 Size and shape of herpes viruses may vary considerably late in infection: Typically, they comprise the core (DNA), capsid, tegument, and envelope with spikes (glycoproteins) well seen in high-pressure frozen and freeze-substituted cells or extracellular space. Image by E.M. Schraner. Bars = 100 nm.

membranes probably has contributed to the unrealistic idea that capsids of herpes viruses acquire their envelope and tegument by budding at small Golgi derived vesicles (Mettenleiter, 2004). In reality herpes virus capsids bud at any site of the Golgi complex (Fig. 4) (Leuzinger *et al.*, 2005; Wild *et al.*, 2005). Herpes virus capsids also bud at Golgi derived vacuoles (Homman-Loudiyi *et al.*, 2003; Stannard *et al.*, 1996), nuclear membranes and membranes of the endoplasmic reticulum (Leuzinger *et al.*, 2005; Wild *et al.*, 2005). Fusion of membrane e.g. in exocytosis is a rapid process taking milliseconds (Knoll *et al.*, 1991), and has very rarely been shown so far in conventionally processed cells (Fig. 13A). Fusion between viral envelope, which derive from cell membranes, with liposomes takes about 1 min (Kanaseki *et al.*, 1997). The discrepancy in time of fusion during exocytosis and simulated viral cell entry by fusion of influenza virus with liposomes is probably due to the fact that during exocytosis membranes approach from the cytoplasmic side whereas in fusion during cell entry the viral envelope approaches the external layer of the plasma membrane or of endosomal membranes with its external layer. Since the cellular fusion machinery (Peters *et al.*, 2004) acts from the cytoplasmic layer a viral fusion machinery has to be evolved to act at the external layer during viral entry. Interestingly, viral entry via fusion of the viral envelope with the plasma membrane has not been shown convincingly to date. Using the setup described under plunge-freezing might help to discover the entry process of enveloped viruses (Fig. 8).

Fusion is required somewhere during entry of enveloped viruses. However, fusion of the viral envelope with cell membranes does probably not take place during viral egress. The dominant process in egress is budding. The morphology of budding differs distinctly form that of fusion (Fig. 13). Despite the distinct features budding was often referred to as fusion in herpes virology obviously to support an assumption: e.g. identical features are addressed as fusion (Klupp *et al.*, 1998) and as budding (Fuchs *et al.*, 1997), respectively. Viruses may bud at RER

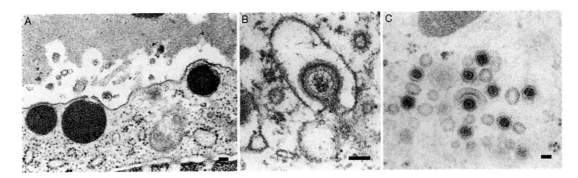

Fig. 13 (A) Close apposition and fusion of secretory granules in a mammary epithelial cell after conventional fixation. (B) Budding of a bovine herpesvirus 1 capsid into a vacuole after conventional fixation. (C) Budding of a bovine herpesvirus 1 capsid at a membrane probably derived from the Golgi complex after high-pressure freezing and freeze-substitution. Bars = 100 nm.

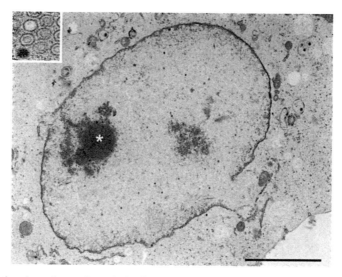

Fig. 14 Section through a nucleus of a bovine herpes virus 1 infected MDBK cell with protrusions of nuclear matrix through impaired nuclear pores containing capsids into the cytoplasmic matrix, accumulation of viral proteins (asterisk), and accumulation of empty and DNA filled (black) capsids (inset). Images by E.M. Schraner. Bar = 5 μm.

membranes, e.g., herpes virus (Leuzinger et al., 2005), vaccinia virus (Griffiths et al., 2001; Sodeik et al., 1993), and probably iridovirus (Cobbold et al., 1996), at Golgi membranes, e.g. herpes virus, or at the plasma membrane, e.g. vaccinia virus (Meiser et al., 2003). In contrast to fusion, budding is a slow process of which mechanistic is poorly understood. It can be easily visualized even by slow fixation methods (Fig. 13B). Budding may got stuck because the amount of membranes needed to engulf a capsid is not sufficient (Fig. 13C). However, from such images it must not be absolutely concluded that the amount of membranes is not sufficient since the membrane hit in a given section plane might be connected to the Golgi complex as has been nicely shown by (Leuzinger et al., 2005).

While budding of herpes viruses can readily be demonstrated at Golgi membranes budding at nuclear membranes are cumbersome to find, possibly because they do not occur often at least late in infection as demonstrated by freeze-fracture images (Fig. 10B). They show that not only the number of budding events but also the number of nuclear pores reduces considerably when infection proceeds making the nuclear envelope a highly interesting field in virus research. Impaired nuclear pores (Fig. 14) are likely to be the gateway for capsids to gain access to the cytoplasm (Wild et al., 2005).

Herpes viruses derived by budding at the inner nuclear membrane have been shown decades ago (Schwartz and Roizman, 1969) to be transported into adjacent RER cisternae. The fate of these virions is not yet clear. It is assumed that the viral envelope fuses somewhere along its way from the perinuclear space into RER cisternae either with the outer nuclear membrane or with the RER membrane

(Skepper *et al.*, 2001). If this idea was true it would represent a biological phenomenon because membrane starts to fuse as soon as they come into close apposition to each other. Therefore, it is difficult to understand how virions can be transported through narrow cisternae without immediate fusion. It is more realistic to assume that the envelope of herpes viruses is protected to fuse so that it can be transported within RER cisternae. If this idea is true virions will be intraluminally transported into Golgi cisternae provided that RER membranes continue into Golgi membranes as has been repeatedly shown (Van Herp *et al.*, 2005; Wild *et al.*, 2002).

This short excursion into herpesvirus envelopment ought to demonstrate that inconsistent if not unrealistic ideas were generated partly due to improper tissue processing, examination of cells after prolonged incubation, and to ignoring cell biologic fundamentals. Electron microscopy is an excellent tool to discover virus particles and to explore viral structures. It is the ultimate instrument to illuminate viral entry and viral egress.

V. Summary

Electron microscopy is a powerful tool to visualize single virus particles and to study virus-cell interactions. For visualizing single virus particles, three methods are available: negative staining, shadowing, and embedding in vitreous ice. Negative staining is rapid and reliable and is the first method to be used in emergencies. A simple method is described including preparation *per se* and preparation of supporting films onto which viruses are adsorbed for staining. Shadowing may be used to verify artificial changes introduced by negative stains. Embedding in ice allows investigation of viral structures in the frozen hydrated state. Images can be used for three-dimensional reconstruction. Ice embedding can be combined with negative staining.

Virus multiplication involves cell entry, replication of the genome, morphogenesis, and intracellular transport to the cell periphery for release. The prerequisite to study cell entry, morphogenesis, and intracellular transport is good preservation of cellular ultrastructure, good contrast, and high resolution considering space and time. Conventional fixation, e.g., fixation with glutaraldehyde and osmium tetroxide at 4 °C or at room temperature, results in substantial loss of lipids and loss of integrity of cell membranes. Both loss of lipids and membrane integrity can be avoided to a large extent by rapid freezing followed by freeze-substitution for both transmission and scanning electron microscopy, or followed by freeze-etching for scanning electron microscopy. The simplest way to freeze cells is by plunging cells grown as monolayer onto sapphire disks into propane/ethane cooled close to liquid nitrogen temperature. The most sophisticated method is freezing under high pressure that yields the highest rate of successful freezing i.e., freezing of cells without formation of ice crystals perturbing cellular ultrastructure. An alternative way is freezing by propelling liquid propane from two sides onto the specimen.

Cells can be prefixed with low concentrations of glutaraldehyde, which is advisable when dealing with cells infected with pathogens. To achieve high resolution and sufficient contrast for transmission electron microscopy, cells can be substituted in acetone containing low concentrations of glutaraldehyde and osmium tetroxide. After substitution at −90 °C, the temperature must be risen to +2 °C to obtain good contrast. Interestingly, good preservation and visibility of cellular ultrastructure is also achieved by substitution with acetone containing pure epon. Freeze-substituted cells can also be dried and coated with platinum for high resolution scanning electron microscopy to study viruses on the cell surface, e.g., during cell entry. Alternatively, frozen cells may be freeze-etched, platinum coated, and examined in the frozen hydrated state by scanning electron microscopy. Employment of these methodologies may give more detailed and more reliable insights into viral morphogenesis and intracellular transportation as discussed on the basis of herpes virus replication.

Acknowledgments

I thank Elisabeth M. Schraner for her excellent technical assistance, Claudia Senn and Monika Engels for providing viruses and cells, Andres Käch, Federal School of Technology, Zürich, Paul Walther, University of Ulm for their help in freeze-etching and cryo-FESEM and Bruno M. Humbel, University of Utrecht, for electron tomography.

References

Adrian, M., Dubochet, J., Fuller, S. D., and Harris, J. R. (1998). Cryonegative staining. *Micron* **29**, 145–160.

Adrian, M., Dubochet, J., Lepault, J., and McDowall, A. W. (1984). Cryo-electron microscopy of viruses. *Nature* **308**, 32–36.

Bailey, S. M., Chiruvolu, S., Longo, M. L., and Zasadzinski, J. A. (1991). Design and operation of a simple environmental chamber for rapid freezing fixation. *J. Electron Microsc. Tech.* **19**, 118–126.

Baker, T. S., Newcomb, W. W., Booy, F. P., Brown, J. C., and Steven, A. C. (1990). Three-dimensional structures of maturable and abortive capsids of equine herpesvirus 1 from cryoelectron microscopy. *J. Virol.* **64**, 563–573.

Baker, T. S., Olson, N. H., and Fuller, S. D. (1999). Adding the third dimension to virus life cycles: three-dimensional reconstruction of icosahedral viruses from cryo-electron micrographs [erratum appears in Microbiol. Mol. Biol. Rev. 2000 Mar;64(1):237]. *Microbiol. Mol. Biol. Rev.* **63**, 862–922.

Cobbold, C., Whittle, J. T., and Wileman, T. (1996). Involvement of the endoplasmic reticulum in the assembly and envelopment of African swine fever virus. *J. Virol.* **70**, 8382–8390.

Cyrklaff, M., Risco, C., Fernandez, J. J., Jimenez, M. V., Esteban, M., Baumeister, W., and Carrascosa, J. L. (2005). Cryo-electron tomography of vaccinia virus. *PNAS* **102**, 2772–2777.

Dalen, H., Lieberman, M., LeFurgey, A., Scheie, P., and Sommer, J. R. (1992). Quick-freezing of cultured cardiac cells in situ with special attention to the mitochondrial ultrastructure. *J. Microsc.* **168**, 259–273.

Dubochet, J., Adrian, M., Richter, K., Garces, J., and Wittek, R. (1994). Structure of intracellular mature vaccinia virus observed by cryoelectron microscopy. *J. Virol.* **68**, 1935–1941.

Eppenberger-Eberhardt, M., Aigner, S., Donath, M. Y., Kurer, V., Walther, P., Zuppinger, C., Schaub, M. C., and Eppenberger, H. M. (1997). IGF-I and bFGF differentially influence atrial natriuretic factor and alpha-smooth muscle actin expression in cultured atrial compared to ventricular adult rat cardiomyocytes. *J. Mol. Cell Cardiol.* **29,** 2027–2039.

Fuchs, W., Klupp, B. G., Granzow, H., and Mettenleiter, T. C. (1997). The UL20 gene product of pseudorabies virus functions in virus egress. *J. Virol.* **71,** 5639–5646.

Giddings, T. H. (2003). Freeze-substitution protocols for improved visualization of membranes in high-pressure frozen samples. *J. Microsc.* **212,** 53–61.

Griffiths, G., Roos, N., Schleich, S., and Locker, J. K. (2001). Structure and assembly of intracellular mature vaccinia virus: Thin-section analyses. *J. Virol.* **75,** 11056–11070.

Grunewald, K., Desai, P., Winkler, D. C., Heymann, J. B., Belnap, D. M., Baumeister, W., and Steven, A. C. (2003). Three-dimensional structure of herpes simplex virus from cryo-electron tomography. *Science* **302,** 1396–1398.

Hahn, E., Wild, P., Hermanns, U., Sebbel, P., Glockshuber, R., Haner, M., Taschner, N., Burkhard, P., Aebi, U., and Muller, S. A. (2002). Exploring the 3D molecular architecture of *Escherichia coli* type 1 pili. *J. Mol. Biol.* **323,** 845–857.

Harris, J. R. (2007). Negative staining of thinly spread biological samples. *In* "Electron Microscopy Methods and Protocols," Vol. **369,** pp. 107–142. Humana Press, Totowa, NJ.

Hawes, P., Netherton, C. L., Mueller, M., Wileman, T., and Monaghan, P. (2007). Rapid freeze-substitution preserves membranes in high-pressure frozen tissue culture cells. *J. Microsc.* **226,** 182–189.

Hazelton, P. R., and Gelderblom, H. R. (2003). Electron microscopy for rapid diagnosis of infectious agents in emergent situations. *Emerging Infect. Dis.* **9,** 294–303.

Hermann, R., and Muller, M. (1991). High resolution biological scanning electron microscopy: a comparative study of low temperature metal coating techniques. *J. Electron Microsc. Techn.* **18,** 440–449.

Hohenberg, H., Mannweiler, K., and Muller, M. (1994). High-pressure freezing of cell suspensions in cellulose capillary tubes. *J. Microsc.* **175,** 34–43.

Homman-Loudiyi, M., Hultenby, K., Britt, W., and Soderberg-Naucler, C. (2003). Envelopment of human cytomegalovirus occurs by budding into Golgi-derived vacuole compartments positive for gB, Rab 3, trans-golgi network 46, and mannosidase II. [erratum appears in J. Virol. 2003 Jul;77(14): 8179]. *J. Virol.* **77,** 3191–3203.

Jimenez, N., Humbel, B. M., van Donselaar, E., Verkleij, A. J., and Burger, K. N. (2006). Aclar discs: A versatile substrate for routine high-pressure freezing of mammalian cell monolayers. *J. Microsc.* **221,** 216–223.

Johnson, R. P., and Gregory, D. W. (1993). Viruses accumulate spontaneously near droplet surfaces: A method to concentrate viruses for electron microscopy. *J. Microsc.* **171,** 125–136.

Kanaseki, T., Kawasaki, K., Murata, M., Ikeuchi, Y., and Ohnishi, S. (1997). Structural features of membrane fusion between influenza virus and liposome as revealed by quick-freezing electron microscopy. *J. Cell Biol.* **137,** 1041–1056.

Kellenberger, E., and Bitterli, D. (1976). Preparation and counts of particles in electron microscopy: application of negative stain in the agar filtration method. *Microsc. Acta* **78,** 131–148.

Klupp, B. G., Baumeister, J., Dietz, P., Granzow, H., and Mettenleiter, T. C. (1998). Pseudorabies virus glycoprotein gK is a virion structural component involved in virus release but is not required for entry. *J. Virol.* **72,** 1949–1958.

Knoll, G., Braun, C., and Plattner, H. (1991). Quenched flow analysis of exocytosis in Paramecium cells: time course, changes in membrane structure, and calcium requirements revealed after rapid mixing and rapid freezing of intact cells. *J. Cell Biol.* **113,** 1295–1304.

Kuhn, D., and Wild, P. (1992). The influence of buffers during fixation on the appearance of smooth endoplasmic reticulum and glycogen in hepatocytes of normal and glycogen-depleted rats. *Histochemistry* **97,** 5–11.

Leuzinger, H., Ziegler, U., Fraefel, C., Schraner, E. M., Glauser, D., Held, I., Ackermann, M., Müller, M., and Wild, P. (2005). Herpes simplex virus 1 envelopment follows two diverse pathways. *J. Virol.* **79,** 13047–13059.

Matsko, N., and Mueller, M. (2005). Epoxy resin as fixative during freeze-substitution. *J. Struct. Biol.* **152**, 92–103.

Meiser, A., Sancho, C., and Krijnse Locker, J. (2003). Plasma membrane budding as an alternative release mechanism of the extracellular enveloped form of vaccinia virus from HeLa cells. *J. Virol.* **77**, 9931–9942.

Mettenleiter, T. C. (2004). Budding events in herpesvirus morphogenesis. *Virus Res.* **106**, 167–180.

Mueller, M. (1992). The integrating power of cryofixation-based electron microscopy in biology. *Acta Microsc.* **1**, 37–46.

Murk, J. L., Posthuma, G., Koster, A. J., Geuze, H. J., Verkleij, A. J., Kleijmeer, M. J., and Humbel, B. M. (2003). Influence of aldehyde fixation on the morphology of endosomes and lysosomes: quantitative analysis and electron tomography. *J. Microsc.* **212**, 81–90.

Naldinho-Souto, R., Browne, H., and Minson, T. (2006). Herpes Simplex Virus Tegument Protein VP16 Is a Component of Primary Enveloped Virions. *J. Virol.* **80**, 2582–2584.

Osborn, M., Webster, R. E., and Weber, K. (1978). Individual microtubules viewed by immunofluorescence and electron microscopy in the same PtK2 cell. *J. Cell Biol.* **77**, R27–R34.

Peters, C., Baars, T. L., Buhler, S., and Mayer, A. (2004). Mutual control of membrane fission and fusion proteins [see comment]. *Cell* **119**, 667–678.

Reynolds, E. S. (1963). The use of lead citrate at high pH as an electron-opaque stain in electron microscopy. *J. Cell Biol.* **17**, 208–212.

Roberts, I. M. (2002). Iso-butanol saturated water: A simple procedure for increasing staining intensity of resin sections for light and electron microscopy. *J. Microsc.* **207**, 97–107.

Sawaguchi, A., Yao, X., Forte, J. G., and McDonald, K. L. (2003). Direct attachment of cell suspensions to high-pressure freezing specimen planchettes. *J. Microsc.* **212**, 13–20.

Schraner, E. M., Engels, M., Bienkowska-Szewczyk, K., Loepfe, E., Walther, P., Mueller, M., and Wild, P. (2004). Shape and size of *in situ* cryo-fixed bovine herpesvirus type 1. Proceedings of 13th European Microscopy Congress, Antwerp, Belgium. **Vol. 3**, pp. 101–102. Belgian Society for Microscopy, Liége, Belgium.

Schwartz, J., and Roizman, B. (1969). Concerning the egress of herpes simplex virus from infected cells: Electron and light microscope observations. *Virology* **38**, 42–49.

Skepper, J. N., Whiteley, A., Browne, H., and Minson, A. (2001). Herpes simplex virus nucleocapsids mature to progeny virions by an envelopment–deenvelopment–reenvelopment pathway. *J. Virol.* **75**, 5697–5702.

Sodeik, B., Doms, R. W., Ericsson, M., Hiller, G., Machamer, C. E., van' t Hof, W., van Meer, G., Moss, B., and Griffiths, G. (1993). Assembly of vaccinia virus: Role of the intermediate compartment between the endoplasmic reticulum and the Golgi stacks. *J. Cell Biol.* **121**, 521–541.

Stannard, L. M., Himmelhoch, S., and Wynchank, S. (1996). Intra-nuclear localization of two envelope proteins, gB and gD, of herpes simplex virus. *Arch. Virol.* **141**, 505–524.

Studer, D., Michel, M., and Muller, M. (1989). High pressure freezing comes of age. *Scanning Microsc. Suppl.* **3**, 253–268; discussion 268–269.

Sung, H. W., Hsu, H. L., Shih, C. C., and Lin, D. S. (1996). Cross-linking characteristics of biological tissues fixed with monofunctional or multifunctional epoxy compounds. *Biomaterials* **17**, 1405–1410.

Van Herp, F., Coenen, T., Geurts, H. P., Janssen, G. J., and Martens, G. J. (2005). A fast method to study the secretory activity of neuroendocrine cells at the ultrastructural level. *J. Microsc.* **218**, 79–83.

Walther, P., Wehrli, E., Hermann, R., and Müller, M. (1995). Double layer coating for high resolution low temperature SEM. *J. Microsc.* **179**, 229–237.

Walther, P., and Ziegler, A. (2002). Freeze substitution of high-pressure frozen samples: the visibility of biological membranes is improved when the substitution medium contains water. *J. Microsc.* **208**, 3–10.

Weibull, C., and Christiansson, A. (1986). Extraction of proteins and membrane lipids during low temperature embedding of biological material for electron microscopy. *J. Microsc.* **142**, 79–86.

Weibull, C., Christiansson, A., and Carlemalm, E. (1983). Extraction of membrane lipids during fixation, dehydration and embedding of *Acholeplasma laidlawii*-cells for electron microscopy. *J. Microsc.* **129**, 201–207.

Weibull, C., Villiger, W., and Carlemalm, E. (1984). Extraction of lipids during freeze-substitution of *Acholeplasma laidlawii*-cells for electron microscopy. *J. Microsc.* **134,** 213–216.

Wild, P., Engels, M., Senn, C., Tobler, K., Ziegler, U., Schraner, E. M., Loepfe, E., Ackermann, M., Mueller, M., and Walther, P. (2005). Impairment of nuclear pores in bovine herpesvirus 1-infected MDBK cells. *J. Virol.* **79,** 1071–1083.

Wild, P., Gabrieli, A., Schraner, E. M., Pellegrini, A., Thomas, U., Frederik, P. M., Stuart, M. C., and Von Fellenberg, R. (1997). Reevaluation of the effect of lysozyme on Escherichia coli employing ultrarapid freezing followed by cryoelectronmicroscopy or freeze substitution. *Microsc. Res. Tech.* **39,** 297–304.

Wild, P., Krahenbuhl, M., and Schraner, E. M. (1989). Potency of microwave irradiation during fixation for electron microscopy. *Histochemistry* **91,** 213–220.

Wild, P., Schraner, E. M., Adler, H., and Humbel, B. (2001a). Enhanced resolution of membranes in cultured cells by cryoimmobilization and freeze-substitution. *Microsc. Res. Tech.* **53,** 313–321.

Wild, P., Schraner, E. M., Adler, H., and Humbel, B. M. (2001b). Enhanced resolution of membranes in cultured cells by cryoimmobilization and freeze-substitution. *Microsc. Res. Tech.* **53,** 313–321.

Wild, P., Schraner, E. M., Cantieni, D., Loepfe, E., Walther, P., Mueller, M., and Engels, M. (2002). The significance of the Golgi complex in envelopment of bovine herpesvirus 1 (BHV-1) as revealed by cryobased electron microscopy. *Micron* **33,** 327–337.

Wild, P., Schraner, E. M., Peter, J., Loepfe, E., and Engels, M. (1998). Novel entry pathway of bovine herpes virus 1 and 5. *J. Virol.* **72,** 9561–9566.

Zhou, Z. H., Chen, D. H., Jakana, J., Rixon, F. J., and Chiu, W. (1999). Visualization of tegument-capsid interactions and DNA in intact herpes simplex virus type 1 virions. *J. Virol.* **73,** 3210–3218.

INDEX

A

Accelerating voltage, 114–115
Actin, 209
Actin filaments, 258
Actin membrane skeleton, 219–221, 223, 226
Acyl-CoA synthetase, 195
Adipocytes, 197
Adipophilin, 200
Ag-NOR-staining, 440–441
Aldehyde fixation, 134–136
Anaglyphs, 217, 221–222, 473
Anchored-protein picket model, 230
Antibody labeling, 124–125
Anti-Nup 116 antibodies, 383
Apoptotic cell death, 39–40
Atomic force microscopy
    contact mode, 282
    of dynamins, 251
    electron microscopy and, 282–284
    equipment, 294
    of intermediate filaments, 282–284
    sample preparation for, 282
    substrates for, 294
    tapping mode, 282
Average cell volume, 74

B

Backscatter electrons, 114, 128
Bacteria, negative staining of, 480–485
Bal-Tec HPM010, 13–15
Biomarkers, 47
Bismuth staining for proteins, 439
Bovine serum albumin solution, 143
Brass hats, 7
Budding of viruses, 520–521
Buffers, 121

C

Cacodylate buffer, 360
Caenorhabditis elegans
    description of, 10, 156, 413
    embryos
        encapsulation of, 416
        high pressure freezing with freeze substitution, 419–423
        immunogold staining of lamina proteins in, 423–427
        infiltration and embedding, 420–423
        microwave fixation of, 416–423
        preparation of, 416–423
    nuclear lamina in
        description of, 413–414
        embryo preparation, 416–423
Calcium-induced calcium release, 191
Carbon-coated copper girds, 132–134, 481
Cardiac muscle cell, 191–192
Cavalieri estimation, 76, 78
Caveolin, 195, 197
Ce-lamin
    assembly of, 415
    description of, 413–414
    filaments, *in vitro* assembly of, 414–416
    protein expression and purification, 414
Cell monolayers grown on sapphire coverslips
    freeze substitution of, 169–174, 507
    high pressure freezing of
        description of, 166
        discussion of, 174
        instrumentation, 167
        materials, 167
        procedures, 167–169
        safety issues, 177
        sapphire glass for, 177
Cell volume, 74
CELLocate grid, 85–86
Cellular electron microscopy, 311–314
Charging, 115–117
Chemical cross-linking, 49–50
Chemical fixatives, 49–50
Chromium coating, 116, 406
Chromosomes
    anaglyphs, 473
    condensation of, 462
    decondensation of, 462, 468–470
    description of, 454
    discovery of, 454

Chromosomes (cont.)
  DNA staining in, with platinum blue, 465–467
  dynamic matrix model of, 462
  field-emission scanning electron microscopy visualization of, 104
  formaldehyde-fixed, 459–460
  isolation of, 458
  kinetochore region of, 104
  labeling methods for, 454
  meiosis, 463–464
  metal impregnation of, 464
  mitotic, in plants
    locations of, 454–456
    spindle fibers, 461
    ultrastructure of, 460–461
  proteins
    immunogold labeling of, 470–472
    staining of, 467–468
  scanning electron microscopy of
    analysis, 464–472
    DNA staining, 457
    DNaseI treatment, 457
    drop/cryo technique, 456, 458–459
    formaldehyde fixation, 457
    materials and methods, 456–458
    preparation of chromosomes, 458–460
    protein staining, 457
    proteinase K, 457
    in situ hybridization, 457–458
Collagen fibrils
  axial structure of, 328–329
  cartilage, 333
  corneal, 330–331
  description of, 320
  electron microscopy of, 327–334
  extracellular channels containing, 334–336
  fibripositors, 336–340
  formation of, 327–328
  heterotypic, 332
  histology of, 321
  isolation of, from tissue, 329–330
  N,N-bipolar, 329
  self-assembly of, 327–328
  three-dimensional reconstruction, 325
  tomographic reconstruction of, 338
  type IX, 334
Confocal light microscopy, 390
Connexins, 194
Continuum normalization, 33–35
Cooling rate, 153–155
Corneal collagen fibrils, 330–331
Correlated light and electron microscopy

cytoskeleton visualization using
  cell relocation, 262–265
  description of, 258–259
  evaluation of, 269–270
  film retrieval, 263
  fixation, 260–262, 265
  live cell microscopy and fixation, 260–262
  materials and methods, 259
  negative staining, 262–265
  patterned thin films for, 260
  results of, 266–269
  structural features, 266
of negative stained samples of bacteria, 484–485
Correlative fluorescence and electron microscopy of intermediate filaments, 284–293
Correlative microscopy, 54–55
Correlative video-light EM
  cell fixation for, 85–86
  critical parameters for, 94–95
  immunolabeling for
    horseradish peroxidase, 88–95
    materials, 87
    nanogold, 87–88
  methods, 85–87
  rationale for using, 84–85
  stages of, 84–85
  troubleshooting of, 94–95
Critical point drying, 119, 125–126, 404–407
Cryo EM
  biological specimen preparation for
    chemical cross-linking, 49–50
    embedding in resin for sectioning, 50
    overview of, 46–47
    vitrification, 48
  correlative microscopy, 54–55
  definition of, 45
  dynamins, 244–249
  microtubules studied using, 307–311
CryoField emission scanning electron microscopy, 514–516
Cryofixation
  of cells
    for scanning electron microscopy, 24–26
    for transmission electron microscopy, 26–27
  description of, 142
  treatment before, 23
Cryo-immobilization, 49
Cryoimmunogold labeling
  aldehyde fixation of cells for, 134–136
  carbon-coated copper girds for, 132–134

cryosectioning for, 138–140
description of, 132
embedding samples for, 136–138
Formvar-coated copper girds for, 132–134
Cryopreparation, 23
Cryoprotection, 52
Cryosectioning
  description of, 138–140, 354
  for electron probe X-ray microanalysis of cell physiology, 28–29
  static electricity caused by, 29
  tools necessary for, 29
Cytochemistry
  acrylic sections, 439
  Ag-NOR-staining for nucleoli, 440–441
  bismuth staining for proteins, 439
  blockade reaction, 435
  controls, 439
  DNA staining with osmium ammine, 433–434
  EDTA staining method for nuclear nucleoproteins, 435–437
  Epon sections, 437–438
  Feulgen reaction, 435
  freeze-fracture
    features of, 181–182
    fracture-label, 186–187
    label-fracture, 187–189
    lipid droplets, 194–201
    methods, 185–191
    nomenclature, 185–186
    principles of, 182–183
    problems associated with, 183–184
    protein organization in plasma membrane, 191–192
    sectioned replica technique, 184
    steps involved in, 183
  goal for, 433
  high resolution autoradiography, 441
  immunocytochemistry, 441–443
  intranuclear substrates targeted using enzyme-colloidal gold complexes, 446
  molecular *in situ* hybridization, 443–444
  nucleic acids identified by enzymatic reactions, 444–445
  overview of, 432
  reagent preparation, 437
  terbium staining for RNA, 437
Cytoplasmic surface of plasma membrane
  actin filaments laterally bound to, 231–232
  actin-based membrane skeleton, 219–221
  immunolabeling of proteins on, 215
  interface structure of membrane skeleton on, 223–226
  methods for exposing, 212–215
  three-dimensional structure of, 219–231
  undercoat surface on, using electron tomography, 222–223
Cytoskeleton
  correlated light and electron microscopy for visualization of
    cell relocation, 262–265
    description of, 258–259
    evaluation of, 269–270
    film retrieval, 263
    fixation, 260–262, 265
    live cell microscopy and fixation, 260–262
    materials and methods, 259
    negative staining, 262–265
    patterned thin films for, 260
    results of, 266–269
    structural features, 266
  "cortical," 209
  plasma membrane and, 208

D

Decoration artifact, 406
Deep-etching, 215–216
Densities, 70–74
Depolymerizing microtubules, 309
Desmin, 280–281
Desmocollins, 348
Desmogleins, 348
Desmoplakin, 348
Desmosomes
  conventional electron microscopy of
    materials for, 360–361
    methods, 351–353
  definition of, 347–348
  extracellular core domain of, 348
  freeze-fracture studies of, 349
  immunogold labeling
    cryopreservation for, 353–356, 361–363
    description of, 353–357
    equipment, 363–364
    quantification of gold particles, 358–360
    solutions for, 363
  protein components of, 348
  ultrastructure of, 348–349
Dilution buffers, 123
Dimethylaminoethanol, 490
Disectors, 78

DNA staining
  with osmium ammine, 433–434
  with platinum blue, 465–467
Dry fracturing, 397–398
Drying, 46
Dynamic matrix model, 462
Dynamin(s)
  atomic force microscopy of, 251
  cofactor proteins, 253
  cryo-EM of, 244–249
  functions of, 237–238
  MxA, 243, 251
  oligomeric structures of, 240
  oligomerization of, 243
  pleckstrin-homology domain, 238, 252
  proline-rich domain, 238, 259
  rotary shadowing of, 249–250
  scanning transmission electron microscopy of, 251
  self-assembly of, 240–241, 252
  transmission electron microscopy of, 244–249
  types of, 238–239
  visualization of, 240–251
Dynamin 1, 243
Dynamin–lipid tubes, 241–243, 247–248
Dynamin-related protein, 237
Dystrophin, 231

E

E surface, 185
EDTA staining method for nuclear nucleoproteins, 435–437
Electron energy-loss spectroscopy, 20
Electron microscopy *(See also specific discussion)*
  atomic force microscopy and, 282–284
  description of, 302
  desmosomes, 351–353
Electron probe X-ray microanalysis of cell physiology
  analysis, 30–32
  applications of, 39
  continuum normalization, 33–35
  equipment for, 37–38
  mapping techniques, 37
  methods, 22–37
  overview of, 20–21
  qualitative and quantitative information, 32–37
  rationale for, 21
  scanning electron microscopy analysis, 31
  equipment, 37
  specimens
    cryosectioning, 28–29
    freeze-drying, 29
    preparation of, 22–27
    semithick, 35
    standards, 36
    thin window detectors, 35–36
    water content estimation, 36–37
  summary of, 38–40
  transmission electron microscopy analysis, 32
    equipment, 37
  treatment post cryofixation, 27–30
  X-ray detection systems, 38
Electron tomography
  description of, 146
  three-dimensional reconstruction of membrane skeleton using, 218–219
  virus cell cultures, 513
Embedding samples, for cryoimmunogold labeling, 136–138
EMPACT, high pressure freezing with, 178–179
Endocytosis, 238
Endoplasmic membrane, 196
Energy dispersive detection systems, 38
Energy dispersive spectroscopy, 20
Epithelial protein lost in neoplasm, 231
Etching, 215–216
Eukaryotic flagella, 313
Extracellular matrix
  characteristics of, 320–321
  procollagen molecules, 322
  rotary shadowing electron microscopy of, 322
  three-dimensional reconstruction, 322–327
Ezrin/Radixin/Moesin proteins, 231

F

Feulgen reaction, 435
Fibripositors, 336–340
Fibrosis, 320
Field emission scanning electron microscopy
  chromosome visualization using, 104
  description of, 110, 369
  immuno-gold labeling for, 118
  nucleus imaging using, 391–392, 407–408
  sample preparation for, 377–379
  viruses, 513–516
"First-order" stereology, 79

Fixation
    aldehyde, 134–136
    description of, 117–118
    materials for, 120–121
Fluorescence microscopy, 98
Fluorescent antibodies, 55
Formaldehyde fixative, 143
Formaldehyde stock solution, 143
Formaldehyde/glutaraldehyde fixative, 143
Formvar-coated copper girds, 132–134, 481
Fractionator, 79
Fracture face, 185
Fracture-label, 186–187
Freeze etching, 209
Freeze substitution (*See also* High pressure freezing)
    alternative approaches to, 53–54
    Caenorhabditis elegans, 419–420
    of *Caenorhabditis elegans*, 10
    of cell monolayers grown on sapphire coverslips, 169–174, 507
    description of, 4, 51, 158, 312, 510–512
    epoxy resin embedding and, 170–172
    of frozen material, 54
    ice damage, 16
    for immunolabeling, 425–426, 512
    Langerhans blood cells, 160
    Lowicryl embedding and, 172–174
    preservation issues, 16
    safety issues, 12
    samples, 159
    specimen preparation, 52
    sucrose for cryo-protection before, 52
    vitrification followed by, 48
    water in medium of, 10
    of yeast, 8–10
Freeze-drying, 29
Freeze-fracture cytochemistry
    applications of
        lipid droplets, 194–201
        protein organization in plasma membrane, 191–192
    features of, 181–182
    fracture-label, 186–187
    label-fracture, 187–189
    methods, 185–191
    nomenclature, 185–186
    principles of, 182–183
    problems associated with, 183–184
    sectioned replica technique, 184
    steps involved in, 183
Freeze-fracture electron microscopy, 182

Freeze-fracture replica immunolabeling
    description of, 182, 184, 188–191
    lipid droplet studies, 194–201
    milk fat globule secretion studies, 199

G

G factor, 36
G-actin molecules, 211
Gap-junctional channels, 193–194
Gelatin, 144
Glutaraldehyde, 53, 353
Gold labeling
    density of, 65–69
    description of, 110
    distribution of, 62–66
    high accelerating voltages for, 115
    micrograph method of estimating, 62–63
    oocyte imaging, 400–403
    over linear profiles, 66
    over profile areas, 65, 67
    particles, 118–119, 184
    platinum replicas, 216
    quantification of gold particles, 358–360
    scanning method of estimating, 63, 65
    specificity, 69
Golgi complex, 503
Green fluorescent protein
    cell culture and transfection, 99–100, 103
    cell preparation on silicon chips, 100–101
    overview of, 98–99
    rationale for using, 99
    real-time light microscopy, 103–104
Grid correction factor, 34
GTP, 300
GTPase effector domain, 238

H

HeLa cells, 105–106
Herpes virus, 502, 515, 519, 521
High pressure freezing (*See also* Freeze substitution)
    artifact induced by, 11
    with Bal-Tec HPM010, 13–15
    Caenorhabditis elegans, 419
    of cell monolayers grown on sapphire coverslips
        description of, 166
        discussion of, 174
        instrumentation, 167

High pressure freezing (*cont.*)
    materials, 167
    procedures, 167–169
    safety issues, 177
    sapphire glass for, 177
  close-to-native ultrastructural preservation by, 151–162
  commercial machines for, 156–158, 166
  description of, 4, 51, 147
  with EMPACT, 178–179
  fluorescent dyes surviving after, 159
  ice formation, 152–153
  instrumentation, 4–7
  Langerhans blood cells, 160
  limitations to small samples, 158
  low-temperature embedding and, 340
  materials, 4–7, 156–158
  methods, 158–159
  process of, 156
  safety issues, 12, 177
  sample holders, 7
  samples, 159
  solutions for, 166
  viruses, 509–510
  yeast sample filtration, 7
High resolution autoradiography, 443
High resolution scanning electron microscopy, 405–407
Hop diffusion, 230
Horseradish peroxidase, for immunolabeling, 88–95

I

Ice damage, from freeze substitution, 16
Immunocytochemistry
  description of, 441–443
  specimen preparation for, 53
  Tokuyasu cryosectioning technique for, 50–51
Immunodetection, 470
  description of, 492–494
  using scanning electron microscopy, 495–496
Immunogold labeling (*See also* Gold labeling)
  of chromosomal proteins, 470–472
  cryo (*See* Cryoimmunogold labeling)
  cryopreservation for, 353–356, 361–363
  description of, 132
  desmosomes
    cryopreservation for, 353–356, 361–363
    description of, 353–357
    equipment, 363–364
    quantification of gold particles, 358–360
    solutions for, 363
  double, 357
  of epidermis, 349
  high accelerating voltages for, 115
  lamina proteins in *C. elegans* embryos, 423–427
  materials for, 141
  negative staining and, for bacteria, 482
  proteins on cytoplasmic surface of plasma membrane, 215
  PVA embedding for, 356
  reagents and solutions for, 143–145
  summary of, 146–148
  technique for, 140–143
Immunolabeling
  freeze substitution for, 425–426, 512
  freeze-fracture replica
    description of, 182, 184, 188–191
    lipid droplet studies, 194–201
    milk fat globule secretion studies, 199
  of HeLa cells, 105–106
  horseradish peroxidase, 88–95
  materials for, 87
  nanogold, 87–88
  nucleus imaging, 399, 403–404
  protocol for, 403–404
  Tokuyasu cryosectioning technique for, 50–51
  "wet chamber" for, 102
  yeast nuclei isolation and visualization, 379–382
Immuno-SEM
  accelerating voltage for, 114–115
  chromium coating for, 116
  coating for, 116
  equipment for, 122
  materials, 120–122
  procedure for
    antibody labeling, 124–125
    attach sample to silicon chip, 122–124
    coating, 126
    critical point drying, 125–126
    fixation, 124–125
    image processing, 127–128
    imaging, 126–127
    processing, 125–126
  sample preparation for
    description of, 102–103
    fixation, 117–118
    imaging, 120

labeling, 118–119
metal coating, 120
sample processing, 119
Infectious process, at cellular level
immunodetection, 492–494
methods and materials for studying
description of, 480
negative staining of isolated
bacteria, 480–485
rapid processing, 492
resin embedding of bacteria-infected cell
cultures, 485–491
resin embedding of infected tissue, 491
overview of, 479–480
scanning electron microscopy of, 494–496
Intermediate filaments
atomic force microscopy of, 282–284
correlative fluorescence and electron
microscopy of, 284–293
desmosomes, 347
discovery of, 274–275
forms of, 274
materials, 293–294
neurofilaments, 293
nucleus position and, 276
plasma membrane and, 276
rationale for studying, 277
structure of, 275
vimentin
cell lines to study expression of, 284–286
desmin and, co-assembly of, 280–281
ectopic expression of, 287–291
ectopically expressed, 281–282
low-angle rotary metal shadowing
of, 278
NLS-, 289
nuclear expression of, 290
preparation of, 293
visualization of, 278–293
YFP-lamin A, 291–293
ISH, 470–471

J

Jet freezing, 510

K

K-function, 79
Kinesins, 310
Knock-down techniques, 497

L

Label-fracture, 187–189
Labeling (*See also* Immunolabeling)
preferential, 67–68
statistical assessment of, 67–69
Lamin-associated proteins, 413
Lamins
assembly of, 415–416
description of, 276, 412
Langerhans blood cells, 160
Lead citrate stain, 360–361
Length density, 71–73
Linear trace length, 67
Lipid droplets, 194–201
Lowicryl embedding, 172–174
Lowicryl polymerization, 179
Lowicryl resins, 48, 50
Low-temperature embedding and high pressure
freezing, 340
L-type calcium channels, 191–192

M

Mannitol, 26
Meiosis, 463–464
Membrane skeleton
actin-based, 219–221, 223
anaglyphs for viewing of, 217, 221–222
characteristics of, 210–211
of cytoplasmic surface of plasma membrane
deep-etching, 215–216
immunolabeling of proteins, 215
mesh size, 226–231
methods for exposing, 212–215
platinum replicas, 216
rapid freezing, 215
stereo views, 217
of human erythrocyte, 211–212
mesh size, 226–231
structure of, 211–212
three-dimensional reconstruction of, using
electron tomography, 218–219
Metal coating, for charging, 116, 120
Metal mirror freezing, 509
Metal shadowing, 305–307
Methacrylates, 50
Methyl cellulose, 144
Microtubule(s)
depolymerizing, 309
description of, 299–301
electron microscopy of

Microtubule(s) (cont.)
  cellular, 311–314
  cryo-electron microscopy, 307–311
  description of, 302
  metal shadowing for, 305–307
  negative stain, 303–305
  functions of, 300
  organization of, 301
  polymerization of, 302–303
  regulation of, 300–301
  stabilization of, 302
  structure of, 300–301
  tubulin, 302–303
Microtubule-associated proteins, 300–301, 314
Milk fat globule secretion, 199
Millipore filter, 12
Mitofusins, 237
Mitotic chromosomes, in plants
  locations of, 454–456
  spindle fibers, 461
  ultrastructure of, 460–461
Molecular *in situ* hybridization, 445–446
MxA, 243, 251
Myocytes, 192–193
Myosin, 209
Myosin-I, 231

N

nanogold, for immunolabeling, 87–88, 119
Negative stain electron microscopy, 303–305
Negative staining
  of bacteria, 480–485
  of viruses, 501–505
Neurofilaments, 293
Nuclear envelope
  definition of, 391
  description of, 98
  field emission electron scanning electron microscopy imaging of, 407
  oocyte, 393–400
  scanning electron microscopy of, 391–393
  structure of, 412
  yeast nuclei, 368, 380
Nuclear lamina
  in *Caenorhabditis elegans*
    description of, 413–414
    embryo preparation, 416–423
  definition of, 391
  structure of, 412–413
Nuclear nucleoproteins, 435–437
Nuclear pore complexes
  definition of, 98
  description of, 391
  field emission electron scanning electron microscopy imaging of, 407
  oocyte, 393–400
  yeast nuclei, 368, 380
Nucleator, 78
Nucleic acids, 444–445
Nucleolus organizing regions, 461
Nucleus
  Ag-NOR-staining of, 440–441
  confocal light microscopy of, 390
  field emission electron scanning electron microscopy imaging of, 391–392, 407–408
  microscopy of, 390–391
  scanning electron microscopy of
    colloidal gold, 400–403
    critical point drying, 404–407
    description of, 391–393
    high resolution, 405–407
    immunolabeling, 403–404
    methods, 393–400
    rationale for, 393
Number density, 73
Number estimation, 75, 75–79
Nup37, 99
Nup43, 99

O

Oocytes, 393–400
OPA1/Mgm1, 237–238
Osmium ammine, 433–434
Osteogenesis imperfecta, 320

P

P surface, 185
Packing density, 71
Paclitaxel, 302, 304
Pappus theorem, 76
Paraformaldehyde fixative, 144, 360
Parapoxvirus, 501
PAT proteins, 195, 198–199
Perichromatin fibrils, 436
PHEM buffer, 144
Phosphatidyl serine, 241
Pick-up loop, 93
Plakoglobin, 348
Plakophilin, 348

Index

Plasma membrane
  compartmentalization of, 229
  cytoplasmic surface of
    actin filaments laterally bound to, 231–232
    actin-based membrane skeleton, 219–221
    immunolabeling of proteins on, 215
    interface structure of membrane skeleton on, 223–226
    methods for exposing, 212–215
    three-dimensional structure of, 219–231
    undercoat surface on, using electron tomography, 222–223
  cytoskeleton and, 208
  intermediate filament connection to, 276
  membrane skeleton
    actin-based, 219–221
    mesh size, 226–231
    on cytoplasmic surface, 223–226
  molecule diffusion, 228
  protein organization in, 191–192
  rapid freezing of, 215
Platinum blue, 465–467
Plectin, 276
Plunge freezing, of cells, 507–509
Polyadenylate nucleotidyl transferase, 445
Polybed 812, 488
Prefixes, 121
Preservation fixes, 122
Procollagen, 322–323
Protein A gold, 141
Protein organization in plasma membrane, 191–192
Protein staining in chromosomes, 467–468
Proteinase K, 457, 469

Q

Quantities
  densities, 70–74
  number, 78–79
  overview of, 69–70
  volume, 74–78
Quantum dots, 482–483

R

Real-time light microscopy, 103–104
Resin embedding
  bacteria-infected cell cultures, 485–491
  description of, 50–51
  of infected tissue, 491
  media for, 489
Ribosomes, 382–383
"Rip-off" protocol, 213
Rotary shadowing, 249–250, 322
Rotator method, of volume estimation, 76
Rough endoplasmic reticulum, 522

S

Saccharomyces cerevisiae, 3–17, 370, 380, 388
  (*See also* Yeast, nuclei)
Sapphire coverslips, cell monolayers grown on
  freeze substitution of, 169–174, 507
  high pressure freezing of
    description of, 166
    discussion of, 174
    instrumentation, 167
    materials, 167
    procedures, 167–169
    safety issues, 177
    sapphire glass for, 177
Sarcoplasmic reticulum, 191
Scanning electron microscopy
  (*See also specific discussion*)
  applications of, 110
  charging, 115–117
  cryofixation of cells for, 24–26
  description of, 109–110
  dynamins analyzed using, 251
  electron types, 113
  external side of cell imaged using, 110
  with field emission electron sources, 110
  high-resolution, 110
  image formation, 111–117
  immunodetection using, 495–496
  processing for, 104
  signal generation, 111–112
  transmission electron microscopy vs., 111–112
Schizosaccharomyces pombe, 3–17, 368, 386
"Second-order" stereology, 79
Semiconductor nanocrystals, 482
Semithick specimens, 35
Semithin sections, 487
Sendai virus, 500
Serial sections, 90–91
Shadowing, 505
Silicon chips, 122–124, 373
Silver staining, 440–441
Sodium dodecylsulphate, 188
Spatial analysis, 79
Spheroplasts, 374, 376–377, 381
Spindle fibers, 461

Spindle pole body, 11
Spurr's resin, 17, 489
Stereomicroscope, 91
Sucrose, 26, 52
Surface density, 71–74
Surface estimation, 71
Systematic uniform random sampling, 60

T

Tannic acid coating, 467
Terbium staining, 437
Terminal deoxynucleotidyl transferase, 444
Thin window detectors, 35–36
Thiocarbohydrazide, 115
Tilt series, 324–327
Tokuyasu cryosectioning technique, 50–51
Transmembrane proteins, 230
Transmission electron microscopy
  cryofixation of cells for, 26–27
  drawbacks of, 322
  image formation, 111–117
  life science applications of, 151
  objectives of, 151
  sample preparation for, 152
  scanning electron microscopy vs., 111–112
  signal generation, 111–112
Triton X-100 solutions, 213
Tubulin, 302–303

U

Ultrastructural cytochemistry
  acrylic sections, 439
  Ag-NOR-staining for nucleoli, 440–441
  bismuth staining for proteins, 439
  blockade reaction, 435
  controls, 439
  DNA staining with osmium ammine, 433–434
  EDTA staining method for nuclear nucleoproteins, 435–437
  Epon sections, 437–438
  Feulgen reaction, 435
  goal for, 433
  high resolution autoradiography, 441
  immunocytochemistry, 441–443
  intranuclear substrates targeted using enzyme-colloidal gold complexes, 446
  molecular *in situ* hybridization, 443
  nucleic acids identified by enzymatic reactions, 444–445

  overview of, 432
  reagent preparation, 437
  terbium staining for RNA, 437
"Unroofing," 213
Uranyl acetate, 144, 360, 420, 502
Uranyl oxalate, 145
Utrophin, 231

V

Virus membrane interactions, 519–520
Viruses
  budding of, 520–521
  cell cultures, 506–516
  collapse of, during air-drying, 518
  CryoField emission scanning electron microscopy, 514–516
  description of, 499–500
  envelope of, 518
  field emission scanning electron microscopy of, 513–516
  fusion, 520–521
  herpes, 502, 515, 519, 521
  materials for studying, 517–518
  multiplicative processes, 522
  negative staining of, 501–505
  parapoxvirus, 501
  Sendai, 500
  shadowing, 505
  structural changes in, 518
Vitrification, 48, 153–155
Volume density, 73
Volume estimations, 74–78
Volume fraction, 70

W

Wavelength dispersive spectroscopy, 20

X

Xanthine oxidoreductase, 200
Xenopus laevis, 369
X-ray microanalysis (*See* Electron probe X-ray microanalysis of cell physiology)

Y

Yeast
  freeze substitution of, 8–10
  nuclei, isolation and visualization of

buffer solutions for, 371
chemicals for, 370–371
equipment for, 372–373
field emission electron scanning electron microscopy, 377–379
fixatives for, 371
immunolabeling, 379–382
materials and instrumentation for, 370–372
nuclear isolation, 375–377
nuclear pore complexes, 380
overview of, 368–369
procedures, 372–385
ribosome removal, 382–383
silicon chips, 373
spheroplast centrifugation, 375
troubleshooting, 384–385
yeast growth, 373–375
resin embedding of, 16
Yeast cells, 161
YFP-lamin A, 291–293

Z

ZAF, 33

VOLUMES IN SERIES

Founding Series Editor
DAVID M. PRESCOTT

Volume 1 (1964)
Methods in Cell Physiology
Edited by David M. Prescott

Volume 2 (1966)
Methods in Cell Physiology
Edited by David M. Prescott

Volume 3 (1968)
Methods in Cell Physiology
Edited by David M. Prescott

Volume 4 (1970)
Methods in Cell Physiology
Edited by David M. Prescott

Volume 5 (1972)
Methods in Cell Physiology
Edited by David M. Prescott

Volume 6 (1973)
Methods in Cell Physiology
Edited by David M. Prescott

Volume 7 (1973)
Methods in Cell Biology
Edited by David M. Prescott

Volume 8 (1974)
Methods in Cell Biology
Edited by David M. Prescott

Volume 9 (1975)
Methods in Cell Biology
Edited by David M. Prescott

Volume 10 (1975)
Methods in Cell Biology
Edited by David M. Prescott

Volume 11 (1975)
Yeast Cells
Edited by David M. Prescott

Volume 12 (1975)
Yeast Cells
Edited by David M. Prescott

Volume 13 (1976)
Methods in Cell Biology
Edited by David M. Prescott

Volume 14 (1976)
Methods in Cell Biology
Edited by David M. Prescott

Volume 15 (1977)
Methods in Cell Biology
Edited by David M. Prescott

Volume 16 (1977)
Chromatin and Chromosomal Protein Research I
Edited by Gary Stein, Janet Stein, and Lewis J. Kleinsmith

Volume 17 (1978)
Chromatin and Chromosomal Protein Research II
Edited by Gary Stein, Janet Stein, and Lewis J. Kleinsmith

Volume 18 (1978)
Chromatin and Chromosomal Protein Research III
Edited by Gary Stein, Janet Stein, and Lewis J. Kleinsmith

Volume 19 (1978)
Chromatin and Chromosomal Protein Research IV
Edited by Gary Stein, Janet Stein, and Lewis J. Kleinsmith

Volume 20 (1978)
Methods in Cell Biology
Edited by David M. Prescott

Advisory Board Chairman
KEITH R. PORTER

Volume 21A (1980)
Normal Human Tissue and Cell Culture, Part A: Respiratory, Cardiovascular, and Integumentary Systems
Edited by Curtis C. Harris, Benjamin F. Trump, and Gary D. Stoner

Volume 21B (1980)
Normal Human Tissue and Cell Culture, Part B: Endocrine, Urogenital, and Gastrointestinal Systems
Edited by Curtis C. Harris, Benjamin F. Trump, and Gray D. Stoner

Volume 22 (1981)
Three-Dimensional Ultrastructure in Biology
Edited by James N. Turner

Volume 23 (1981)
Basic Mechanisms of Cellular Secretion
Edited by Arthur R. Hand and Constance Oliver

Volume 24 (1982)
The Cytoskeleton, Part A: Cytoskeletal Proteins, Isolation and Characterization
Edited by Leslie Wilson

Volume 25 (1982)
The Cytoskeleton, Part B: Biological Systems and *In Vitro* Models
Edited by Leslie Wilson

Volume 26 (1982)
Prenatal Diagnosis: Cell Biological Approaches
Edited by Samuel A. Latt and Gretchen J. Darlington

Series Editor
LESLIE WILSON

Volume 27 (1986)
Echinoderm Gametes and Embryos
Edited by Thomas E. Schroeder

Volume 28 (1987)
***Dictyostelium discoideum:* Molecular Approaches to Cell Biology**
Edited by James A. Spudich

Volume 29 (1989)
Fluorescence Microscopy of Living Cells in Culture, Part A: Fluorescent Analogs, Labeling Cells, and Basic Microscopy
Edited by Yu-Li Wang and D. Lansing Taylor

Volume 30 (1989)
Fluorescence Microscopy of Living Cells in Culture, Part B: Quantitative Fluorescence Microscopy—Imaging and Spectroscopy
Edited by D. Lansing Taylor and Yu-Li Wang

Volume 31 (1989)
Vesicular Transport, Part A
Edited by Alan M. Tartakoff

Volume 32 (1989)
Vesicular Transport, Part B
Edited by Alan M. Tartakoff

Volume 33 (1990)
Flow Cytometry
Edited by Zbigniew Darzynkiewicz and Harry A. Crissman

Volume 34 (1991)
Vectorial Transport of Proteins into and across Membranes
Edited by Alan M. Tartakoff

Selected from Volumes 31, 32, and 34 (1991)
Laboratory Methods for Vesicular and Vectorial Transport
Edited by Alan M. Tartakoff

Volume 35 (1991)
Functional Organization of the Nucleus: A Laboratory Guide
Edited by Barbara A. Hamkalo and Sarah C. R. Elgin

Volume 36 (1991)
***Xenopus laevis:* Practical Uses in Cell and Molecular Biology**
Edited by Brian K. Kay and H. Benjamin Peng

Series Editors
LESLIE WILSON AND PAUL MATSUDAIRA

Volume 37 (1993)
Antibodies in Cell Biology
Edited by David J. Asai

Volume 38 (1993)
Cell Biological Applications of Confocal Microscopy
Edited by Brian Matsumoto

Volume 39 (1993)
Motility Assays for Motor Proteins
Edited by Jonathan M. Scholey

Volume 40 (1994)
A Practical Guide to the Study of Calcium in Living Cells
Edited by Richard Nuccitelli

Volume 41 (1994)
Flow Cytometry, Second Edition, Part A
Edited by Zbigniew Darzynkiewicz, J. Paul Robinson, and Harry A. Crissman

Volume 42 (1994)
Flow Cytometry, Second Edition, Part B
Edited by Zbigniew Darzynkiewicz, J. Paul Robinson, and Harry A. Crissman

Volume 43 (1994)
Protein Expression in Animal Cells
Edited by Michael G. Roth

Volume 44 (1994)
Drosophila melanogaster: **Practical Uses in Cell and Molecular Biology**
Edited by Lawrence S. B. Goldstein and Eric A. Fyrberg

Volume 45 (1994)
Microbes as Tools for Cell Biology
Edited by David G. Russell

Volume 46 (1995)
Cell Death
Edited by Lawrence M. Schwartz and Barbara A. Osborne

Volume 47 (1995)
Cilia and Flagella
Edited by William Dentler and George Witman

Volume 48 (1995)
Caenorhabditis elegans: **Modern Biological Analysis of an Organism**
Edited by Henry F. Epstein and Diane C. Shakes

Volume 49 (1995)
Methods in Plant Cell Biology, Part A
Edited by David W. Galbraith, Hans J. Bohnert, and Don P. Bourque

Volume 50 (1995)
Methods in Plant Cell Biology, Part B
Edited by David W. Galbraith, Don P. Bourque, and Hans J. Bohnert

Volume 51 (1996)
Methods in Avian Embryology
Edited by Marianne Bronner-Fraser

Volume 52 (1997)
Methods in Muscle Biology
Edited by Charles P. Emerson, Jr. and H. Lee Sweeney

Volume 53 (1997)
Nuclear Structure and Function
Edited by Miguel Berrios

Volume 54 (1997)
Cumulative Index

Volume 55 (1997)
Laser Tweezers in Cell Biology
Edited by Michael P. Sheetz

Volume 56 (1998)
Video Microscopy
Edited by Greenfield Sluder and David E. Wolf

Volume 57 (1998)
Animal Cell Culture Methods
Edited by Jennie P. Mather and David Barnes

Volume 58 (1998)
Green Fluorescent Protein
Edited by Kevin F. Sullivan and Steve A. Kay

Volume 59 (1998)
The Zebrafish: Biology
Edited by H. William Detrich III, Monte Westerfield, and Leonard I. Zon

Volume 60 (1998)
The Zebrafish: Genetics and Genomics
Edited by H. William Detrich III, Monte Westerfield, and Leonard I. Zon

Volume 61 (1998)
Mitosis and Meiosis
Edited by Conly L. Rieder

Volume 62 (1999)
Tetrahymena thermophila
Edited by David J. Asai and James D. Forney

Volume 63 (2000)
Cytometry, Third Edition, Part A
Edited by Zbigniew Darzynkiewicz, J. Paul Robinson, and Harry Crissman

Volume 64 (2000)
Cytometry, Third Edition, Part B
Edited by Zbigniew Darzynkiewicz, J. Paul Robinson, and Harry Crissman

Volume 65 (2001)
Mitochondria
Edited by Liza A. Pon and Eric A. Schon

Volume 66 (2001)
Apoptosis
Edited by Lawrence M. Schwartz and Jonathan D. Ashwell

Volume 67 (2001)
Centrosomes and Spindle Pole Bodies
Edited by Robert E. Palazzo and Trisha N. Davis

Volume 68 (2002)
Atomic Force Microscopy in Cell Biology
Edited by Bhanu P. Jena and J. K. Heinrich Hörber

Volume 69 (2002)
Methods in Cell–Matrix Adhesion
Edited by Josephine C. Adams

Volume 70 (2002)
Cell Biological Applications of Confocal Microscopy
Edited by Brian Matsumoto

Volume 71 (2003)
Neurons: Methods and Applications for Cell Biologist
Edited by Peter J. Hollenbeck and James R. Bamburg

Volume 72 (2003)
Digital Microscopy: A Second Edition of Video Microscopy
Edited by Greenfield Sluder and David E. Wolf

Volume 73 (2003)
Cumulative Index

Volume 74 (2004)
Development of Sea Urchins, Ascidians, and Other Invertebrate Deuterostomes: Experimental Approaches
Edited by Charles A. Ettensohn, Gary M. Wessel, and Gregory A. Wray

Volume 75 (2004)
Cytometry, 4th Edition: New Developments
Edited by Zbigniew Darzynkiewicz, Mario Roederer, and Hans Tanke

Volume 76 (2004)
The Zebrafish: Cellular and Developmental Biology
Edited by H. William Detrich, III, Monte Westerfield, and Leonard I. Zon

Volume 77 (2004)
The Zebrafish: Genetics, Genomics, and Informatics
Edited by William H. Detrich, III, Monte Westerfield, and Leonard I. Zon

Volume 78 (2004)
Intermediate Filament Cytoskeleton
Edited by M. Bishr Omary and Pierre A. Coulombe

Volume 79 (2007)
Cellular Electron Microscopy
Edited by J. Richard McIntosh

Volume 80 (2007)
Mitochondria, 2nd Edition
Edited by Liza A. Pon and Eric A. Schon

Volume 81 (2007)
Digital Microscopy, 3rd Edition
Edited by Greenfield Sluder and David E. Wolf

Volume 82 (2007)
Laser Manipulation of Cells and Tissues
Edited by Michael W. Berns and Karl Otto Greulich

Volume 83 (2007)
Cell Mechanics
Edited by Yu-Li Wang and Dennis E. Discher

Volume 84 (2007)
Biophysical Tools for Biologists, Volume One: *In Vitro* Techniques
Edited by John J. Correia and H. William Detrich, III

Volume 85 (2008)
Fluorescent Proteins
Edited by Kevin F. Sullivan

Volume 86 (2008)
Stem Cell Culture
Edited by Dr. Jennie P. Mather

Volume 87 (2008)
Avian Embryology, 2nd Edition
Edited by Dr. Marianne Bronner-Fraser

Edwards Brothers Malloy
Thorofare, NJ USA
December 15, 2014